Planning and Design
of Airports

Planning and Design of Airports

Robert Horonjeff
Late Professor of Transportation Engineering
Institute of Transportation Studies
University of California, Berkeley

Francis X. McKelvey
Professor of Civil and Environmental Engineering
Michigan State University

Fourth Edition

Boston, Massachusetts Burr Ridge, Illinois
Dubuque, Iowa Madison, Wisconsin New York, New York
San Francisco, California St. Louis, Missouri

Library of Congress Cataloging-in-Publication Data

Horonjeff, Robert.
    Planning and design of airports / Robert Horonjeff and Francis
X. McKelvey.—4th ed.
      p.   cm.
    Includes bibliographical references and index.
    ISBN 0-07-045345-4
    1. Airports—Planning. 2. Airports—Design and construction.
I. McKelvey, Francis X. II. Title.
TL725.3.P5H6   1993
629.136—dc20   93-23138
                                         CIP

McGraw-Hill

A Division of The McGraw-Hill Companies

7 8 9   BKM BKM   9 0 9 8 7 6 5 4 3 2 1

ISBN 0-07-045345-4

The sponsoring editor for this book was Larry S. Hager, the editing supervisor was Frank Kotowski, Jr., and the production supervisor was Donald F. Schmidt. It was set in Century Schoolbook by McGraw-Hill's Professional Book Group composition unit.

INTERNATIONAL EDITION

Copyright © 1994. Exclusive rights by McGraw-Hill, Inc. for manufacture and export. This book cannot be re-exported from the country to which it is consigned by McGraw-Hill. The International Edition is not available in North America.

When ordering this title, use ISBN 0-07-113352-6

To our students at the University of California and Michigan State University who continue to make a substantial impact on the field of airport planning and design.

Contents

Preface

The technological and legislative developments related to the air transportation industry in the 1980s and early 1990s are of such significance as to necessitate an updating of the Third Edition of *The Planning and Design of Airports* published in 1983. This edition attempts to follow the framework and philosophy of previous editions and draws substantially from the latest information available from the Federal Aviation Administration, International Civil Aviation Organization, International Air Transportation Association, Boeing Commercial Airplane Company, McDonnell-Douglas Corporation, and many other governmental, industrial, and trade organizations. The basic principles, techniques, and methodologies appropriate to the planning and design of airports are incorporated in this book and the reader is encouraged to access the publications referenced herein for further explanations of the use and limitations of their application to airport problems. The reader should be aware that the technology, regulations, standards, and specifications associated with airport planning and design change frequently and, therefore, the latest references should be consulted prior to applying the principles contained herein to airport problems.

This edition has updated and expanded the statistical data and legislation associated with air transport. It contains information about the recent developments in environmental, fiscal, and revenue analysis, aircraft characteristics and performance, and air traffic control and navigation. The chapters on airport configuration, airside capacity and delay, geometric design, terminal planning and design, airport lighting, marking, and signing, and the environmental aspects of airport design have been completely rewritten to reflect the recent and significant developments in these areas. A substantial number of solved example problems have been included throughout the book. The reference section at the end of each chapter has been updated and expanded to reflect the current literature appropriate to the field.

An appendix has also been included which contains problems appropriate for assignment to students in the academic environment.

The author is deeply indebted to Dick Horonjeff who was kind enough to substantially rewrite the section of Chap. 15 dealing with airport noise. The assistance of Mel Cullen in preparing a critique and reviewing the chapter on air traffic control is greatly appreciated. The contributions of Abid Kanafani, Hanan Kivett, and Carl Monismith in preparing material for earlier editions of this book is also acknowledged. The support of Michigan State University in providing the resources needed to complete the manuscript is gratefully acknowledged. The assistance of so many organizations which provided original drawings, materials, photographs, reports, and other resources is appreciated. Many organizations and professionals assisted substantially in this work and an effort has been made to recognize them in the acknowledgment section.

Finally, the support and understanding of my wife, Betty, and three children, Gene, Michael, and Keri, in providing the impetus and strength to complete this work are immeasurable.

Frank McKelvey

Acknowledgments

The assistance of the following professionals and organizations in preparing the material for the fourth edition of this book is deeply appreciated and gratefully acknowledged.

David Schlothauer	Aviation Planning Associates
Jack Renton	CH2MHill
Donald G. Andrews	Department of Aviation, City of Houston
Pat Beardsley	Federal Aviation Administration
Walter Frucht	
Larry Kiernan	
Gene S. Mercer	
Elisha Novak	
Harry Smetana	
Richard Horonjeff	Harris Miller Miller Hanson
Stuart S. Holder	Landrum and Brown, Inc.
Bradley T. Jacobsen	
Douglas Goldberg	
Mike Hanlon	
Marc Schoen	McDonnell-Douglas Company
Malcolm G. Cullen	MITRE Corporation
John C. Hayward	Michael Baker, Jr., Inc.
G. John Kurgan	
James A. Moorcroft	
Brian P. Reed	Reynolds, Smith and Hills, Inc.
Howard W. Yaws	
Ramon Ricondo	Ricondo and Associates
Doug Trezise	
George Perinis	Tasso Katselas Associates
Larry Jenney	Transportation Research Board
Walter Janse Van Rensburg	Scott-DeWaal
Zale Anis	Volpe Transportation Systems Center

The Nature of Civil Aviation

The Air Age

On December 17, 1903, near Kitty Hawk, North Carolina, a bicycle repairer by the name of Orville Wright propelled himself through the air a distance of 120 ft. This was the first powered flight in a heavier-than-air aircraft known to humans; it was the equivalent of 0.23 passenger-mile, the first such statistic that could be recorded in aviation history. In contrast to this humble beginning, the commercial airlines in the United States in 1990 carried more than 450 million passengers almost 350 billion passenger-miles, which is nearly 10 times the number of passenger-miles logged by railroads and buses. In 1990, more than 1 billion passengers were recorded worldwide, and these passengers flew more than 1.1 trillion passenger-miles. In addition to passenger traffic, there has been a substantial increase in the carriage of mail and freight by air. Cargo traffic has been estimated at about 16 billion ton-miles by U.S. airlines and almost 44 billion ton-miles worldwide in 1990.

The magnitude of the impact of the commercial air transportation industry on the economy of the United States can be appreciated by the facts that the total operating revenue received by the domestic air carriers was over $64 billion in 1988 and that the scheduled airlines employed over 514,000 people in 1988 [24]. The investment in the civil aviation industry consists of commercial aircraft and their supporting facilities and equipment; general aviation aircraft used for business, agricultural purposes, aerial photography, patrol, advertising, and other uses; airports; and airways. It is difficult to assess the total investment in monetary terms, but in 1988 the investment in aircraft alone exceeded $60 billion. It is projected that from 1992 to 2000, U.S. airlines will invest nearly $15 billion annually in new, more efficient commercial aircraft [13]. The total capital investment required in U.S. airports between 1990 and 2015 has been estimated at $148 billion [25].

The availability of air transport, however, has done more than provide a carrier service. It has affected our economic way of life, it has made changes in our social and cultural viewpoints, and it has had a hand in shaping the course of political history.

The sociological changes brought about by air transportation are perhaps as important as those it has brought about in the economy. People have been brought closer together and so have reached a better understanding of interregional problems. Industry has found new ways to do business. The opportunity for more frequent exchanges of information has been facilitated, and air transport is enabling more people to enjoy the cultures and traditions of distant lands.

Air Transport and the National Economy

The growth of the U.S. economy has been closely associated with the development of transportation. The amount of money spent for transportation services by all modes of transportation has correlated closely with the level of economic activity as measured by the *gross domestic product* (GDP) or national income. The expenditures for domestic intercity travel have averaged 5 to 6 percent of the GDP. Estimates of the GDP in the future vary, but if the average annual increase in GDP is about 2.4 percent in future years [15], by the year 2000, the real GDP will reach $6147 billion, in terms of 1987 dollars. Civil aviation contributions to the GDP have consistently grown at a faster pace than the economy as a whole.

Total intercity travel consists of two major parts: private automobile travel, which accounts for about 80 percent of the total, and public transportation or common-carrier travel (bus, rail, and air), which accounts for the remaining portion. In the last decade (1980 to 1990) there has continued to be a steady rise in both private automobile and public transportation travel. Since 1950, however, there has taken place a dramatic reduction in rail travel which has been more than offset by the phenomenal growth in air travel. From 1950 to 1990, the air share of public transportation travel rose from about 14 percent to more than 90 percent. Air travel now accounts for over 17 percent of the total intercity traffic in the United States. These relationships are shown in Table 1-1.

The steady rise in consumer income has had a favorable impact on air travel. As incomes have grown, larger proportions have been spent on transportation and other items related to the enjoyment of life. Consumer expenditures have been substantially greater for air travel

TABLE 1-1   Domestic Intercity Passenger-Miles by Mode of Travel and Type of
Service, 1950–1990 (Billions)

	1950	1960	1970	1980	1990
Total	503.6	781.0	1180.8	1482.5	2004.6
Private transport	439.1	708.4	1035.1	1239.7	1622.5
Auto	438.3	706.1	1026.0	1225.0	1610.2
Air	0.8	2.3	9.1	14.7	12.3
Public transport	64.5	72.6	145.7	242.8	382.1
Airlines	9.3	31.7	109.5	204.4	345.8
Railroads	32.8	21.6	10.9	11.0	13.3
Motor buses	22.7	19.3	25.3	27.4	23.0
Percentages					
Private	87.2	90.7	87.7	83.6	80.9
Public	12.8	9.3	12.3	16.4	19.1
Auto	87.0	90.4	86.9	82.6	80.3
Airlines of total	1.7	4.1	12.3	16.5	17.3
Airlines of public	14.4	43.7	75.2	84.2	90.5

SOURCE: Eno Foundation for Transportation [32].

than for other modes of transportation. This trend clearly indicates
that expenditures for air travel will continue to rise if disposable
income continues to increase and people have more time for leisure.

Growth of Air Transport and Future Trends

Although some ninety years have elapsed since the first successful
flight at Kitty Hawk, air transport is essentially a post–World War II
development.

United States air carrier passenger traffic

The stature of the United States in the world market for air transportation is clearly indicated by the fact that the commercial airlines
in domestic service accounted for about 37 percent of the passengers
and about 29 percent of the passenger-miles flown in the entire world
in 1990. The average annual growth rate in terms of passengers carried from 1960 to 1990 was about 7 percent. Estimates of future
growth rates vary from 4 to 8 percent annually through the year
2000. Thus domestic air carrier traffic, in terms of passengers and
passenger-miles, is expected to increase by approximately 40 percent
in the 10-year period from 1990 to 2000. This is shown in Table 1-2.

The relationship between domestic and international air traffic by
the U.S. commercial air carriers is shown in Fig. 1-1. In 1990, international passenger enplanements represented about 8 percent of total
passenger enplanements, and international revenue passenger-miles

TABLE 1-2   U.S. Scheduled Air Carriers and Regional-Commuter
Revenue Passengers Enplaned and Passenger-Miles Flown in
Domestic Service

Year	Revenue passengers (thousands)	Revenue passenger-miles (millions)
1940	2,500	1,100
1950	17,300	8,000
1960	56,300	30,600
1970	153,200	103,800
1980	287,900	205,400
1981	274,700	199,200
1982	286,100	209,500
1983	308,200	226,000
1984	334,000	240,700
1985	370,100	268,800
1986	404,700	297,400
1987	441,200	325,800
1988	441,200	329,900
1989	443,600	333,200
1990	456,700	344,900
1995*	518,500	399,400
2000*	633,900	493,700

*Estimated.

SOURCE: Federal Aviation Administration [15].

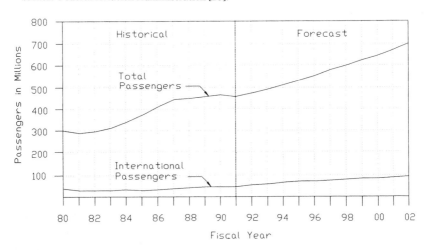

Figure 1-1   U.S. commercial air carriers scheduled passenger enplanements
(*Federal Aviation Administration [15]*).

represented about 25 percent of the revenue passenger-miles for
these carriers. It expected that these percentages will increase to
about 10 and 31 percent, respectively, by the year 2000.

It is interesting to examine the pattern of commercial air travel in
terms of trip length. While it is common knowledge that air trans-

portation is the predominant mode for long distance trips, it is interesting to note that in 1988 nearly 38 percent of the domestic passenger trips on the scheduled airlines and regional carriers were not more than 500 mi and about 52 percent did not exceed 750 mi. The distribution of scheduled flights in the United States in 1988 indicates that about 24 percent of the domestic flights scheduled by the national and major airlines were not more than 250 mi and 55 percent were not more than 500 mi. This information indicates that a substantial portion of air travel can be classified as *short-haul*, a term generally used to indicate flights of distances under 500 mi. In 1990 the average passenger trip length by the scheduled domestic air carriers in the United States was 755 mi and by the small regional and commuter carriers was 183 mi.

While the automobile will undoubtedly remain the principal means of short-haul travel for some time to come, it is useful to compare air transport to the other competing modes of intercity travel. From Table 1-3 it is clear that for domestic round-trips up to 500 mi, travel by automobile accounts for about 47 percent of the total person-trips, or 61 percent of all person-trips by automobile [33]. However, this leaves a considerable percentage of trips less than 500 mi round-trip which are completed by using other modes of travel. It is felt that the potential growth of air travel for one-way trips in excess of 250 mi is substantial, provided efficient and economical short-haul aircraft and adequate ground facilities can be developed. This potential, together with the severe airspace and airfield congestion that is occurring at terminals in large metropolitan areas, has stimulated the development of smaller capacity turboprop aircraft and the use of satellite airports to service the short-haul markets.

The geographic distribution of domestic passenger traffic is also of interest. Air travel is highly concentrated in relatively few metropoli-

TABLE 1-3   Domestic Person-Trips in the United States in 1989

Round-trip length, mi	Percentage of trips	Travel mode, %				
		Auto	Air	Bus	Train	Other
200–299	24	23	0			
300–399	17	15	1			
400–499	12	9	1			
500–599	7	6	1			
600–999	14	11	2			
1000–1999	15	9	6			
2000 and over	12	4	8			
Totals*	100	77	19	2	1	1

*Totals may not add up due to rounding error.

SOURCE: U.S. Travel Data Center [33].

tan areas. The Federal Aviation Administration (FAA) classifies airports in a community on the basis of the percentage of annual national passengers enplaned in that community. A large air traffic hub airport is defined as an airport in a community which enplaned 1.00 percent or more of the total enplaned passengers; a medium air traffic hub airport is an airport in a community which enplaned between 0.25 and 0.99 percent of the total enplaned passengers; a small air traffic hub is an airport in a community which enplaned between 0.05 and 0.24 percent of the total enplaned passengers; and a nonhub airport is an airport in a community which enplaned less than 0.05 percent of the total enplaned passengers. In 1990 the 29 large hub airports enplaned 68 percent of all air carrier passengers. The large, medium, and small hub airports accounted for 98 percent of all passengers enplaned in the United States. In 1990, 17 airports accounted for one-half of all passengers, while 50 airports accounted for 83 percent of all passenger traffic. The 100 largest airports in the United States enplaned almost 96 percent of the passengers. These data are shown in Table 1-4. This trend is expected to continue as the population of the United States continues to concentrate into several distinct regions throughout the country.

TABLE 1-4   Ranking of Passenger Enplanements, All Services at Major Airports in the United States, 1990

Cumulative percentage of passengers enplaned	Rank	Airport name
5.78	1	Chicago O'Hare International
10.80	2	Dallas–Fort Worth International
15.79	3	Atlanta Hartsfield International
20.40	4	Los Angeles International
23.44	5	San Francisco International
26.43	6	Kennedy International
29.07	7	Denver Stapleton International
31.59	8	Miami International
33.95	9	New York LaGuardia
36.25	10	Boston Logan International
38.53	11	Newark International
40.78	12	Phoenix Sky Harbor
42.96	13	Detroit Metropolitan
45.11	14	Honolulu International
47.19	15	St. Louis International
49.20	16	Minneapolis–St.Paul
51.13	17	Las Vegas International
75.35	37	Chicago Midway
80.07	44	Indianapolis
83.35	50	Bradley International
95.53	100	Long Island MacArthur Field

SOURCE: Federal Aviation Administration [16].

United States air cargo traffic

The term *air cargo* includes mail, express, and freight. Although quite insignificant in the total intercity domestic cargo market, less than 0.2 percent in 1980 [24], air cargo has developed at a very rapid pace, increasing more than 7 times during the 30-year period from 1960 to 1990, as shown in Table 1-5. The large increase in the shipment of mail since 1965 is due to the policy of the U.S. Postal Service of sending nonpriority mail by air whenever space is available. A significant portion of the air cargo ton-miles is handled in the cargo areas of passenger aircraft, approximately 59 percent in 1990, the remainder in all-cargo aircraft. Although projections of air cargo growth vary considerably, annual growth rates averaged 18 percent in the 1960s, slowed to a little less than 4 percent in the 1970s, and grew at a little less than 9 percent in the 1980s. The traffic in the 1990–2000 period might be expected to nearly double if the average rate of annual growth from the last 20 years occurs. Likewise, it has been predicted that the tonnage of air cargo carried in all freighter aircraft will increase from about 41 percent in 1990 to about 44 percent in 2005 [13]. Aircraft such as the Boeing 747-200F have a lift capacity in excess of 100 tons and provisions for end-loading of bulk cargo rather than side-loading. This offers a potential for lowering costs. As air cargo volumes expand, the continued growth in the use of containerized freight will tend to further reduce costs.

Small regional and commuter air carriers

A *commuter* air carrier is defined as an air carrier certified in accordance with FAR part 135 [5] or FAR part 121 [10] that operates air-

TABLE 1-5   Revenue Ton-Miles of Cargo Flown in Domestic and International Service by U.S. Air Carriers (Millions)

Year	Mail	Express and freight	Total
1955	61	96	157
1960	241	703	944
1965	483	1,820	2,303
1970	1,470	3,514	4,984
1975	1,097	4,495	5,892
1980	1,313	5,677	6,990
1985	1,653	7,284	8,936
1990	1,994	14,313	16,307
1995*	2,425	20,075	22,500
2000*	2,950	28,150	31,100

*Reflects a projection of an average annual increase of 4.0 percent for mail and 7.0 percent for freight through the year 2000.

SOURCE: Federal Aviation Administration [15].

craft with a maximum of 60 seats and that provides at least five scheduled round-trips per week between two or more points, or that carries mail. In 1980 there were over 300 commuter and small regional airline operators of over 1300 aircraft, which on the average had about 13 seats. In 1990 there were 151 commuter and small regional airlines operating in excess of 1900 aircraft which had on average 22 seats. The route lengths of the commuter carriers are increasing with new and more efficient aircraft. In 1990 about 30 percent of the flights were more than 200 mi [8] whereas in 1979 only 11 percent were more than 200 mi [11]. It has been estimated that the commuter industry carried more than 42 million passengers in 1990 over 7.6 billion passenger-miles [8]. Recent data indicate that the annual growth rate for commuter carriers averaged nearly 15 percent through the 1980s largely due to airline deregulation and code-sharing agreements with commercial air carriers. The commuter carrier activity has been concentrated in the large metropolitan regions where commuter airlines have offered an extension of the airline service offered by the large air carriers to outlying parts of the region through air carrier hub airports. Recent trend data for the commuter airline industry are shown in Table 1-6.

The passage of the airline deregulation legislation in 1978 has had a very favorable impact upon the growth and stability of commuter airline operations. The legislation made these carriers eligible for fed-

TABLE 1-6   Commuter and Small Regional Airline Statistics, 1970 to 1990

Year	Enplaned passengers (thousands)	Passenger-miles (thousands)	Average trip length, mi	Number of aircraft	Average seats
1970	4,270	399,000	98	741	10.6
1975	7,243	698,000	110	948	13.0
1980	14,810	1,920,000	129	1,339	13.9
1981	15,400	2,090,000	136	1,463	15.1
1982	18,550	2,610,000	141	1,573	15.6
1983	21,820	3,240,000	149	1,545	18.1
1984	26,140	4,170,000	160	1,747	18.4
1985	26,000	4,410,000	173	1,745	19.2
1986	28,360	4,470,000	158	1,806	18.4
1987	31,788	5,000,000	158	1,841	19.7
1988	35,188	6,040,000	173	1,801	20.5
1989	37,360	6,770,000	181	1,907	21.8
1990	42,099	7,610,000	183	1,917	22.1
1995*	50,700	10,064,000	205	2,099	26.4
2000*	69,000	14,359,000	216	2,229	31.8

*Forecast FAA aviation forecasts 1992–2003.

SOURCES: Commuter Airline Association [11], Regional Airline Association [8], Federal Aviation Administration [15].

eral loan guarantees for the purchase of aircraft and extended subsidies to operators of essential air services. Over the last several years, the commuter industry has been realigning route structures to increase aircraft utilization and fuel efficiency and to capitalize on the opportunity to provide feeder service from airports in smaller communities to the larger airports serviced by the national and major air carriers. As a result, commuter airlines have been able to offer air service in many markets where domestic air carrier operations were not viable. Many communities now receive greater schedule frequency to more markets than was possible in the past with the high-capacity aircraft used by the larger carriers in these low-density markets. Of the 792 airports receiving scheduled air service in North America in 1990, 552 received service exclusively by small regional and commuter airlines [8]. In response to the growth shown in the commuter market in recent years, several aircraft manufacturers continue to plan the introduction of aircraft designed to meet the particular needs of the commuter operators. Presently, commuter carriers are authorized to utilize aircraft with seating capacities up to 60 passengers on their routes, a significant increase from the restrictions placed upon seating (19 passengers) and maximum gross takeoff weight (12,000 lb) by the Civil Aeronautics Board in the past.

The growth of the commuter industry in the 1980s was largely the result of solving the problems encountered by the industry in attaining suitable working relationships and ownership positions with the larger air carriers, particularly with respect to joint fare agreements, code-sharing, and shared reservation systems. With many large airports in this country reaching saturation in terms of airport facility use, provisions for the commuter carriers to continue to gain access to large airports are essential. Advances in the air traffic control system and on-board instrumentation for instrument flight operations in commuter aircraft allow these operators to use approach and departure paths and specifically designated runways separate from the larger jet transports. Such developments will significantly improve the ability of commuter airlines to gain access to airports and increase the performance reliability of these airlines in the future.

Impact of airline deregulation

Some of the more significant developments in the early years of deregulation included the introduction of a great variety of discount fares for the discretionary traveler, adjustment of fares to more closely reflect the cost of service provided, realignment of the route systems of the airlines, strong growth and apparent acceptance of commuter carriers, and development and growth of "low-cost airlines"

offering basic air transportation service with few of the amenities traditionally associated with air carrier service [14, 27, 34]. Most of
these "low-cost airlines" no longer exist, since they were acquired
through mergers with or consolidations by the larger air carriers.

Service patterns to small communities have shifted since deregulation [27]. The large airlines have abandoned linear route systems in
favor of hub-and-spoke route systems. The structural change in
the route networks of large carriers is clearly shown by comparing
Fig. 1-2, which indicates a typical airline linear route structure in
1979, and Fig. 1-3, which indicates a typical airline hub-and-spoke
structure in 1989.

Less service is being provided between small communities, but
more service is being provided between large communities and small
communities. The majority of cities receiving air service in 1978

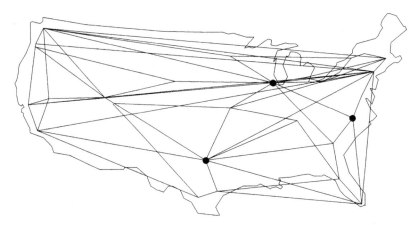

Figure 1-2 Typical airline linear route structure in 1979.

Figure 1-3 Typical airline hub-and-spoke route structure in 1989.

received more frequent air service in 1989. Although in the early years of deregulation there was an expansion in the number of carriers providing service, the later years saw considerable consolidation of the industry through mergers, acquisitions, and bankruptcies. In 1989 about 90 percent of the revenue passenger-miles provided by the domestic air carriers were provided by only eight carriers [27]. In 1990 the largest 50 regional and commuter airlines served 96.4 percent of the total passengers carried by the regional and commuter airlines [8]. The total number of daily departures has risen significantly since deregulation, the number of competitive markets has increased, and the number of single-plane services between cities has increased [27]. The growth experienced by the regional and commuter carriers has outpaced that of the large domestic carriers.

Comparative data on scheduled airline passenger service for 1979 and 1988 are shown in Figs. 1-4 through 1-6 [27]. The total number of city pairs served increased by about 20 percent from 1979 to 1988. Fig. 1-4 shows a comparison of the cumulative percentage of markets

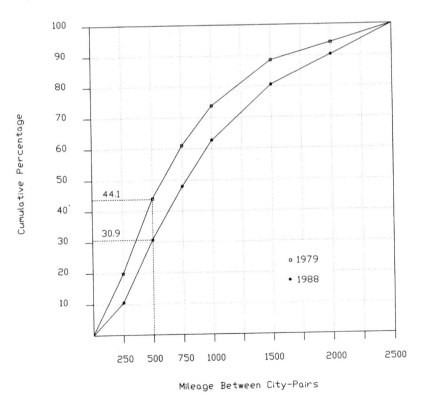

Figure 1-4 Comparison of city-pair markets by mileage block served by U.S. air carriers in 1979 and 1988 (*U.S. Department of Transportation [27]*).

(city pairs) served by the scheduled air carriers in 1979 and 1988. As may be observed, in 1979 these carriers served a greater percentage of longer-stage-length markets than in 1988. In 1979 about 44 percent of the markets served were under 500 mi while this value decreased to about 31 percent in 1988. The total number of scheduled flights increased by about 27 percent from 1979 to 1988. However, the percentage of scheduled flights by the larger air carriers with stage length under 500 mi decreased from about 63 percent in 1979 to about 55 percent in 1988, as shown in Fig. 1-5. The total number of passengers served increased by about 40 percent from 1979 to 1988. However, the percentage of passengers served by the larger air carriers on trips under 500 mi decreased from about 44 percent in 1979 to about 38 percent in 1988, as shown in Fig. 1-6. These changes since deregulation are due to the larger airlines providing longer flight stage lengths from hub airports in which a greater portion of the shorter-stage-length flights are provided by commuter or small

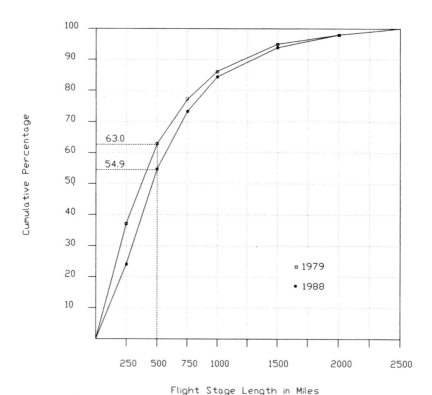

Flight Stage Length in Miles

Figure 1-5 Comparison of the airline passenger flights by mileage block served by U.S. air carriers in 1979 and 1988 (*U.S. Department of Transportation [27]*).

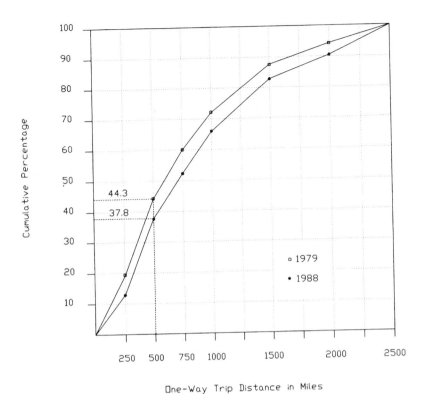

One-Way Trip Distance in Miles

Figure 1-6 Comparison of the total airline passenger trips by mileage block served by U.S. air carriers in 1979 and 1988 (*U.S. Department of Transportation [27]*).

regional air carriers, many of which have code-sharing agreements with the larger air carriers.

Worldwide air commerce

Although international air transport was inaugurated in the mid-1930s, rapid growth did not begin until 1950. Since that time, the average annual growth rate in the number of worldwide passengers was nearly 14 percent in the 1960s, slightly less than 7 percent in the 1970s, and slightly less than 5 percent in the 1980s. Worldwide passenger traffic is shown in Table 1-7. Projections of world traffic growth vary considerably, but if the estimated growth rate of 5 percent per year from 1990–2000 occurs, it has been predicted that an increase to almost 1.9 billion passengers in the year 2000 may occur, nearly a 63 percent growth in passenger traffic during the 1990 to 2000 period.

TABLE 1-7   World Civil Air Transport Revenue of Passengers Carried and
Passenger-Miles Flown on Domestic and International Routes

Year	Passengers (thousands)	Passenger-miles (millions)
1950	31,000	18,000
1960	106,000	68,000
1970	386,000	289,000
1975	534,000	433,000
1980	734,000	665,000
1981	752,000	695,000
1982	766,000	710,000
1983	798,000	739,000
1984	848,000	794,000
1985	899,000	849,000
1986	960,000	902,000
1987	1,027,000	987,000
1988	1,082,000	1,060,000
1989	1,118,000	1,107,000
1990	1,159,000	1,177,000
2000*	1,888,000	1,917,000

*Reflects a projection of an average annual increase of 5.0 percent through the year 2000.
SOURCE: International Civil Aviation Organization [18].

World air cargo has expanded at a very rapid pace—even more so than domestic cargo in the United States. Statistical information on world air cargo is provided in Table 1-8. The average yearly growth rate from 1980 to 1990 was almost 8 percent, which is considerably higher than the passenger growth of 5 percent during the same period. Cargo traffic by the U.S. airlines represents about one-third of the total world air cargo. Estimates of the annual growth of world air cargo ton-miles from 1990 to 2000 indicate that an annual growth rate of 6.3 percent is likely, which will mean that in the year 2000 the world cargo ton-miles will be nearly twice that in 1990. The international portion of the world air cargo revenue ton-miles is expected to increase faster than the domestic portion [13].

The geographic distribution of world air transport is also of interest. For statistical purposes, the International Civil Aviation Organization (ICAO) has divided the world into six regions: Asia and Pacific, Europe, North America, Latin America and Caribbean, Africa, and Middle East. Table 1-9 indicates the regional percentage distribution of total ton-kilometers performed on scheduled services from 1968 to 1990. Note that over 70 percent of all traffic is generated in North America and Europe. The relative growth of traffic in the Asian and Pacific region is quite dramatic, paralleling the growth of importance of this area in the political, social, and economic sectors during the same period. The principal regions of projected relative growth

TABLE 1-8   Revenue Ton-Miles of International and
Domestic Cargo Flown in the World in Scheduled Service
(Millions)

Year	Mail	Freight	Total
1955	281	996	1,277
1960	457	1,628	2,085
1965	831	3,740	4,571
1970	1,876	6,895	8,771
1975	1,981	13,236	15,217
1980	2,529	18,047	20,576
1981	2,598	19,852	22,450
1982	2,405	21,562	23,967
1983	2,646	24,003	26,649
1984	2,946	27,121	30,067
1985	3,007	27,237	30,244
1986	3,111	29,527	32,638
1987	3,200	33,068	36,268
1988	3,302	36,480	39,782
1989	3,452	39,146	42,598
1990	3,610	40,363	43,973
2000*	5,344	79,400	84,744

*Reflects a projection of an average annual increase of 4.0 percent
for mail and 7.0 percent for freight through the year 2000.

SOURCE: International Civil Aviation Organization [18].

TABLE 1-9   Regional Percentage Distribution of Total Ton-Kilometers Performed
on Scheduled Service

Year	Asia and Pacific	Europe	North America	Latin America and Caribbean	Africa	Middle East
1968	8.4	19.8*	63.8	4.3	2.0	1.5
1972	8.4	35.9	47.6	4.4	2.0	1.7
1976	12.4	36.5	41.0	4.8	2.5	2.8
1980	15.5	35.0	38.6	5.5	2.6	2.8
1984	17.6	33.0	38.0	4.9	2.8	3.7
1988	19.8	31.0	39.3	4.6	2.2	3.0
1990	19.8	31.9	38.5	4.7	2.2	2.9
2000†	22.8	25.1	41.5	5.6	2.3	2.7

*Data for 1968 do not include the U.S.S.R. since these data were not available.

†Forecast by Boeing Commercial Airplane Group.

SOURCE: International Civil Aviation Organization [18], Boeing Commercial Airplane
Group [13].

through the year 2000 are the Asian and Pacific region and North America. A significant reduction in the proportion of worldwide traffic in the European region is projected due in part to the formation of the European Economic Community (EEC) and the impact of the removal of barriers for airline travel within the EEC.

General aviation

General aviation is the term used to designate all flying done other than by the commercial airlines. For statistical purposes, general aviation in the United States is usually divided into business flying (i.e., transportation not for hire), commercial flying, instructional flying, and personal flying. Some concept of the size of the general aviation activities can be gained from the fact that in 1990 general aviation accounted for about 35 times as many airplanes, accumulated nearly twice the mileage, and flew almost 2.5 times the number of hours as the scheduled airlines. It also accounted for 61 percent of all civil aircraft operations at airports with FAA control towers [16].

A comparison of the air carrier fleet with the general aviation fleet is presented in Table 1-10. In 1990 about 68 percent of the air carrier fleet was turbojet, this percentage being dramatically reduced over the past 15 years due to the addition of smaller airlines in scheduled service as a direct result of deregulation of the industry. About 78 percent of the general aviation fleet consists of single-engine aircraft. The dominant uses of general aviation aircraft are for business, commercial, instructional, and personal purposes, as shown in Table 1-11. From 1980 to 1990, the use of general aviation aircraft in terms of

TABLE 1-10   U.S. Civil Air Fleet

Calendar year ending	Air carrier			General aviation		
	Jet	Piston, turboprop, rotary wing	Total	Single-engine	Multiengine	Total*
1960	202	1,933	2,135	69,000	7,000	77,000
1965	725	1,400	2,125	81,000	11,000	95,000
1970	2,136	543	2,679	111,000	16,000	134,000
1975	2,114	381	2,495	137,000	24,000	168,000
1980	2,531	1,277	3,808	168,000	32,000	211,000
1985	3,164	1,514	4,678	164,000	34,000	211,000
1990	4,148	1,935	6,083	165,000	33,000	212,000
1995†	4,658			164,500	33,500	214,200
2000†	5,548			167,400	35,400	221,600

*Total includes other aircraft such as helicopters.

†Forecast.

SOURCE: Federal Aviation Administration [15, 16].

TABLE 1-11   Hours Flown in General Aviation by Type of Use (Thousands)

Year	Total hours	Business h	Business %	Commercial h	Commercial %	Instructional h	Instructional %	Personal h	Personal %	Other h	Other %
1940	3,200	314	10	387	12	1,529	48	970	30	135	1
1951	8,451	2,950	35	1,584	19	1,902	23	1,880	22	57	1
1960	13,121	5,699	44	2,365	18	1,828	14	3,172	24	75	1
1965	16,733	5,857	35	3,348	20	3,346	20	4,016	24	106	1
1970	26,028	7,182	28	4,582	18	6,798	26	6,936	27	530	2
1975	34,165	9,545	28	6,480	19	8,174	24	9,244	27	722	2
1980	38,300	12,300	32	7,300	19	9,600	25	8,400	22	700	2
1985	31,500	8,900	28	6,200	20	7,600	24	8,200	26	600	2
1990	32,000	7,600	24	6,400	20	7,400	23	9,700	30	900	3
1995*	37,000										
2000*	39,300										

*Only total hours flown are forecast.
SOURCE: Federal Aviation Administration [15, 16].

hours flown has decreased by 16 percent. The number of licensed pilots and general aircraft deliveries also decreased during this period. These decreases are due in large measure to fuel cost increases, product liability costs, economic factors, repeal of the investment tax credit, and air traffic congestion. It is expected that this trend will continue through the year 2000. Although the general aviation fleet has not changed materially in recent years, there has been a marked increase in the number of aircraft equipped for instrument flying and in the number of pilots with instrument certification. At present, general aviation represents about 41 percent of all instrument operations monitored by the FAA at all air traffic control facilities [16].

Airports

The number of airports in the United States has grown steadily during the last decade, as shown in Table 1-12. Nearly 68 percent are not open to the public due principally to private ownership without the granting of public access. However, note that of the 17,451 airports on

TABLE 1-12   Number of Airports in the United States

	1960	1970	1980	1990
Total airports on record with FAA	6,881	11,340	15,161	17,451
Publicly owned	2,780	4,260	4,785	4,169
With paved runways	1,893	3,717	5,833	7,694
With scheduled airline service	575	518	628	568
National airport system plan	2,770	3,236	3,159	3,285

SOURCE: Federal Aviation Administration [16, 22].

record in 1990, only 38 percent had paved runways, and only 568 were being served by scheduled passenger airline service. In 1990 there were 29 large air traffic hub airports, 42 medium air traffic hub airports, 64 small air traffic hub airports, and 433 nonhub airports [16]. Only 400 airports had sufficient traffic to justify FAA control towers. Finally, only 3285 airports satisfied the criteria to be included in the National Plan of Integrated Airport Systems [22]. Should commuter travel continue to expand at its present rate, many more airports will undoubtedly be served by scheduled airlines.

In order that the reader may gain an appreciation of the magnitude of traffic at some of the busiest airports in the world, statistical data concerning annual aircraft operations and annual passengers handled are presented in Table 1-13. It is interesting to note the fairly large

TABLE 1-13   Passenger Volumes and Aircraft Operations at Selected Airports in the United States and the World for Calendar Years 1980 and 1990

Airports	Total aircraft operations		Total passengers	
	1980	1990	1980	1990
United States				
Chicago O'Hare	724,155	810,865	43,653,000	59,936,137
Dallas–Ft. Worth	463,447	731,036	21,951,355	48,515,464
Hartsfield-Atlanta	612,525	790,502	40,180,000	48,024,566
Los Angeles	523,961	679,861	33,038,000	45,810,221
San Francisco	359,146	430,253	22,248,000	31,059,820
John F. Kennedy	307,527	302,038	26,796,000	29,786,657
Denver Stapleton	486,888	484,040	20,849,000	27,432,989
Miami	369,563	480,987	20,505,000	25,837,445
Honolulu	375,408	407,048	14,036,350	23,367,770
Boston Logan	296,851	424,568	14,722,000	22,935,844
LaGuardia	316,811	354,229	17,459,000	22,753,812
Detroit Metro	267,280	387,848	9,759,404	22,585,156
Newark	196,781	379,432	9,223,260	22,255,002
Phoenix Sky Harbor	376,421	498,752	6,585,854	21,718,068
Minneapolis–St. Paul	282,027	379,785	9,024,472	20,381,314
Europe				
London-Heathrow	294,600	390,485	27,472,000	42,964,200
Frankfurt	237,765	324,387	16,873,000	28,912,145
Paris-Orly	183,850	193,451	15,866,557	24,329,700
Osaka International	127,612	130,550	16,433,450	23,511,611
Paris-De Gaulle	90,000	235,350	8,841,000	22,506,107
London-Gatwick	143,500	203,200	9,707,400	21,185,400
Toronto-Pearson	253,559	353,682	13,707,238	20,304,271
Tokyo Narita	62,000	125,187	7,200,000	19,264,650
Hong Kong	74,070	122,945	6,813,279	18,687,525
Rome-Fiumicino	146,508	176,456	11,434,000	17,835,885
Amsterdam Schiphol	185,386	245,350	9,401,025	16,185,810
Stockholm Arlanda	89,514	257,606	8,532,200	14,821,692
Bangkok International	54,763	125,998	4,589,826	14,329,337

SOURCE: Airports Association Council International [35].

TABLE 1-14   Profile of Aircraft Operations at Selected Air Carrier Airports in the United States, Fiscal Year 1980

Airport	Annual	Total aircraft operations			
		Average day	Peak day	Average hour	Peak hour
Chicago O'Hare International	734,555	2012	2639	84	178
Los Angeles International	534,414	1464	1742	61	133
Hartsfield-Atlanta International	609,466	1670	1869	70	145
John F. Kennedy International	311,777	854	1167	36	128
San Francisco International	371,222	1017	1198	42	95
Denver Stapleton International	485,695	1331	1390	55	169
LaGuardia	319,891	876	1125	37	92
Miami International	376,820	1032	1340	43	111
Washington National	354,717	972	1289	41	108
Boston Logan International	340,896	934	1219	39	112

SOURCE: Federal Aviation Administration [28].

increase in the number of passengers handled from 1980 to 1990 and the fact that the increase in the number of aircraft operations has been less significant and even decreased in some instances. This latter observation may be attributed to the greater availability and use of wide-bodied aircraft for passenger travel during this period.

Although the data in Table 1-14 are not current, it is useful to compare the relationships among the annual, average and peak day, and average and peak hour aircraft operations for a sample of airports in the United States for fiscal year 1980. It is informative to note the relationships between the average and peak traffic on both a daily and an hourly basis. Similarly, a comparison of the level of total aircraft operations and air carrier operations on an annual and busy-hour basis for selected air traffic hub and general aviation airports in fiscal year 1980 is shown in Table 1-15. Observe that air carrier traffic is a significant portion of total aircraft operations at only the large hub airports and that the busy-hour operations at other hub and general aviation airports can be quite large.

Historical Review of the Federal Role in Aviation

Commercial air transportation started in the United States with the carriage of mail. As early as 1911, the Post Office Department showed an interest in the transportation of mail by air, and from then on the department did much to encourage civil aviation. Attempts to obtain federal appropriations for airmail began in 1912 but met with little success until 1916, when an appropriation for experimental purposes was made. In 1918 the first airmail route in the United States was established between Washington, D.C. and New York City. At the

TABLE 1-15   Annual and Busy-Hour Aircraft Operations at Selected United States
Air Traffic Hub and General Aviation Airports, Fiscal Year 1980

Airport	Annual		Busy hour	
	Total	Air carrier	Total	Air carrier
Large hub airports				
Chicago O'Hare International	734,555	577,671	178	141
Hartsfield-Atlanta International	609,466	541,440	145	134
Los Angeles International	534,414	419,506	133	88
Denver Stapleton International	485,695	316,664	169	105
Miami International	376,820	283,457	111	84
Washington National	354,717	204,560	108	56
Boston Logan International	340,896	204,159	112	59
Medium hub airports				
Memphis International	377,603	143,294	115	48
Salt Lake City International	285,104	83,723	155	27
Milwaukee-Mitchell Field	247,290	85,011	136	35
Portland International	219,404	75,827	107	37
Orlando Jetport	157,535	115,936	67	42
San Diego International	155,914	69,513	54	18
Greater Cincinnati	119,088	59,773	68	25
Small hub airports				
San Jose Municipal	415,543	47,255	183	19
Daytona Beach Regional	292,534	15,030	164	9
Wichita Mid-Continent	230,128	37,439	169	37
Sacramento Metro	170,733	36,749	106	20
Grand Rapids Kent County	167,922	27,334	126	11
Richmond Byrd International	163,998	33,813	75	17
Knoxville McGhee-Tyson	143,575	19,598	71	7
General aviation				
Fort Worth Meacham	400,722		284	
Pontiac	274,686		220	
Seattle Boeing Field	408,207		247	
Tamiami	422,867		236	
Teterboro	308,413		231	
Torrance Municipal	370,398		284	
Van Nuys	567,055		310	

SOURCE: Federal Aviation Administration [28].

start of this service the flying operations were conducted by the War Department, but later that year the Post Office Department took over the entire operation with its own equipment and pilots. Service was inaugurated between New York City and Chicago in 1919 and was extended to San Francisco in 1920.

The Post Office Department, having demonstrated the practicality of moving mail by air, desired to turn over the operation to private enterprise. By 1925 the development work of the government had reached a stage where private operation seemed feasible. Accordingly, legislation permitting the Post Office Department to contract with private operators for the carriage of mail by air was provided by the Air Mail Act of 1925 (Kelly Act). It was not until 1926, however, that a number of contract routes were opened. Some of the early contractors were the Ford Motor Company, Boeing Air Transport (predecessor of United Airlines), and National Air Transport.

Air Commerce Act of 1926

Thus far there was no legislation which fostered and encouraged the orderly development of air transportation. Oddly enough, the first year of the carriage of mail also saw the passage of the first federal law dealing with air commerce, the Air Commerce Act of 1926 (Public Law 64-254). Although this law provided regulatory measures, it did more to aid and encourage civil aviation than to regulate. The principal provisions of this act were as follows:

1. All aircraft owned by U.S. citizens operating in common-carrier service or in connection with any business must be registered.

2. All aircraft must be certificated and operated by certified pilots.

3. Authority was given to the Secretary of Commerce to establish air traffic rules.

4. The Secretary of Commerce was authorized to establish, operate, and maintain lighted civil airways.

In drafting the legislation, Congress relied considerably upon the precedents in maritime law. An analogy was utilized between the role of government in meeting water navigation needs and the role of government in air navigation. In water navigation these services included the signing, lighting, and marking of channels; safety inspection of ships and operating personnel; assistance in the development and improvement of ports and waterways; and laws concerning the operations of the industry. The provision of docks and terminal facilities was the responsibility of local government or the private sector of the economy. Therefore, the legislation was adopted in such a framework which held that airports were analogous to the docks of waterborne transportation [31].

Aid for airports was specifically prohibited. This policy remained in effect until the passage of the Civil Aeronautics Act of 1938. The Air Commerce Act had one serious disadvantage, namely, that control of air transportation was divided among several government agencies. The airmail contracts were let by the Post Office Department, airmail rates were subject to regulation by the Interstate Commerce Commission, and matters having to do with registration, certification, and airways were vested in the Bureau of Air Commerce in the

Department of Commerce. The result of this divided jurisdiction was a lack of coordination in the efforts of government to develop the air transportation industry.

Civil Aeronautics Act of 1938

The Air Commerce Act of 1926 had been passed before the carriage of mail and passengers had developed into a substantial business enterprise. The failure of this legislation to provide adequate economic control led to wasteful and destructive competitive practices. The carriers had little security in their routes and therefore could not attract private investors and develop traffic volumes sufficient to achieve economic stability. These particular weaknesses in the existing legislation led to the enactment of the Civil Aeronautics Act of 1938 (Public Law 76-706). This act defined in a precise manner the role of the federal government in respect to the economic phases of air transport. It created one independent agency to foster and regulate air transport in lieu of the three agencies operating under the Air Commerce Act. This new agency was called the Civil Aeronautics Authority (not to be confused with the Civil Aeronautics Administration). It consisted of a five-member Authority, a three-member Air Safety Board, and an Administrator. The five-member Authority was principally concerned with the economic regulation of air carriers; the Air Safety Board was an independent body for the investigation of accidents; and the concerns of the Administrator dealt primarily with construction, operation, and maintenance of the airways.

During the first year and a half of its existence, a number of organizational difficulties arose within the Civil Aeronautics Authority. As a result, President Franklin D. Roosevelt, acting within the authority granted to him in the Reorganization Act of 1939 (55 Stat. 561), reorganized the Civil Aeronautics Authority and created two separate agencies, the Civil Aeronautics Board and the Civil Aeronautics Administration. The five-member Authority remained as an independent agency and became known as the Civil Aeronautics Board; the Air Safety Board was abolished, and its functions were given to the Civil Aeronautics Board; and the Administrator became the head of an agency within the Department of Commerce known as the Civil Aeronautics Administration (CAA). The duties of the original five-member Authority were unchanged, but certain responsibilities, such as accident investigation, were added because of the abolition of the Air Safety Board. The Administrator, in addition to retaining the functions of supervising construction, maintenance, and operation of the airways, was required to undertake the administration and enforcement of safety regulations and the administration of the laws with regard to aircraft operation. Subsequently, the Administrator became directly responsible to the Secretary of Commerce (1950). The Civil Aeronautics Act, like its predecessor, the Air Commerce Act, authorized the federal government to establish, operate, and maintain the airways; but again, authorization for actively aiding airport development was lacking. The act, however, authorized the expenditure of federal funds for the construction of landing areas, provided the Administrator certified "that such landing area was reasonably necessary for use in air commerce or in the interests of national defense." The act also directed the Administrator to make a survey of airport needs in the United States and to report to Congress about the desirability of federal participation and the extent to which the federal government should participate.

In accordance with the requirements of the act, the Civil Aeronautics Authority conducted a detailed survey of the airport needs of the nation. An advisory committee was appointed, composed of representatives of interested federal agencies (both military and civil), state aviation officials, airport managers, airline representatives, and others. A report was submitted to Congress on March 23, 1939 (House Document 245, 76th Congress, 1st session). Some of the more important recommendations in this report were as follows:

1. Development and maintenance of an adequate system of airports and seaplane bases should be recognized in principle as a matter of national concern.

2. Such a system should be regarded, under certain conditions, as a proper object of federal expenditure.

3. In passing upon applications for federal expenditure on airport development or improvement, the highest preference should be given to airports which are important to the maintenance of safe and efficient operation of air transportation along the major trade routes of the nation; and to those rendering special service to the national defense.

4. At such times as the national policy includes the making of grants to local units of government for public works purposes, or any work relief activity, a proportion of the funds involved should be allocated to airport purposes. Such purposes should be given preference as rendering an important service to the localities concerned and at the same time being of particular importance to the nation's commerce and defense.

5. Whenever emergency public works programs may be terminated, or when such programs may be curtailed to a degree not enabling adequate airport development to continue, or when Congress for other reasons may determine that federal assistance for airports should be continued through annual appropriations for that purpose, based upon annual reports (which should include a review of the gen-

eral status of the nation's airport system and of the work recently done or currently in course of being done) and recommendations for future work in the interest of developing and maintaining a system adequate to national needs, expenditures at these periods should be limited to projects of exceptional national interest.

6. In connection with such public works or work relief programs as normally involve joint contributions by the federal government and by local government, there should be a provision of supplementary funds to enable the federal government to increase its share of the total expense, in any proportion justified by the importance of the project.

7. All applications for Federal airport grants from such a supplementary appropriation should be presented through agencies of state government.

8. In deciding upon the wisdom and propriety of granting any such applications, and the priority that should be given to them, consideration should be given to the aeronautical policy of the state in question, with reference to such matters as the state's policy in protecting the approaches to airports; the state's policy in respect to the employment of any taxes collected on the fuel used in aircraft; and any measures taken by the state to ensure the proper maintenance of airports and the maintenance of reasonable charges for the services given them.

9. The detailed plans for the location and development of any airport with respect to which there is federal contribution of any kind should be subject to the approval of the federal agency charged with the establishment of civil airways, landing areas, and necessary air navigation facilities.

10. There should be no direct federal contribution to the cost of maintaining airports, other than federal airports; except that the Administrator of the Civil Aeronautics Authority may, in accordance with the Civil Aeronautics Act and so far as available funds permit, assume the cost of operating any lighting equipment and other air navigation facility as a part of the cost of operation of the federal airways system.

The airport survey submitted in 1939 was brought up to date with new studies completed in 1940. Continuing studies were made through the war years. While first importance was attached to the military requirements, care was taken whenever possible to anticipate the needs of postwar civil aviation. During the war years the federal government, through the CAA, spent $353 million for the development of military landing areas in the continental United States. This does not include funds spent by the military agencies. During the same period the CAA spent $9.5 million for the development of landing areas in the United States solely for civil purposes.

At the end of World War II, over 500 airports constructed for the military by the CAA were declared surplus and were turned over to cities, counties, and states for airport use. The interest in adequate airport facilities by various political subdivisions of government continued. The needs were made known to Congress by various interests. As a result, the House of Representatives passed a resolution (H.R. 598, 78th Congress) directing the CAA to make a survey of the "need for a system of airports and landing areas throughout the United States" and to report back to Congress.

The results of this survey were completed in 1944 (House Document 807, 78th Congress, 2d session) and contained the following principal recommendations:

1. Congress should authorize an appropriation to the office of the Administrator of Civil Aeronautics not to exceed $100 million annually to be used in a program of federal aid to public agencies for the development of a nationwide system of public airports adequate to meet the present and immediate future needs of civil aeronautics. The Administrator should be authorized to allocate such funds for any construction work involved in constructing, improving, or repairing an airport, including the construction, alteration, and repair of airport buildings other than hangars and the removal, lowering, marking, and lighting of airport obstructions; for the acquisition of any lands or property interest necessary either for any such construction or to protect airport approaches; for making field surveys; preparing plans and specifications; supervising and inspecting construction work, and for any necessary federal expenses in the administration of this program.

2. Such a program should be conducted in cooperation with the state and other nonfederal public agencies on a basis to be determined by Congress. The Federal contribution should be determined by Congress in passing the necessary enabling legislation. A good precedent for the proportionate sharing of costs exists in the public roads program which has operated satisfactorily for many years on a 50-50 basis.

3. Any project for which federal aid is requested must meet with the approval of the Administrator of Civil Aeronautics as to the scope of development and cost and must conform to Civil Aeronautics Administration standards for location, layout, grading, drainage, paving, and lighting. All work thereon is subject to the inspection and approval of the Civil Aeronautics Administration.

4. To participate in the federal aid program, a state shall:
   a. Establish and empower an official body equipped to conduct its share of the program.

b. Have legislation adequate for the clearing and protection of airport approaches and such other legislation as may be necessary to vest in its political subdivisions all powers necessary to enable them to participate through the state as sponsors of airport projects.

c. Have no special tax on aviation facilities, fuel, operations, or businesses, the proceeds of which are not used entirely for aviation purposes.

d. Ensure the operation of all public airports in the public interest, without unjust discrimination or unreasonable charges.

e. Ensure the proper operation and maintenance of all public airports within its jurisdiction.

f. Make airports developed with federal aid available for unrestricted use by U.S. government aircraft without charge other than an amount sufficient to cover the cost of repairing damage done by such aircraft.

g. Require the installation at all airports for which federal funds have been provided for a standard accounting and fiscal reporting system satisfactory to the Administrator.

5. Sponsors of projects are required to enter into contracts with the Civil Aeronautics Administration ensuring the proper maintenance and protection of airports developed with federal aid and their operation in the public interest.

Federal Airport Act of 1946

In 1944 an airport development bill was introduced in the House of Representatives (H.R. 5024), but no action was taken on it. After extensive hearings in both houses of Congress, the Senate passed an airport bill (S. 2) in 1945, and later that year the House passed a bill (H.R. 3615). The language in these two bills differed in several respects. One of the principal differences was the method employed in channeling funds to the municipalities. The Senate bill provided that funds be channeled to the municipalities through appropriate state aviation organizations unless a state failed to have an appropriate agency to handle the matter. The House bill permitted channeling of funds either through the state or directly to a municipality or other political subdivision of government. The substitute bill agreed to in conference conformed more nearly to the House language. Another difference had to do with the size of the discretionary fund, which, instead of being apportioned among the states by a fixed formula, would be available for use at the sole discretion of the Administrator. The House bill provided 25 percent of the total appropriation for airport development as a discretionary fund; the Senate bill, 35 percent. The compromise reached in conference retained the House version. Other differences which were worked out in conference concerned whether the costs of the acquisition of land and interest in airspace should be eligible for federal aid, project sponsorship requirements, and the reimbursement for damage to public airports caused by federal agencies.

The conference report was approved by Congress, and the Federal Airport Act of 1946 was enacted (Public Law 79-377).

Federal Aviation Act of 1958

For a number of years there had been a growing concern about the division of responsibility in aviation matters among different agencies of the federal government. Unlike highway or other forms of transport, aviation is unique in its relation to the federal government. It is the only mode whose operations are conducted almost wholly within federal jurisdiction, and it is subject to little or no regulation by states or local authorities. Thus, the federal government bears virtually complete responsibility for the promotion and supervision of the industry in the public interest. The military interest and the entire national defense concept are also intimately related to aviation.

Recognizing that the demands on the federal government in the years ahead would be substantial, the director of the Bureau of the Budget requested a review of aviation facilities problems in 1955. A report was issued later that year, recommending that a study of "long-range needs for aviation facilities and aids be undertaken" and that such a study be made under the direction of an individual of national reputation.

President Dwight D. Eisenhower accepted these recommendations and appointed Edward P. Curtis as his Special Assistant for Aviation Matters in 1957 [23]. Curtis was charged with the responsibility of preparing a comprehensive aviation facilities plan which would "provide the basis for the timely installation of technically adequate aids, for optimum coordination of the efforts of the civil and military departments, and for effective participation by state and local authorities and the aircraft operators in meeting facilities requirements." Curtis completed his report and submitted it to the President. In this report Curtis stated that "it has become evident that the fundamental reason for our previous failures lies with the inability of our governmental organizations to keep pace with the tremendous growth in private, commercial, and military aviation which has occurred in the last twenty years." Curtis recommended the consolidation of all aviation functions, other than military, into one independent agency responsible only to the President. However, the report recognized that to "develop new management structures and policy, to coordinate proposals within the executive branch and to obtain legislation implementing a new permanent organization might be as long as two or three years." The most urgent matter requiring attention was in the area of air traffic control. The collision of two aircraft over the Grand Canyon in 1956 provided the impetus for rapid legislative action for remedying midair collisions. Curtis recommended that, as an interim measure, there be created an Airways Modernization Board whose function was to "develop, modify, test, and evaluate systems, procedures, facilities, and devices, as well as define the performance characteristics thereof, to meet the needs for safe and efficient navigation and traffic control of all civil and military aviation except for those needs of military agencies which are peculiar to air warfare and primarily of military concern, and select such systems, procedures, facilities, and devices which will best serve such needs and will promote maximum coordination of air traffic control and air defense systems." The board was to consist of the Secretary of Commerce, the Secretary of Defense, and an independent chairperson.

Congress was receptive to this recommendation and passed the Airways Modernization Act of 1957 (Public Law 85-133), establishing the board for a 3-year term.

In the meantime, there were more midair collisions and reports of near misses were given wide circulation. Costly disagreements between the CAA and the military on the type of navigational aids to be used on the airways no doubt also spurred congressional action. As a result, instead of taking 2 or 3 years to create a single aviation agency as was predicted, Congress acted favorably on the legislation within a year of the passage of the Airways Modernization Act. This legislation is known as the Federal Aviation Act of 1958 (Public Law 85-726). This law superseded the Civil Aeronautics Act of 1938 but not the Federal Airport Act of 1946.

The principal provisions of the law relating to organizational changes are as follows:

1. The Federal Aviation Agency was created as an independent agency with an Administrator directly responsible to the President. The agency incorporated the functions of the Civil Aeronautics Administration and the Airways Modernization Board, both of which were abolished.

2. The Civil Aeronautics Board was retained as an independent agency including all its functions except its safety rule-making powers, which were transferred to the Federal Aviation Agency.

Creation of the Department of Transportation

For many years it had been argued that there had been a proliferation of federal activities with regard to transportation. For example, the

Bureau of Public Roads was part of the Department of Commerce whereas the Federal Aviation Agency was an independent agency. It was felt by different transport interests that there was a lack of coordination and effective administration of the transportation programs of the federal government, resulting in a lack of a sound national transportation policy. It is interesting to note that the first legislative proposal in this direction dates back to 1874. However, in recent years, the involvement of the federal government in the development of the transportation systems of the nation has been enormous, requiring much more coordination among federal transport activities than ever before. With this as a background, a Cabinet-level Department of Transportation (DOT) was created, headed by the Secretary of Transportation (Public Law 89-670). The department began to function on April 1, 1967.

The agencies and functions transferred to the Department of Transportation related to air transportation included the Federal Aviation Agency in its entirety and the safety functions of the Civil Aeronautics Board, including the responsibility for investigating and determining the probable cause of aircraft accidents, and its appellate safety functions involving review on appeal of the suspension, modification, or denial of certificates or licenses. The name of the Federal Aviation Agency was changed to the Federal Aviation Administration (FAA). The Administrator is still appointed by the President but reports directly to the Secretary of Transportation.

The National Transportation Safety Board (NTSB) was established by the same act which created the Department of Transportation to determine "the cause or probable cause of transportation accidents and reporting the facts, conditions, and circumstances relating to such accidents" for all modes of transportation. Although created by the act which established the DOT, the board in carrying out its functions is "independent of the Secretary and other offices and officers of the Department." The board consists of five members appointed by the President and annually reports directly to Congress.

The creation of the DOT did not alter any legislation in the Federal Aviation Act of 1958, with the exception of the transfer of the safety functions from the Civil Aeronautics Board to the National Transportation Safety Board. In the act of establishing the new department, however, there was a statutory requirement to establish an Office of Noise Abatement to provide policy guidance with respect to interagency activities related to the reduction of transportation noise. With the introduction of jet aircraft in 1958, the complaints against aircraft noise increased significantly. As a result, in 1968 the Federal Aviation Act was amended by Congress (Public Law 90-411) to require noise abatement regulations. Its principal purpose was to establish noise levels which aircraft manufacturers cannot exceed in the development of new aircraft.

Airport and Airway Development Act of 1970

In the mid-1960s, as air traffic was expanding at a fairly rapid pace, air traffic delays getting into and out of major airports began to increase rapidly. Along with the delays in the air, there was congestion on the ground in parking areas, on access roads, and in terminal buildings. It was evident that to reduce congestion, substantial financial resources would be required for investment in airway and airport improvements. For airports alone it was estimated that $13 billion in new capital improvements would be required for public airports in the 1970–1980 period. The amount of money authorized by the Federal Airport Act of 1946 was insufficient to assist in financing such a vast program. The normal and anticipated sources of revenue available to public airports were also not sufficient to raise the required funds for capital expenditures. It was argued that much of the congestion in the air at major airports was due to a lack of funds to modernize the airways system. Funds for airport development came from the budget of the FAA, authorized by Congress each year, and not from the Federal Airport Act. The deficiencies in airport and airway development were documented in several reports. It was the consensus of industry and government that the only way to provide the funds needed for airports and airways was through increased or new taxes imposed upon the users of the air transport system. It was also argued that the revenues from these taxes should be specifically earmarked for aviation and should not go to the general fund. The concept of establishing a trust fund similar to that of the national highways program was agreed upon. Finally, after much debate in Congress, the Airport and Airway Development Act of 1970 and the Airport and Airway Revenue Act of 1970 were enacted (Public Law 91-258).

As finally passed, the act was divided into two sections: Title I detailed the airport assistance programs and established a financing program for airport grants, airways hardware acquisition, and research and development; Title II created the Airport and Airway Trust Fund and established the pattern of aviation excise taxes which would provide the resources upon which the Title I capital programs would depend through 1980. The excise taxes adopted consisted of a tax on domestic passenger tickets, a head tax on international passenger departures, a flowage tax on all fuel used by general aviation, and tax on all air cargo waybills. In addition, an annual aircraft registration tax was levied on all aircraft (commercial and general aviation) plus an annual weight surtax for all aircraft weighing in excess of 2500 lb. Finally, revenues from existing taxes on aircraft tires and tubes were transferred from the Highway Trust Fund to the Airport and Airway Trust Fund.

The amount of these excise taxes was changed most recently in the Omnibus Budget Reconciliation Act of 1990 (Public Law 101-508) and currently consist of a 10 percent tax on domestic passenger tickets, a $6 head tax on international passenger departures, a flowage tax of 17.5 cents per gallon on all fuel used by general aviation, and a 6.25 percent tax on all air cargo waybills.

There were significant changes from the Federal Airport Act:

1. The provision of funds to local agencies for airport system planning and master planning

2. The emphasis on airports served by air carriers and general aviation airports to relieve congested air carrier airports

3. The provision of funds for commuter service airports

4. The requirement that the FAA issue airport operating certificates to ensure that airports were adequately equipped for safe operations

5. Provision of requirements to ensure that airport projects did not adversely affect the environment and were consistent with long-range development plans of the area in which the project was proposed

6. Provision for terminal facility development in non-revenue-producing public areas

7. The requirement that the FAA develop a national airport system plan (NASP)

To be eligible for federal aid, airport ownership was required to be vested in a public agency and the airport must be included in the *National Airport System Plan* (NASP). This plan was reviewed and revised as necessary to keep it current. It was prepared by the FAA and submitted to Congress by the Secretary of Transportation. The plan specified, in terms of general location and type of development, the projects considered necessary to provide a system of public airports adequate to anticipate and meet the needs of civil aeronautics. These projects included all types of airport development eligible for federal aid under the act and were not limited to any classes or categories of public airports. The plan was based on projected needs over 5- and 10-year periods.

Because of the mounting public concern for the enhancement of the environment, the act specifically stated that authorized projects provide for the protection and enhancement of the natural resources and the quality of the environment of the nation. The Secretary of Transportation was required to consult with the Secretary of the

Interior and the Secretary of Health, Education, and Welfare regarding the effect of certain projects on natural resources and whether "all possible steps have been taken to minimize such adverse effects." The act required that airport sponsors provide the "opportunity for public hearings for the purpose of considering the social, economic, and environmental effect on any project involving the location of an airport, the location of a runway, and a runway extension." In addition, the National Environmental Policy Act of 1969 (Public Law 91-190), supported by a Presidential Executive Order (11514, March 5, 1970), required the preparation of detailed environmental impact statements for all major airport development actions significantly affecting the quality of the environment. The environmental impact statement was required to include the probable impact of the proposed project on both the human and the natural environment, including impact on ecological systems such as wildlife, fish, and marine life, and any probable adverse environmental effects which could not be avoided if the project were implemented. The act also stipulated that no airport project involving "airport location, a major runway extension, or runway location" could be approved for federal funding unless the governor of the state in which the project was located certified to the Secretary of Transportation that there was reasonable assurance that the project would comply with applicable air and water quality standards. Finally, the project had to be consistent with the plans of other agencies for development of the area, and the airport sponsor had to assure the government that adequate housing was available for any displaced people. The Aviation Safety and Noise Abatement Act of 1979 (Public Law 96-193) amended this legislation to place increased emphasis on reducing the noise impact of airports. Thus one of the principal differences between the Federal Airport Aid Program and the Airport Development Aid Program was the emphasis on environmental protection in the latter.

The Airport and Airway Development Act made no mention of specific standards for determining airports to be included in the NASP. It did state, however, that

> The Plan shall set forth, for at least a ten-year period, the type and estimated cost of an airport development considered by the Secretary to be necessary to provide a system of public airports adequate to anticipate and meet the needs of civil aeronautics, to meet the requirements in support of the national defense as determined by the Secretary of Defense, and to meet the special needs of the Postal Service. In formulating and revising the plan, the Secretary shall take into consideration, among other things, the relationship of each airport to the rest of the transportation system in the particular area, to the forecasted technological developments in aeronautics, and to developments forecasted in the other modes of transportation.

With this and other policy guidelines, the FAA developed entry criteria which described a broad and balanced airport system. For example, the 1980 NASP included about 3600 airport locations, indicating that the federal interest in developing a basic airport system extended well beyond the major airports with scheduled airline service. In an effort to provide a safe and adequate airport for as many communities as possible, NASP criteria were developed to include the general aviation airports which serve smaller cities and towns.

The NASP airport entry criteria evolved from both policy and legislative considerations and focused on two broad categories of airports: those with scheduled service and those without significant scheduled service in the general aviation and reliever category. Airports with scheduled service were included in NASP because of their use by the general public, legislative provisions which specifically designated airports to receive development funds, and their use by CAB-certified carriers. Commuter airports were identified in the NASP starting in 1976, when legislation was enacted which designated them as a type of air carrier airport and provided them with special development funds. About 70 percent of the airports in the NASP were general aviation locations which met the criteria of viability because of the number of based aircraft or aircraft activity and which provided reasonable access for aircraft owners and users to their community. Reliever airports have been included as a separate NASP category since the 1960s, when Congress designated special funding for the purpose of relieving congestion in large metropolitan areas by providing additional general aviation capacity.

Airline Deregulation Act of 1978

The Airline Deregulation Act (Public Law 95-504) was passed by Congress in October 1978. This legislation eliminated the statutory authority for the economic regulation of the passenger airline industry in the United States. It provided that the Civil Aeronautics Board (CAB) would be abolished in 1985. The legislation was intended to increase competition in the passenger airline industry by phasing out federal authority to exercise regulatory controls between 1978 and 1985.

The principal provisions of this legislation were as follows:

1. It required the CAB to place maximum reliance on competition in its regulation of interstate airline service, while continuing to ensure the safety of air transportation; to maintain service to small communities; and to prevent practices which were deemed anticompetitive in nature.

2. It required CAB approval of airline acquisitions, consolidations, mergers, purchases, and operating contracts. The burden to prove

that an action was anticompetitive was placed upon the party challenging that action.

3. It permitted carriers to change rates within a range of reasonableness from the standard industry fare without prior CAB approval. The CAB was authorized to disallow a fare change if it considered the change predatory.

4. It provided interstate carriers an exemption from state regulation of rates and routes.

5. It required the CAB to authorize new routes and services that were consistent with the public convenience and necessity.

6. It allowed carriers to be granted operating rights to any route on which only one other carrier was providing service and on which other airlines were authorized to provide service but were not actually providing a minimum level of service. If more than one airline was providing service on this route, the CAB was required to determine if the granting of additional route authority was consistent with the public convenience and necessity before allowing additional carriers to service the route. An airline not providing the specified minimum level of service on a route (dormant authority) could begin providing such service and retain its authority. Otherwise, the CAB was required to revoke the unused authority.

7. It provided for an automatic market entry program, whereby airlines could begin service on one additional route each year during the 1979–1981 period without formal CAB approval. Each carrier was also permitted to protect one of its existing routes each year by declaring it as ineligible for automatic market entry by another airline.

8. It authorized the CAB to order an airline to continue to provide "essential air transportation service" and, for a 10-year period, to provide subsidies or seek other willing carriers to ensure the continuation of essential service.

9. It required the CAB to determine within one year of the enactment of the legislation what it considered to be "essential air transportation service" for each point being serviced at the time of enactment of the legislation.

10. It required the CAB and the Department of Transportation to determine mechanism by which the state and local governments should share the cost of subsidies from the federal government to preserve small-community air service and to make policy recommendations to Congress on this matter.

11. It exempted from most CAB regulations all commuter aircraft weighing less than 18,000 lb and carrying fewer than 56 passengers.

12. It made commuter and intrastate air carriers eligible for the federal loan guarantee program.

13. It provided that the domestic route authority of the board cease in 1981; its authority over domestic fares, acquisitions, and mergers

would cease in 1983; and the board would be abolished (sunset) in 1985.

14. It provided that after the board was abolished, the local service carrier subsidy program was to be transferred to the Department of Transportation; the foreign air transportation authority of the board was to be transferred to the Transportation and Justice departments, in consultation with the State Department; and the mail subsidy program was to be transferred to the U.S. Postal Service.

The Airport and Airway Improvement Act of 1982

The Airport and Airway Improvement Act (Title V of the Tax Equity and Fiscal Responsibility Act of 1982, Public Law 97-248) was enacted by Congress to provide continued funding for airport planning and development under a single program called the *Airport Improvement Program* (AIP). This law also authorized federal funding for noise compatibility planning and to implement noise compatibility programs contained in the Noise Abatement Act of 1979 (Public Law 96-193).

This act directed the Secretary of Transportation to prepare, publish and revise every 2 years a national airport system plan—the National Plan of Integrated Airport Systems (NPIAS)—for the development of public-use airports in the United States. The NPIAS is closely coordinated with the FAA's 10-year plan to improve the air traffic control system and airway facilities. It sets forth the type and estimated costs of eligible airport development considered necessary to provide a safe, efficient, and integrated system of public-use airports. It further considers the needs of civil aeronautics without limitation to the requirements of any classes or categories of public-use airports. It takes into consideration, among other things, the relationship of each airport to the rest of the transportation system in a particular area, the forecast of technological developments in aeronautics, and the development forecast in other modes of transportation.

Projects eligible for funding under this legislation were required to be at airports included in the NPIAS, and these projects were restricted to planning, development, and noise compatibility projects at or associated with public-use airports. Further discussion of the specific provisions of the airport improvement program is found in Chap. 2.

State Role in Aviation and Airports

State interest in aviation began as early as 1921, when Oregon established an agency to handle matters concerning aviation. Since that time, virtually all states have established aeronautical agencies as commissions, departments, bureaus, boards, or divisions. Their responsibilities vary considerably and include channeling federal aid funds, planning state airport systems, providing state aid to local airport authorities, constructing and maintaining navigational equipment, investigating small aircraft accidents, enforcing safety regulations, and licensing airports.

Despite the growing concern of the states in airport development and aviation planning, their participation in the past has had little resemblance to the pattern associated with highway development. The states have always played a leading role in the development of roads and streets within their boundaries, whereas in airport development this has not been true. The reason for this pattern can be best explained by looking into the background of the entry of the states into aviation.

The majority of airports for civil aviation served by air carriers are municipally owned and operated. In a large number of states, these airports were in operation prior to the formation of a state aeronautical agency. From its inception, air transportation, because it is inherently of an interstate nature, became a matter of federal concern. The federal government provided much technical and financial assistance to municipalities. During World War II a great number of municipalities were the recipients of federal aid from the Civil Aeronautics Administration through the Defense Landing Area Program. Thus, in the early stages of airport development in this country, a fairly close relationship was established between the federal government and the municipalities. This relationship was furthered when the Federal Airport Act of 1946 authorized the CAA to issue grants directly to municipalities, as long as such a procedure was not opposed to state policy. In the meantime, the majority of states were doing very little in the way of providing funds to municipalities for airports. While significant increases in state aid for airport development have occurred in the last several years, the amount of federal aid has been substantially higher.

In those states where monetary aid is made available for airport development, the plans, specifications, and design for airport construction are generally reviewed by the state aeronautical agency.

There is no doubt that participation by the states in airport development is assuming a more significant role with the emergence of recent legislation in Congress to channel funds directly to state aeronautical agencies through block grant programs. The public concern for environmental control has resulted in legislation being passed at the state level, in addition to federal statutes, aimed at the control of aircraft noise and pollution. As general aviation and commuter activities continue to grow, the states will have to share the burden with the federal government in providing facilities for these activities, enforcing safety regulations, and other matters.

Aviation Organizations and Their Functions

The organizations directly involved in U.S. and international air carrier transportation and general aviation activity have an important influence on airport development as well as aircraft operations. These organizations and their functions can be classified into four groups: international government agencies, federal agencies, state agencies, and industry or trade organizations.

International government agencies

International Civil Aviation Organization. Perhaps the most important international agency concerned with airport development is the International Civil Aviation Organization (ICAO), which is now a specialized agency of the United Nations with headquarters in Montreal, Canada. One hundred sixty-one nations were members of ICAO in 1990.

The ICAO concept was formed during a conference of 52 nations held in Chicago in 1944. This conference was by the invitation of the United States to consider matters of mutual interest in the field of air transportation. The objectives of ICAO, as stated in its charter, are to develop the principles and techniques of international air transportation so as to

1. Ensure the safe and orderly growth of international civil aviation throughout the world

2. Encourage the arts of aircraft design and operation for peaceful purposes

3. Encourage the development of airways, airports, and air navigation facilities for international aviation

4. Meet the needs of the peoples of the world for safe, regular, efficient, and economical air transport

5. Prevent economic waste by unreasonable competition

6. Ensure that the rights of contracting states are fully respected and that every contracting state has a fair opportunity to operate international airlines

7. Avoid discrimination between contracting states

8. Promote safety of flight in international air navigation

9. Promote generally the development of all aspects of international civil aeronautics

The ICAO has two governing bodies: the Assembly and the Council. The Council is a permanent body responsible to the Assembly and is composed of representatives from 30 countries. The Council is the working group for the organization. It carries out the directives of the Assembly and discharges the duties and obligations specified in the ICAO charter.

To the airport planner and designer, perhaps the most important document issued by ICAO is *Aerodromes, Annex 14 to the Convention on International Civil Aviation* [1]. Annex 14 contains the international design standards and recommended practices which are applicable to nearly all airports serving international air commerce. In addition to Annex 14, ICAO publishes a great deal of technical and statistical information concerning international air transport [19].

Federal agencies of the U.S. government

There are currently two agencies at the federal level which are concerned primarily with air transportation: the Federal Aviation Administration and the National Transportation Safety Board. The Department of Transportation supports the activities of the FAA in many aspects of aviation and develops national policy with respect to aviation as part of its function of integrating the planning and operations of all modes of transport. The National Aeronautics and Space Administration (NASA) has assumed a basic and applied research role in augmenting the activities of the other agencies concerned with aviation.

The Civil Aeronautics Board no longer exists, but its functions during the period when the commercial air transportation industry was regulated by the Civil Aeronautics Board are discussed because of its impact on the evolution of the U.S. air transportation industry. The Civil Aeronautics Board was composed of five members, appointed by the President. In general, the board performed two chief functions, namely, the regulation of the economic aspects of U.S. air carrier operations, both domestic and international, and cooperation and assistance in the establishment and development of international air transportation. The Civil Aeronautics Board was phased out of existence in 1985, and some of its powers have been transferred to other agencies of the federal government.

Federal Aviation Administration. The Federal Aviation Administration is headed by the chief executive known as the administrator who is appointed by the President. The FAA performs the following functions:

1. It encourages the establishment of civil airways, landing areas, and other air facilities.
2. It designates federal airways and acquires, establishes, operates, and conducts research and development, and maintains air navigation facilities along such civil airways.

3. It makes provision for the control and protection of air traffic moving in air commerce.

4. It undertakes or supervises technical development work in the field of aeronautics and the development of aeronautical facilities.

5. It prescribes and enforces the civil air regulations for safety standards, including:

   a. Effectuation of safety standards, rules, and regulations

   b. Examination, inspection, or rating of pilots and other flight personnel, aircraft engines, air navigation facilities, aircraft, and air agencies

   c. Issuance of various types of safety certificates

6. It provides for aircraft registration

7. It requires notice and issues orders with respect to hazards to air commerce.

8. It issues airport operating certificates to airports serving air carriers.

The FAA develops, directs, and fosters the coordination of a national system of airports (the National Plan of Integrated Airport Systems [22]), an aviation system capacity enhancement plan (the Aviation System Capacity Plan [9]), and a plan to modernize and significantly upgrade the air traffic control system (the National Airspace System Plan [21]). Also the FAA directs the federal-aid airport program. In this connection the FAA performs the following functions:

1. It provides consultation and advisory assistance on airport planning, design, construction, management, operation, and maintenance to government, professional, industrial, and other individuals and agencies.

2. It develops and establishes standards, government planning methods, and procedures; airport and seaplane base design and construction; and airport management, operation, and maintenance.

3. It collects and maintains an accurate record of all available airport facilities in the United States.

4. It directs, formulates, and keeps current a national plan (NPIAS) for the development of an adequate system of airports in cooperation with federal, state, and local agencies. And it determines and recommends the extent to which portions or units of that system should be developed or improved.

5. It develops and recommends principles, for incorporation in state and local legislation, to permit or facilitate airport development, regulation, and protection of approaches through zoning or property acquisition.

6. It secures compliance with statutory and contractual requirements relative to airport operation practices, conditions, and arrangements.

7. It develops and recommends policies, requirements, and procedures governing the participation of states, municipalities, and other public agencies in federal-aid airport projects; and it secures adherence to such policies, requirements, and procedures.

National Transportation Safety Board. The National Transportation Safety Board consists of five members who are appointed by the President. The NTSB performs the following functions:

1. It investigates certain aviation, highway, marine, pipeline, and railroad accidents, and it reports publicly on the facts, conditions and circumstances, and cause or probable cause of such accidents.

2. It recommends to Congress and to federal, state, and local agencies measures to reduce the incidence of transportation accidents.

3. It initiates and conducts transportation safety studies and investigations.

4. It establishes procedures for reporting accidents to the board.

5. It assesses accident investigation techniques and issues recommendations for improving accident investigation procedures.

6. It evaluates the adequacy of the procedures and safeguards used for the transportation of hazardous materials.

7. It reviews, on appeal, the suspension, amendment, modification, revocation, or denial of certain operating certificates, documents, or licenses issued by the Federal Aviation Administration or the U.S. Coast Guard.

State agencies

As mentioned earlier, the states are involved in varying degrees in the many aspects of aviation including airport financial assistance, flight safety, enforcement, aviation education, airport licensing, accident investigation, zoning, and environmental control. Because of the interstate nature of air transportation, the federal government has preempted the legislative and administrative controls since the early days of aviation. However, for those aviation activities which occur wholly within the borders of a state, there have been formed regulatory agencies at the state level to oversee that these activities are operated in the best interests of the state.

Industry and trade organizations

Many groups are involved in the technical and promotional aspects of aviation. The following is a partial list of those groups primarily concerned with the airport aspects of aviation.

1. Aerospace Industries Association of America (AIA), the national trade association of companies in the United States, is engaged in the research, development, and manufacture of aerospace systems. It is located in Washington, D.C.

2. The Aircraft Owners and Pilots Association (AOPA) is an association of owners and pilots of general aviation aircraft. It is headquartered in Frederick, Maryland, a suburb of Washington, D.C.

3. The Air Line Pilots Association (ALPA) is an association of airline pilots. It is headquartered in Herndon, Virginia, a suburb of Washington, D.C.

4. The Airports Council International (ACI) is an association of over 400 large airports and airport authorities throughout the world. It is based in Geneva, Switzerland. The North American region of this organization, which was formerly called the Airport Operators Council International (AOCI), is headquartered in Washington, D.C.

5. The Air Transport Association of America (ATA) is an association of scheduled domestic and international airlines in the United States. The headquarters is in Washington, D.C.

6. The American Association of Airport Executives (AAAE) is an association of managers of public and private airports. It is located in Alexandria, Virginia, a suburb of Washington, D.C.

7. The General Aviation Manufacturers Association (GAMA) is an association promoting the interests of general aviation. It is located in Washington, D.C.

8. The Helicopter Association International (HAI) is an association which represents the interests of manufacturers and users of helicopters and which promotes the use of helicopters. It is located in Alexandria, Virginia.

9. The International Air Transport Association (IATA) is an association of scheduled carriers in international air transportation. This organization is headquartered in Montreal, Canada.

10. The National Association of State Aviation Officials (NASAO) is an association of the representatives of state aviation agencies. Its headquarters is in Silver Springs, Maryland, a suburb of Washington, D.C.

11. The Regional Airline Association (RAA) is an association of small regional and commuter aircraft operators promoting the needs of this segment of the air transportation industry. It was formerly called the Commuter Airline Association of America (CAAA). It is located in Washington, D.C.

References

1. *Aerodromes, Annex 14 to the Convention on International Civil Aviation*, vol. 1, *Aerodrome Design and Operations*, 1st ed., International Civil Aviation Organization, Montreal, Canada, 1990.
2. *Aerospace Facts and Figures*, Aerospace Industries Association of America, Inc., Washington, 1980.
3. *Airline Capital Requirements in the 1980's*, Economics and Finance Department, Air Transportation Association of America, Washington, September 1979.
4. *Airline Deregulation*, M. A. Brenner, J. O. Leet, and E. Schott, Eno Foundation for Transportation, Inc., Westport, Conn., 1985.
5. *Air Taxi Operators and Commercial Operators, Federal Aviation Regulations*, pt. 135, Federal Aviation Administration, Washington, 1978.
6. *Air Transport Facts and Figures*, Air Transportation Association of America, Washington, annual.
7. *Air Transportation*, 10th ed., Robert M. Kane, Kendall/Hunt Publishing Company, Dubuque, Iowa, 1990.
8. *Annual Report of the Regional Airline Association*, Regional Airline Association, Washington, 1991.
9. *Aviation System Capacity Plan 1991–1992*, Report no. DOT/FAA/ASC-91-1, Department of Transportation, Federal Aviation Administration, Washington, 1991.
10. *Certification and Operations: Domestic, Flag and Supplemental Air Carriers and Commercial Operators of Large Aircraft*, pt. 121, Federal Aviation Regulations, Federal Aviation Administration, Washington, 1980.
11. *Commuter Air*, 1981 yearbook edition, Commuter Airline Association of America, Washington, April 1981.
12. *Commuter Air Carrier Traffic Statistics*, 12 months ended June 30, 1980, Civil Aeronautics Board, Washington.
13. *Current Market Outlook*, Boeing Commercial Airplane Group, Seattle, March 1992.
14. *Developments in the Deregulated Airline Industry*, D. R. Graham and D. P. Kaplan, Office of Economic Analysis, Civil Aeronautics Board, Washington, June 1981.
15. *FAA Aviation Forecasts, Fiscal Years 1992–2003*, Federal Aviation Administration, Washington, February 1992.
16. *FAA Statistical Handbook of Civil Aviation*, Federal Aviation Administration, Washington, 1990.
17. *Hearings before the Subcommittee on Aviation of the Committee on Commerce, Science, and Transportation*, U.S. Senate, Washington, August 1980.
18. *ICAO Journal*, International Civil Aviation Organization, Montreal, Quebec, Canada, monthly.
19. *ICAO Publications and Audio Visual Training Aids*, Catalog, International Civil Aviation Organization, Montreal, Quebec, Canada, 1992.
20. *National Airport System Plan, 1978–1987*, Federal Aviation Administration, Department of Transportation, Washington.
21. *National Airspace System Plan*, Federal Aviation Administration, Washington, 1989.
22. *National Plan of Integrated Airport Systems (NPIAS), 1990–1999*, Federal Aviation Administration, Department of Transportation, Washington, 1991
23. *National Requirements for Aviation Facilities: 1956–1975*, final report prepared by the Aeronautical Research Foundation for Edward P. Curtis, Special Assistant to the President for Aviation Facilities Planning, Washington, 1957.

24. *National Transportation Statistics,* annual report, Research and Special Programs Administration, Department of Transportation, Washington, July 1990.
25. *National Transportation Strategic Planning Study,* Department of Transportation, Washington, 1990.
26. *Report on Airline Service, Fares, Traffic, Load Factors, and Market Share,* a staff study, fourteenth in a series, Civil Aeronautics Board, Washington, 1981.
27. *Secretary's Task Force on Competition in the U.S. Domestic Airline Industry,* Department of Transportation, Washington, February 1990.
28. *Terminal Area Air Traffic Relationships,* fiscal year 1980, Federal Aviation Administration, Washington.
29. *Terminal Area Forecasts,* Federal Aviation Administration, Washington, annual.
30. *The Changing Airline Industry: A Status Report through 1979,* Comptroller General of the United States, General Accounting Office, Washington, September 1980.
31. *The Federal Turnaround on Aid to Airports 1926–38,* Federal Aviation Administration, Department of Transportation, Washington, 1973.
32. *Transportation in America,* Frank A. Smith, Eno Foundation for Transportation, Inc., Waldorf, Md., annual.
33. *Travel Market Closeup 1989,* National Travel Survey Tabulations, U.S. Travel Data Center, Washington, 1990.
34. *Winds of Change, Domestic Air Transport since Deregulation, Special Report 230,* Transportation Research Board, National Research Council, Washington, 1991.
35. *Worldwide Airport Traffic Report,* Airports Council International, Washington, annual.

2

Airport Financing

Airports are owned and operated by both private interests and public agencies. In 1990 there were 17,451 airports in the United States of which more than two-thirds were privately owned. However, only a small portion of the privately owned fields are paved, whereas a majority of the publicly owned airports are paved. Of the total number of airports in 1990 in the United States, 5598 airports were open to the public, called public-use airports, and of these 4169 were publicly owned.

In the early years of aviation, airport ownership was vested almost entirely in private hands. State and federal financial participation in airport development was virtually nonexistent. The depression of the 1920s witnessed a collapse in private investments in airports and gave rise to public ownership. Although they represent less than 25 percent of the total airports in the country, the bulk of aviation activities today are carried out at publicly owned airports.

Although public ownership in airports is vested in a number of different types of government subdivisions, the majority of airports are owned and operated by municipalities. The next most prevalent ownership is by counties, followed by special airport districts and airport authorities. Although both the federal and state governments may have provisions which affect the airport, it is the decision of the airport owner, called the airport sponsor, which ultimately determines the development of the airport.

Airport improvements are financed in a variety of ways. Among these are federal grants, state grants, taxes levied by local governments, airport user fees, general obligation bonds, and revenue bonds. In a few instances, capital improvements of a minor nature have been financed from accumulated surpluses from airport revenues. It has been estimated by the Federal Aviation Administration

(FAA) that about one-third of the costs associated with the airport improvements identified in the current National Plan of Integrated Airport Systems (NPIAS) will be funded through federal revenues and the remainder through state and local revenues [14].

The Federal Airport Act of 1946 was the first legislation in the United States which made provisions for federal grants to airports under the Federal Aid to Airports Program (FAAP). In this legislation, the maximum federal grant for an eligible project was for approximately 50 percent of the total project cost. This amount has varied over the years in subsequent airport funding legislation, such as the Airport Development Aid Program (ADAP) and the Airport Improvement Program (AIP), and has been authorized for up to 90 percent of the total project cost for certain types of projects.

Virtually all the states participate financially in airport development, either in conjunction with or independently of the federal aid program. Where state grants are made in conjunction with the federal program, the state usually contributes one-half of the local government share of the project cost.

Local financing has generally been accomplished through the levying of taxes and user fees or through the sale of general obligation or revenue bonds. For the smaller airports which have not reached financial maturity, funds for capital improvements have usually been taken from the general tax funds of local governments. This is particularly true if the amount required is small. The bulk of the improvements at the larger airports have been financed by the sale of general obligation and revenue bonds.

Federal Participation in Airport Funding

Until 1933, airports for civil use were developed mainly through investments by municipalities and private sources. The depression was largely responsible for the first substantial federal participation in the development of civil airports. The bulk of the funds provided from 1933 until the beginning of World War II were through work relief programs. The first such program was under the Civil Works Administration (CWA). In the fall of 1933, the CWA provided more than $15 million for airport construction, with most of the money going to smaller communities.

In 1934, the Civil Works Administration was succeeded by the Federal Emergency Relief Administration (FERA). This agency provided over $17 million for the development of 943 airport projects.

The administration of federal aid for airports was taken over in 1935 by the Works Progress Administration (WPA). The WPA spent $323 million for airport construction in the United States. It was under this program that contributions by municipalities were encouraged and a pattern of cost sharing emerged. The local contributions amounted to about $110 million.

Another federal program contributing to airport development in the 1930s was the Public Works Administration (PWA) which made loans or grants amounting to almost $29 million, primarily to municipalities.

Thus, up to World War II the federal government expended a total of about $384 million for airport development under the four programs of CWA, FERA, WPA, and PWA. It must be recognized that these were primarily work relief programs and provided no basis for federal support in times of a normal economy.

During World War II the federal government, through the Civil Aeronautics Administration, expended $353 million for the development of landing areas for military use. While first importance in this program was attached to military requirements, the needs of postwar civil aviation were considered in the location and construction of these facilities. During the same period the federal government, through the CAA, spent over $9 million for the development of airports solely for civil use. These two programs are referred to as DLA (Defense Landing Area) and DCLA (Development of Civil Landing Areas). The DLA and DCLA programs were independent of the airports constructed by the War and Navy Departments. After the war some 500 military airports were declared surplus and turned over to cities, counties, and states. This is the principal reason for public ownership today being mainly vested in local authorities.

Federal Airport Act of 1946

At the end of World War II, interest was renewed in establishing a federal program for monetary aid for airport development. A resolution was introduced in Congress (H.R. 598, 78th Congress) requiring the Civil Aeronautics Administration to make a survey of airport needs and prepare a report on the subject. These recommendations formed the basis of the Federal Airport Act of 1946 (Public Law 79-377). Appropriations of $500 million over a 7-year period were authorized for projects within the United States plus $20 million for projects in Alaska, Hawaii, Puerto Rico, and the Virgin Islands. In 1950 the 7-year period was extended for an additional 5 years (Public Law 81-846). However, annual appropriations approved by Congress were much less than the amounts authorized by the act.

The original act provided that a project shall not be approved for federal aid unless "sufficient funds are available for that portion of the project which is not to be paid by the United States."

Local governments often required 2 to 3 years to make arrangements for raising funds. Most of the larger projects are financed locally through the sale of bonds. This method of financing requires legislation at the local level and, in some cases, also at the state level. General obligation bonds normally require approval by the electorate. Programs to inform the public on the need for airport improvement must be carefully planned and executed. Thus, after the completion of these events, local governments frequently found that sufficient federal funds were not appropriated to match local funds, and the projects were delayed. Another complaint of local governments had been that Congress failed to fulfill its obligation, since the amount appropriated by Congress fell far short of the amount authorized by the Federal Airport Act. These deficiencies as well as other matters were incorporated in a bill (S. 1855). Hearings on the bill were held before the Subcommittee of the Committee on Interstate and Foreign Commerce of the U.S. Senate in 1955. Representatives of the Council of State Governments, the American Municipal Association, the National Association of State Aviation Officials, airport and industry trade associations, and individuals were unanimous in the feeling that air transportation had reached a stage of maturity where many airports were woefully inadequate and greater financial assistance from the federal government would be required to meet the current needs of aviation. After much debate, the bill was approved by the President (Public Law 84-211).

This amending act made no change in the basic policies and purposes expressed in the original act. There were no changes in the requirements with respect to the administration of the grants authorized, such as the distribution and apportionment of funds, eligibility of the various types of airport construction, sponsorship requirements, etc. The primary purpose of the act was to substitute for the procedure of authorizing annual appropriations for airport projects, provisions granting substantial annual contract authorization in specific amounts over a period of 4 fiscal years. Airport sponsors were thus furnished assurance that federal funds would be available at the time projects were to be undertaken.

This law provided $40 million for fiscal year 1956 and $60 million for each of fiscal years 1957, 1958, and 1959 for airport construction in the continental United States. It also provided $2.5 million in fiscal year 1956 and $3 million for the 3 succeeding fiscal years for airport construction in Alaska, Hawaii, Puerto Rico, and the Virgin Islands. Besides the $42.5 million made available in fiscal year 1956 by Public Law 84-211, Congress approved an additional appropriation of $20 million for airport projects.

In 1958, the 85th Congress passed a bill (S. 3502) proposing to extend the Federal Airport Act for 4 years at an annual funding rate of $100 million. This bill was vetoed by the President in September 1958 with the veto statement (S. 3502 veto statement), which stated in part:

> I am convinced that the time has come for the federal government to begin an orderly withdrawal from the airport grant program. This conclusion is based, first, on the hard fact that the government must now devote the resources it can make available for the promotion of civil aviation programs which cannot be assumed by others, and second, on the conviction that others should begin to assume the full responsibility for the cost of construction and improvement of civil airports.

In the 86th Congress, much debate involved a significant increase in the federal airport program. Two bills, one introduced in the House ($297 million over a 4-year period) and the other in the Senate ($465 million for a 4-year period), together with the President's recommended bill for a 4-year program of $200 million, were finally merged into a 2-year continuation of the existing aid program at $63 million per year (Public Law 86-72).

Significant changes were the removal of the territorial status from Hawaii and Alaska which had been admitted as states and the exclusion of automobile parking and certain portions of airport building improvement costs as allowable costs.

A 3-year continuation of the federal aid program providing $75 million annually was enacted by the 87th Congress (Public Law 87-255). The provisions of the bill were very similar to those of Public Law 86-72. One new feature of the legislation was that it provided the Administrator of the FAA with a discretionary fund of $7 million from the $75 million for developing general aviation airports to relieve congestion at high-density commercial airports. Thus this legislation started the reliever airport program. The Federal Airport Act was again amended in 1964, authorizing the expenditure of $75 million for fiscal years 1965 to 1967 (Public Law 88-280). The final amendment was accepted in 1966, authorizing the continuation of the expenditure of $75 million for fiscal years 1968 to 1970 (Public Law 89-647). This marked the end of the Federal Airport Act, as the Airport and Airway Development Act of 1970 became law in 1970.

During the 24 years of airport funding under the Federal Airport Act, a total of $1.2 billion was appropriated by the federal government for improvements at 2316 airports involving almost 8000 projects.

Airport and Airway Development Act of 1970

Because of the rapidly growing requirements for modernizing the air traffic control system and airport expansion, neither the federal government nor the local authorities were able to fund capital improvements badly needed for the growth of aviation. The FAA needed more money in its budget to accelerate the implementation of a program to modernize the air traffic control system. The $75 million authorized annually by the Federal Aviation Act, together with local matching funds, fell far short of the needs of the local airport authorities to meet the current and projected growth of airport traffic. Cities were unable to raise sufficient funds at the local level to meet the rising costs of airport construction. The Federal Aviation Act of 1946 was supported by general rather than user tax revenues; therefore it had to compete annually with other government programs for scarce federal dollars. The increased competition for fewer dollars resulted in delay and postponement of airport construction throughout the nation. As a result of this situation, aviation organizations representing airport owners, airlines, pilots, and general aviation aircraft owners joined in pressing for more funds for airports and airways and the establishment of a trust fund similar to that for the national highway program. Several bills were introduced in the House and Senate to enact legislation which would remedy the deficiencies in the Federal Airport Act of 1946 (S. 1637, S. 2437, S. 2651). After much debate including the question of whether airport terminal buildings should be included in the legislation (initially they were not), Public Law 91-258 was signed into law in May 1970. As stated in Chap. 1, the law consisted of two parts, referred to as Titles I and II. Title I was known as the Airport and Airway Development Act of 1970, which replaced the Federal Airport Act of 1946. Title II was known as the Airport and Airway Revenue Act of 1970, and it provided the excise taxes required to furnish the resources necessary to carry out the Title I programs through 1980. These excise taxes provided the revenues which funded the programs under the act and were deposited in the Airport and Airway Trust Fund established by the act.

Title I of the original act provided for $250 million annually for the "acquisition, establishment, and improvement of air navigational facilities" and security equipment required by the sponsor for fiscal years 1971 through 1980. For airport assistance, the Airport Development Aid Program (ADAP) initially authorized a total of $2.5 billion for the 10-year period. The act further specifically authorized $250 million annually through fiscal year 1973 and $275 million each for fiscal years 1974 and 1975 for airports served by air carriers and general aviation airports which relieve high-density air carrier airports. Also authorized were $30 million annually through fiscal year 1973 and $35 million each for fiscal years 1974 and 1975 for all other general aviation airports (Public Law 93-44). Later amendments (Public Law 94-353) raised the program level to range from $500 to $610 million annually through 1980. The Aviation Safety and Noise Abatement Act (Public Law 96-193) further raised the final-year program to $667 million. These amendments provided $435 to $539 million annually for airports serving all segments of aviation and $65 to $95 million annually for general aviation airports. The act also authorized the issuance of planning grants for the preparation of airport system plans and airport master plans. The Planning Grant Program (PGP) was designed to promote the effective location and development of publicly owned airports and to develop a national airport system plan. System plans were prepared by state and regional agencies to formulate air transportation policy, determine facility requirements needed to meet forecast aviation demand, and establish a framework for detailed airport master planning. Airport master plans, which were developed by the airport owner, focused on the nature and extent of the development required to meet the future aviation demand at specific facilities.

The funds that were authorized for airport development for the several classes of airports were apportioned to air carrier and general aviation airports. In the final amended form of the law, two-thirds of the air carrier and commuter service funds were made available to air carrier airports based upon the number of annual enplaned passengers. These were termed *entitlement funds*. The remaining monies were placed in a discretionary fund, of which $15 million annually was apportioned to commuter service airports. Air carrier airports serving aircraft heavier than 12,500 lb were authorized to receive not less than $150,000 nor more than $10 million annually. Airports serving aircraft weighing less than 12,500 lb were to receive not less than $50,000 annually. Of the funds appropriated for general aviation and general aviation reliever airports, $15 million annually was apportioned to reliever airports. Seventy-five percent of the remaining funds was allocated to the states on the basis of population and area; 1 percent was allocated to Puerto Rico, Guam, the Virgin Islands, American Samoa, and the Trust Territories of the Pacific Islands and was distributed by the Secretary of Transportation.

The maximum federal grant for any specified project varied from 50 to 90 percent of the total eligible project costs over the life of the act, depending upon the type of project being considered for funding. A maximum of 75 percent of the allowable project cost was allowed for airports in areas that enplaned 0.25 percent or more of the total annual passengers enplaned by air carriers certified by the Civil Aeronautics Board. These airports were called the large and medium air traffic hub airports. A maximum of 90 percent of the allowable project cost was allowed for airports in areas that enplaned less than 0.25 percent of the total annual passengers, for the small air traffic hub and nonhub airports, and for general aviation airports.

The original act specifically prohibited the use of federal funds for automobile parking facilities or airport buildings except those parts "intended to house facilities or activities directly related to the safety of persons at the airport." However, the 1976 amendments to the act (Public Law 94-353) provided federal funding for the non-revenue-producing public areas of terminal facilities required for the processing of passengers and baggage. In this case the federal share was limited to 50 percent of the project costs, and the airport could not spend more than 60 percent of its enplanement funds on such development.

There was no state apportionment for planning grant funds. The Secretary of Transportation prescribed the regulations governing the award and administration of these grants. When the program first began, the federal government provided up to two-thirds of the cost of planning grant projects. However, the 1976 amendments to the act increased this share to 75 percent of the cost of airport system plans, 90 percent of the cost for master plans at general aviation airports, and a range from 75 to 90 percent of the cost for master plans at air carrier airports, depending upon the number of enplaned passengers.

In administering the Airport and Airways Development Act, the FAA established, in detail, the types of improvement which were eligible for federal aid under the act. In general, items that were eligible included land acquisition, paving and grading, lighting and electrical work, utilities, roads, removal of obstructions to air navigation, fencing, fire and rescue equipment, snow removal equipment, terminal-area development, and physical barriers and landscaping for noise attenuation.

Separate buildings for airport emergency, snow removal, and fire-fighting equipment were eligible. Administration buildings serving air commerce or general aviation were not eligible. Buildings used exclusively for the handling of cargo were also ineligible. The non-revenue-producing public-use areas of terminal facilities became eligible for funding in 1976 (Public Law 94-353) if these areas were "directly related to the movement of passengers and baggage in air commerce within the boundaries of the airport."

Roads and streets were eligible if they were within the boundary of the airport and were needed for the operation and maintenance of the airport, or were directly related to the movement of passengers and baggage.

Each eligible project was evaluated separately. It was rated on the basis of the aeronautical necessity of the airport, volume and character of traffic, and type of work included in the project. These ratings established a priority score for each increment of work in any one state and were used to program the funds allocated to that state. The ratings were based on such factors as safety, efficiency, and convenience. In the development of priorities, the views of the state aviation agency were solicited. An environmental impact statement was necessary for certain projects. This statement was required to include the effect of the project on the ambient noise level, animal and bird life, air and water pollution, recreation areas, areas of unique interest or scenic beauty, and displacement of significant numbers of people.

A community interested in obtaining federal aid contacted the Airports District Office (ADO) of the FAA in the geographic area in which the airport was located. In states which required that all federal aid be channeled through the state, submission of a request for aid was made through the state aeronautical agency.

If the project qualified for aid, if there were sufficient funds, and if the project was found by the Secretary of Transportation to be acceptable from the standpoint of its economic, social, and environmental effects on the community, then the sponsor was notified that a tentative allocation of funds had been made for part of or all the items listed in the request. The tentative allocation was an indication that funds had been placed in reserve pending the completion of arrangements for necessary financing, land acquisition, and preparation of plans, specifications, and contract documents.

Upon submitting detailed plans and specifications with a project application and upon approval of the FAA, the sponsor secured bids from contractors and made recommendations for the award of the contracts. At this time the sponsor also formally accepted federal aid and the obligations connected therewith by executing what was known as a "grant agreement." The execution of the grant agreement legally bound the community to fulfill the obligations (sponsors' assurances) set forth in the project application. Some of the sponsor's important obligations were as follows:

1. The sponsor would operate the airport for the use and benefit of the public, on fair and reasonable terms without unjust discrimination.

2. It would keep the airport open to all types, kinds, and classes of aeronautical use without discrimination between such types, kinds, and classes.

3. It would operate and maintain in a safe and serviceable condition the airport and all facilities thereon which are necessary to serve aeronautical uses other than facilities owned or controlled by the United States.

4. It would make every effort to maintain clear approaches to the runways.

5. It would not charge government-owned or military aircraft for the use of runways and taxiways unless the use was substantial.

6. These obligations would remain in effect not more than 20 years.

The authority to issue grants under this act expired in 1981. During the 11-year period under this legislation, 8809 grants totaling

$4.5 billion were approved for airport planning and development at over 1800 airports. This was about 4 times greater than the total amount provided by the Federal Airport Act of 1946. Over 6700 of these grants were made under the Airport Development Aid Program, and almost 2000 of these grants were made under the Planning Grant Program.

The Airport and Airway Improvement Act of 1982

In 1982 Congress enacted the Airport and Airway Improvement Act (Title V of the Tax Equity and Fiscal Responsibility Act of 1982, Public Law 97-248). This act continued to provide funding for airport planning and development under a single program called the Airport Improvement Program (AIP). The act also authorized funding for noise compatibility planning and to implement noise compatibility programs contained in the Noise Abatement Act of 1979 (Public Law 96-193). It required that to be eligible for a grant, the airport must be included in the National Plan of Integrated Airport Systems (NPIAS). The NPIAS, the successor to the National Airport System Plan (NASP), is prepared by the FAA and published every 2 years; it identifies public-use airports considered necessary to provide a safe, efficient, and integrated system of airports to meet the needs of civil aviation, national defense, and the U.S. Postal Service.

The Airport and Airway Improvement Act has been amended several times, resulting in significant changes in the provisions of the act and in the appropriations authorized under the act. These amendments are included in the Continuing Appropriations Act of 1982 (Public Law 97-276), the Surface Transportation Assistance Act (Public Law 97-424), the Airport and Airway Safety and Capacity Expansion Act of 1987 (Public Law 100-223), the Airway Safety and Capacity Expansion Act of 1990 (Public Law 101-508), and the Airport and Airway Safety, Capacity, Noise Improvement and Intermodal Transportation Act of 1992 (Public Law 102-581). This legislation, as amended, significantly increased the level of federal funding for airports to an aggregate total of more than $14 billion for 1982 through 1992.

Projects eligible for funding under this legislation were restricted to planning, development, and noise compatibility projects at or associated with public-use airports, including heliports and seaplane bases, which were defined as airports open to the public and publicly owned, or privately owned but designated by the FAA as a reliever airport, or privately owned and having scheduled service and at least 2500 annual enplanements.

Airports were defined in five categories: commercial service airports, primary airports, cargo service airports, reliever airports, and other airports. Commercial service airports are publicly owned airports which enplaned at least 2500 passengers annually and received scheduled service. Primary service airports are commercial service airports which enplaned at least 10,000 passengers annually. Cargo service airports are airports served by aircraft providing air transportation of property only, including mail, with an aggregate annual aircraft landed weight in excess of 100,000,000 lb. Reliever airports are in metropolitan areas designated by the FAA as having the function of relieving congestion at large commercial service airports by providing alternative landing areas for general aviation aircraft and which provided more general aviation access to the community. Other airports are the remaining airports, commonly referred to as general aviation airports.

The allocation of funds under the AIP is also defined in the legislation such that these funds are distributed between apportioned and discretionary funds. As amended by the Airport and Airway Safety, Capacity, Noise Improvement and Intermodal Transportation Act of 1992, of the total funds available not more than 44 percent is apportioned as entitlements to primary airports and 3.5 percent is apportioned as entitlements to cargo service airports. Additionally, 12 percent of the total funds is apportioned for states and insular areas, there is a separate apportionment for airports in Alaska, and 2.5 percent is apportioned to the Military Airport Program for current and former military airfields, to enhance the capacity of the national transportation system by enhancement of airport and air traffic control systems in major metropolitan areas. The remaining funds are designated as discretionary funds, which are required to be used so that of the total funds available, a minimum of 10 percent is to be used for reliever airports, 12.5 percent for noise compatibility projects, 2.5 percent for nonprimary commercial service airports, and 0.5 percent for integrated system plans for states, regions, or metropolitan areas. Of the remaining discretionary funds, 75 percent is to be used for projects to preserve and enhance capacity, safety, and security and projects carrying out noise compatibility planning programs at primary and reliever airports.

The federal share of the costs associated with integrated airport system planning was limited to 90 percent. For individual airports, the federal share of planning and airport development project costs was limited to 75 percent at primary airports and 90 percent at all other airports. There are approximately 70 large primary airports. The federal share of the cost of noise compatibility projects was limited to 80 percent. The federal share of non-revenue-producing public-area terminal development costs at large, medium, and small hub commercial service airports was limited to 75 percent. The federal share of both revenue-producing and non-revenue-producing public areas in terminal buildings and non-revenue-producing parking lots at nonhub commercial service airports was limited to 85 percent.

The formula for the apportionment of funds to primary airports in fiscal year 1991 was set at $7.80 for each of the first 50,000 enplaned passengers, $5.20 for each of the next 50,000 enplaned passengers, $2.60 for each of the next 400,000 enplaned passengers, and $0.65 for each enplaned passenger in excess of 500,000 enplaned passengers. In fiscal year 1992, no primary airport could receive less than $400,000 nor more than $22 million. The total appropriation for fiscal year 1992 was increased to $2.05 billion.

Selection for the NPIAS

The NPIAS is closely coordinated with the FAA's plan to improve the air traffic control system and airway facilities, the National Airspace System Plan [13]. It sets forth the type and estimated costs of eligible airport development considered necessary to provide a safe, efficient, and integrated system of public-use airports to meet the needs of civil aviation, the requirements of national defense, and the special needs of the U.S. Postal Service (USPS). It further considers the needs of civil aeronautics without limitation to the requirements of any classes or categories of public-use airports. It takes into consideration, among other things, the relationship of each airport to the rest of the transportation system in a particular area, the forecast of technological developments in aeronautics, and the development forecast in other modes of transportation.

Federal involvement in airports is intended to provide the public with access to the national air transportation system. Commonly stated as an objective in national airport planning is the federal intent to provide a balanced transportation system. This means taking into account the diverse needs of different communities and the various segments of aviation and coordinating planned airport development with plans for air traffic, approach and navigational aids, and other components of the air transportation system.

The NPIAS is formulated in conjunction with state and metropolitan-area system plans. A number of objectives are common in all system planning efforts. The primary objective is to provide the public with reasonable access to safe and adequate airports. Quantifying reasonable access is challenging, and several approaches have been considered. An analysis prepared for the FAA in 1983 indicated that 97.3 percent of the population resides within 20 mi of an airport that is included in the NPIAS.

The viability of an airport is also considered in the entry criteria. Viability means that once an airport is built, the community will maintain, if not further improve, the airport. Ten based aircraft have been used as a rule of thumb to identify communities which have the activity to support a public airport, provided they do not already have reasonable access to another airport. When there are less than 10 based aircraft registered in a community but there is evidence of substantial itinerant use, a special justification may be used as the basis for including the community in the NPIAS.

The following criteria are used to select and classify locations for inclusion in the NPIAS:

1. *Commercial service airports* are public airports that enplane 2500 or more annual passengers and receive aircraft offering scheduled passenger service.
2. *Primary airports* are commercial service airports that have more than 10,000 annual enplaned passengers.
3. *Cargo service airports* are served by aircraft providing air transportation of property only, including mail, with an aggregate annual aircraft landed weight in excess of 100,000,000 lb.
4. *Reliever airports* provide substantial capacity or instrument training relief to a commercial service airport that serves a metropolitan statistical area with a population of at least 250,000 persons or has at least 250,000 annual enplaned passengers and operates at 60 percent of its capacity; or would be operated at such a level before being relieved by one or more reliever airports; or is subject to restrictions that limit activity that would otherwise reach 60 percent of capacity. An airport not meeting these criteria may be included in NPIAS as a reliever airport if it is so designated in a state, regional, or metropolitan system plan and the FAA concurs with that portion of the plan.

    Evidence of providing substantial capacity or instrument training relief requires a current activity level or, in the case of a new airport or an airport that is scheduled for major improvement, a forecast activity level of at least 50 based aircraft or 25,000 itinerant operations, or 35,000 local operations; or a heliport if it has one-half of these activity levels; or the FAA regional administrator has determined that the airport is a desirable location for instrument training activity.
5. *General aviation airports* that receive U.S. mail service, are listed as a scheduled stop by an air carrier transporting mail pursuant to a current contract with the USPS, or are otherwise designated by USPS and served through a public airport are included in the plan. General aviation civil public-use airports with military activity where a unit of the Air National Guard or a reserve component

of the U.S. armed forces is permanently based on or adjacent to the airport and operates permanently assigned aircraft in activities directly related to the mission of the unit are also included in the plan.

General aviation airports that were included in predecessor plans if subject to a current compliance obligation resulting from accepting prior federal grants, unless the airport is omitted from an accepted state airport system plan or a metropolitan airport system plan and there is clearly no longer a continuing system role for the airport, are also included in the plan. Existing general aviation airports that are included in an accepted state airport system plan or metropolitan airport system plan may be included in the plan if they serve a community located 30 min or more average ground travel time from the nearest existing or proposed plan airport and have at least 10 based aircraft. An existing or proposed general aviation airport not meeting the above criteria may be included in the plan if it meets all the following requirements:

a. It is included in a state airport system plan or a metropolitan airport system plan.

b. It serves a community more than 30 min from the nearest existing or proposed airport included in the plan.

c. It is forecast to have 10 based aircraft during the short-range planning period of 5 years.

d. There is an eligible sponsor willing to undertake ownership and development of the airport.

An existing or proposed general aviation airport not meeting the other criteria may be included on the basis of a special justification. This may be a favorable benefit-cost analysis or a written justification in the case of remote areas, islands, or recreational areas.

6. *Public-use heliports* that do not meet other criteria are included in the plan if they make a significant contribution to public transportation. Helicopter landing areas are included in the plan if they have at least four based rotorcraft or 800 annual itinerant operations, or 400 itinerant operations by air taxi rotorcraft. Preferably, the heliports are included in the state or metropolitan airport system plan. Private-use heliports or special service heliports that are primarily intended to provide community services such as police patrol, traffic surveillance, or air ambulance transportation are not included.

The NPIAS for the 1990–1999 period contains 3692 airports, 3285 existing airports, and 407 proposed airports. Of the 3285 existing airports, 568 are commercial service airports, 396 of which are primary airports and 172 are other commercial service airports. In addition, there are 285 reliever airports and 2432 general aviation airports. Of the 407 proposed airports, 17 are commercial service airports, 71 are reliever airports, and 319 are general aviation airports [14].

Over the 10-year period from 1982 through 1991, the Airport Improvement Program appropriated a total of 11,285 grants totaling $10.9 billion for airport planning and improvement [20]. The total apportionment of funds and the total number of grants issued by airport category under the Airport Improvement Program from 1982 through 1991 are shown in Table 2-1. The apportionment of funds to various work categories is shown in Table 2-2.

Passenger facility charges

Perhaps the most significant feature of the Airway Safety and Capacity Expansion Act of 1990 (Public Law 101-508) was the authorization for the imposition of passenger facility charges (PFCs) at commercial service airports. The law allowed a public agency that controlled a commercial service airport to submit an application to the FAA to impose a $1, $2, or $3 charge on air carrier enplaned passengers to finance eligible airport-related projects. The law limited the PFC to no more than two charges on each leg of a round-trip at airports at which passengers enplaned aircraft. The revenue gained from such charges may be used only to finance the allowable costs of the approved projects that the public agency controls. This includes all or a portion of the total project cost, bond-associated debt service and financing costs, and nonfederal share of the costs of projects funded under the federal airport grant program. Therefore, revenues

TABLE 2-1   AIP Allocations by Program Category for Fiscal
Years 1982 to 1991

	Grants	Allocations (billions)
Program Category		
Primary airports	4,108	$7.34
Other commercial service airports	1,126	0.53
Reliever airports	1,445	1.25
General aviation airports	4,090	1.61
Systems planning	510	0.07
State block grant program	6	0.09
Total	11,285	$10.89

SOURCE: Federal Aviation Administration [20].

TABLE 2-2   AIP Allocations by Work Increment for Fiscal Years
1982 to 1991

Work increment	Allocations (billions)	Percentage
Landing area		
Runways	$2.42	22.2
Taxiways	1.86	17.1
Aprons	1.61	14.8
Buildings		
Terminals	0.51	4.7
Other	0.08	0.7
Roadways	0.63	5.8
Navigation	0.62	5.7
Safety and security	0.52	4.8
Land		
Noise	0.75	6.9
Other	0.94	8.6
Noise control	0.35	3.2
Planning	0.21	1.9
State block grants	0.09	0.8
Miscellaneous	0.30	2.8
Total	$10.89	100.0

SOURCE: Federal Aviation Administration [20].

can be used for airport development, airport planning, terminal development, noise compatibility planning, noise compatibility implementation measures, and construction of gates and related areas at which passengers enplane or deplane and other areas directly related to the movement of passengers and baggage in air commerce within the boundaries of the airport.

The legislation also provided that at those commercial service airports in areas which enplaned at least 0.25 percent of the national annual enplanements in any year, i.e., the large and medium hub airports, the entitlement funds apportioned to the airport based upon passenger enplanements would be reduced by 50 percent of the revenue obtained through the PFCs, but these reductions of entitlement funds would not exceed 50 percent of the apportioned funds. The legislation directed that 25 percent of the revenues obtained through a reduction in these apportioned funds is to be placed in the AIP discretionary fund, of which one-half is to be used for small hub airports, and 75 percent is to be used to establish a Small Airports Fund. One-third of the revenues in the Small Airport Fund is to be distributed to general aviation airports and two-thirds to nonhub commercial service airports.

The specific requirements imposed upon airports requesting authority to impose passenger facility charges are contained in FAR part 158 [15].

State Participation in Financing Airport Improvements

Airport development and aviation planning are of major concern in most states. Virtually all states provide some financial assistance for airport improvements. Increases in state financial support have been dramatic during the past several years. Data show that expenditures by states for airports have risen from $59 million in 1966 [11] to over $860 million in 1990 [22].

State sharing in airport development varies depending on whether federal aid is involved and whether this aid is channeled through a state aeronautical agency. Most states require that federal funds be channeled through the state to local sponsors. In those states where this is a requirement, the state normally contributes one-half of the sponsor's share of the project costs, which amounts to approximately one-quarter of the total project cost. If no federal aid is involved, the state often contributes one-half of the project cost. In other states where revenues are obtained from user taxes, formulas for apportioning the revenues are established.

Several states obtain the revenues to finance aviation and airport improvement projects from a variety of sources, including [19]

1. The general fund

2. Highway or transportation funds or bond issues

3. State-owned airport revenue bonds

4. Fuel taxes on general aviation and air carrier fuel

5. Sales taxes on the purchase of general aviation or air carrier fuel

6. Aircraft registration fees

7. Pilot registration fees

8. Airport licensing fees

9. Airline flight property taxes

10. Dedicated property tax millages

Some idea of the amount provided by the states for airport improvements in fiscal years 1980 and 1990 is shown in Table 2-3. It should be emphasized that 65 percent of the funds made available by the

TABLE 2-3   State Airport Development and Aviation Funds for Fiscal Years 1980 and 1990 ($1000)

	1980	1990		1980	1990
Alabama	637	895	Montana	365	557
Alaska	71,019	69,213	Nebraska	1,626	8,167
Arizona	4,122	11,870	Nevada		50
Arkansas	643	1,358	New Hampshire	981	929
California	4,960	5,672	New Jersey	113	653
Colorado		177	New Mexico	365	1,194
Connecticut	5,970	18,073	New York	4,843	15,040
Delaware	67	47	North Carolina	7,000	7,058
Florida	4,366	44,877	North Dakota	2,159	675
Georgia	2,040	2,922	Ohio	2,700	3,255
Hawaii	85,787	496,508	Oklahoma	828	1,320
Idaho	624	978	Oregon	662	66
Illinois	11,260	5,728	Pennsylvania	15,286	15,494
Indiana	2,517	1,168	Rhode Island	6,419	7,680
Iowa	1,134	2,363	South Carolina	5,920	3,622
Kansas	200	202	South Dakota	417	825
Kentucky	2,451	974	Tennessee	2,143	9,318
Louisiana	4,851	318	Texas	2,600	1,924
Maine	3,821	1,294	Utah	304	6,716
Maryland	21,602	68,771	Vermont	566	1,750
Massachusetts	323	2,581	Virginia	5,423	15,589
Michigan	2,280	6,528	Washington	966	1,181
Minnesota	8,895	9,363	West Virginia	1,303	85
Mississippi	183	742	Wisconsin	922	6,747
Missouri	231	1,419	Wyoming	1,570	1,620
			Total	$306,520	$864,500

SOURCE: National Association of State Aviation Officials [19, 22].

states in 1990 was provided by Alaska and Hawaii. This is because in these states most publicly owned airports are owned and operated by the state.

Local Sources of Funds

Municipalities have provided the major portion of the funds for airport improvements. These funds have come from three principal sources: taxes for the support of local government as a whole, sale of general obligation bonds, and sale of revenue bonds. In the early years of aviation, the general tax fund was the principal source of local funds, and it still is for small airports. The taxes levied are not earmarked specifically for airport use but are the kind that are normally imposed to operate most of the affairs of local government. As long as the amount of funding is relatively small, this method of financing has not met with much opposition from the citizens in whose political jurisdiction the airport is located.

As aviation grew and the amounts of funds required became large in relation to other community expenditures, drawing funds from the general tax fund became impractical, and municipalities had to resort to the sale of bonds. Initially these were mostly of the general obligation type. There were several reasons for resorting to this type of bond financing. First, obligating the entire resources of a local government to back the bonds resulted in much more favorable interest rates than could be obtained by any other form of financing. Second, the projected revenues to be derived from the facility to be developed were often insufficient to utilize any other method of financing.

As air transport continued to grow and mature, the requirements for airport capital improvements also grew substantially. At the same time, communities were faced with an increased demand for schools, streets, sewage disposal, and other public services. In many cases cities either had reached or were reaching the statutory limit on the amount of general obligation bonds which they could issue, and the cities desired to reserve whatever remaining margin of bonding capacity they had to carry out needed improvements that did not have the revenue potential of airports.

To overcome the limitation of financing through the sale of general obligation bonds, many communities are raising funds through the sale of revenue bonds whenever possible. In general, only those airports which generate sufficient revenue over the term of the bond issues are candidates for revenue bonds. Agreements between the airlines and the airport are the most common form of supporting the issuance of revenue bonds. In contrast, most general aviation airports and the smaller air carrier airports do not generate sufficient revenue to support revenue bonds. Consequently, in these cases major capital improvements are usually undertaken with general obligations and combinations of federal and state grants.

Revenue bond financing is gaining impetus. This is partially due to the fact that the investor has gained confidence in the stability of air transportation and does not consider the investment as much of a risk as it might have been in the earlier years of aviation. While interest rates are generally higher for revenue bonds than they are for general obligation bonds, the differential between the two has decreased considerably since the first issuance of revenue bonds. Revenue bond financing is most successful for those components of the airport which are good revenue producers, such as terminal buildings and parking garages. Revenues from runways and taxiways, on the other hand, are normally insufficient to finance these specific facilities solely through the sale of revenue bonds.

Some facilities at an airport, such as hangars, hotels, and shopping centers, have been financed through private debt. The tenant constructs the facility on airport property being leased from the airport owner. One advantage of this type of airport financing is that it relieves the community of all capital investment in the facility except for utilities and access roads or taxiways. However, it does require the community to commit the land on the airport for 25 to 30 years, the period normally required in a ground lease if the tenant is to secure private financing.

Privatization of Airports

Privatization is a mechanism by which some level of airport management, operation, or ownership is transferred from the public sector to the private sector. Its purpose is to introduce market competition into the operation of airports and to relieve government of the financial burden of providing the large investments required to maintain and operate a system of airports. Proposals exist for the private involvement in airports in several ways including the outright transfer of airport ownership, the leasing of the airport to private sector management firms, and the private development, ownership, and operation of a segment of an airport such as the terminal buildings. Several examples of the various types of airport privatization ventures presently exist [4].

Although in the United States the privatization of airports has been considered in only a limited number of isolated situations, in other countries privatization of airports is a significant issue because of the absence of a specifically designated revenue base dedicated to financing aviation system improvements such as the Airport and Airway Trust Fund. In these countries, investments in aviation facilities compete with other public-sector programs for limited funds. Given the significant level of financial investment required to maintain an adequate aviation system and the pressures to limit overall government expenditures, privatization is viewed as a mechanism to finance aviation system improvements and operations with limited public-sector involvement.

Proponents of airport privatization often argue that economic market forces and profit motivation can stimulate private-sector investments, resulting in the development of both new airports and increased capacity at existing airports. Furthermore, it is argued that privatization can lead to cost savings in the management and operation of the airport because of private-sector profit motivation, which tends to lower costs and increase productivity. It is also argued that private-sector management and operation of airports can lead to revenue enhancement through market pricing strategies being employed for airport airside and landside services. Market pricing strategies, such as marginal cost pricing, often lead to a more efficient utilization of airport resources and generate the revenues necessary to increase capacity in those aspects of airport operations which need capacity enhancement. Furthermore, market pricing can be used to increase revenue from the various commercial enterprises offering services at the airport.

Proposals for the privatization of airports require the careful consideration of several factors, the most fundamental of which is that an airport is essentially a monopoly upon which airport users, i.e., airlines and other aircraft operators, passengers, and shippers, are highly dependent. For this reason regulatory safeguards must be implemented with privatization that maintain freedom of access and nondiscrimination among different groups of airport users, ensure conformity with operating standards and agreements, and satisfy both national and local air transportation policy [3].

Financial Planning

The financial plan for the capital improvements at an airport requires a detailed analysis of projected traffic, costs, and revenues. At larger airports, significant portions of the capital costs of a project are recovered through revenues from the airlines, concessionaires, and other tenants. The remainder is recovered through capital grants from federal and state sources. Since the airlines become long-term tenants obligated to pay user fees and rents for the facilities utilized in conducting their operations, an evaluation of the conceptual alternatives in the planning phases of a project should only be undertaken with direct input from these users. The financial feasibility of a program is in a large measure determined by the magnitude and reasonableness of the charges and rents paid by airport users and tenants. The financing of general aviation airports continues to be a major problem since the revenue base available is usually insufficient to support significant capital improvements. Therefore, it is imperative that the benefits of such airports to the community at large be carefully analyzed so that other sources of financing can be demonstrated economically viable.

The determination of the financial feasibility of a project is initiated with an agreement between the airport owners and airport users defining the fiscal policies which will govern the setting of rates and charges for the airport users. The basic process consists of a series of steps, as defined below [21]:

1. Allocation of the capital costs of the project to the various cost centers is established by airport management and airport users.

These cost centers are usually categorized as the airfield area, hangar and other operational support building areas, terminal area, concessions area, and other areas of the airport.

2. The net annual costs of the capital construction program are projected, and these costs are assigned to the various cost centers. These costs are amortized over the period specified in the agreement between management and users.

3. Projection and allocation of the net annual administrative, operating, and maintenance costs to each of the cost centers are based upon a knowledge of past cost experience and projections of these anticipated costs for the new facility.

4. Conversion of the total annual capital and administrative, operating, and maintenance costs to a schedule of the fees and rents to be paid by the users of the facilities utilizes available forecasts of aircraft activity, passenger enplanements, parking usage, and other relevant indices of projected airport activity.

Recovery of capital costs requires that the total capital investment in each cost center be determined. Projected costs in the airfield area for runways, taxiways, apron ramps, and land acquisition and improvement and in the terminal area for terminal building construction, land, and terminal support facilities must be ascertained and assigned to the relevant cost center. It is essential that airport support facilities such as access roads, service roads, sanitary and storm sewer systems, electrical and mechanical systems, communication and security services, emergency medical services, and crash, fire, and rescue services be properly apportioned to the appropriate cost centers to eliminate imbalances in the determination of the facility cost-center revenue requirements, which may result in unreasonable rates and charges. For projects where bond issues are utilized to finance portions of the capital costs, the annual cost of debt service (i.e., principal, interest, and the required reserve, called coverage) over the recovery period must be included and assigned to the relevant cost-center category.

The anticipated costs of airport administration, operation, and maintenance are assigned to each cost center on an annual basis. These costs generally include all direct costs for salaries, materials, supplies, and outside services and related indirect costs.

Terminal costs are divided by the terminal area. Often the location and degree of finishing in ticketing facilities, baggage facilities, office space, and car rental space are taken into consideration to establish rental rates for the terminal building tenants.

Concession and other airport revenues will normally be applied against the appropriate cost center and the net revenue recovery requirements determined. Forecasts of landing weights are used to determine the landing fees charged to the airlines to recover airfield costs. Often the cost of the apron area is isolated as a separate cost center, and ramp fees are established based upon the gate frontage required by the airlines.

Concession area costs are derived from rentals paid and from charging the concessionaire a percentage of the gross receipts. Usually the most significant concession cost center at a large airport is the parking facility. Since the capital costs of a parking structure are considerable, these costs are usually assigned to the terminal-area cost center. The administrative, operating, and maintenance costs of these facilities are recovered through a percentage of the gross receipts charge. A breakdown of the items included in the various revenue categories recommended by the Airport Operators Council International (AOCI), now the Airports Council International (ACI), is shown in Table 2-4.

The percentage distribution of operating revenue sources for a sampling of air carrier airports is shown in Table 2-5. This table indicates the percentage of total operating revenue receipts which could be expected from various revenue source items in each cost-center category. Based upon data collected from a sample of general aviation airports, it was found that such airports needed a strong industrial and commercial base in their service area and total operations in excess of 100,000 annually for the airport to recover operating expenses through operating revenues [7]. In these cases, a sampling of such general aviation airports demonstrated the operating revenue distributions shown in Table 2-6.

TABLE 2-4   Airport Report Revenue Classifications

Landing area	Airline-leased areas
Landing fees	Ground or land rentals
Parking ramp fees	Cargo terminals
Fuel flowage fees	Office rentals
Terminal-area concessions	Ticket counters
Advertising	Operations and maintenance areas
Bus transportation	Hangars
Car rentals	Other leased areas
Coin-operated devices	Fixed-base operators
Flight insurance	Ground or land rentals
Hotel	Cargo terminals
Limousine and taxi service	Industrial buildings
Parking	Other buildings
Personal services	Fuel and servicing
Restaurants	Other operating areas
Specialty shops, stores, and facilities	Sale of insurance
Miscellaneous	Other

SOURCE: Airport Operators Council International, Inc. [1].

TABLE 2-5   Percentage Distribution of Operating Revenue Sources at Air Carrier Airports by Annual Enplaned Passengers

	Annual enplaned passengers				
Revenue category	Greater than 2,000,000	500,000 to 2,000,000	250,000 to 500,000	125,000 to 250,000	Less than 125,000
Airfield area	27.6	32.1	31.9	30.2	22.2
Air carrier landing fees	25.1	22.5	19.0	16.5	8.9
Other landing fees	0.4	0.6	1.0	1.2	2.3
Fuel and oil sales	0.9	4.6	10.0	10.6	8.4
Airline catering fees	1.1	3.3			
Aircraft parking	0.1	1.1	1.9	1.9	2.6
Hangar and building area	11.4	11.6	13.6	20.2	43.9
Terminal area	12.8	13.3	18.7	14.8	10.5
Systems and services	4.3	3.1	4.0	4.0	4.0
Concessions	43.9	39.9	31.8	30.8	19.4
Airport parking	19.7	15.5	11.5	11.0	2.1
Car rental	8.2	10.3	10.2	8.0	6.8
Restaurants and lounges	4.8	5.5	5.9	3.5	2.5
Advertising	0.6	1.2	0.9	1.5	1.2
Ground transportation	1.9	1.9	0.7	1.9	0.2
Flight insurance	2.3	2.4	0.7	0.7	0.5
Hotel and motel	1.8	1.6	0.1	1.5	0.1
Miscellaneous	4.6	1.5	1.8	2.7	6.0
Total	100.0	100.0	100.0	100.0	100.0

SOURCE: Federal Aviation Administration [7].

TABLE 2-6   Percentage Distribution of Operating Revenue Sources at General Aviation Airports

Revenue category	Percentage
Airfield area	18.7
Hangar and building area	61.6
Terminal area	0.7
Systems and services	8.3
Concessions	10.7
Total	100.0

SOURCE: Federal Aviation Administration [7].

Rate setting

The airport is designed to service two distinct groups: the airlines and the commercial entities serving them, and the passengers and those retail enterprises which service them [16]. The airport leases its facilities to the airlines, concessionaires, industries, general aviation, and airport support services. The airlines lease ground for aircraft storage and space for ticket counters, operations, maintenance, and baggage handling, and they pay fees for landing and ramp rights. Cargo and hangar facilities are also utilized by the airlines at specific locations. Concessionaires rent space within the terminal and are charged on the basis of the amount and quality of space rented and a percentage of receipts. Those franchisees outside the terminal, such as taxicab companies, are charged in various ways. One method is based upon the number of passengers enplaned at the airport, and another is a fixed rate based upon the number of times the airport is utilized for the service.

The method for determining rates varies from airport to airport. The most common method is the *residual cost approach.* In such an approach, the total annualized costs of the airport are reduced by the amount of all nonairline revenues, and the remainder is proportioned among the airlines based upon the level of activity measures. Those costs apportioned to the terminal area are divided by the gross terminal area to determine space rental charges. Those costs apportioned to the airfield are divided by the total annual gross landing weight of the carriers at the airport to determine landing fees. This type of cost recovery approach essentially guarantees that the airlines will provide the revenues necessary to cover airport costs. It also places the airlines in a unique position relative to airport management in that the airlines have a vested interest in maximizing nonairline revenues to minimize their costs.

Other approaches to rate setting attempt to classify airport expenses into distinct cost centers and to apportion the cost of each among users through equitable rates, or to assign the expenses associated with certain cost centers directly to users and to group other expenses to be shared by all users. This type of rate setting is called the *compensatory cost method.*

The test of the validity of any rate setting scheme, however, lies in its ability to reflect rates which are reasonable and justifiable to the airlines, concessionaires, and other tenants. An illustration of the mechanics of rate setting and the determination of measures to assess financial feasibility are contained in Example Problem 2-1.

Example Problem 2-1 Estimate the rates and charges required to support the capital costs of airport development shown in Table 2-7. Use the compensatory cost method of determining rates and charges. The capital costs are financed by issuing 6 percent bonds which are repaid in 20 years. Financial considerations require a 1.25 coverage factor on revenues to support the repayment of airport development bonds. The interest earned at an annual rate of 6 percent on the accumulated balance in the capital recovery fund is to be used to decrease the rates and charges.

The airport is being developed for an initial annual enplaned passenger level of 2 million, and it is expected that the annual demand will increase to 4 million enplaned passengers in 20 years. Air carrier demand is expected to increase from 48,000 to 96,000 annual operations, and total aircraft operations are expected to increase from 150,000 to 250,000 annual operations over the 20-year

TABLE 2-7   Airport Capital Development Costs for Example
Problem 2-1

Airfield	$72,265,000
Apron area and concourses	46,510,000
Main terminal building	50,000,000
Parking facilities	15,400,000
Airport access roads	4,260,000
Land acquisition	10,980,000
Total	$199,415,000

period. Assume that the increases in annual passengers and aircraft operations are constant in each year rather than the growth rate of annual passengers and operations being constant. The air carrier aircraft mix using the airport is expected to consist of 25 percent Boeing 767-200 and 75 percent McDonnell-Douglas MD-87 aircraft. A typical general aviation aircraft will have a maximum certified landing weight of 3000 lb. The airport will be developed with 4 wide-bodied gates to accommodate the Boeing 767-200 and 12 narrow bodied gates to accommodate the MD-87. The main terminal building will have an area of 250,000 ft^2 exclusive of the concourses housing the aircraft gates.

The characteristics of these aircraft relevant to this problem, i.e., the maximum certified landing weight, the aircraft wingspan, and the average passenger capacity, can be found in Table 3.1. Therefore, the Boeing 767-200 is found to have a maximum certified landing weight of 272,000 lb, a wingspan of 156 ft 1 in, and an average passenger capacity of 236 passengers. The MD-87 has a maximum certified landing weight of 130,000 lb, a wingspan of 107 ft 10 in, and an average passenger capacity of 135 passengers.

The rates and charges to the airport users must recover the capital development costs, including bond interest, over the capital recovery period of the bonds, which is 20 years.

In general, airfield costs are recovered through landing fees, apron area and concourse costs through ramp charges, terminal building costs through square-foot rental charges, and parking facility costs through parking rates. The costs for the ground access system and the land acquisition are usually allocated to the other charges to recover these costs. Sometimes the ground access system costs are allocated to either the terminal building or parking charges or to both. In this problem, the ground access system and land acquisition costs total $15,240,000 and represent about 8 percent of the total project cost. Therefore, all rates and charges would be increased by 8 percent to cover these costs.

First, the determination of landing fees is made. Generally, landing fees are charged to only the commercial air carriers at an airport. Since the air carrier demand will double over the 20-year span, the total number of air carrier aircraft landings using the airport over a 20-year period will be about 720,000. Assuming that the air carrier mix over the project period is constant, the average landing weight of the air carrier aircraft is expected to be $0.25 \times 272,000 + 0.75 \times 130,000 = 165,500$ lb. Therefore, the total landed weight of all air carrier aircraft over the project life is found to be $720,000 \times 165,500 = 119,160,000,000$ lb.

If bonds are issued to finance the project, the bond interest payments for 20 years are equal to $72,265,000 \times 0.06 \times 20 = $86,718,000$. The required average annual contribution to the capital recovery fund to repay the face value of the bonds, considering the interest earned on the accumulated surplus, is

$2,413,200. Therefore, the air carrier landing fee must generate a total revenue of $86,718,000 + 20 × $2,413,200 = $134,982,000.

To recover the airfield cost through air carrier landing fees, the average rate per 1000 lb of landed weight is found to be $134,982,000 ÷ 119,160,000 = $1.13. The landing fee that must be assessed to air carrier aircraft to recover the airfield costs is computed for the Boeing 767-200 as $1.13 × 272 = $307.36 and for the MD-87 as $1.13 × 130 = $146.90. (The reader should examine the cash flow for each year to determine that the required revenue is attained through the landing fee computed. If this is done, it will be found that the landing fee must be increased to about $1.133 per thousand lb to realize sufficient revenue to pay the bond interest and retire the bonds at the end of 20 years. In any situation in which rates are based on a varying demand, a cash flow analysis should be performed to verify the rates and charges, as average values of demand typically yield either too little or too much revenue.)

It is instructive to look at the impact of a policy which charges general aviation aircraft landing fees on the cost to air carriers. Since there are initially 102,000 annual general aviation operations which increase to 154,000 annual operations in the design year, general aviation aircraft over the project life will conduct about 1,280,000 landings. General aviation aircraft at this airport average about 64 percent of the aircraft fleet. Therefore, the average landed weight of all aircraft using the airport is equal to 0.64 × 3000 + 0.36 × 165,500 = 61,500 lb. The total landed weight of all aircraft over the project life is then found to be 2,000,000 × 61,500 = 123,000,000,000 lb.

To recover the airfield cost through landing fees assessed to all aircraft, the average rate per 1000 lb of landed weight is found to be equal to $134,982,000 ÷ 123,000,000 = $1.10. This is not a significant reduction from the case where only air carrier aircraft were charged landing fees. The landing fee that must be assessed to air carrier aircraft to recover the airfield costs is computed for the Boeing 767-200 as $1.10 × 272 = $299.20 and for the MD-87 as $1.10 × 130 = $143.00. For general aviation aircraft the landing fee is $1.10 × 3 = $3.30.

Clearly there is very little benefit to air carrier aircraft from a policy which charges landing fees to all aircraft using the airport. Furthermore, the design of the airfield is significantly impacted by the presence of air carrier aircraft, and these costs are appreciably higher than if the airfield were designed for only general aviation aircraft. Collecting general aviation aircraft landing fees also presents an operational problem for airport management, and the cost of collecting the fee will very likely exceed the fee collected. Airport management must determine the best method of charging general aviation aircraft for airport use if landing fees are not assessed to these users.

Next a determination of ramp fees for air carrier aircraft is made. Normally, ramp fees are charged to the airlines to recover the cost of the apron and concourse system. The ramp fee is based upon the wingspan of the gate design aircraft.

In this problem, the average wingspan of the gate design aircraft is computed as 0.25 × 156.1 + 0.75 × 107.84 = 120 ft. The total wingspan of the aircraft occupying the 16 gates is then 16 × 120 = 1920 ft. The cost of the apron and concourse development was $46,510,000. The interest payments on the bonds over 20 years will be $55,812,000. The required average annual contribution to the capital recovery fund to repay the face value of the bonds, considering the interest earned on the accumulated surplus, is $1,553,000. Therefore, the air carrier landing fee must generate a total revenue of $55,812,000 + 20 × $1,553,000 = $86,872,000. Therefore the cost per foot of gate over the life of the project is $86,872,000 ÷ 1920 = $45,250. This means that each of the Boeing 767-200 gates would cost $45,250 \times 156.1 = \$7,063,500$ and that each of the MD-87 gates would cost $45,250 \times 107.84 = \$4,879,800$. This is a significant cost to the airline leasing the gate. Since there are 720,000 landings over a 20-year period, there will be 45,000 aircraft occupancies at each gate. The cost per aircraft for gate use is then $156.97 for a Boeing 767-200 and $108.44 for an MD-87.

The cost of the main terminal building is usually recovered through square-footage rental charges. However, typically only about 50 percent of the space in a terminal building is rentable. Therefore, the space in the main terminal building which is rentable is $0.50 \times 250,000 = 125,000$ ft^2. The cost of the main terminal building is then recovered based upon this area. The cost of the main terminal building is $50 million. The interest payments on the bonds over 20 years will be $60 million. The required average annual contribution to the capital recovery fund to repay the face value of the bonds, considering the interest earned on the accumulated surplus, is $1,359,300. Therefore, the main terminal building area charges must generate a total revenue of $60,000,000 + 20 \times \$1,359,300 = \$87,186,000$. Therefore, the cost per square foot to recover terminal building costs is then $87,186,000 \div 125,000 = \$698$ over the life of the project. Assuming a 20-year project life, this becomes $35 per square foot per year.

Since typically about 40 percent of the terminal building area is rented by the airlines, this represents a total annual cost to the airlines of $35 \times 0.40 \times 250,000 = \$3.5$ million. This represents a lifetime cost of $70 million to the airlines, which is significant since the airlines normally operate from many airports.

Parking charges are used to recover the cost of the parking facility. This requires that one know the number of parkers and average parking duration. If it is assumed that vehicle occupancy rates are 2.5 passengers per vehicle, then 4 million annual enplaned passengers translate to 1.6 million annual vehicles with enplaning passengers in the design year. Since over time the enplaning and deplaning passengers are about the same, this means a total of 3.2 million vehicles on the ground access system during the design year. Typically 70 percent are passenger cars, and of these typically 30 percent park. The number of vehicles parking in the design year is then $0.70 \times 0.30 \times 3,200,000 = 672,000$.

The total number of vehicles parking in the parking facilities over the project life is then about 10,080,000. The parking facility cost is $15,400,000. The interest payments on the bonds over 20 years will be $18,480,000. The required average annual contribution to the capital recovery fund to repay the face value of the bonds, considering the interest earned on the accumulated surplus, is $514,300. Therefore, the terminal area charges must generate a total revenue of $18,480,000 + 20 \times \$514,300 = \$28,766,000$. The average parking rate required is $28,766,000 \div 10,080,000 = \$2.85$ per vehicle. This requires that time-related parking rates be established based upon the average vehicle parking time to realize a revenue of $2.85 per parked vehicle.

As noted earlier, one method of recovering land acquisition and ground access system costs is to proportionally increase the other rates for the effect of these costs. In this problem all the above rates would be increased by 8 percent for this purpose, since these costs are about 8 percent of the total project development costs. Additionally, bonding agencies usually require that the airport demonstrate that its rate structure will realize actual revenues which are 1.25 times the required revenues to retire airport debt. This is called bond *coverage*. Therefore, each of the rates and charges evaluated above must be increased in this problem by $0.08 + 0.25 = 0.33$, or 33 percent, for these purposes.

Based upon the cost recovery and allocation factors discussed above, and to account for the allocation of ground access system and land acquisition costs as well as the impact of bond interest and coverage, the final rates and charges can be determined. A Boeing 767-200 aircraft would be assessed a landing fee of $409 and a ramp fee of $209. An MD-87 aircraft would be assessed a landing fee of $195 and a ramp fee of $144. Tenants would be charged a rental fee of $47 per square foot per year. Parking charges would average about $3.79. The total number of enplaned passengers serviced over this 20-year project life is 60,500,000. The total number of deplaned passengers would also be about 60,500,000. Including the bond coverage requirements, the average cost per passenger for landing fees and ramp charges amounts to $2.44, and the total airport development cost per enplaned passenger is equal to $449,282,000 ÷ 60,500,000 = $7.43. Each of these metrics is an indicator of the financial viability of the development project. In both cases these are very reasonable values.

The reader should examine the cash flow for each year to determine that the required revenue is attained through the fees calculated. In any situation in which rates are based on a varying demand, a cash flow analysis should be performed to verify the rates and charges, as average values of demand typically yield either too little or too much revenue.

It should be emphasized that airport rates and charges include not only capital development costs but also the operating and maintenance costs associated with the airport. These charges are usually reevaluated each year by the airport.

The compensatory cost method of determining rates and charges was used in Example Problem 2-1. It is likely that very different rates and charges would be realized if the residual cost method were used. This is shown in Example Problem 2-2.

Example Problem 2-2   Estimate the landing fees required to support the capital costs of the airfield development in Example Problem 2-1 if a passenger facility charge is imposed at the rate of $2 per enplaned passenger for 10 years. This passenger facility charge is dedicated to airfield development. Solve this problem using the residual cost method of determining rates and charges and assuming that the passenger facility charge is the only additional revenue received by the airport.

As before, the airport is being developed for an initial annual enplaned passenger level of 2 million, and it is expected that the annual demand will increase to 3 million enplaned passengers in 10 years. Therefore, the total revenue gained from the passenger facility charge is $2 × 25,000,000 = $50,000,000. The net cost of airfield development, including interest over the 20-year period, becomes $117,802,100 − $50,000,000 = $67,802,100. The average landing fee to support airfield development over a capital recovery period of 20 years will be determined.

Since the air carrier demand doubles over the 20-year span, the total number of air carrier aircraft landings using the airport over a 20-year period will still be about 720,000. The average landing weight of the air carrier aircraft is still equal to 165,500 lb, and the total landed weight of all air carrier aircraft over the project life is still 119,160,000,000 lb.

To recover the net airfield development cost through air carrier landing fees, the average rate per 1000 lb of landed weight is then found to equal $67,802,100 ÷ 119,160,000 = $0.57. Therefore, the landing fee that must be assessed to air carrier aircraft to recover the airfield costs is computed for the

Boeing 767-200 as $0.57 \times 272 = 155.04 and for the MD-87 as $0.57 \times 130 = 74.10. As in Example Problem 2-1, to account for land acquisition and ground access system costs, and coverage, these landing fees must be increased by 33 percent. Therefore, a Boeing 767-200 aircraft would be assessed a landing fee of $206, and an MD-87 aircraft would be assessed a landing fee of $99. The cash flow is shown in Table 2-8.

As may be observed, the residual cost method of determining rates and charges results in a decrease in cost to the airlines.

Evaluation of the financial plan

Criteria for measuring the financial effectiveness of an airport plan are usually determined by considering various evaluative measures, including these [21]:

1. The effectiveness of functional areas, as measured by the ratios of the amount of public space, revenue space, airline exclusive space, and concession space to the total space within the terminal building

2. The relative effectiveness of areas within the terminal building, as indicated by the ratio of airline exclusive space to the number of gates and the ratio of the ramp area to the total building area

3. An evaluation of annual costs and revenues for various items in each of the cost-center categories, as shown by the cost and revenue per enplanement, per operation, per 1000 lb of aircraft landing weight, and per square foot of building space

4. The effectiveness of the schedule plan of the airline, as indicated by the number of departures per gate and enplaned passengers per unit of airline exclusive space

The final determination of the most effective plan is made through the process of discussion and negotiation between airport managers and users. Various assumptions are made concerning the allocation of costs and revenues between cost centers until a consensus is reached. At this point airline lease agreements and concession policies are developed which result in long-term commitments by the airlines and tenants to the airport project. In the final analysis, airport expansion plans must address not only the needs for changes in physical facilities but also the economic, environmental, and financial feasibility associated with such development. It can be expected in this age of limited financial resources, with energy and aircraft equipment needs foremost in the management of airlines, that a clear determination of the feasibility of airport expansion projects will be required before long-term commitments for support by the airline will be made.

TABLE 2-8 Landing Fee Cash Flow Analysis with Passenger Facility Charge for Example Problem 2-2

Year	Aircraft landing demand	Enplaned passengers demand	Landing fee revenue	PFC revenue	Total revenue	Bond interest	Capital recovery fund	
							Deposit	Accumul.
1	24000	2000000	$2260068	$4000000	$6260068	$4335900	$1924168	$1924168
2	25263	2111111	2379019	4222222	6601241	4335900	2265341	4304959
3	26526	2222222	2497970	4444444	6942415	4335900	2606515	7169772
4	27789	2333333	2616921	4666667	7283588	4335900	2947688	10547646
5	29053	2444444	2735873	4888889	7624761	4335900	3288861	14469366
6	30316	2555556	2854824	5111111	7965935	4335900	3630035	18967563
7	31579	2666667	2973775	5333333	8307108	4335900	3971208	24076825
8	32842	2777778	3092726	5555556	8648282	4335900	4312382	29833816
9	34105	2888889	3211677	5777778	8989455	4335900	4653555	36277400
10	35368	3000000	3330628	6000000	9330628	4335900	4994728	43448772
11	36632	3100000	3449579	0	3449579	4335900	−886321	45169378
12	37895	3200000	3568531	0	3568531	4335900	−767369	47112172
13	39158	3300000	3687482	0	3687482	4335900	−648418	49290484
14	40421	3400000	3806433	0	3806433	4335900	−529467	51718445
15	41684	3500000	3925384	0	3925384	4335900	−410516	54411036
16	42947	3600000	4044335	0	4044335	4335900	−291565	57384134
17	44211	3700000	4163286	0	4163286	4335900	−172614	60654568
18	45474	3800000	4282237	0	4282237	4335900	−53663	64240180
19	46737	3900000	4401189	0	4401189	4335900	65289	68159879
20	48000	4000000	4520140	0	4520140	4335900	184240	72433711
Total	720000	60500000	$67802100	$50000000	$117802078	$86718000	$31084078	$72433711

References

1. *AOCI Uniform Airport Financial Statement,* Airport Operators Council International, Inc., Washington,
2. *Airport Administration and Management,* John R. Wiley, Eno Foundation for Transportation. Inc., Westport, Conn., 1986.
3. *Airport Economics Manual,* 1st ed., Doc. no. 9562, International Civil Aviation Organization, Montreal, Quebec, Canada, 1991.
4. *Airport Finance,* Norman Ashford and Clifton A. Moore, Van Nostrand Reinhold, New York, 1992.
5. *Airport Planning and Management,* D. F. Smith, J. D. Odegard, and W. Shea, Wadsworth Publishing Company, Belmont, Calif., 1984.
6. Airport Revenues and Expenses, *Airport Economic Planning,* G. P. Howard, ed., M.I.T. Press, Cambridge, Mass., 1974.
7. *Economics of Airport Operation,* calendar year 1972, J. A. Neiss, Federal Aviation Administration, Washington, April 1974.
8. *Eleventh Annual Report of Operations under the Airport and Airway Development Act,* fiscal year ended September 30, 1980, Department of Transportation, Federal Aviation Administration, Washington, 1981.
9. *Fort Lauderdale–Hollywood International Airport Economic Feasibility,* preliminary report, Aviation Planning Associates, Inc., Cincinnati, Ohio, April 1981.
10. *General Aviation and the Airport and Airway System: An Analysis of Cost Allocation and Recovery,* National Business Aircraft Association, Inc., Washington, April 1981.
11. *Hearings before the Subcommittee on Aviation of the Committee on Commerce,* U.S. Senate, Washington, July 1969.
12. *National Airport System Plan, 1978–1987,* Federal Aviation Administration, Department of Transportation, Washington.
13. *National Airspace System Plan,* Federal Aviation Administration, Washington, 1989.
14. *National Plan of Integrated Airport Systems (NPIAS), 1990–1999,* Federal Aviation Administration, Department of Transportation, Washington, 1991.
15. *Passenger Facility Charges,* pt. 158, Federal Aviation Regulations, Federal Aviation Administration, Washington, June 1991.
16. *Planning for Airport Access: An Analysis of the San Francisco Bay Area,* Conference publication 2044, National Aeronautics and Space Administration, Ames Research Center, Moffett Field, Calif., May 1978.
17. *Reauthorizing Programs of the Federal Aviation Administration, Future Capacity Needs and Proposals to Meet Those Needs,* Subcommittee on Aviation, Committee on Public Works and Transportation, House of Representatives Rep. no. 101-37, Washington, 1990.
18. *Airport and Airway System Act of 1981,* Senate Rep. no. 97-97, Committee on Commerce, Science, and Transportation, U.S. Senate, Washington, May 15, 1981.
19. *State Funding of Airport and Aviation Programs,* National Association of State Aviation Officials, Washington, 1981.
20. *Tenth Annual Report of Accomplishments under the Airport Improvement Program,* fiscal year 1991, Federal Aviation Administration, Washington, 1992.
21. *The Apron-Terminal Complex,* The Ralph M. Parsons Company, Federal Aviation Administration, Washington, September 1973.
22. *The States and Air Transportation: Expenditures and Tax Revenues,* Center for Aviation Research and Education, National Association of State Aviation Officials, Silver Spring, Md., 1991.

3

Aircraft Characteristics
Related to Airport Design

A general knowledge of aircraft is essential in planning facilities for their use. Aircraft presently used in airline operations have capacities ranging from less than 20 to approximately 600 passengers, and it is expected that over the next several years aircraft will be operating in commercial service with capacities which may approach 800 to 1000 passengers. General aviation aircraft, on the other hand, have a transportation function similar to that of the private automobile.

To present a perspective of the variety of aircraft which make up the airline fleet, Table 3-1 has been prepared to summarize the principal characteristics of air carrier aircraft in terms of size, weight, capacity, and runway length requirements. This list is by no means complete but does include the principal aircraft in use or in production. In a similar manner, some of the characteristics of typical general aviation and short-haul commuter or regional aircraft, including those used for corporate purposes, are shown in Table 3-2. It is important to recognize that such characteristics as operating empty weight, passenger capacity, and runway length can only be approximated in a very general way in such tabulations because many variables affect these quantities. Main landing gear dimensions and typical tire inflation pressures for some of the more common air carrier aircraft are shown in Table 3-3. Figure 3-1 illustrates the definition of the principal dimensions shown in Tables 3-1 and 3-2.

The characteristics shown in Tables 3-1 and 3-2 are essential for the planning and design of airports. The aircraft weight is important for determining the thickness of the runway, taxiway, and apron pavements, and it affects the takeoff and landing runway length requirements at an airport. The wingspan and the fuselage length influence the size of parking aprons, which in turn influence the configuration of the terminal buildings. Size also dictates the width of runways and taxiways and the distances between these traffic ways, and size affects

TABLE 3.1   Characteristics of Principal Transport Aircraft

Aircraft	Manufacturer	Wingspan	Length	Wheel base	Wheel track
A-300-600	Airbus Industrie	147'01"	175'06"	61'01"	31'06"
A-310-300	Airbus Industrie	144'00"	153'01"	49'11"	31'06"
A-320-200	Airbus Industrie	111'03"	123'03"	41'05"	24'11"
A-340-200	Airbus Industrie	197'10"	195'00"	62'11"	16'09"
B-727-200	Boeing	108'00"	153'02"	63'03"	18'09"
B-737-200	Boeing	93'00"	100'02"	37'04"	17'02"
B-737-300	Boeing	94'09"	109'07"	40'10"	17'02"
B-737-400	Boeing	94'09"	119'07"	46'10"	17'02"
B-737-500	Boeing	94'09"	101'09"	36'04"	17'02"
B-747-100	Boeing	195'08"	231'10"	84'00"	36'01"
B-747-200B	Boeing	195'08"	231'10"	84'00"	36'01"
B-747-300	Boeing	195'08"	231'10"	84'00"	36'01"
B-747-400	Boeing	213'00"	231'10"	84'00"	36'01"
B-747SP	Boeing	195'08"	184'09"	67'04"	36'01"
B-757-200	Boeing	124'10"	155'03"	60'00"	24'00"
B-767-200	Boeing	156'01"	159'02"	64'07"	30'06"
B-767-300	Boeing	156'01"	180'03"	74'08"	30'06"
B-777-200	Boeing	199'11"	209'01"	84'11"	36'00"
DC-8-73	McDonnell-Douglas	148'05"	187'05"	77'06"	20'10"
DC-9-32	McDonnell-Douglas	95'04"	119'04"	53'02"	16'04"
DC-9-51	McDonnell-Douglas	93'04"	133'07"	60'11"	16'00"
MD-81	McDonnell-Douglas	107'10"	147'10"	72'05"	16'08"
MD-87	McDonnell-Douglas	107'10"	130'05"	62'11"	16'08"
MD-90-30	McDonnell-Douglas	107'10"	152'07"	77'02"	16'08"
DC-10-10	McDonnell-Douglas	155'04"	182'03"	72'05"	35'00"
DC-10-30	McDonnell-Douglas	165'04"	182'03"	72'05"	35'00"
DC-10-40	McDonnell-Douglas	165'04"	182'03"	72'05"	35'00"
MD-11	McDonnell-Douglas	170'06"	201'04"	80'09"	35'00"
L-1011-500	Lockheed	164'04"	164'03"	61'08"	36'00"
BAe111-500	British Aerospace	93'06"	107'00"	41'05"	14'03"
F-28-4000	Fokker	82'03"	97'02"	33'11"	16'07"
Concorde	Aerospatiale/BAC	83'10"	205'05"	59'08"	25'04"
Ilyushine-62	Former U.S.S.R.	141'09"	174'04"	80'05"	22'04"
Tupolev-154M	Former U.S.S.R.	123'03"	157'02"	62'01"	37'09"

the required turning radius on pavement curves. The passenger capacity has an important bearing on facilities within and adjacent to the terminal building. The length of the runway influences, to a large extent, the land area required at an airport. Many of the values provided in Tables 3-1 and 3-2 are only approximate. For more precise values, the appropriate references, such as those listed in this chapter, should be consulted since the various types of aircraft are equipped differently for the specific market applications of the airlines.

An examination of Tables 3-1 and 3-2 reveals some interesting information. The maximum takeoff weight of principal airline aircraft

TABLE 3.1   (Continued)

Maximum structural takeoff weight, lb	Maximum landing weight, lb	Operating empty weight,* lb	Zero fuel weight, lb	No. and type of engines†	Payload, passengers	Runway length‡
363,765	304,240	197,192	286,600	2 TF	247–375	7,600
330,690	271,170	169,842	249,120	2 TF	200–280	7,575
158,730	134,480	84,171	125,662	2 TF	138–179	5,630
558,900	399,000	270,700	372,500	4 TF	262–375	7,600
184,800	150,000	101,773	138,000	3 TF	145–189	8,600
100,000	95,000	59,900	85,000	2 TF	97–136	5,600
124,500	114,000	69,400	105,000	2 TF	128–149	6,300
138,500	121,000	73,170	113,000	2 TF	146–189	7,300
115,500	110,000	69,030	102,500	2 TF	108–149	5,100
710,000	564,000	358,000	526,500	4 TF	452–480	9,500
775,000	564,000	381,150	526,500	4 TF	452–480	12,200
710,000	564,000	390,300	536,500	4 TF	565–608	7,700
800,000	574,000	396,142	535,000	4 TF	400	8,800
630,000	450,000	325,660	410,000	4 TF	297–331	7,000
220,000	198,000	128,380	184,000	2 TF	186–239	5,800
315,000	272,000	176,650	250,000	2 TF	216–255	6,000
345,000	300,000	186,378	278,000	2 TF	261–290	8,000
535,000	445,000	299,550	420,000	2 TF	305–375	8,700
355,000	258,000	166,500	231,000	4 TF	196–269	10,000
121,000	110,000	57,190	98,500	2 TF	115	5,530
121,000	110,000	64,675	98,500	2 TF	139	7,100
140,500	128,000	77,888	118,000	2 TF	155–172	7,250
149,500	130,000	74,880	112,000	2 TF	130–139	7,600
156,000	142,000	86,588	130,000	2 TF	158–172	6,800
430,000	363,500	240,171	335,000	3 TF	270–399	9,000
572,000	403,000	267,197	368,000	3 TF	255–380	9,290
555,000	403,000	270,213	368,000	3 TF	255–399	14,500
602,500	430,000	285,846	400,000	3 TF	323–410	9,800
510,000	368,000	245,400	338,000	3 TF	246–330	9,200
119,048	109,127	66,260	87,080	2 TF	86–104	6,900
73,000	69,500	38,900	62,000	2 TF	85	5,200
408,000	245,000	175,000	200,000	4 TJ	108–128	11,300
363,760	231,500	157,520	208,550	4 TF	168–186	10,830
220,460	176,366	121,915	163,140	3 TF	162–180	8,200

*Approximate only; depends upon seating configuration.   †TF = turbofan, TJ = turbojet.
‡At sea level, standard day, no wind, level runway.

SOURCE: Manufacturers' data

varies from 73,000 to 800,000 lb. For small general aviation aircraft, maximum takeoff weight ranges from 1600 to 5500 lb, while small regional, commuter, and corporate aircraft vary from 6600 to nearly 100,000 lb. The maximum number of passengers carried by airline aircraft varies from about 80 to over 600. On the other hand small general aviation airplanes seat from 2 to 6 people, and small regional, commuter, and corporate aircraft seat from less than 10 to nearly 130 persons depending on the configuration of the interior of these air-

TABLE 3-2 Characteristics of General Aviation and Short-Haul Passenger Aircraft

Aircraft	Wingspan	Fuselage length	Wheel track	Maximum takeoff weight, lb	Maximum number of seats	Number and type of engines*	Runway length, ft
Aerospatiale N262	74'02"	63'03"	10'03"	23,480	29	2 TP	2,690
ATR-42-300	80'07"	74'05"	13'05"	36,815	42–50	2 TP	3,576
ATR-72	88'09"	89'02"	13'05"	47,400	64–74	2 TP	4,620
BAe ATP	100'06"	85'04"	27'09"	50,550	64–72	2 TP	4,800
BAe 146-100	86'05"	85'10"	15'06"	84,000	94	4 TF	4,000
BAe 146-300	86'05"	101'08"	15'06"	95,000	128	4 TF	5,600
Beech SuperKingAir	54'06"	43'09"	17'02"	12,500	9–14	2 TP	2,580
Beech 1900C	54'06"	57'10"	17'02"	16,600	19	2 TP	3,260
BN2B-Islander	49'00"	35'08"	11'10"	6,600	9	2 P	1,155
Canadiar RJ-100ER	69'07"	87'10"	25'00"	51,000	50	2 TF	5,265
CASA C-212-300	66'07"	53'00"	10'02"	16,975	25	2 TP	2,680
Cessna 182 Skylane	35'10"	28'01"	8'00"	2,950	4	1 P	1,350
Cessna 310	37'06"	29'07"	12'00"	5,500	6	2 P	1,790
Cessna 402C	44'01"	36'05"	18'00"	6,850	10	2 P	2,195
Cessna Citation II	51'09"	47'03"	17'07"	13,300	6–10	2 TF	2,990
deHavilland DHC6	65'00"	51'08"	12'06"	12,500	21	2 TP	1,500
deHavilland DHC7	93'00"	80'08"	23'06"	44,000	52	4 TP	2,260
deHavilland DHC8-300	90'00"	84'03"	25'10"	41,100	50–56	2 TP	3,500

Dornier 228-212	55'08"	54'04"	14,109	19–20	2 TP	2,250
Dornier 328-100	68'10"	68'08"	27,557	30–33	2 TP	3,300
Embraer EmB 120	64'11"	65'08"	25,353	30	2 TP	4,600
Fairchild Metro III	57'00"	59'05"	16,500	19–20	2 TP	4,640
Fokker F27	95'02"	77'04"	45,000	44	2 TP	3,240
Fokker 50	92'01"	82'10"	45,900	50	2 TP	4,450
Fokker 100	92'02"	106'08"	98,000	107	2 TF	5,645
Gulfstream IV	77'10"	87'10"	71,780	14–19	2 TF	5,280
Jet Falcon 20T	54'03"	60'00"	29,100	28	2 TF	4,430
Jetstream Super31	52'00"	47'02"	16,204	18–19	2 TP	4,725
Piper PA-28 Archer	35'00"	23'09"	2,550	4	1 P	1,660
Piper PA-32 Saratoga	36'02"	27'08"	3,600	6	1 P	1,760
Piper Cheyenne IIIA	47'08"	43'05"	11,200	8–11	2 TP	2,380
Piper PA-28 Arrow 2	35'05"	24'08"	2,750	4	1 P	1,525
SAAB 340B	70'04"	64'09"	28,500	37	2 TP	4,170
SAAB 2000	81'03"	88'09"	47,000	50	2 TP	4,920
Shorts 330	74'08"	58'01"	22,900	30	2 TP	3,420
Shorts 360	74'10"	70'10"	27,100	36–39	2 TP	4,280

*TP = turboprop, P = piston, TF = turbofan.

SOURCE: Manufacturers' data; *Jane's All the World's Aircraft* [37].

TABLE 3-3   Main Landing Gear Dimensions for Typical Transport Aircraft

Main landing gear configuration		Aircraft type	Dimensions, in					Typical inflation pressures, lb/in²
			X	Y	Z	U	V	
(a)		B-727	34.0					168
		B-737	30.5					148
		MD-81	28.1					170
(b)		A-300	36.5	55.0				181
		A-310	36.5	55.0				172
		A-320	30.7	39.5				149
		B-707-120B	34.0	56.0				170
		B-707-320B	34.6	56.0				180
		B-720B	32.0	49.0				145
		B-757	34.0	45.0				161
		B-767	45.0	56.0				183
		Concorde	26.4	65.7				184
		DC-8-61	30.0	55.0				188
		DC-8-62	32.0	55.0				187
		DC-8-63	32.0	55.0				196
		DC-10-10	54.0	64.0				173
		L1011-500	52.0	70.0				184
(c)		B-747-100	44.0	58.0	121.1	141.0		192
		B-747-200	44.0	58.0	121.1	141.0		204
		B-747-400	44.0	58.0	121.1	141.0		195
		B-747-SP	43.3	54.0	121.1	141.0		205
(d)		A-340	55.0	78.0	39.0	211.0	38.0	
		DC-10-30	54.0	64.0	30.0	216.0	37.5	157 *
		DC-10-40	54.0	64.0	30.0	216.0	37.5	165 †

SOURCE: Manufacturers' data.

*Tire pressure of 134 lb/in² supports 16 percent of total weight.

†Tire pressure of 140 lb/in² supports 16 percent of total weight.

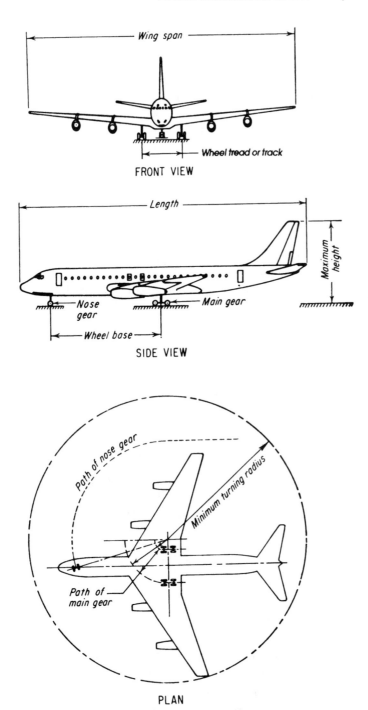

Figure 3-1 Definition of terms related to aircraft dimensions.

craft. Runway lengths for typical airline aircraft vary from about 5000 to over 12,000 ft, but note that it is not valid to assume that the larger weight of an aircraft, the longer the runway length required. For large aircraft, especially, the trip length has a profound influence on takeoff weight and hence the required runway length. Therefore, in the analysis of runway length requirements, an estimate of the trip length is very important. Runway lengths for small general aviation aircraft seldom exceed 2000 ft, while for commuter and corporate aircraft this figure is on the order of 5000 ft.

In Tables 3-1 and 3-2, aircraft are referred to according to the type of propulsion and thrust-generating medium. The term *piston engine* applies to all propeller-driven aircraft powered by gasoline-fed reciprocating engines. Most small general aviation aircraft are powered by piston engines. The term *turboprop* refers to propeller-driven aircraft powered by turbine engines. A few twin-engine general aviation aircraft and a few of the earlier airline aircraft are powered in this manner. The term *turbojet* has reference to those aircraft which are not dependent on propellers for thrust, but which obtain the thrust directly from a turbine engine. The early jet airline aircraft, particularly the Boeing 707 and the DC-8, were powered by turbojet engines, but these were discarded in favor of turbofan engines principally because the latter are far more economical. When a fan is added in the front or rear of a turbojet engine, it is referred to as a *turbofan*. Most fans are installed in front of the main engine. A fan can be thought of as a small-diameter propeller driven by the turbine of the main engine. Nearly all airline transport aircraft are now powered by turbofan engines for the reason just cited. Current technological advances in engines are concentrated toward the development of *prop-fan* engines for short- and medium-haul aircraft and ultrahigh-bypass-ratio turbofan engines for long-haul aircraft. These engine technologies are expected to reduce fuel consumption by 25 to 35 percent and are likely to be introduced on commercial aircraft in the 1990s. Various manufacturers are testing such engines which are variously termed the *unducted fan* (UDF) engines and the *ultrahigh-bypass-ratio* (UHB) turbofan engines.

Trends in Size, Speed, and Productivity of Transport Aircraft

The size, speed, and productivity of aircraft have grown enormously in the last four decades. The introduction of jet transports in the late 1950s was a tremendous step forward in the areas of speed, weight, and productivity. Since that time, aircraft grew gradually larger until the introduction of the Boeing 747 in the latter part of 1969, which was a quantum jump in size. Yet there has only been a very nominal increase in speed in subsonic jet aircraft. With the introduction of the supersonic transport, there was again a quantum jump in speed. In general, the speed of the subsonic jets is about twice that of the piston-engine aircraft of the late 1950s, while the speed of the first supersonic transport aircraft is again about twice the speed of the subsonic jets. The trend in cruising speed of transport aircraft is shown in Fig. 3-2. A perspective of the increase in speed in the last 50 years can also be gained by comparing the speeds of various transport aircraft with the speed of the DC-3, as shown in Table 3-4.

During the last four decades, we have witnessed an increase in not only the speed of aircraft but also the size and weight. The trend in maximum weight of transport aircraft is shown in Fig. 3-3. Trends in wingspan and overall length are shown in Figs. 3-4 and 3-5.

Note from Fig. 3-3 that in about 50 years the weight of transport aircraft has increased from slightly over 20,000 to nearly 800,000 lb, a 40-fold increase. New technology is not needed to produce aircraft with weights of 1,000,000 lb or more. Demand for travel is the princi-

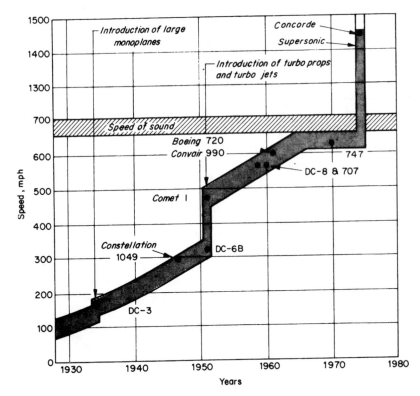

Figure 3-2 Trends in the speed of transport aircraft.

TABLE 3-4   Speed of DC-3 Compared to Speed of Other
Aircraft

Aircraft	Cruising speed, mi/h	Ratio of speed of aircraft to speed of DC-3
DC-3	185	1.0
DC-4	240	1.3
DC-6	305	1.7
DC-9	530	2.9
DC-8	570	3.1
DC-10	600	3.2
B-747	600	3.2
Concorde	1450	6.9

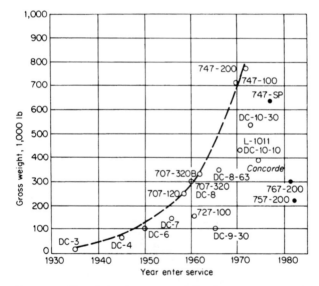

Figure 3-3 Trends in the gross weight of transport aircraft
(*Aerospace Industries Association of America, Inc. [28]*).

pal factor that should determine the growth in aircraft size; however, airport facilities may be a limiting factor in growth. The wingspan and overall length of an aircraft are, in general, a function of aircraft weight; i.e., the heavier an aircraft, the longer it is and the greater its wingspan. However, the growth in overall length is dependent upon the extent to which multideck aircraft are built. If these aircraft become common, the lengths will be shorter. For supersonic aircraft, the ratio of length to wingspan is larger than that for subsonic aircraft of equivalent weight.

The productivity of aircraft in ton-miles/h has increased enormously since the days of the DC-3. The *average productivity* is the product of the average load and the cruising speed. The average productivity

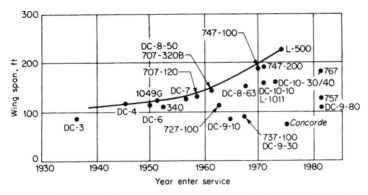

Figure 3-4 Trends in wingspan of transport aircraft (*Aerospace Industries Association of America, Inc. [28]*).

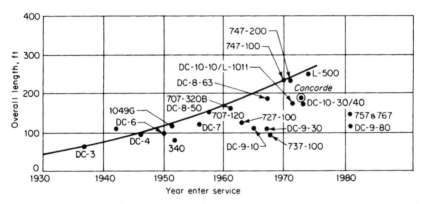

Figure 3-5 Trends in overall length of transport aircraft (*Aerospace Industries Association of America, Inc. [28]*).

of a DC-3, introduced in 1935, is 350 ton-miles/h; for a DC-7, introduced in 1953, it is 2700 ton-miles/h; for a 707-120, introduced in 1959, it is 8500 ton-miles/h; and for the 747, introduced in 1969, it is 38,500 ton-miles/h. Thus, the increase in productivity in a little less than 40 years has been over 100-fold. There are not likely to be further significant increases in productivity for subsonic aircraft in the near future because market forces do not seem to indicate a need for greater payload capabilities. The supersonic transport, because of its great speed, also has high productivity. However, the first models with their smaller load-carrying capability are not as productive as the 747. The productivity of one large jet transport between New York and London can be equal to or greater than that of a large ocean liner.

The revenue-producing lift capability of transport aircraft, referred to as *payload,* has also grown enormously over four decades, as shown

TABLE 3-5   Typical Payload of Transport
Aircraft

Aircraft type	Payload, tons
DC-3	2.4
DC-4	7.0
DC-7	9.0
DC-9-15	12.4
B-737-100	14.0
B-737-200	17.6
A-320	18.3
MD-81	20.1
B-737-400	21.4
DC-8-61	22.0
B-757-200	29.5
A-310	36.1
B-747-SP	42.2
B-767-300	44.2
A-300	44.7
DC-10-10	47.4
A-330	50.1
A-340	51.7
MD-11	57.1
B-777	60.3
B-747-400	69.5
B-747-200	73.0
B-747-100	84.3

in Table 3-5. Piston-engine transports of the middle and late 1950s—
the DC-7 and the Constellation—carried nearly 4 times more than
the DC-3. The first jet transports, the 707-120 and DC-8, in the late
1950s and early 1960s were able to carry about twice the load of piston-engine transports of the time. Through the 1960s the growth in
payload was gradual until the introduction of the wide-bodied aircraft
in the late 1960s and 1970s, when a very large jump in payload took
place.

The distance that aircraft can fly without refueling, the *range*, has
also greatly increased since the introduction of the DC-3. The DC-6
could fly about 4 times farther than the DC-3. The DC-6 was followed
by the DC-7, which was able to fly about twice the distance of the DC-
6. The DC-7 provided the first regular nonstop coast-to-coast travel in
the United States. Then came the jet transports. The early versions
were not able to fly as far as the DC-7, but as engine technology
improved and turbofan engines were introduced in the 1960s, the
range of the jet transport exceeded that of the DC-7. This made possible nonstop flights from the west coast of the United States to Europe.
From this point on there was a modest increase in range. Today there
are several aircraft which have the capability of flying about 7000 mi

(statute), making possible nonstop flights from the west coast of the United States to Europe or the Far East, or from Europe to Africa. The range of these aircraft is so large that it is doubtful that any significant increases in range will take place in the next decade.

To the airport designer the matter of runway length is important. Given the DC-3 as a starting point, runway lengths for piston-engine aircraft kept increasing as the DC-4, DC-6, and DC-7 were introduced. A further increase in runway length was required when jet transports were introduced into service. Since that time, there has been a leveling off in runway length requirements, as shown in Fig. 3-6. In this figure the lengths are shown relative to the DC-7. The lower curve in Fig. 3-6 represents recent trends in runway length requirements. To meet the runway length limitations existing at many major airports due to the inability to obtain additional land for increased runway length, both engine technology and wing technology have been utilized

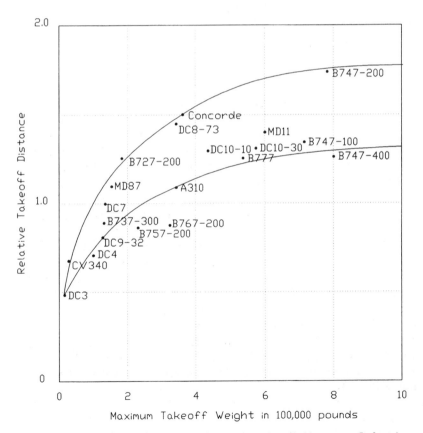

Figure 3-6 Trends in runway length of transport aircraft (*Aerospace Industries Association of America, Inc. [28]*).

to limit runway length requirements with increasing aircraft weight. As indicated in Table 3.1, runway lengths for jet transports vary widely depending on the trip length. For the longest intercontinental trips, a runway length on the order of 12,000 ft is generally adequate.

Characteristics of Transport Aircraft

Engine types and thrust

Jet engines can be classified into two general categories: turbojet and turbofan. A turbojet engine consists of a compressor, combustion chamber, and turbine at the rear of the engine. A turbofan is essentially a turbojet engine to which has been added large-diameter blades, usually located in front of the compressor. These blades are normally referred to as the *fan*. A single row of blades is referred to as *single-stage*; two rows of blades as *multistage*. In dealing with turbofan engines, reference is made to the *bypass ratio*. This is the ratio of the mass airflow through the fan to the mass airflow through the core of the engine, or the turbojet portion. In turbofan engines, the airflow through the core of the engine, the inner flow, is hot and very compressed and is burned in it. The airflow through the fan, the outer flow, is compressed much less and exits from the engine without burning into an annulus around the inner core. Fan engines are quieter than turbojet engines, and the development of quiet propulsive integrated power plants in modern turbofan has included extensive acoustic lining development both in the inlet and in the fan exhaust [38]. Bypass ratios of engines on aircraft other than the very large ones range from 1.1 to 1.4. For the very large wide-bodied aircraft the bypass ratio is on the order of 5 to 6. In high-bypass-ratio engines, 60 to 70 percent of the thrust is derived from the fan. Engines are usually rated in lb of thrust in a static position at sea level. If provision is made to use water during takeoff, the engine is referred to as *wet*; if no water is used, it is known as *dry*. Adding water to the engine allows it to operate at higher temperatures than normal, providing slightly more thrust. This larger amount of thrust is available only during takeoff. A typical thrust-versus-temperature relationship is shown in Fig. 3-7. The engine in this figure would be rated at 43,500 lb dry and 45,500 lb wet. Some dry ratings of typical engines at sea level are given in Table 3-6. When an aircraft is cruising, it is utilizing from one-fifth to one-fourth of the static sea-level thrust rating listed in the table.

Performance

One index of engine performance, insofar as consumption of fuel is concerned, is known as the *specific fuel consumption*. It is expressed

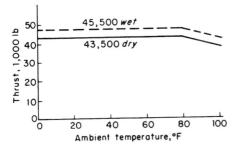

Figure 3-7 Thrust versus temperature for Pratt and Whitney JT9D-3A engine at sea level.

TABLE 3-6 Turbojet Aircraft Engines

Engine	Manufacturer	Thrust, lb	Aircraft
JT3D-3B	Pratt and Whitney	18,000	B-707, DC-8
JT8D-7	Pratt and Whitney	14,000	B-727, B-737, DC-9
JT8D-219	Pratt and Whitney	21,700	MD-80
JT9D-7	Pratt and Whitney	45,600	B-747
JT9D-59A	Pratt and Whitney	53,000	DC-10, A-300
PW2040	Pratt and Whitney	40,900	B-757
PW4052	Pratt and Whitney	52,200	B-767
PW4056	Pratt and Whitney	56,750	B-747, B-767
PW4084	Pratt and Whitney	84,000	B-777
RB211-22B	Rolls-Royce	42,000	L-1011
RB211-524G	Rolls-Royce	58,000	B-747
RB211-524H	Rolls-Royce	63,000	B-767
RB211-535E4B	Rolls-Royce	43,100	B-757
RB211 Trent668	Rolls-Royce	67,950	MD-11, MD-12X
CF6-45A2	General Electric	46,500	B-747
CF6-50C1	General Electric	52,500	DC-10, A-300
CF6-80A2	General Electric	50,000	B-767
CF6-80C2A3	General Electric	60,200	A-300, MD-11
CF6-80E1A3	General Electric	72,000	A-330
GE-90	General Electric	86,800	B-777
CFM56-3C	CFM International	23,500	B-737
CFM56-5B1	CFM International	30,000	A-321
CFM56-5C-2	CFM International	31,200	A-340
V2500-A1	International Aero	25,000	A-320
V2530-A5	International Aero	30,000	A-321
V2522-D5	International Aero	22,000	MD-90

in terms of pounds of fuel per hour per pound of thrust. In aviation, fuel consumption is expressed in lb rather than gallons. This is because the volumetric expansion and contraction of fuel with changes in temperature can be misleading in the amount of fuel which is available. Each gallon of jet fuel weighs about 6.7 lb.

TABLE 3-7   Performance Characteristics of Typical Jet Aircraft
Engines

Aircraft	Takeoff weight, 1000 lb	Engine	Bypass ratio	Specific fuel consumption
MD-87	140	JT8D-217C	1.7	0.736
MD-81	140	JT8D-209	1.8	0.724
B-757	220	RB211-535C	4.3	0.646
DC-9-41	114	JT8D-15	1.0	0.630
B-747-100	520	JT9D-3A	5.2	0.624
B-747-200	823	RB211-524D4	4.3	0.617
B-727-200	185	JT8D-9	1.0	0.595
A-320	159	V2527-A5	4.8	0.575
B-747-400	800	RB211-524H	4.3	0.570
B-767-300	345	PW4056	4.9	0.537
A-310	291	CF6-80A	4.7	0.376
MD-11	602	CF6-80C2	5.1	0.329
A-330	467	CF6-80E1	5.1	0.324
B-777	506	GE90	9.0*	0.310*

*Estimated.

The specific fuel consumption for a particular type of aircraft is a function of its weight, altitude, and speed. Some typical values are given in Table 3-7 merely to illustrate the fuel economy of a turbofan engine, particularly at high bypass ratios. It will be observed that significant gains in specific fuel consumption have been made with modern aircraft engines such as the RB211, CF6, and PW4000 engines. Table 3-8 gives the approximate average consumption of fuel for typical aircraft.

Fuel consumption improvements in the last two decades have been significant, and pressures continue to exist for further improvements in this area. New engines, such as the CFM56, CF6, RB211-524D, and PW4000, as well as derivatives of current engines have resulted in significant fuel economy gains. Figure 3-8 illustrates the overall trend in specific fuel consumption improvements. It is likely that advanced turboprop engines, the prop-fan engines, will be introduced in the future to take advantage of their fuel economy in applications where speeds below 0.8 Mach are possible [26].

TABLE 3-8   Average Fuel Consumption of Typical Jet Aircraft

Aircraft	Engine	Fuel consumption, lb/h
B-737-200	JT8D-9	9,300
B-757	RB211-535C	9,600
B-727-200	JT8D-9	11,000
DC-9-50	JT8D-17	11,000
DC-10-10	CF6-6D	17,400
L-1011-385	RB211.22B	19,700
B-747-200	JT9D-7A	26,000

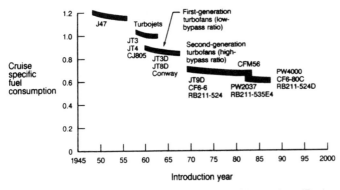

Figure 3-8 Trends in specific fuel consumption of jet engines (*Boeing Commercial Airplane Group [38]*).

An indication of the differences in fuel consumption attained by the various types of passenger aircraft in the different trip modes is given in Table 3-9. It should be pointed out, however, that the data are indicators of only fuel consumption, not productivity. Those aircraft which burn the higher rates of fuel generally are capable of greater speeds and have greater passenger capacity, as indicated in Table 3-10. Finally, Table 3-11 indicates the time spent in the various modes of an aircraft trip and may be combined with Table 3-9 to obtain estimates of fuel consumption for various trip lengths. As may be observed in Fig. 3-9, the fuel consumption in gallons per available seat-mile decreases with increasing route segment length.

Aircraft operating costs

Prior to 1973, fuel prices were relatively stable. However, in the middle to late 1970s, fuel costs rose very rapidly. In the early to mid-1980s these costs decreased and have become generally stable since then. The historical trends in and the projections for the price of oil and the price of jet fuel for U.S. airlines are shown in Fig. 3-10.

TABLE 3-9 **Average Fuel Consumption of Passenger Aircraft in Various Trip Modes, gal/h**

Aircraft type	Taxi or idle	Takeoff	Climb-out	Cruise	Approach and land
Jumbo jet	1,053	10,335	8,677	3,576	3,154
Long-range jet	528	6,567	5,428	1,879	2,508
Medium-range jet	436	3,980	3,335	1,136	1,550
Turboprop	299	1,450	1,326	582	695
STOL (short takeoff and landing)	45	182	152	85	76
General aviation turboprop	44	111	103	61	62

SOURCE: California Department of Transportation [34].

TABLE 3-10   Average Fuel Consumption at Best Cruising Speed for Passenger Aircraft

Aircraft type	Average number of seats	Best cruising speed, mi/h	Fuel consumption Cruise, gal/ seat-mile	Fuel consumption Noncruise, gal/seat
Jumbo jet	315	575	0.020	7.2
Long-range jet	140	565	0.024	10.1
Medium-range jet	90	565	0.022	10.1
Turboprop	85	360	0.019	3.0
STOL	19	190	0.028	1.8
General aviation turboprop	10	300	0.020	3.0

SOURCE: California Department of Transportation [34].

TABLE 3-11   Average Time in Operating Mode for Passenger Aircraft, h

Aircraft type	Taxi or idle Departure	Taxi or idle Arrival	Takeoff	Climb-out	Approach and land
Jumbo jet	0.32	0.12	0.012	0.04	0.07
Long-range jet	0.32	0.12	0.012	0.04	0.07
Medium-range jet	0.32	0.12	0.012	0.04	0.07
Turboprop	0.32	0.12	0.010	0.04	0.08
STOL	0.32	0.12	0.010	0.04	0.08
General aviation turboprop	0.32	0.12	0.010	0.04	0.08

SOURCE: California Department of Transportation [34].

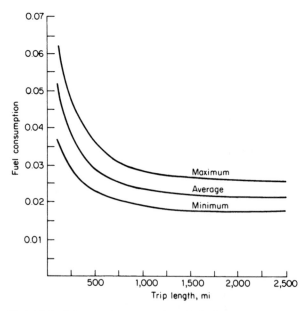

Figure 3-9 Average passenger aircraft fuel consumption in gallons per seat-mile as a function of route distance (*California Department of Transportation [34]*).

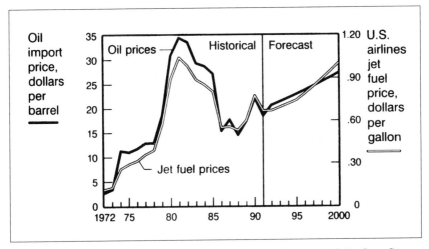

Figure 3-10   Trends in oil and aviation fuel prices (*Boeing Commercial Airplane Group* [29]).

TABLE 3-12   **Trends in the Percentage Distribution of Operating Costs for U.S. Airlines**

	1970	1980	1990	2000
Fuel	12.2	31.0	17.7	13.0
Crew	13.1	11.0	9.0	9.1
Maintenance	13.7	9.0	10.2	8.4
Aircraft investment	12.3	5.0	8.0	12.9
Indirect costs	48.7	44.0	55.1	56.6

SOURCE: Boeing Commercial Airplane Group [29].

Table 3-12 indicates the trends in the distribution of total operating costs for U.S. airlines. The variations in fuel costs have resulted in significant changes in the operating economics of aircraft. In 1970, fuel costs were about 12 percent of the total operating costs of a U.S. airline aircraft, in 1980 they rose to about 31 percent of these costs, and in 1990 they were reduced to slightly less than 18 percent of these costs. The trends in and projections of total aircraft operating costs and labor costs per available seat-mile are shown in Fig. 3-11. As may be observed, these costs have demonstrated a decreasing trend since 1980 and are forecast to remain constant or to decrease slightly through the year 2000. Of course, the price of jet fuel has a significant impact on the total operating costs per available seat-mile. To counteract the impact of escalating fuel prices extensive research has been undertaken to investigate the impact of such technological factors as drag reduction, engine efficiency, structural weight changes, and avionics on operating cost [27].

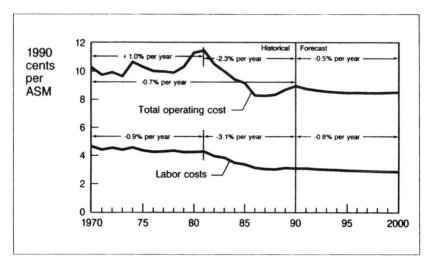

Figure 3-11 Trends in aircraft operating costs and labor costs per available seat-mile (*Boeing Commercial Airplane Group [29]*).

Knowledge of the relative magnitude of the operating expenses of the various aircraft in service today allows one to examine the alternatives for service by particular aircraft in specific markets. The operating and performance characteristics of aircraft are suited to a particular range of applications; and when aircraft are utilized in a route context for which they were not designed, considerable diseconomies can result. For example, the Boeing 737 was designed for the short-haul lower-density markets whereas the Boeing 747 was designed for the long-haul heavily traveled markets. The ability to project the types of aircraft which will be used in the several markets served by an airport is essential to the airport planning and design process.

Noise

The major sources of noise in an engine are the machinery noise and the primary jet noise. Machinery noise is noise generated primarily by the moving parts of the engine such as the fan, compressor, and turbine blades. The fan and compressor blade noise is radiated forward of the engine while the turbine blade noise is radiated aft of the engine. The parts of a jet engine are shown in Fig. 3-12. The primary jet noise is generated by the mixing of the high-velocity exhaust gas from the main body of the engine with the ambient air. The fan exhaust also generates noise, but during high levels of thrust, such as takeoff, the overall noise level from the primary jet is reduced by the presence of the fan exhaust. This is due to the fact that the veloc-

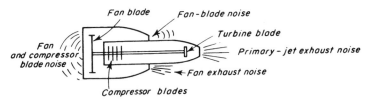

Figure 3-12   Sources of noise in jet engines.

ity of the fan exhaust is less than that of the primary jet; therefore it provides for a less abrupt mixing of the primary jet with the ambient air.

The dominant source of noise during takeoff is the primary jet, whereas on approach the dominant source is machinery noise. Acoustical lining of the inlet and ducts for the fan exhaust and removal of the inlet guide vanes have done a great deal to reduce machinery noise. Significant reductions in primary jet noise can be accomplished only by reducing the velocity of the primary jet. Several types of nozzle noise suppressors have been developed to reduce primary jet noise, including corrugated, multitube, and teeth types, but reductions in noise have been marginal.

Since one of the most serious problems facing aviation is noise, a substantial effort is being made by the federal government and engine manufacturers to reduce engine noise [7, 26]. It is interesting to note that, since the introduction of commercial jet aircraft in 1958, there has been a reduction of approximately 25 EPNLdB, where EPNLdB is the effective perceived noise level in decibels, which is quite significant. An indication of the advancements made in reducing sideline noise levels may be found in Fig. 3-13. The sideline noise levels shown in Fig. 3-13 are normalized to a total airplane sea-level static thrust and show that technology has made it possible to substantially reduce noise for a given amount of thrust and to dramatically increase thrust without significantly increasing noise levels. Prospects of producing substantially quieter engines continue to be promising. A more serious problem, however, was the requirement to reduce the noise of existing low-bypass-ratio engines which powered most of the earlier jet aircraft, such as the DC-8, DC-9, and the Boeing 707, 727, and 737. Some of these aircraft were fitted with "hush kits," but many have been retired from service.

Components of aircraft weight

If the information in Tables 3-1 and 3-2 is to be useful for planning purposes, the planner must understand the basic components which make up the weight of an aircraft during takeoff and landing, since

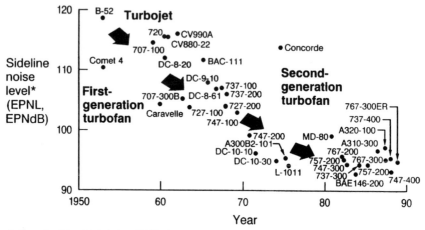

Figure 3-13 Trends in sideline noise level (*Boeing Commercial Airplane Group [38]*).

weight is one of the major factors which govern the runway length. A number of different weights which are referenced in airline operations are discussed below.

The *operating empty weight* (OEW) is the basic weight of the aircraft including the crew and all the gear required for flight but not including the payload and fuel. The operating empty weight is not a constant for passenger aircraft but varies with the seating configuration. The *zero-fuel weight* (ZFW) is the weight above which all additional weight must be fuel, so that when the aircraft is in flight, the bending moments at the junction of the wing and fuselage do not become excessive. The *payload* refers to the total revenue-producing load. This includes the weight of passengers and their baggage, mail, express, and cargo. The *maximum structural payload* is the maximum load which the aircraft is certified to carry, whether this load be passengers, cargo, or a combination of both. Theoretically, the maximum structural payload is the difference between the zero-fuel weight and the operating empty weight. The maximum payload actually carried is usually less than the maximum structural payload because of space limitations. This is especially true for passenger aircraft, in which seats and other items consume a lot of space. The *maximum ramp weight* is the maximum weight authorized for ground maneuver including taxi and run-up fuel. As the aircraft taxis between the apron and the end of the runway, it burns fuel and consequently loses weight. The *maximum structural takeoff weight* is the maximum weight authorized at brake release for takeoff. It excludes taxi and run-up fuel and includes the operating empty weight, trip and reserve fuel, and payload. The difference between the maximum structural takeoff weight and the maximum ramp weight is very nominal, only a few thousand pounds.

The *maximum structural landing weight* is the structural capability of the aircraft in landing. The main gear is structurally designed to absorb the forces encountered during landing; the larger the forces, the heavier the gear must be. Normally the main gears of transport aircraft are structurally designed for a weight less than the maximum structural takeoff weight. This is so because an aircraft loses weight en route by burning fuel. This loss in weight is considerable if the journey is long, being in excess of 80,000 lb for large jet transports. It is therefore not economical to design the main gear of an aircraft to support the maximum structural takeoff weight during landing, since this situation will rarely occur. If it does occur, as in the case of aircraft malfunction just after takeoff, the pilot must jettison or burn off sufficient fuel prior to returning to the airport so as not to exceed the maximum landing weight. For short-range aircraft such as the DC-9, the main gear is designed to support, in a landing operation, a weight nearly equal to the maximum structural takeoff weight. This is so because the distances between stops are short, and therefore a large amount of fuel is not consumed between stops.

A portion of the weight of an aircraft at the start of takeoff is composed of fuel. The fuel requirements can be segregated into two components: the amount of fuel needed to make the trip and the reserves required by regulations set forth by the government in the Federal Aviation Regulations (FAR), part 121 [23].

The fuel for the trip depends on the distance to be traveled, speed, meteorological conditions, altitude at which the aircraft is flown, and payload. The fuel reserve is dependent on the distance to the alternate airport, the amount of specified waiting time to land, and, for international flights, the trip length.

Thus the aircraft weight is composed of the operating empty weight and the three variables of payload, trip fuel, and fuel reserve.

On landing, the weight of an aircraft is the sum of the operating empty weight, payload, and fuel reserve, assuming that the aircraft lands at its destination and is not diverted to an alternate airport. This landing weight cannot exceed the maximum structural landing weight of the aircraft. The takeoff weight is the sum of the landing weight and the trip fuel. This weight cannot exceed the maximum structural takeoff weight of the aircraft.

An approximate estimate of the distribution of the components of aircraft weight is given in Table 3-13. Note that as the range of an aircraft is increased, the proportion of trip fuel to takeoff weight increases, while the proportion of payload decreases.

TABLE 3-13   Average Distribution of Weight Components for Passenger Turbine-Powered Aircraft, % of Takeoff Weight

Aircraft type	Operating empty weight	Payload	Trip fuel	Fuel reserve
Short range	66	24	6	4
Medium range	59	16	21	4
Long range	44	10	42	5

Payload and range

The question that is often asked is, How far can an aircraft fly? The distance it can fly is referred to as the *range*. A number of factors influence the range of an aircraft; among the most important is payload. Normally as the range is increased, the payload is decreased, with a weight tradeoff occurring between fuel to fly to the destination and the payload which can be carried. The relationship between payload and range is illustrated in Fig. 3-14. Point A, the range at maximum payload, designates the farthest distance R_a that an aircraft can fly with a maximum structural payload. To fly a distance of R_a and carry a payload of P_a, the aircraft has to take off at its maximum structural takeoff weight; however, its fuel tanks are not completely filled. Point B, the range at maximum fuel, represents the farthest distance R_b that an aircraft can fly if its fuel tanks are completely filled at the start of the journey. The corresponding payload that can be carried is P_b. To travel the distance R_b, the aircraft must take off at its maximum structural takeoff weight. Therefore to extend the

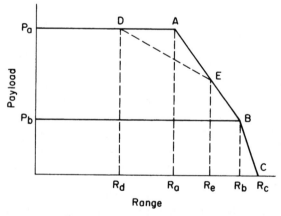

Figure 3-14 General representation of the relationship between payload and range.

distance of travel from R_a to R_b, the payload has to be reduced in favor of adding more fuel. Point C represents the maximum distance that an aircraft can fly without any payload. Sometimes this is referred to as the *ferry range*, and it is used, if necessary, for delivery of aircraft. To travel this distance R_c, the maximum amount of fuel is necessary, but since there is no payload, the takeoff weight is less than maximum. In some cases the maximum structural landing weight may dictate how long an aircraft can fly with a maximum structural payload. If this is the case, line DE represents the tradeoff between payload and range which must occur, since the payload is limited by the maximum structural landing weight. The shape of the payload-versus-range curve would then follow line $DEBC$ instead of ABC. Payload versus range depends on a number of factors such as meteorological conditions en route, flight altitude, speed, fuel, wind, and amount of reserve fuel. For performance comparison of different aircraft in an approximate way, the payload range curves are usually shown for standard day, no wind, and long-range cruise. For this purpose, approximate values of payload versus range are shown for some typical transport aircraft in Table 3-14 and Figs. 3-15, 3-16, and 3-17. Note the wide variation in payload and range demonstrated for short-, medium-, and long-range aircraft.

The actual payload, particularly in passenger aircraft, is normally less than the maximum structural payload even when the aircraft is completely full. This is due to the limitation in the use of space when passengers are carried. For computing payload, passengers and their baggage are normally considered as 200-lb units.

The aircraft manufacturers publish payload-versus-range diagrams in aircraft characteristics manuals for each aircraft which may be used for airport planning purposes. These diagrams are most useful in airport planning for determining the most probable weight characteristics of aircraft flying particular stage lengths between airports. Figure 3-18 shows a payload-versus-range diagram for the Boeing 757-200 aircraft.

Example Problem 3-1 shows the typical computations which can be made relative to the payload-versus-range diagram.

Example Problem 3-1. The weight characteristics of a commercial aircraft are given in Table 3-15. It is necessary to plot the payload-versus-range diagram for this aircraft. We assume that the regulations governing the use of this aircraft require only a 1.25-h reserve in route service. The aircraft has an average route speed of 540 m/h and an average fuel burn of 22.8 lb/mi.

To plot payload versus range, the range at maximum payload and the range at maximum fuel must be found. The range at maximum payload is found by determining the maximum fuel which can be placed on this aircraft for takeoff at the maximum structural takeoff weight (MSTOW).

TABLE 3-14   Payload-versus-Range Information for Typical Transport Aircraft*

Aircraft	P_a	R_a	P_b	R_b	R_c	R_d	P_e	R_e
A-300-600	95.4	2200	64.0	3500				4300
A-310-200	68.0	1600	26.0	3800				4300
A-320-100	43.0	1100	35.0	1800				2300
B-727-200	37.5				2200	950	23.0	1800
B-737-200	35.2	1200	28.2	1800				2500
B-737-300	33.9	1900	25.5	2800				3600
B-747-100	168.5	2600	22.0	5800				6200
B-747-200	140.4	4800	89.0	5800				7200
B-747-400	138.9	5500	74.0	7300				8000
B-757-200	58.9	2300	39.0	3400				4300
B-767-200	73.4	2300	26.4	4900				5200
B-767-300ER	92.9	4100	50.0	6400				7500
B-777-200	121.1	3100	26.0	6800				7200
DC-9-32	30.1	950	10.0	2650				2900
DC-9-51	36.5	750	23.0	1650				2000
DC-8-73	64.8	4000	25.0	5800				5800
DC-10-10	94.0	2100	46.0	3800				4200
DC-10-30	101.8	3800	46.0	5500				6100
MD-81	40.2	800	24.0	2150				2600
MD-87	38.8	1400	28.0	2400				3000
MD-11	114.1	4600	55.0	6600				7500

*Weights are in 1000 lb, and distances are in nautical miles. The weights and ranges are approximate and may vary based upon the particular configuration of the aircraft model and the engines with which it is equipped.

SOURCE: Data from aircraft manufacturers.

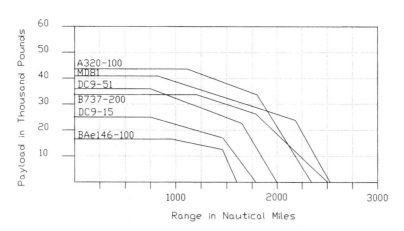

Figure 3-15   Payload versus range for typical short-range transport aircraft.

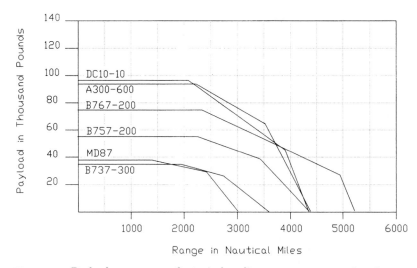

Figure 3-16 Payload versus range for typical medium-range transport aircraft.

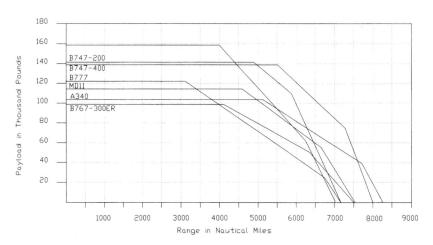

Figure 3-17 Payload versus range for typical long-range transport aircraft.

$$\text{MSTOW} = \text{OEW} + \text{max. structural payload} + \text{allowable fuel}$$

$$220{,}000 = 125{,}513 + 57{,}000 + \text{allowable fuel}$$

$$\text{Allowable fuel} = 37{,}487 \text{ lb}$$

$$\text{Reserve fuel} = 1.25 \times 540 \times 22.8 = 15{,}390 \text{ lb}$$

$$\text{Allowable fuel} = \text{route fuel} + \text{reserve fuel}$$

$$37{,}487 = \text{route fuel} + 15{,}390$$

$$\text{Route fuel} = 22{,}097 \text{ lb}$$

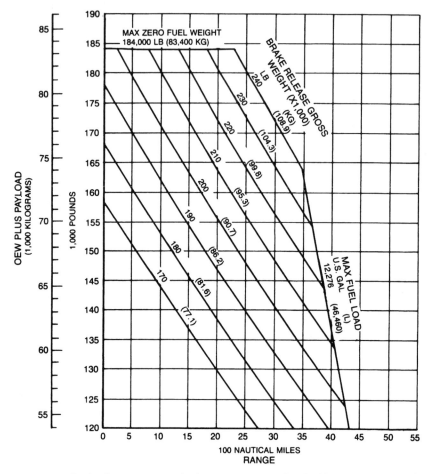

Figure 3-18 Payload versus range for long-range cruise for the Boeing 757-200 with RB211-535C engines (*Boeing Commercial Airplane Group [19]*).

TABLE 3-15   Aircraft Weight Characteristics for Example Problem 3-1, lb

Maximum structural takeoff weight	220,000
Maximum structural landing weight	198,000
Zero-fuel weight	182,513
Operating empty weight	125,513
Maximum structural payload	57,000
Fuel capacity	75,400

Now the range at maximum structural payload can be found by computing the route distance that this route fuel will take this aircraft.

$$\text{Range} = \frac{22,097}{22.8} = 969 \text{ mi}$$

Therefore, this aircraft can fly only 969 mi at maximum structural payload. The landing weight at the destination cannot exceed the maximum structural landing weight. The actual landing weight (LW) at the destination is the take-off weight (MSTOW) minus the route fuel. Therefore,

$$\text{LW} = \text{MSTOW} - \text{route fuel}$$

$$\text{LW} = 220,000 - 22,097 = 197,903 \text{ lb}$$

This is less than the maximum structural landing weight.

Next the range at maximum fuel must be computed. This is done by first subtracting the reserve fuel from the fuel capacity of the aircraft to determine the route fuel available and then computing the route range of the aircraft.

$$\text{Fuel capacity} = 75,400 \text{ lb}$$

$$\text{Max. route fuel} = 75,400 - 15,390 = 60,010 \text{ lb}$$

$$\text{Range at max. fuel} = \frac{60,010}{22.8} = 2632 \text{ mi}$$

Therefore, this aircraft can fly a maximum route length of 2632 mi, but the payload must be restricted. The allowable payload is found by subtracting the operating empty weight (OEW) and the weight of the fuel at capacity from the maximum structural takeoff weight (MSTOW).

$$\text{MSTOW} = \text{OEW} + \text{max. fuel} + \text{allowable payload}$$

$$220,000 = 125,513 + 75,400 + \text{allowable payload}$$

$$\text{Allowable payload} = 19,087 \text{ lb}$$

Finally, the ferry range is computed by determining the range the aircraft can fly by using all the fuel available including the reserve fuel. Therefore,

$$\text{Ferry range} = \frac{75,400}{22.8} = 3307 \text{ mi}$$

These points may now be plotted to produce a payload-versus-range diagram such as shown in Fig. 3-19.

Turning radii

For determining aircraft positions on the apron adjacent to the terminal building and establishing the paths of aircraft at other locations in the airport, it is important to understand the geometry of movement of an aircraft. Turning radii are a function of the nose gear steering angle. The larger the angle, the smaller the radii. From the center of rotation, the distances to the various parts of the aircraft, such as the wing tips, nose, or tail, create a number of radii. The largest radius is

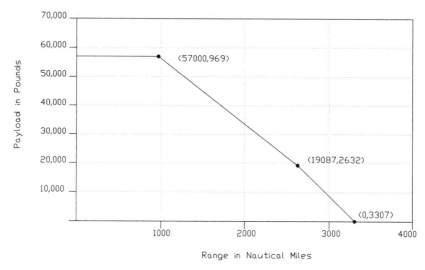

Figure 3-19 Payload versus range for Example Problem 3-1.

the most critical from the standpoint of clearance to buildings or adjacent aircraft. The minimum turning radius corresponds to the maximum nose gear steering angle specified by the aircraft manufacturer. The maximum angles vary from 60° to 80°. The center of rotation can be easily determined by drawing a line through the axis of the nose gear at whatever steering angle is desired. The intersection of this line with a line drawn through the axes of the two main gears is the center of rotation. For aircraft with more than two main gears, such as the Boeing 747, the axis is drawn midway between the gears. This was illustrated in Fig. 3-1. Some of the newer large aircraft have the capability of swiveling the main gear when making sharp turns. The effect of the swivel is to reduce the turning radius.

Minimum turning radii are not used in practice very often because the maneuver produces excessive tire wear and in some instances results in scuffing of the pavement surface. Accordingly lesser angles on the order of 50° are more proper. Minimum turning radii for some typical transport aircraft are given in Table 3-16.

Static weight on the main gear and the nose gear

The distribution of the load between the main gears and the nose gear depends on the type of aircraft and the location of its center of gravity. For any gross weight there is a maximum aft and forward center-of-gravity position to which the aircraft can be loaded for flight in order to maintain stability. Thus the distribution of weight

TABLE 3-16   Minimum Turning Radii* for Typical Passenger
Aircraft

Aircraft	Maximum Steering angle, deg	Wing tips	Radius, ft† Nose	Tail
A-300-600	65	104.9	87.7	108.4
A-310	65	98.0	75.6	94.9
A-320	70	72.2	60.0	71.9
A-340	78	130.6	109.9	120.4
B-727-200	78	71.0	79.5	80.0
B-737-200	78	56.0	51.0	56.0
B-737-300	78	58.0	55.0	63.0
B-737-400	78	59.0	61.0	67.0
B-737-500	78	57.0	50.0	60.0
B-747-200	70	113.0	110.0	125.0
B-747-400	70	157.0	117.0	96.0
B-747-SP	70	113.0	93.0	97.0
B-757-200	65	92.0	84.0	91.0
B-767-200	65	112.0	85.0	98.0
B-767-300	65	116.0	96.0	108.0
B-777	70	135.0‡	106.0	126.0
DC-8-63	67	110.4	99.0	109.7
DC-8-62	61	111.2	83.8	99.0
DC-9-32	82	55.5	61.2	64.0
DC-10-10	68	112.4	104.6	101.0
DC-10-30	68	118.1	105.0	100.8
MD-11	70	121.5	113.8	102.0
MD-81	82	65.9	80.7	74.3
MD-87	82	64.5	71.1	66.6

*Note that the wing tip radius is not always the largest of the three radii
listed.

†From center of rotation.

‡Folded wing radius is 113 ft.

between the nose and main gears is not a constant. An illustration of
the variation in main landing gear weight is given in Fig. 3-20. For
the design of pavements, it is normally assumed that 5 percent of the
weight is supported on the nose gear and the remainder on the main
gears. Thus if there are two main gears, each gear supports 47.5 percent of the total weight. For example, if the takeoff weight of an aircraft is 300,000 lb, each main gear is assumed to support 142,500 lb.
If the main gear has four tires, it is assumed that each tire supports
an equal fraction of the weight on the gear, in this case, 35,625 lb.

Wing-tip vortices

Whenever the wings lift an aircraft, vortices form near the ends of the
wings. The vortices are made up of two counterrotating cylindrical air
masses about a wingspan apart, extending aft along the flight path.

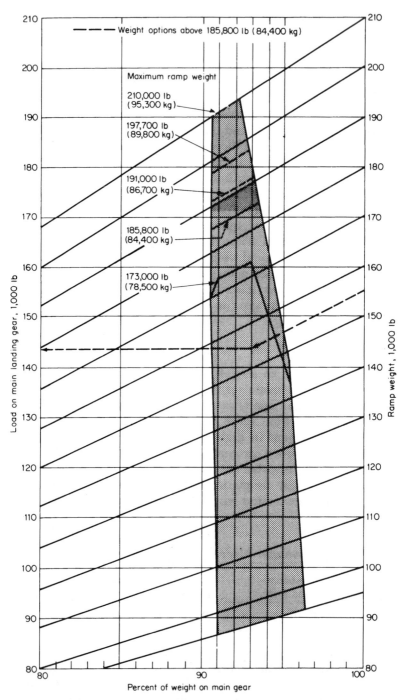

Figure 3-20 Landing gear pavement loading for the Boeing 727-200 aircraft (*Boeing Commercial Airplane Company [14]*).

The velocity of the wind within these cylinders can be hazardous to other aircraft encountering them in flight. This is particularly true if a lighter aircraft encounters a vortex generated by a much heavier aircraft. The tangential velocities in a vortex are directly proportional to the weight of the aircraft and inversely proportional to the speed. The more intense vortices are therefore generated when the aircraft is flying slowly near an airport [52]. The winds created by vortices are often referred to as *wake turbulence* or *wake vortex*.

Once vortices are generated, they move downward and drift laterally in the direction of the wind. The rate at which vortices settle toward the ground is dependent to some extent on the weight of an aircraft: The heavier the vehicle, the faster the vortex will settle. About one wingspan height above the ground, the vortices begin to move laterally away from the aircraft, as shown in Fig. 3-21. The duration of a vortex is dependent to a great extent on the velocity of the wind. When there is very little or no wind, a vortex can persist for longer than 2 min. It was observed that during flight the vertical displacement of the wake tended to level off about 1000 ft below the flight path of a heavy aircraft [8].

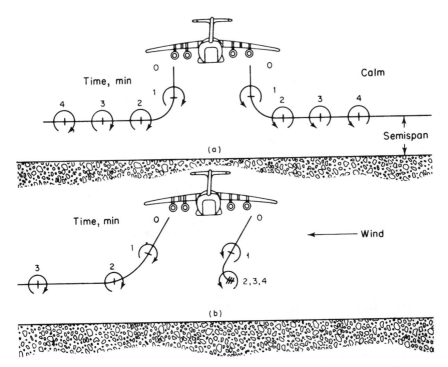

Figure 3-21 Vortex movement near the ground showing the effect of wind (*National Aeronautics and Space Administration [8]*).

The vortex hazard to small general aviation aircraft has been known for some time. However, when wide-bodied aircraft were introduced, it was found that they could create a hazard to smaller airline aircraft as well. This prompted the federal government through NASA and the FAA to undertake an accelerated flight test program to determine the safe distance that lighter aircraft should keep behind a heavy aircraft. The results of the test program led the FAA to divide airline jet aircraft into two classes: large and heavy. Large aircraft have a maximum takeoff weight of less than 300,000 lb; all other aircraft are classified as heavy. This classification places the DC-9, B-737, B-727, B-720, and B-707-120B in the large category. On the other hand, the 707-320B, DC-8-61, DC-8-62, and DC-8-63, and all wide-bodied jets are considered heavy. The minimum headway between a large aircraft and a preceding heavy one was increased as a result of the tests. The spacing between two heavy aircraft following each other was also increased. The headway was not altered for two large aircraft following each other. Likewise the spacing between parallel runways was increased for simultaneous operations in good weather. The specific separation rules are described in Chap. 4 on air traffic control. Since the size of headways has a profound influence on the runway capacity, considerable attention is being devoted to the wake vortex problem. NASA has a comprehensive research effort underway to give a better analytical description of vortex behavior and concurrently to develop methods for its detection, early dissipation, or prevention.

Methods of avoiding hazardous wake vortices, particularly for use by operators of small general aviation aircraft, have been published [8]. Among the more important suggestions is to select a flight path which will always be above the path of a heavy aircraft.

Effect of Aircraft Performance on Runway Length

The factors which have a bearing on runway length may be grouped into three general categories:

1. Performance requirements imposed by the government on aircraft manufacturers and operators

2. The environment at the airport

3. Those items which establish the operating takeoff and landing gross weights for each aircraft type

Performance requirements imposed by the government

Transport aircraft are licensed and operated under the code of regulations known as the Federal Aviation Regulations (FAR). This code is promulgated by the federal government in coordination with industry. The regulations govern the aircraft gross weights at takeoff and landing by specifying performance requirements which must be met in terms related to the runway lengths available.

The regulations pertaining to turbine aircraft consider three general cases in establishing the length of a runway necessary for safe operations:

1. A normal takeoff in which all engines are available and sufficient runway is required to accommodate variations in liftoff techniques and the distinctive performance characteristics of these aircraft

2. Takeoff involving an engine failure in which sufficient runway is required to allow aircraft to continue the takeoff despite the loss of power or else to brake to a stop

3. Landing in which sufficient runway is required to allow for normal variation in landing technique, overshoots, poor approaches, and the like

The regulations pertaining to piston-engine aircraft retain in principle the above criteria, but the first criterion is not used. This particular regulation is aimed toward the everyday, normal takeoff maneuver, since engine failure occurs rather infrequently with turbine-powered aircraft. The runway length needed at an airport by a particular type and weight of turbine-powered aircraft is established by one of these three cases—whichever yields the longest length.

In the regulations for both piston-engine aircraft and turbine-powered aircraft, the word *runway* refers to the full-strength pavement (designated FS in upcoming equations). Thus, in the discussion which follows, the terms *runway* and *full-strength pavement* are synonymous. In any discussion of the effect of the regulations on the runway length, however, note that the current regulations for turbine-powered aircraft do not require a runway for the entire takeoff distance, while the regulations for piston-engine aircraft normally do.

To indicate why there is a difference in the two regulations with regard to the length of full-strength pavement, it is necessary to examine in more detail the regulations pertaining to turbine-powered transports.

The three cases referred to above, as defined by the current turbine-powered transport regulations FAR part 25 [12] and part 121 [23], are illustrated in Fig. 3-22.

The landing case (Fig. 3-22a) is probably easiest to explain. The regulations state that the *landing distance* LD needed for each aircraft using the airport must be sufficient to permit the aircraft to come to a full stop, the *stop distance* SD, within 60 percent of this distance, assuming that the pilot makes an approach at the proper speed

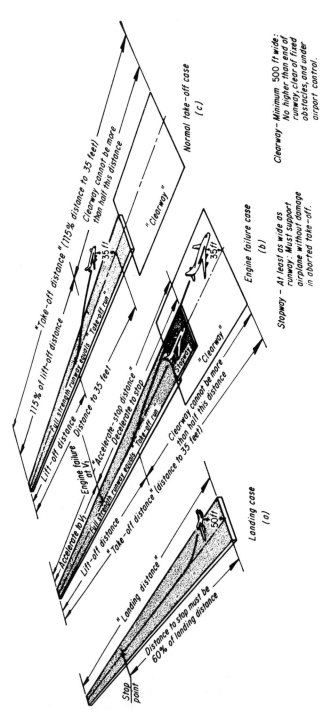

Figure 3-22 Summary of turbine-powered aircraft performance regulations affecting runway length requirements.

and crosses the threshold of the runway at a height of 50 ft. The landing distance must be of full-strength pavement. The landing distance for piston-engine aircraft is defined in exactly the same manner.

The normal case, all-engine takeoff (Fig. 3-22c), defines a *takeoff distance* TOD which, for a specific weight of aircraft, must be 115 percent of the actual distance that the aircraft uses to reach a height of 35 ft (designated D35). Not all of this distance has to be of full-strength pavement; however, pilot techniques would never vary so greatly as to need all the distance before lifting off the ground. What is necessary is that all this distance be free from obstructions to protect against an overshooting takeoff. Consequently the regulations permit the use of a clearway CL for part of this distance. A clearway is a rectangular area beyond the runway not less than 500 ft wide and not longer than 1000 ft, centrally located about the extended centerline of the runway and under the control of the airport authorities. The clearway is expressed in terms of a clearway plane, extending from the end of the runway with an upward slope not exceeding 1.25 percent above which no object nor any portion of the terrain protrudes, except that threshold lights may protrude above the plane if their height above the end of the runway is not greater than 26 in and if they are located to each side of the runway. Up to one-half the difference between 115 percent of the distance to reach the point of liftoff, the *liftoff distance* LOD, and the takeoff distance may be clearway. The remainder of the takeoff distance must be full-strength pavement and is identified as the *takeoff run* TOR.

For the engine-failure case (Fig. 3-22b), the regulation specifies that the takeoff distance required is the actual distance to reach a height of 35 ft (D35) with no percentage applied, as in the all-engine takeoff case. This recognizes the rarity of engine failure. The regulations again permit the use of a clearway, in this case up to one-half the difference between the liftoff distance and the takeoff distance, with the remainder being full-strength pavement. The regulations for piston-engine aircraft normally require full-strength pavement for the entire takeoff distance.

As mentioned previously, the engine-failure case requires that sufficient distance also be available to stop the airplane rather than to continue the takeoff. This distance is referred to as the *accelerate-stop distance* DAS. For piston-engine aircraft, only full-strength pavement was normally used for this purpose. The regulations for turbine-powered aircraft, however, recognize that an aborted takeoff is relatively rare and permit use of lesser-strength pavement, known as *stopway* SW, for that part of the accelerate-stop distance beyond the takeoff run. The stopway is an area beyond the runway, not less wide than the width of the runway, centrally located about the extended centerline of the runway, and designated by the airport authorities for use in decelerating the aircraft during an aborted takeoff. To be considered as such, the stopway must be capable of supporting the airplane during an aborted takeoff without inducing structural damage to the aircraft.

The required *field length* FL is generally made up of three components: the full-strength pavement FS, the partial-strength pavement or stopway SW, and the clearway CL.

The above regulations for turbine-powered aircraft may be summarized for each case in equation form to find the required field length.

Normal takeoff: Case 1

$$FL_1 = FS_1 + CL_{1\,max} \tag{3-1}$$

where

$$TOD_1 = 1.15(D35_1) \tag{3-1a}$$

$$CL_{1\,max} = 0.50[TOD_1 - 1.15(LOD_1)] \tag{3-1b}$$

$$TOR_1 = TOD_1 - CL_{1\,max} \tag{3-1c}$$

$$FS_1 = TOR_1 \tag{3-1d}$$

Engine-failure takeoff: Case 2

$$FL_2 = FS_2 + CL_{2\,max} \tag{3-2}$$

where

$$TOD_2 = D35_2 \tag{3-2a}$$

$$CL_{2\,max} = 0.50(TOD_2 - LOD_2) \tag{3-2b}$$

$$TOR_2 = TOD_2 - CL_{2\,max} \tag{3-2c}$$

$$FS_2 = TOR_2 \tag{3-2d}$$

Engine-failure aborted takeoff: Case 3

$$FL_3 = FS + SW \tag{3-3}$$

where

$$FL_3 = DAS \tag{3-3a}$$

Landing: Case 4

$$FL_4 = LD \tag{3-4}$$

where

$$LD = \frac{SD}{0.60} \qquad (3\text{-}4a)$$

$$FS_4 = LD \qquad (3\text{-}4b)$$

To determine the required field length and the various components of length such as full-strength pavement, stopway, and clearway, the above equations must each be solved for the critical design aircraft at the airport. This will result in finding each of the following values:

$$FL = \max (TOD_1, TOD_2, DAS, LD) \qquad (3\text{-}5)$$

$$FS = \max (TOR_1, TOR_2, LD) \qquad (3\text{-}6)$$

$$SW = DAS - \max (TOR_1, TOR_2, LD) \qquad SW_{min} = 0 \qquad (3\text{-}7)$$

$$CL = \min (FL - DAS, CL_{1\,max}, CL_{2\,max}) \qquad (3\text{-}8)$$

$$CL_{min} = 0, \qquad CL_{max} = 1000 \text{ ft}$$

If operations are to take place on the runway in both directions, as is the usual case, the field length components must exist in each direction. Example Problem 3-2 illustrates the application of these requirements for a hypothetical aircraft.

Example Problem 3-2. Determine the runway length requirements according to the specifications of FAR parts 25 and 121 for a turbine-powered aircraft with the following performance characteristics:

Normal takeoff:

> Liftoff distance = 7000 ft
>
> Distance to 35-ft height = 8000 ft

Engine failure:

> Liftoff distance = 8200 ft
>
> Distance to 35-ft height = 9100 ft

Engine-failure aborted takeoff:

> Accelerate-stop distance = 9500 ft

Normal landing:

> Stop distance = 5000 ft

From Eqs. (3-1) for a normal takeoff

$$TOD_1 = 1.15(D35_1) = 1.15 \times 8000 = 9200 \text{ ft}$$

$$CL_{1\,max} = 0.50\,[TOD_1 - 1.15(LOD_1)]$$
$$= 0.50[9200 - 1.15(7000)] = 575 \text{ ft}$$
$$TOR_1 = TOD_1 - CL_{1\,max} = 9200 - 575 = 8625 \text{ ft}$$

From Eqs. (3-2) for an engine-failure takeoff,

$$TOD_2 = D35_2 = 9100 \text{ ft}$$
$$CL_{2\,max} = 0.50(TOD_2 - LOD_2) = 0.50(9100 - 8200) = 450 \text{ ft}$$
$$TOR_2 = TOD_2 - CL_{2\,max} = 9100 - 450 = 8650 \text{ ft}$$

From Eq. (3-3) for an engine-failure aborted takeoff,

$$DAS = 9500 \text{ ft}$$

From Eqs. (3-4) for a normal landing,

$$LD = \frac{SD}{0.60} = \frac{5000}{0.60} = 8333 \text{ ft}$$

By using the above quantities in Eqs. (3-5) through (3-8), the actual runway component requirements become

$$FL = \max (TOD_1, TOD_2, DAS, LD)$$
$$= \max (9200, 9100, 9500, 8333)$$
$$= 9500 \text{ ft}$$
$$FS = \max (TOR_1, TOR_2, LD)$$
$$= \max (8625, 8650, 8333) = 8650 \text{ ft}$$
$$SW = DAS - \max(TOR_1, TOR_2, LD)$$
$$= 9500 - \max (8625, 8650, 8333)$$
$$= 9500 - 8650 = 850 \text{ ft}$$
$$CL = \min (FL - DAS, CL_{1\,max}, CL_{2\,max})$$
$$= \min (9500 - 9500, 575, 450) = 0$$

The sketch in Fig. 3-23 shows the required runway field length and the components of this field for operations in both directions. Observe that for this case there is a displaced threshold at each end of the runway since the stopway is not available for normal aircraft operations.

The results of Example Problem 3-2 may be used to illustrate the declared-distance concept [1, 9]. *Declared distances* are the distances

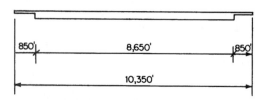

Figure 3-23 Illustration of runway length components for Example Problem 3-2.

which the airport owner declares are available and suitable for satis-
fying the takeoff run, takeoff distance, accelerate-stop distance, and
landing distance requirements of aircraft. These declared distances
must take into consideration the provision of the runway safety area,
runway obstacle-free area, and runway protection zone. In the application of the declared-distance concept to Example Problem 3-2, we
assume that sufficient area is available for the provision of the runway safety area, runway obstacle-free area, and runway protection
zone without reducing the dimensions of the full-strength pavement
or stopway.

The *takeoff run available* TORA is the runway length declared
available and suitable for the ground run of an aircraft during takeoff. For Example Problem 3-2, TORA is 8650 ft. The takeoff distance
available TODA is the takeoff run available plus the length of any
remaining runway and clearway beyond the far end of the takeoff run
available. For Example Problem 3-2, TODA is 9500 ft. The *accelerate-
stop distance available* ASDA is the amount of runway plus stopway
declared available and suitable for the acceleration and deceleration
of an aircraft during an aborted takeoff. For Example Problem 3-2,
the ASDA is also 9500 ft. The *landing distance available* LDA is the
runway length available and suitable for landing an aircraft. For
Example Problem 3-2, the LDA is 8650 ft.

Clearly both the takeoff distance and the accelerate-stop distance
will depend on the speed that the aircraft has achieved when an
engine fails. The speed at which engine failure is assumed to occur is
selected by the aircraft manufacturer and is referred to as the *critical
engine-failure speed* V_1. If an engine actually fails prior to this selected
speed, the pilot brakes to a stop. If the engine fails at a speed greater
than this speed, the pilot has no choice but to continue the takeoff.

Since, for piston-engine aircraft, full-strength pavement was normally used for the entire accelerate-stop distance and the takeoff distance, it was general practice to select V_1 so that the distance
required to stop from the point where V_1 was reached was equal to
the distance (from the same point) to reach a specified height above
the runway. The runway length established on this basis is referred
to as the *balanced field* or *balanced runway* and results in the shortest runway. For turbine-powered aircraft, the selection of V_1 on this
basis will not necessarily result in the shortest runway if a clearway
or a stopway is provided.

For this reason it is necessary to understand the interrelationships
between V_1 and the various components of the takeoff distance and
the accelerate-stop distance. These are illustrated in Fig. 3-24. At
increasingly higher V_1 speeds, the takeoff distance becomes shorter
and shorter because the aircraft has the benefit of all-engine accelera-

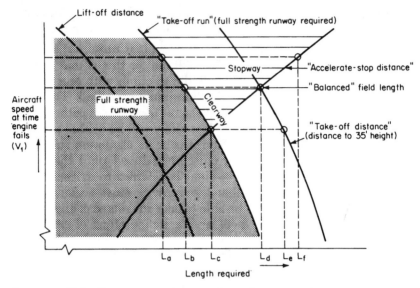

Figure 3-24 Takeoff runway requirements for turbine-powered aircraft in the engine-failure case.

tion for greater portions of the takeoff roll, but the accelerate-stop distance is correspondingly increased.

Although Fig. 3-24 is included primarily to illustrate the clearway and stopway concepts, the *balanced field length* is also shown for comparison and corresponds to the crossover of the takeoff and stopping requirements at point D.

When clearways and stopways are utilized, several alternatives are possible:

1. If the same V_1 is selected as in the balanced-field concept, the lengths of the clearway and stopway are equal. This would mean that the full-strength runway, denoted by L_b in Fig. 3-24, could be shortened a distance equal to the clearway, but a stopway would have to be constructed to the field length L_d.

2. Select V_1 so as to balance the accelerate-stop distance with the takeoff run. In this case the full-strength runway becomes L_c, which is less than L_d, and no stopway is required. Thus the runway has been shortened, but a clearway is required from L_c to L_e.

3. Select a fairly high V_1 in order to reduce the takeoff distance; but when this is done, the accelerate-stop distance is increased greatly. For this case the length of the runway is denoted by L_a, but a stopway must be provided to point L_f because of the greater accelerate-stop distance. This alternative may be advantageous to the

operator at airports where there are obstacles near the ends of the runway.

Thus, one can see that the regulations pertaining to turbine-powered aircraft offer a number of alternatives to the aircraft operator. It should be emphasized that the takeoff distance and the takeoff run for the engine-failure case must be compared with the corresponding distance for the normal all-engine takeoff case. The longer distance always governs. A further discussion of these concepts is presented by the International Civil Aviation Organization (ICAO) [1].

From the presentation thus far, clearly the length of the runway is intimately associated with the heights, speeds, and other requirements specified in the performance regulations.

Both aircraft operators and airport managers are interested in clearways, because clearways will, for a fixed available length of runway, allow the operator additional gross takeoff weight with less expense to airport management than building full-strength pavement would require.

Thus far we assumed that if sufficient length of runway were provided, an aircraft could take off at its maximum structural takeoff weight. However, this may not be the case, particularly if it is quite hot and the airport is located at higher elevations. The reason is that federal regulations state that the aircraft must be able to climb with one engine inoperative at gradients not less than the minimum specified. This performance must be demonstrated with no obstacles in the flight path. On hot days and at higher elevations, an aircraft most likely will be unable to climb at the minimum specified gradients when it is at its maximum structural takeoff weight. Therefore, the weight must be reduced until the climb requirements are met. The resulting weight is referred to as the *climb-limited weight*. It is this weight that the runway must accommodate because an aircraft operator will not be able to take advantage of any longer runway. As an example, at sea level and a temperature of 80°F, the maximum allowable takeoff weight of a Boeing 747-200B is 785,000 lb. This corresponds to its maximum structural takeoff weight. With the same temperature but an elevation of 2000 ft, the maximum allowable takeoff weight is reduced to 747,000 lb, which corresponds to the climb-limited weight. If at an elevation of 2000 ft there were no climb restrictions, the length of runway required for 785,000 lb, the maximum structural takeoff weight, would be about 12,300 ft. However, the runway length for the climb-limited weight of 747,000 lb is about 10,900 ft. Therefore any length greater than 10,900 ft could not be used by the aircraft operator when the temperature is 80°F.

Thus far no mention has been made of obstacles, sometimes referred to as *obstructions*, in the flight path. If there are any and they are sufficiently high, the allowable takeoff weight may have to be further reduced to provide adequate clearance of the flight path above these obstacles. The corresponding weight is often referred to as the *obstacle-limited weight.*

Climb and obstacle clearance requirements are too complex to describe in detail in this text, but a summary of the requirements is shown in Fig. 3-25. The details of these requirements are found in FAR part 25 [12]. The climb regulations pertain to the takeoff flight path which extends from a point where the aircraft has reached a height of 35 ft above the pavement surface with one engine inoperative to a point where a height of 1500 ft is reached. The takeoff flight path is divided into several segments, referred to as the first, second, third, and final segments. Sometimes the third and final segments are referred to as the *transition* segment. Note that for each segment there are several minimum climb gradients specified depending on the number of engines on the aircraft. The largest gradients are in the second segment, and so this segment is normally the critical one for determining the climb-limited weight. Note that the second segment begins at the point when the landing gear is retracted and ends when the aircraft reaches a height of 400 ft above the end of the runway. In the transition segment, the aircraft operator has the option of

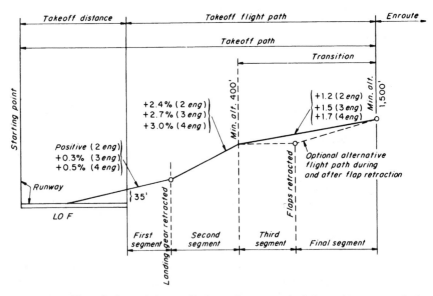

Figure 3-25 Obstacle-free minimum climb gradients required for turbine-powered aircraft with one engine inoperative.

climbing the aircraft in several ways as long as the average gradient is not less than that specified in Fig. 3-25.

If there are obstacles in the flight path and they are high enough, the minimum gradients that are shown in Fig. 3-25 will no longer be applicable. It is necessary to determine a flight path that will clear the obstacles in a manner specified by the federal regulations. These regulations require that the net takeoff flight path with one engine inoperative clear the obstacle by a distance of not less than 35 ft. The net takeoff flight path is computed by reducing the actual flight path by 0.8 percent for two-engine aircraft, 0.9 percent for three-engine aircraft, and 1.0 percent for four-engine aircraft.

Using the previous illustration with the Boeing 747-200B as an example, suppose there was an obstacle 120 ft high that is 3400 ft from the takeoff end of the runway. The obstacle-limited weight for this condition might be 740,000 lb, slightly less than the climb-limited weight of 747,000 lb. The corresponding runway length to accommodate 740,000 lb is about 10,600 ft. However, if it were possible to lower or remove the obstacle, the aircraft would still be limited to its climb-limited weight of 747,000 lb. The removal of the obstacle reduces the weight penalty from 45,000 to 38,000 lb. Beyond this there is nothing that can be done to further reduce the penalty in weight other than to adjust V_1 to the extent that sufficient runway is available. This example points out the importance of selecting airports sites that are relatively free of obstacles in the flight path and are as close to sea level as possible, particularly if it is necessary for aircraft to fly at or near their maximum structural takeoff weights while atmospheric temperatures are high.

Environs at the Airport

Thus far it has been shown how the requirements imposed by the government can influence runway length. Certain conditions at the airport can do so, too. The more important conditions are temperature, surface wind, runway gradient, altitude of the airport, and condition of the runway surface. The effect of these conditions on runway length can only be approximated. However, orders of magnitude can be beneficial for planning, and they are therefore presented here in this context.

Temperature

The higher the temperature, the longer the runway required, because high temperatures reflect lower air densities, resulting in lower output of thrust. The increase in runway length is not linear with temperature. The rate of increase at high temperatures is greater than at low temperatures. The increase can be specified in terms of the percentage of runway length at 59°F, which is the standard temperature at sea level. For estimating purposes, the approximate variations of runway length with temperature which apply to turbine-powered transport between 59 and 90°F can be used. In this temperature range, the average increase in runway length varies from about 0.42 to 0.65 percent per degree Fahrenheit above the runway requirements at 59°F. Therefore, at 85°F the runway length would be increased by an average range of about 11 to 18 percent over the length required at 59°F.

Surface wind

The greater the headwind, the shorter the runway; and the greater the tailwind, the longer the runway. For computing the takeoff weight, federal regulations allow operators to use only one-half of the reported headwind. If there is a tailwind, 1.5 times the reported tailwind is used for computation of the takeoff weight, up to a maximum allowable tailwind of 10 kn. The effect of wind varies depending in part on temperature and aircraft weight. Nevertheless, approximations are useful for planning. A 5-kn headwind reduces the takeoff length by about 3 percent, whereas a 5-kn tailwind increases this length by about 7 percent. For airport planning purposes, it is desirable to use no wind, particularly if light winds occur at the airport sight.

Runway gradient

An uphill gradient requires more runway length than a level or downhill gradient, the specific amount depending on the elevation of the airport and the temperature. Studies indicate that the relationship between uniform gradient and an increase or decrease in runway length is nearly linear [55]. For turbine-powered aircraft this amounts to 7 to 10 percent for each 1 percent of uniform gradient. Airport design criteria limit the gradient to a maximum of 1.5 percent. Information provided by aircraft manufacturers in flight manuals is based on a uniform gradient, yet most runway profiles are not uniform. In the United States, aircraft operators are allowed to substitute an average uniform gradient, which is a straight line joining the ends of the runway, as long as no intervening point along the actual path profile lies more than 5 ft above or below the average line. Fortunately most runways meet this requirement. For airport planning purposes only, the FAA uses an effective gradient. The *effective gradient* is the difference in elevation between the highest and lowest points on the actual runway profile divided by the length of the runway. Studies indicate that within the degree of accuracy required for airport planning, there is not very much difference between the average uniform gradient and, effective gradient.

Altitude

All other things being equal, the higher the altitude of the airport, the longer the runway required. This increase is not linear but varies with weight and temperature. At high altitudes the rate of increase is higher than at low altitudes. For planning purposes an increase from sea level of 7 percent per 1000 ft of altitude will suffice for most airport sites except those that experience very hot temperatures or are located at high altitudes. Then the rate of increase can be as much as 10 percent. Manufacturer-provided performance data are in terms of pressure altitude not geographic altitude. The former is described in a later section. Unless the airport site is very unusual, the two altitudes can be considered equal for the purposes of airport planning.

Condition of runway surface

Slush or standing water on the runway has an undesirable effect on aircraft operation. Slush is equivalent to wet snow. It has a slippery texture which makes braking extremely poor. Being a fluid, slush is displaced by tires rolling through it, causing a significant retarding force, especially on takeoff. The retarding forces can get so large that aircraft can no longer accelerate to takeoff speed. In the process slush is sprayed on the aircraft, which further increases the resisting forces on the vehicle and can cause damage to some parts. Considerable experimental work has been conducted by NASA and the FAA on the effect of standing water and slush. As a result of these tests, jet operations are limited to no more than 0.5 in of slush or water. Between 0.25 and 0.5 in, the takeoff weight must be reduced substantially to overcome the retarding force of water or slush. It is therefore important to provide adequate drainage on the surface of the runway for removal of water and a means for rapidly removing slush. Both water and slush result in a very poor coefficient of braking friction. When tires ride on the surface of the water or slush, the phenomenon is known as *hydroplaning*. When the tires hydroplane, the coefficient of friction is on the order of wet ice and so the steering ability is completely lost. Hydroplaning is primarily a function of tire inflation pressure and to some extent the condition and type of grooves in the tires. According to tests made by NASA, the approximate speed at which hydroplaning develops may be determined by the formula [48]

$$V_p = 10p^{0.5} \qquad (3\text{-}9)$$

where V_p is the speed in miles per hour where hydroplaning develops and p is the tire inflation pressure in pounds per square inch. The range of inflation pressures for commercial jet transports varies from 120 to over 200 lb/in². Therefore the hydroplaning speeds would range from 110 to 140 mi/h or more. The landing speeds are in the same range. Therefore hydroplaning can be a hazard to jet operations. Hydroplaning can occur when the depth of water or slush is on the order of 0.2 in or less, the exact depth depending on tire tread design, condition of the tires, and texture of the pavement surface. Smooth tread operating on a smooth pavement surface requires the least depth of fluid for hydroplaning.

To reduce the hazard of hydroplaning and to improve the coefficient of braking friction, runway pavements have been grooved in a transverse direction. The grooves form reservoirs for the water on the surface. The FAA is conducting extensive research to establish standards for groove dimensions and shape [54]. In the past, the grooves normally were 0.25 in wide and deep and were spaced 1 in apart [44].

Computation of Runway Length

The runway length at a particular airport site is based upon the critical aircraft flying the longest nonstop flight segment regularly serviced at the airport. To compute the runway length, the following steps are required:

1. The operating empty weight of the critical aircraft is obtained.

2. The payload for the trip is ascertained.

3. The required fuel reserve is found.

4. The landing weight at the destination is computed as the sum of the operating empty weight, payload, and reserve fuel. This weight should not exceed the maximum structural landing weight of the aircraft.

5. The trip fuel requirements for climb, cruise, and descent are computed.

6. The takeoff weight of the aircraft is obtained by adding the trip fuel requirements to the landing weight. This should not exceed the maximum structural takeoff weight of the aircraft.

7. A determination of the temperature, surface wind, runway gradient, and altitude at the origination airport is made.

8. By using the takeoff weight of the aircraft with the origination airport temperature, surface wind, runway gradient, and elevation, the approved flight manual for the specific aircraft is utilized to determine the runway length requirements.

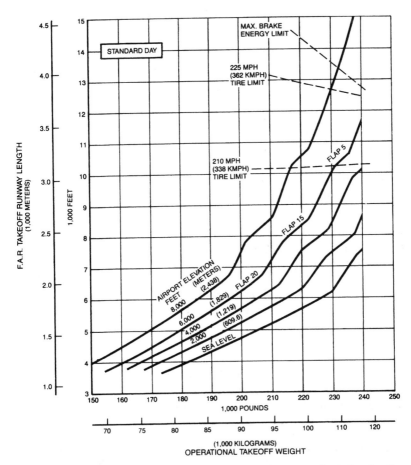

Figure 3-26 Takeoff runway length requirements on a standard day for the Boeing 757-200 aircraft with RB211-535C engines (*Boeing Commercial Airplane Group [19]*).

The aircraft manufacturers publish runway length diagrams for the required takeoff and landing runway in the aircraft characteristics manuals for each aircraft which may be used for airport planning purposes. Figure 3-26 shows the takeoff runway requirements for the Boeing 757-200 aircraft.

Advisory circular techniques for runway requirements

The FAA has published procedures for determining runway length requirements for airport planning purposes in Advisory Circular AC 150/5325-4A [50]. In this publication two types of procedure are available, one based upon aircraft performance curves and the other based upon performance charts. The design runway length is found for the critical aircraft, defined as the aircraft which flies the greatest non-stop route segment from the airport at least 250 times per year and requires the longest runway system. To use these procedures, the following data are necessary:

1. Designation of the critical aircraft
2. The longest nonstop route distance flown by the critical aircraft
3. The takeoff and landing weights of the critical aircraft at the airport
4. The airport elevation
5. The mean daily maximum temperature for the hottest month at the airport
6. The maximum difference in elevation along the runway centerline, i.e., the difference in elevation between the highest and lowest point on the runway centerline

In the chart procedure, discussed here, we assume that all points on the centerline of the runway are at the same elevation, the runway gradient is zero, and, to account for actual conditions, the length found from the charts is increased by 10 ft for each foot of centerline elevation between the highest and lowest points on the centerline of the runway for turbine-powered aircraft. The charts are based upon wet runway conditions, and no adjustment is made for dry conditions. The utilization of the chart procedure is illustrated in Example Problem 3-3.

Example Problem 3-3.   Determine the required runway at an airport for a Lockheed L-1011-385-1 aircraft equipped with RB211-22B engines. There are to be no weight restrictions in landing or takeoff. The FAA advisory circular AC 150/5325-4 appendix 6 procedures are to be used under the following conditions:

Flap settings:
  Takeoff: 22°
  Landing: 42°
Airport environs:
  Elevation: 6300 ft above mean sea level
  Temperature: 93.5°F
  Maximum centerline elevation difference: 20 ft
  Wet runways

Reference to the advisory circular yields the takeoff and landing charts for this aircraft shown in Tables 3-17 and 3-18.

For the landing conditions, the top portion of Table 3-17 allows for the interpolation of the maximum allowable landing weight, as shown in Table 3-19a. Therefore the maximum allowable landing weight of this aircraft at this airport is found to be 350,480 lb.

The required runway landing distance is found by interpolating from the bottom portion of Table 3-17, as shown in Table 3-19b. Therefore the landing run-

TABLE 3-17   Aircraft Landing Performance for Lockheed TriStar L-1011-385-1 with Rolls-Royce RB-221-22B Engine at 42° Flap Setting

Maximum Allowable Landing Weight, 1000 lb

Temper-	Airport elevation, ft								
ature, °F	0	1000	2000	3000	4000	5000	6000	7000	8000
50	358.0	358.0	358.0	358.0	358.0	358.0	358.0	358.0	358.0
55	358.0	358.0	358.0	358.0	358.0	358.0	358.0	358.0	358.0
60	358.0	358.0	358.0	358.0	358.0	358.0	358.0	358.0	358.0
65	358.0	358.0	358.0	358.0	358.0	358.0	358.0	358.0	358.0
70	358.0	358.0	358.0	358.0	358.0	358.0	358.0	358.0	358.0
75	358.0	358.0	358.0	358.0	358.0	358.0	358.0	358.0	357.4
80	358.0	358.0	358.0	358.0	358.0	358.0	358.0	358.0	349.5
85	358.0	358.0	358.0	358.0	358.0	358.0	358.0	355.3	341.6
90	358.0	358.0	358.0	358.0	358.0	358.0	358.0	347.1	333.8
95	358.0	358.0	358.0	358.0	358.0	358.0	352.8	339.0	326.0
100	358.0	358.0	358.0	358.0	358.0	358.0	345.0	331.2	318.4
105	358.0	358.0	358.0	358.0	358.0	351.2	336.8	323.4	310.8
110	358.0	358.0	358.0	358.0	355.6	341.8	328.7	315.9	303.4

Runway Length, 1000 ft

Weight,	Airport elevation, ft								
1000 lb	0	1000	2000	3000	4000	5000	6000	7000	8000
260	5.20	5.31	5.44	5.57	5.70	5.84	5.98	6.12	6.25
270	5.37	5.48	5.60	5.73	5.86	6.00	6.15	6.30	6.44
280	5.52	5.63	5.76	5.89	6.03	6.17	6.32	6.47	6.63
290	5.67	5.79	5.91	6.05	6.19	6.34	6.49	6.65	6.81
300	5.81	5.93	6.07	6.21	6.35	6.50	6.66	6.82	6.99
310	5.94	6.08	6.22	6.36	6.51	6.67	6.83	6.99	7.17
320	6.07	6.22	6.37	6.52	6.67	6.83	6.99	7.16	7.34
330	6.20	6.35	6.51	6.66	6.82	6.98	7.15	7.32	7.50
340	6.33	6.49	6.65	6.81	6.97	7.13	7.30	7.48	7.66
350	6.45	6.62	6.78	6.94	7.10	7.27	7.44	7.62	7.82
360	6.58	6.75	6.90	7.06	7.23	7.40	7.57	7.77	7.97

Airplane Characteristics

Typical operating empty weight plus reserve fuel:	270,300 lb
Average fuel consumption:	34 lb/mi
Typical maximum passenger load at 200 lb per passenger:	51,200 lb
Maximum structural payload:	86,183 lb

SOURCE: Federal Aviation Administration [50].

way length under wet conditions is 7501 ft which is rounded up to 7525 ft. Note that the chart procedure outlined here yields results for wet and slippery runways and does not contain adjustments for a condition where dry runways may be the design condition.

The determination of the required takeoff runway is made from Table 3-18. From the top portion of this table, the maximum allowable takeoff weight is found by interpolation, as shown in Table 3-19c. Therefore the maximum allow-

TABLE 3-18   Aircraft Takeoff Performance for Lockheed TriStar L-1011-385-1 with Rolls Royce RB211-22B Engines at 22°Flap Setting

Maximum Allowable Takeoff Weight, 1000 lb

Tempera-ture, °F	Airport elevation, ft								
	0	1000	2000	3000	4000	5000	6000	7000	8000
50	430.0	428.0	418.0	407.9	397.8	387.7	377.7	367.9	358.3
55	430.0	428.0	418.0	407.9	397.8	387.7	377.7	367.9	358.3
60	430.0	428.0	418.0	407.9	397.8	387.7	377.7	366.7	353.0
65	430.0	428.0	418.0	407.9	397.8	387.7	374.6	360.5	346.8
70	430.0	428.0	418.0	407.9	397.4	382.5	368.0	353.9	340.3
75	430.0	428.0	418.0	405.2	389.9	375.2	360.9	347.0	333.6
80	430.0	428.0	412.5	397.0	382.0	367.5	353.5	339.9	326.7
85	430.0	419.5	403.8	388.5	373.8	359.6	345.8	332.6	319.8
90	426.5	410.4	394.9	379.9	365.4	351.5	338.1	325.2	312.8
95	417.1	401.2	385.9	371.2	357.0	343.4	330.4	317.8	305.7
100	407.7	392.1	377.0	362.6	348.8	335.5	322.8	310.5	298.8
105	398.5	383.1	368.3	354.2	340.7	327.8	315.3	303.4	291.8
110	389.6	374.4	360.0	346.2	333.1	320.4	308.3	296.5	285.1

Reference Factor R

Tempera-ture, °F	Airport elevation, ft								
	0	1000	2000	3000	4000	5000	6000	7000	8000
50	50.7	53.7	57.0	60.6	64.5	68.8	73.6	78.7	84.4
55	51.0	54.1	57.5	61.2	65.1	69.5	74.2	79.4	85.0
60	51.5	54.6	58.0	61.7	65.7	70.1	74.9	80.2	87.4
65	51.9	55.0	58.4	62.1	66.2	70.7	76.2	82.9	90.4
70	52.2	55.4	58.8	62.6	66.7	72.4	78.8	85.8	93.6
75	52.6	55.8	59.2	63.6	69.0	74.9	81.6	88.9	96.9
80	53.0	56.2	60.8	65.8	71.4	77.6	84.5	92.1	100.5
85	54.1	58.3	62.9	68.2	74.0	80.5	87.7	95.6	104.3
90	56.0	60.3	65.2	70.7	76.8	83.5	91.0	99.2	108.3
95	57.9	62.5	67.7	73.4	79.7	86.8	94.5	103.1	112.5
100	60.0	64.9	70.2	76.2	82.8	90.2	98.2	107.2	117.0
105	62.2	67.3	73.0	79.2	86.1	93.7	102.2	111.5	121.7
110	64.5	69.9	75.9	82.4	89.6	97.5	106.3	116.0	126.7

able takeoff weight at this airport is 328,900 lb. Using the middle portion of Table 3-18 yields the R factor as shown in Table 3-19d.

This R factor is then used with the maximum allowable takeoff weight in the bottom portion of Table 3-18 to yield the required takeoff runway distance, as shown in Table 3-19e.

The required takeoff runway is 8370 ft with the elevation of all points on the runway centerline being the same. The correction for the maximum centerline elevation difference is an increase of 10 ft for each foot of difference in centerline elevation between the high and low points on the runway centerline. This correction yields

$$\text{Takeoff distance} = 8370 + 10 \times 20 = 8570 \text{ ft}$$

TABLE 3-18   (Continued)

Weight, 1000 lb	Runway Length, 1000 ft								
	Reference factor R								
	50	60	70	80	90	100	110	120	130
260	4.50	4.50	4.50	4.50	4.85	5.31	5.78	6.24	6.71
270	4.50	4.50	4.50	4.70	5.21	5.73	6.24	6.75	7.27
280	4.50	4.50	4.50	5.04	5.60	6.17	6.73	7.30	7.87
290	4.50	4.50	4.78	5.40	6.01	6.63	7.25	7.88	8.50
300	4.50	4.50	5.10	5.78	6.45	7.13	7.81	8.49	9.17
310	4.50	4.71	5.45	6.18	6.92	7.65	8.39	9.13	9.87
320	4.50	5.01	5.81	6.60	7.40	8.20	9.00	9.80	10.60
330	4.50	5.32	6.19	7.05	7.91	8.78	9.64	10.51	11.37
340	4.72	5.65	6.59	7.52	8.45	9.38	10.31	11.24	12.17
350	5.00	6.00	7.00	8.01	9.01	10.01	11.01	12.01	13.00
360	5.29	6.36	7.44	8.52	9.59	10.66	11.73	12.80	13.86
370	5.59	6.74	7.89	9.05	10.20	11.34	12.49	13.62	14.75
380	5.90	7.13	8.37	9.60	10.82	12.05	13.27	14.48	
390	6.23	7.55	8.36	10.17	11.48	12.78	14.07		
400	6.57	7.97	9.37	10.76	12.15	13.53	14.90		
410	6.93	8.41	9.89	11.37	12.84	14.31			
420	7.30	8.87	10.44	12.00	13.56				
430	7.68	9.34	11.00	12.66	14.30				

SOURCE: Federal Aviation Administration [50].

This is rounded to 8600 ft.

In summary, the required landing distance for the runway is 7525 ft and the required takeoff distance for the runway is 8600 ft.

Runway length requirements for general aviation aircraft

The discussion thus far has been devoted to the primary factors affecting the runway length for transport aircraft. The same factors also affect the runway length required for general aviation aircraft. General aviation aircraft are certified under FAR part 23 [11] in three categories: normal, utility, and acrobatic. The normal category includes aircraft for use in nonacrobatic, nonscheduled passenger and cargo operations. This is the category under which most general aviation aircraft are certified. The utility category involves the same uses as the normal category, but certain acrobatic maneuvers are permitted. Many general aviation aircraft are certified for both normal and utility operations, allowing them to be used for flight instruction. The maximum allowable gross weight, however, is less in the utility than in the normal category. The acrobatic category has no maneuver restrictions. It is a relatively rare category since stunt flying is one of

TABLE 3-19   Determination of Parameters for Computation
of Runway Length for Example Problem 3-3

a. Maximum Allowable Landing Weight MALW

Temperature, °F	Elevation, ft		
	6000	6300	7000
90	358.0		347.1
93.5	354.36	350.48	341.43
95	352.8		339.0

b. Landing Distance

MALW	Elevation, ft		
	6000	6300	7000
350	7.44		7.62
350.5	7.446	7.501	7.628
360	7.57		7.77

c. Maximum Allowable Takeoff Weight MATOW

Temperature, °F	Elevation, ft		
	6000	6300	7000
90	338.1		325.2
93.5	332.71	328.90	320.02
95	330.4		317.8

d. *R* Factor

Temperature, °F	Elevation, ft		
	6000	6300	7000
90	91.0		99.2
93.5	93.45	95.99	101.93
95	94.5		103.1

e. Takeoff Distance

MATOW	*R* factor		
	90	95.99	100
320	7.40		8.20
328.9	7.854	8.370	8.716
330	7.91		8.78

the minor applications of the aircraft. In general, the regulations governing aircraft of less than 6000 lb differ from those for aircraft heavier than 6000 lb. For general aviation aircraft weighing more than 6000 lb, the distance to takeoff and climb over a 50-foot obstacle must be determined. From this requirement the takeoff distance can be interpreted as the distance from start of takeoff roll to attainment of a 50-ft height above the surface of the ground. For landing, the horizontal distance required to bring the aircraft to a stop from a height of 50 ft above the surface must be demonstrated. There is no such requirement for aircraft of less than 6000 lb. Most manufacturers, however, provide the distances for a 50-ft height, and it is these distances that are listed in Tables 3-1 and 3-2 under the data for runway length.

Aircraft Certification for Noise

The growing concern over jet aircraft noise by the public led to the passage of legislation by Congress which required appropriate federal agencies to take steps "to provide for the control and abatement of aircraft noise" (Public Law 90-411). This resulted in the implementation of FAR part 36 [45], which prescribes noise standards for subsonic transport aircraft and for subsonic jet-powered aircraft regardless of category. The effective date of the regulation was December 1, 1969. Originally, the regulation applied only to aircraft with engines having a bypass ratio of 2 or more. This means that all aircraft prior to the introduction of the 747 were exempt from this regulation. Most of these aircraft have noise levels higher than prescribed in FAR part 36.

In establishing FAR part 36, the FAA made it clear that the prescribed noise levels would be further reduced as the reductions became technologically practical and economically reasonable. A noise floor of 80 EPNLdB was proposed for some future date "as an objective to aim for, and to achieve where economically reasonable."

In 1976, FAR part 36 was amended to require that all subsonic aircraft with a maximum certified takeoff weight in excess of 75,000 lb be either retired from the fleet or retrofitted to meet standard prescribed noise levels. These noise levels were called *stage 2* noise limits. A schedule requiring the phasing in, or staging, of compliance with these standards by January 1, 1985, was adopted. Subsequently, FAR part 36 was amended by the Airport Noise and Capacity Act of 1990 (Public Law 101-508) which required that aircraft not meeting the stage 3 noise limits be retired from the airline fleet in the United States by December 31, 1999. Airlines may apply for a 4-year waiver from this requirement provided that 85 percent of their fleet meets the stage 3 noise limits and firm orders are in place for replacement aircraft for those aircraft which do not meet the stage 3 noise limits. In any case, all stage 2 aircraft must be retired from airline fleets by December 31, 2003.

Stage 1 aircraft are aircraft which have not been shown to comply with the takeoff, sideline, or approach noise levels required for stage 2 or stage 3 aircraft. *Stage 2* aircraft are aircraft which have been shown to comply with the stage 2 noise limits but not the stage 3 noise limits. *Stage 3* aircraft are refined as aircraft which comply with the stage 3 noise limits.

The measuring points for noise certification are as follows:

1. For takeoff, 21,325 ft (6500 m) from the start of takeoff roll on the extended centerline of the runway.

2. For approach, 6562 ft (2000 m) from the threshold on the extended centerline of the runway.

3. For sideline, 1476 ft (450 m) from and parallel to the extended runway centerline, where noise after liftoff is greater, except that for an airplane with more than three engines this distance must be 2296 ft (700 m) for the purpose of showing compliance with stage 1 or stage 2 noise limits.

The prescribed noise limits are as follows:

Stage 2 noise limits

1. For approach and sideline, a maximum of 108 EPNLdB for aircraft weights of 600,000 lb or more. This value is reduced by 2 EPNLdB for each halving of the maximum takeoff weight of 600,000 lb down to 102 EPNLdB for aircraft with maximum takeoff weights of 75,000 lb or less.

2. For takeoff, a maximum of 108 EPNLdB for aircraft weights of 600,000 lb or more. This value is reduced by 5 EPNLdB for each halving of the maximum takeoff weight of 600,000 lb down to 93 EPNLdB for aircraft with maximum takeoff weights of 75,000 lb or less.

Stage 3 noise limits

1. For approach, a maximum of 105 EPNLdB for aircraft weights of 617,300 lb or more. This value is reduced by 2.33 EPNLdB for each halving of the maximum takeoff weight of 617,300 lb down to 98 EPNLdB for aircraft with maximum takeoff weights of 77,200 lb or less.

2. For takeoff with aircraft with four or more engines, a maximum of 106 EPNLdB for aircraft weights of 850,000 lb or more. This value is reduced by 4 EPNLdB for each halving of the maximum takeoff weight of 850,000 lb down to 89 EPNLdB for aircraft with maximum takeoff weights of 44,673 lb or less.

3. For takeoff with aircraft with three engines, a maximum of 104 EPNLdB for aircraft weights of 850,000 lb or more. This value is

reduced by 4 EPNLdB for each halving of the maximum takeoff weight of 850,000 lb down to 89 EPNLdB for aircraft with maximum takeoff weights of 63,177 lb or less.

4. For takeoff with aircraft with two or fewer engines, a maximum of 101 EPNLdB for aircraft weights of 850,000 lb or more. This value is reduced by 4 EPNLdB for each halving of the maximum takeoff weight of 850,000 lb down to 89 EPNLdB for aircraft with maximum takeoff weights of 106,250 lb or less.

5. For sideline, a maximum of 103 EPNLdB for aircraft weights of 882,000 lb or more. This value is reduced by 2.56 EPNLdB for each halving of the maximum takeoff weight of 882,000 lb down to 94 EPNLdB for aircraft with maximum takeoff weights of 77,200 lb or less.

The noise levels prescribed above may be exceeded at one or two of the measuring points if the sum of the two exceeding points is not greater than 3 EPNLdB and no exceeding point is greater than 2 EPNLdB. The exceeding points must be completely offset by reductions at other required measuring points.

The precise formulas for the stage 3 noise limits are as follows:

$$(\text{EPNLdB})_a = 98 + \frac{\log W - \log 77.2}{(\frac{1}{2}_{2.33}) \log 2} \tag{3-10}$$

$$(\text{EPNLdB})_{t4} = 89 + \frac{\log W - \log 44.673}{\frac{1}{4} \log 2} \tag{3-11}$$

$$(\text{EPNLdB})_{t3} = 89 + \frac{\log W - \log 63.177}{\frac{1}{4} \log 2} \tag{3-12}$$

$$(\text{EPNLdB})_{t2} = 89 + \frac{\log W - \log 106.25}{\frac{1}{4} \log 2} \tag{3-13}$$

$$(\text{EPNLdB})_s = 94 + \frac{\log W - \log 77.2}{(\frac{1}{2}_{2.56}) \log 2} \tag{3-14}$$

where $(\text{EPNLdB})_a$ = effective perceived noise level in decibels for approach

$(\text{EPNLdB})_{tn}$ = effective perceived noise level in decibels for takeoff with aircraft with n engines

$(\text{EPNLdB})_s$ = effective perceived noise level in decibels for sideline

W = maximum takeoff weight in 1000 lb

The atmospheric conditions for which the noise levels are prescribed are sea level, 77°F, 70 percent relative humidity, and zero wind on the runway [35].

Important Aeronautical Terms

When dealing with matters concerning aircraft operations as they relate to the design of airports, planners encounter a number of aeronautical terms that should be familiar: standard atmosphere, pressure altitude, aircraft speed, crosswind, track, and heading.

Standard atmosphere

The actual characteristics of the atmosphere vary from day to day and from place to place, but for practical convenience for comparing the performance of aircraft, a standard atmosphere has been adopted by agreement. A *standard atmosphere* represents the average conditions found in the actual atmosphere in a particular geographic region. However, remember that it is a fictitious atmosphere of assumed composition. Several different standard atmospheres are in use, but the one most commonly used is the one proposed by ICAO. In the standard atmosphere it is assumed that from sea level to an altitude of about 36,000 ft the temperature decreases linearly. Above 36,000 ft to about 65,000 ft, the temperature remains constant; and above 65,000 ft, the temperature rises. Many conventional jet aircraft fly as high as 41,000 ft. The supersonic transports fly at altitudes on the order of 60,000 ft or more. The layer of the earth's atmosphere from sea level to about 36,000 ft is known as the *troposphere.* Immediately above this layer is the *stratosphere.* In the troposphere the standard atmosphere is defined as follows:

1. The temperature at sea level is 59°F.

2. The pressure at sea level is 29.92126 inHg.

3. The temperature gradient from sea level to the altitude at which the temperature becomes − 69.7°F is 0.003566°F/ft and 0 above.

The following relation establishes the standard pressure in the troposphere up to a temperature of − 69.7°F.

$$\frac{P_0}{P} = \frac{T_0^{\,5.2561}}{T} \tag{3-15}$$

where P_0 = standard pressure at sea level (29.92 inHg)
  P = standard pressure at specified altitude

T_o = standard temperature at sea level (59°F)
T = standard temperature at specified altitude

In Eq. (3-15), the temperature must be expressed in absolute or Rankine units. Absolute zero is equal to -459.7°F. Therefore 0°F is equal to -459.7°R and 59°F is equal to 518.7°R.

By using these criteria, the standard temperature at an altitude of 5000 ft is 41.2°F, and the standard pressure is 24.90 inHg. It is common to refer to *standard conditions* or *standard day*. A standard condition is one in which the actual temperature and pressure correspond to the standard temperature and pressure at a particular altitude. When reference is made to the temperature being "above standard," it means that the temperature is higher than the standard temperature. Table 3-20 contains a partial list of standard temperatures and pressures.

Pressure altitude

Aircraft takeoff performance data are related to the pressure altitude. The reason is that aircraft performance is dependent on the density of the air. When the atmospheric pressure drops, the air becomes less

TABLE 3-20   The Standard Atmosphere*

Altitude, ft	Temperature, °F	Pressure, inHg	Speed of sound, kn
0	59	29.92	661.2
1,000	55.4	28.86	658.9
2,000	51.9	27.82	656.6
3,000	48.3	26.82	654.3
4,000	44.7	25.84	652.0
5,000	41.2	24.90	649.7
6,000	37.6	23.98	647.7
7,000	34.0	23.09	645.1
8,000	30.5	22.23	642.7
9,000	26.9	21.39	640.4
10,000	23.3	20.58	638.0
15,000	5.5	16.89	626.2
20,000	-12.2	13.76	614.1
30,000	-47.8	8.90	589.2
35,000	-65.6	7.06	576.3
37,000	-69.7	6.41	573.3
40,000	-69.7	5.55	573.3
50,000	-69.7	3.44	573.3
60,000	-69.7	2.13	573.3
65,000	-69.7	1.68	573.3
70,000	-67.4	1.32	575.0
80,000	-62.0	0.82	579.0

*Adopted in 1962.

dense, requiring a longer run on the ground to obtain the same amount of lift as on a day when the pressure is high. Thus a reduction in atmospheric pressure at an airport has the same effect on its air density as if the airport had been moved to a higher elevation. Pressure altitude is the altitude corresponding to the pressure of the standard atmosphere. Thus if the atmospheric pressure is 29.92 inHg, the pressure altitude is zero. If the pressure drops to 28.86 inHg, the pressure altitude is 1000 ft. This can be obtained from the formula relating pressure and temperature. If this lower pressure occurred at a sea-level airport, the geographic altitude would be zero but the pressure altitude would be 1000 ft. For airport planning purposes, it is satisfactory to assume that the geographic and pressure altitudes are equal unless the barometric pressures at a particular site are unusually low a great deal of the time.

Aircraft speed

Reference is made to aircraft speed in several ways. It is important to understand the basic difference between ground speed and airspeed. The *ground speed* is the speed of the aircraft relative to the ground. *True airspeed* is the speed of an aircraft relative to the medium in which it is traveling. Thus, if an aircraft is flying at a ground speed of 500 kn in air where the wind is moving in the opposite direction at a speed of 100 kn, the true airspeed is 600 kn. Likewise, if the wind were moving in the same direction and the aircraft maintained a ground speed of 500 kn, the true airspeed would be 400 kn. Thus, the airspeed is the speed that the aircraft wing (airfoil) encounters.

In examining aircraft performance data, the reader will find that reference is made to two airspeeds, namely, *true airspeed* (TAS) and *indicated airspeed* (IAS). The pilot obtains the speed from an airspeed indicator. This indicator works by comparing the dynamic air pressure due to the forward motion of the aircraft with the static atmospheric pressure. As the forward speed is increased, so is the dynamic pressure. The airspeed indicator works on the principle of the pitot tube. From physics we know that the dynamic pressure is proportional both to the square of the speed and to the density of the air. The variation with the square of the speed is taken care of by the mechanism of the airspeed indicator, but not the variation in density. The indicator is sensitive to the product of the density of the air and the square of the velocity. At high altitudes the density becomes smaller, and thus the indicated airspeed is less than the true airspeed.

If the true airspeed is required, it can be found with the aid of tables. As a very rough guide, one can add 2 percent to the indicated speed for each 1000 ft above sea level to obtain the true airspeed.

The indicated airspeed is more important to the pilot than the true airspeed. The concern is with the generation of lift, specifically the stall speed, which is dependent on speed and air density. At high altitudes an aircraft will stall at a higher speed than it does at sea level. But at higher altitudes the airspeed indicator is indicating speeds lower than the true speeds; consequently this is on the safe side, and no corrections are necessary. Thus, an aircraft with a stalling speed of 90 kn will stall at the same indicated airspeed regardless of altitude. This is why aircraft manufacturers always report stalling speeds in terms of indicated airspeed rather than true airspeed. With the introduction of jet transports and high-speed military aircraft, the reference datum for speed is often the speed of sound. The speed of sound is defined as *Mach 1* (after Ernst Mach, Austrian scientist). Thus Mach 3 means 3 times the speed of sound. Most of our current jet transports are *subsonic* (slower than the speed of sound) and cruise at a speed in the neighborhood of 0.8 to 0.9 Mach. Many military aircraft are *supersonic* (faster than the speed of sound). Again the reader is reminded that when the maximum speed of an aircraft is quoted as 0.9 Mach, this is in terms of true airspeed, not ground speed. Such an aircraft can conceivably be traveling at a ground speed higher than the speed of sound, depending on the magnitude of the tailwind.

The speed of sound is not a fixed speed; it depends on temperature, not on atmospheric pressure. As the temperature decreases, so does the speed of sound. The speed of sound at 32°F (0°C) is 742 mi/h (1090 ft/s), at -13°F (-25°C) it is 707 mi/h, and at 86°F (30°C) it is 785 mi/h. In fact, the speed of sound varies 2 ft/s for every change in temperature of 1°C above or below the speed at 0°C. The speed of sound at the altitudes at which jets normally fly is less than 700 mi/h, but at altitudes at which small aircraft normally fly (20,000 ft or less) it is greater than 700 mi/h.

The speed of sound may be computed from:

$$V_{sm} = 33.4T^{0.5} \tag{3-16}$$

$$V_{sf} = 49.04T^{0.5} \tag{3-17}$$

where V_{sm} = speed of sound at some temperature, mi/h
V_{sf} = speed of sound at some temperature, ft/s
T = temperature, °R

For convenience in navigation, aircraft distances and speeds are measured in nautical miles and knots, just like measurement on the high seas. A nautical mile, 1 nmi (6080 ft), is practically equal to 1 minute of arc of the earth's circumference. And 1 kn is 1 nmi/h. A nautical mile is approximately 1.15 statute miles.

Crosswind, track, and heading

As an aircraft approaches a runway, its heading (direction in which the nose is pointing) is dependent on the strength of the wind traveling across the path of the aircraft (crosswind). The approach flight path to the runway is an extension of the centerline of the runway, and it is referred to as the *track*. An aircraft must fly along the track to safely reach the runway. The relation between track, heading, and crosswind is illustrated in Fig. 3-27. In order not to be blown laterally off the track by the wind, the aircraft must fly at an angle x from the track. The magnitude of this angle can be obtained from the following relation:

$$\sin x = \frac{V_c}{V_h} \tag{3-18}$$

where V_c is the crosswind, in miles per hour or knots, and V_h is the true airspeed, in miles per hour or knots.

The *crosswind* V_c is defined as the component of the wind V_w that is at a right angle to the track. Angle x is referred to as the crab angle. Note that the magnitude of the angle is directly proportional to the speed of the wind and inversely proportional to the speed of the aircraft. This means that when the aircraft is moving slowly (as it does when it approaches a runway) and there is a strong crosswind, angle x will be large. The term V_t is the true airspeed along the track and is equal to $V_h \cos x$. To obtain the ground speed along the track, the component of the wind along the track must be subtracted from V_t. In the diagram, the ground speed along the track is equal to V_t minus the wind along the track $V_w \sin x$.

Assume that an aircraft is approaching a runway at a speed of 135 kn and that the crosswind is 25 kn. The crab angle x is 10°10'. To land properly, the pilot must reduce the crab angle to zero just prior to touchdown. If it is reduced too early, the aircraft might be blown laterally off the runway. For this reason it is desirable to orient run-

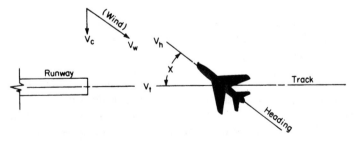

Figure 3-27 Definitions of track, heading, and crab angle for an aircraft.

ways so as to minimize crosswinds. Large transport aircraft are capable of landing on a dry runway with crosswinds as high as 30 kn, but this is certainly undesirable from the standpoint of the pilot, particularly if the runway is wet. For the orientation of runways, the following criterion might be used (see Chap. 7): Runways should be oriented so that 95 percent of the time the crosswind is not greater than 13 kn (15 mi/h). Even this value is too large for small light aircraft [9].

References

1. *Aerodrome Design Manual,* Part 1: Runways, 2d ed., Document 9157-AN/901, International Civil Aviation Organization, Montreal, Canada, 1984.
2. *Airbus Industrie A300 Airplane Characteristics for Airport Planning,* A.AC E00A, Airbus Industrie, Biagnac, France, October 1987.
3. *Airbus Industrie A300-600 Airplane Characteristics for Airport Planning,* D.AC E00A, Airbus Industrie, Biagnac, France, October 1990.
4. *Airbus Industrie A310 Airplane Characteristics for Airport Planning,* B.AC E00A, Airbus Industrie, Biagnac, France, December 1991.
5. *Airbus Industrie A320 Airplane Characteristics for Airport Planning,* Airbus Industrie, Biagnac, France, February 1988.
6. *Airbus Industrie A340 Airplane Characteristics for Airport Planning,* Preliminary, Airbus Industrie, Biagnac, France, July 1991.
7. "Aircraft of the Future," W. E. Parsons and J. A. Stern, Paper no. 800743, *International Air Transportation Meeting,* Society of Automotive Engineers, Warrendale, Pa., May 1980.
8. "Aircraft Wake Turbulence Avoidance," W. A. McGowan, Paper no. 72/6, *12th Anglo-American Aeronautical Conference,* National Aeronautics and Space Administration, Washington, July 1971.
9. *Airport Design,* Advisory Circular AC150/5300-13, Federal Aviation Administration, Washington, 1989.
10. *Air Taxi Operators and Commercial Operators of Small Aircraft,* Part 135, Federal Aviation Regulations, Federal Aviation Administration, Washington, 1978.
11. *Airworthiness Standards: Normal, Utility, and Acrobatic Category Airplanes,* Part 23, Federal Aviation Regulations, Federal Aviation Administration, Washington, 1974.
12. *Airworthiness Standards: Transport Category Airplanes,* Part 25, Federal Aviation Regulations, Federal Aviation Administration, Washington, 1974.
13. *Boeing 707 Airplane Characteristics—Airport Planning,* Document D6-58322, Boeing Commercial Airplane Company, Seattle, Wash., December 1968.
14. *Boeing 727 Airplane Characteristics—Airport Planning,* Document D6-58324-R2, Boeing Commercial Airplane Company, Seattle, June 1978.
15. *Boeing 737-100/200 Airplane Characteristics—Airport Planning,* Document D6-58325, revision D, Boeing Commercial Airplane Group, Seattle, September 1988.
16. *Boeing 737-300/400/500 Airplane Characteristics—Airport Planning,* Document D6-58325-2, revision A, Boeing Commercial Airplane Group, Seattle, July 1990.
17. *Boeing 747 Airplane Characteristics—Airport Planning,* Document D6-58326, revision E, Boeing Commercial Airplane Group, Seattle, May 1984.
18. *Boeing 747-400 Airplane Characteristics—Airport Planning,* Document D6-58326-1, revision B, Boeing Commercial Airplane Group, Seattle, March 1990.
19. *Boeing 757 Airplane Characteristics—Airport Planning,* Document D6-58327, revision D, Boeing Commercial Airplane Group, Seattle, September 1989.
20. *Boeing 767 Airplane Characteristics—Airport Planning,* Document D6-58328 revision F, Boeing Commercial Airplane Group, Seattle, February 1989.
21. *Boeing 777 Airplane Characteristics—Airport Planning,* Preliminary information, Document D6-58329, Boeing Commercial Airplane Group, Seattle, February 1992.

22. *British Aerospace 146 Airplane Characteristics for Airport Planning,* APM 146.1, British Aerospace Limited, Hatfield, Hertfordshire, England, June 1984.

23. *Certification and Operations—Domestic, Flag, and Supplemental Air Carriers and Commercial Operators of Large Aircraft,* Part 121, Federal Aviation Regulations, Federal Aviation Administration, Washington, 1974.

24. *Certification and Operations of Scheduled Air Carriers with Helicopters,* Part 127, Federal Aviation Regulations, Federal Aviation Administration, Washington, 1974.

25. *Commercial Air Transportation in the Next Three Decades,* H. W. Withington, Boeing Commercial Airplane Company, Seattle, 1980.

26. *Commercial Air Transportation 1980's and Beyond,* H. W. Withington, Boeing Commercial Airplane Company, Seattle, November 1980.

27. "CTOL Concepts and Technology Development," D. William Conner, *Astronautics and Aeronautics,* American Institute of Aeronautics and Astronautics, July–August 1978.

28. *CTOL Transport Aircraft Characteristics, Trends, and Growth Projections,* Aerospace Industries Association of America, Inc., Washington, 1979.

29. *Current Market Outlook,* Boeing Commercial Airplane Group, Seattle, March 1992.

30. *DC-8 Airplane Characteristics,* Airport Planning, Rep. DAC-67492, Douglas Aircraft Company, McDonnell-Douglas Corporation, Long Beach, Calif., March 1969.

31. *DC-9 Airplane Characteristics,* Airport Planning, Rep. DAC-67264, Douglas Aircraft Company, McDonnell-Douglas Corporation, Long Beach, Calif., September 1978.

32. *DC-10 Airplane Characteristics,* Airport Planning, Rep. DAC-67803A, Douglas Aircraft Company, McDonnell-Douglas Corporation, Long Beach, Calif., January 1991.

33. *Dimensions of Airline Growth,* Boeing Commercial Airplane Company, Seattle, March 1980.

34. *Energy and Transportation Systems,* Final Report, California Department of Transportation, Sacramento, December 1981.

35. *Environmental Protection,* Annex 16 to the Convention on International Civil Aviation, vol. 1: *Aircraft Noise,* 2d ed., International Civil Aviation Organization, Montreal, Canada, 1988.

36. *High Speed Civil Transport,* Program Review, Boeing Commercial Airplane Group, Seattle, 1990.

37. *Jane's All the World's Aircraft,* Franklin Watts, Inc., New York, annual.

38. *Jet Aviation Development: One Company's Perspective,* John E. Steiner, Boeing Commercial Airplane Group, Seattle, 1989.

39. "Jet Transport Characteristics Related to Airports," R. Horonjeff and G. Ahlborn, *Journal of the Aerospace Transport Division,* vol. 91 AT1, American Society of Civil Engineers, New York, April 1965.

40. *L-1011 Airplane Characteristics,* Airport Planning Document CER-12013, Lockheed California Company, Burbank, Calif., December 1972.

41. *MD-11 Airplane Characteristics for Airport Planning,* Rep. MDC-K0388, McDonnell-Douglas Corporation, Long Beach, Calif., October 1990.

42. *MD-80 Series Airplane Characteristics for Airport Planning,* Rep. MDC-J2904, McDonnell-Douglas Corporation, Long Beach, Calif., February 1992.

43. *MD-90-30 Aircraft Airport Compatibility Brochure,* Rep. MDC-91K0393, McDonnell-Douglas Corporation, Long Beach, Calif., February 1992.

44. *Measurement, Construction and Maintenance of Skid Resistant Airport Pavement Surfaces,* Advisory Circular AC 150/5320-12A, Federal Aviation Administration, Washington, July 1986.

45. *Noise Standards: Aircraft Type and Airworthiness Certification,* Part 36, Federal Aviation Regulations, Federal Aviation Administration, Washington, 1974.

46. *Outlook for Commercial Aircraft 1980–1994,* Douglas Aircraft Company, McDonnell-Douglas Corporation, Long Beach, Calif., June 1980.

47. *Pavement Grooving and Traction Studies,* NASA SP-5073, Proceedings of Conference at Langley Research Center, National Aeronautics and Space Administration, Langley Field, Va., November 1968.

48. "Pneumatic Tire Hydroplaning and Some Effects on Vehicle Performance," W. B. Horne and U. T. Joyner, Paper No. 97UC, *Proceedings of the Society of Automotive Engineers,* New York, 1965.

49. "Runway Grooving for Increasing Traction—The Current Program and an Assessment of Available Results," W. B. Horne and G. W. Brooks, *20th Annual International Air Safety Seminar,* Williamsburg, Va., December 1967.

50. *Runway Length Requirements for Airport Design,* Advisory Circular AC 150/5325-4A, Federal Aviation Administration, Washington, January 1990.

51. *Short-Haul Transport Aircraft: Future Trends,* Aerospace Industries Association of America, Inc., Washington, January 1978.

52. "Simulated Vortex Encounters by a Twin-Engine Commercial Transport Aircraft during Final Approach," E. C. Hastings, Jr., and G. L. Keyser, Jr., Paper No. 800775, *International Air Transportation Meeting,* Society of Automotive Engineers, Warrendale, Pa., May 1980.

53. "Technology Requirements and Readiness for Very Large Vehicles," D. William Conner, *AIAA Very Large Vehicle Conference,* American Institute of Aeronautics and Astronautics, Arlington, Va., April 1979.

54. *The Braking Performance of an Aircraft Tire on Grooved Portland Cement Concrete Surfaces,* S. K. Agrawal and H. Diautolo, Rep. No. FAA-RD-80-78, Federal Aviation Administration Technical Center, Federal Aviation Administration, Atlantic City, January 1981.

55. *The Effect of Variable Runway Slopes on Takeoff Runway Length for Transport Aeroplanes,* ICAO Circular 91-AN/75, International Civil Aviation Organization, Montreal, Canada, 1970.

56. "Trailing Vortex Hazard," W. A. McGowan, Paper No. 680220, *Proceedings of the Society of Automotive Engineers,* New York, April 1968.

57. *Water, Slush, and Snow on the Runway,* Advisory Circular AC 91-6A, Federal Aviation Administration, Washington, May 1978.

4

Air Traffic Control

So that the airport designer may be aware of the importance of *air traffic control* (ATC) in airport planning, a very brief summary of what constitutes air traffic control, how it is managed and operated, and the principal aids to air navigation is presented in this chapter. An appreciation of air traffic control, its capabilities, and its problems will focus attention on the fact that any extensive reorientation of runways on existing airports or the construction of entirely new airports requires consultation with the people controlling traffic and very often an airspace study. This is particularly true in large metropolitan areas where several airports are present and the existing airspace must be shared by several airports. Conflicts in air traffic control procedures can seriously affect the capacity of any single airport or a system of airports in a region. The planning of airports must include provisions for facilities located at airports that support the air traffic control system. The importance of the segregation of airspace in the vicinity of major airports is illustrated in Fig. 4-1, which shows the location of airports in the Chicago terminal control area.

Elements of the Air Traffic Control System

The first attempt to set up rules for air traffic control was made by the International Commission for Air Navigation (ICAN), which was under the direction of the League of Nations. The procedures which the commission promulgated in July 1922 were adopted by 14 countries. Although the United States was not a member of the League of Nations, and therefore did not officially adopt the rules, many of the procedures established by ICAN were used in the promulgation of air traffic procedures in this country.

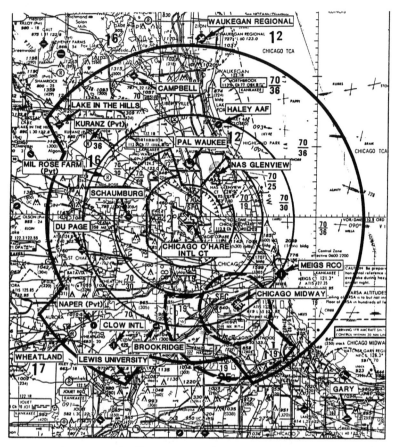

Figure 4-1 Airports in the Chicago terminal control area (*Landrum and Brown, Inc. [7]*).

Construction and operation of the airways system in this country prior to 1926 were controlled by the military and by the Post Office Department. The formal entry of the federal government into the regulation of air traffic came with the passage of the Air Commerce Act of 1926 (Public Law 64-254). This act directed the Bureau of Air Commerce to establish, maintain, and operate lighted civil airways. At present the Federal Aviation Administration (FAA) maintains and operates the U.S. airways system.

Today, the FAA, in providing control and navigational assistance for the movement of air traffic on the airways and in support of approaches to and departures from airports, has established a pattern of integrated radio signal stations, radar, instrument landing systems, air route traffic control centers, terminal radar approach control facilities, airport traffic control towers, flight service stations, continuous weather reporting, and regulations for the use of these facilities. Most navigational aids are owned and operated by the FAA with the exception of a small number of aids provided by private organizations and states at locations not involving a substantial volume of interstate commerce, and certain military facilities operated in conjunction with military airfields.

Definition of air traffic control

The federal government prescribes two types of flight rules for air traffic. These are known as the *visual flight rules* (VFR) and *instrument flight rules* (IFR). In general, VFR are in effect when visual meteorological conditions (VMC) prevail, i.e., when the weather conditions are such that aircraft can maintain safe separation by visual means. On the other hand, IFR are in effect when instrument meteorological conditions (IMC) prevail, i.e., when the visibility or the ceiling (the height of the dominant cloud base) falls below that prescribed under visual meteorological conditions. In IFR conditions, safe separation between aircraft is the responsibility of air traffic control personnel, while in VFR conditions it is the responsibility of the pilot. Thus in VFR conditions there is essentially very little air traffic control, and aircraft separations are maintained by the pilots themselves. Air traffic control monitors flights under VFR and intervenes only when apparent conflicts between aircraft develop. This is passive control. Positive air traffic control is exercised when IFR apply or when aircraft operating under VFR enter controlled airspace. Essentially these rules require the assignment of aircraft to specific routes and altitudes and the maintenance of minimum separation between aircraft.

As the speed of the aircraft and the density of traffic in the airspace increased, there was greater concern over the possibility of midair collisions. This concern was substantiated by the occurrence of several such collisions involving many lives. Accordingly, in certain parts of the airspace, IFR have been prescribed regardless of weather conditions. This is referred to as *positive control airspace*. Positive control airspace usually encompasses the airspace where high-speed jet aircraft operate, and therefore it can include the airspace in the vicinity of a major airport, often designated as a *terminal control area* (TCA) or an *airport radar service area* (ARSA), as well as the airspace at and above 18,000 ft AMSL (above mean sea level) in which jets fly en route from one airport to another. The limits of positive control airspace can be extended by the FAA as necessary to ensure safe operations. The trend is toward more positive control.

Terminal control areas

A terminal control area consists of the controlled airspace extending upward from the surface or higher to specified altitudes, within which all aircraft are subject to operating rules and equipment requirements specified in FAR part 91 [16]. Terminal control areas are usually designated around major airports to improve the safe and efficient movement of aircraft within the TCA. In addition to the major airport, some smaller airports may fall within the TCA, as shown in Fig. 4-1. Terminal control areas are classified as group I, group II, or group III depending on the volume of air traffic within the TCA. A group I TCA represents some of the busiest airport locations in terms of the volume of air traffic and passengers carried. The Chicago TCA is illustrated in Fig. 4-1.

Airport radar service areas

An airport radar service area (ARSA) is regulatory airspace around designated airports wherein ATC provides radar vectoring and sequencing on a continuous basis for all VFR and IFR aircraft. An ARSA is usually established at busy airports that cannot justify a TCA and is established for the safety of aircraft departing from or arriving at the airport. An ARSA consists of the airspace from the airport surface up to an elevation of 4000 ft above ground level from the airport surface within a radius of 5 nmi of the airport and from an elevation of 1200 ft above the ground surface to an elevation of 4000 ft above the ground surface between 5 and 10 nmi radially from the airport. Pilots are required to establish radio contact with ATC before entering an ARSA. ARSA services include the sequencing of all aircraft to the primary airport, standard IFR separations to IFR aircraft, separation, traffic advisories and safety alerts between IFR and VFR aircraft, and mandatory traffic advisories and safety alerts between VFR aircraft. A schematic diagram of an ARSA is shown in Fig. 4-2.

Airspace classifications

Effective September 16, 1993, the FAA has reclassified its airspace designations as follows [9]:

> *Class A Airspace—Positive control areas.* Airspace in which only IFR operations are permitted. Pilots must be instrument-rated and have ATC clearance to enter this airspace. A transponder with altitude reporting capability (mode C) is required. This airspace is designated from 18,000 ft above mean sea level to flight level 600 (60,000-ft pressure altitude) over the 48 contiguous states.

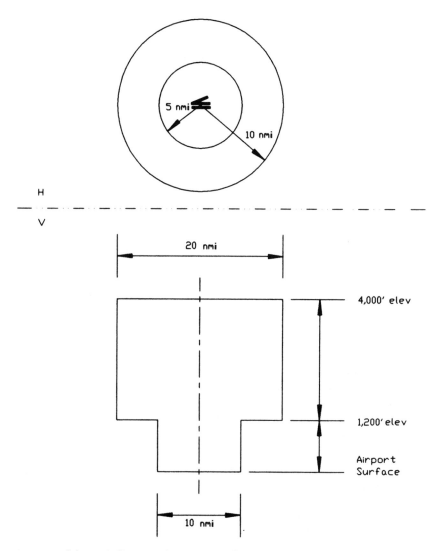

Figure 4-2 Schematic diagram of an airport radar service area.

Class B Airspace—Terminal control areas. Airspace in which both VFR and IFR operations are permitted but an ATC clearance is required to enter the airspace. Pilots must hold a private certificate or a student certificate with the appropriate instructor endorsements. Aircraft must be equipped with a mode C transponder, and must have a two-way radio with appropriate frequencies and a very high-frequency omnirange radio receiver for IFR operations. All aircraft are subject to ATC separation, and VFR aircraft must remain clear of clouds.

Class C Airspace—Airport radar service areas.   Airspace in which entry requires prior two-way radio communications with ATC and a transponder with mode C capability. ATC provides traffic advisories and conflict resolution between all aircraft. VFR aircraft must maintain visual flight rules, that is, 3-mi (statute) visibility, and remain 500 ft below the clouds, 1000 ft above the clouds, and 2000 ft horizontally from the clouds. Special VFR operations may be approved by ATC when basic VFR minima cannot be maintained.

Class D Airspace—Airport traffic areas and control zones at airports with control towers.   Airspace established around airports with operating air traffic control towers. When it is consistent with safety, nearby uncontrolled airports may be included. Pilots must establish and maintain radio contact with ATC prior to entering and while operating in this airspace. The ceiling upper limit will normally be 2500 ft above ground level. However, this upper limit may be different if local conditions warrant. Weather minima are basic VFR.

Class E Airspace—General controlled airspace and control zones at airports without control towers.   The airspace above 14,500 ft above mean sea level, i.e., in the *continental control area* (CCA). Additionally, federal airways and airspace above 1200 ft above ground level or 700 ft above ground level for the transition areas for instrument approaches. Basic VFR minima are 3-mi visibility, 500 ft below the clouds, 1000 ft above the clouds, and 2000 ft horizontally from the clouds. Above 10,000 ft mean sea level, the VFR minima are 5 mi, 1000 ft below the clouds, 1000 ft above the clouds, and 1 mi horizontally from the clouds.

Class G Airspace—Uncontrolled airspace.   The airspace not under the control of any ATC facility.

Conformity with instrument flight rules requires that, prior to departure or en route prior to operational necessity, the pilot file a flight plan with the air route traffic control center. Each filed flight plan indicates the aircraft's destination, desired routes, and desired altitudes. Progress reports are continuously updated by air traffic control as the flight progresses along its planned route.

Airways

Aircraft flying from one point to another follow designated routes. In the United States, these are referred to as *victor airways* and *jet routes*. These routes have evolved over time, as discussed below.

Colored airways.   The airways were initially given a color designation. The trunk lines east and west were green, trunk lines north and south were amber, secondary lines east and west were red, and secondaries north and south were blue. Each of these colored airways was then given a number, such as green 3, red 4, etc. The numbering for the airways began at the Canadian border and the Pacific coast and then progressed to the south and east. These airways were then assigned an altitude level, which for green and red was at odd 1000-ft levels eastbound and at even 1000-ft levels westbound. On the amber and blue airways northbound, odd 1000-ft levels were assigned, and southbound the even 1000-ft levels were assigned. These airways were delineated on the ground by low-frequency/medium-frequency (LF/MF) four-course radio ranges. The colored airways were phased out as aircraft became equipped to use the victor airways.

Victor airways.   Following the development of the LF/MF four-course radio ranges, the routes now known as the victor airways were established. The victor airways are delineated on the ground by *very high-frequency omnirange radio* (VOR) equipment. Each VOR station has a discrete radio frequency to which a pilot could tune a navigational radio and thus be able to maintain a course from one VOR station to the next. The numbering system for these airways is as follows: even numbers east and west, odd numbers north and south. The advantages of the victor airways are that the VOR stations are relatively free of static and it is much easier for a pilot to determine air position relative to a VOR station than with the LF/MF four-course radio range. Victor airways are designated on aeronautical charts as V-1, V-2, etc. The airway includes the airspace within parallel lines 4 mi each side of the centerline of the airway. If two VOR stations delineating an airway are more than 120 mi apart, the airspace included in the airway is as indicated for jet routes.

An illustration of a portion of an air navigation chart showing the victor airways is given in Fig. 4-3.

Jet routes.   With the introduction of commercial jet aircraft in 1958, the altitudes at which these aircraft flew increased significantly. At high altitudes the number of ground stations (VOR stations) required to delineate a specific route is smaller than at low altitudes because the signal is transmitted on a line of sight. Therefore there was no need to clutter the high-altitude routes with all the ground stations required for low-altitude flying. All the routes in the continental United States could be placed on one chart. Jet routes were established. Although in one sense these routes are airways, they are not referred to as such. Today both victor airways and jet routes exist. Thus the jet routes are delineated by the same aids to navigation on the ground (VOR stations) as are victor airways, but fewer stations are used. Victor airways extend from 1200 ft above the terrain to, but not including, 18,000 ft *above mean sea level* (AMSL). Jet routes

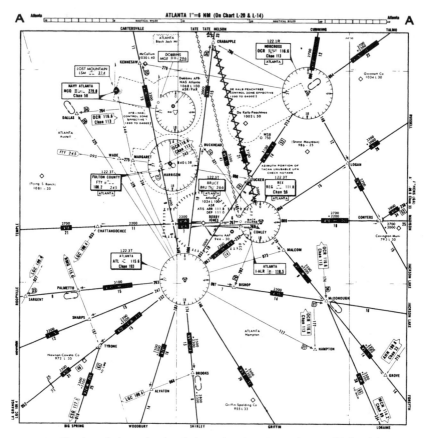

Figure 4-3 Portion of air navigational chart showing victor airways (*Federal Aviation Administration [19]*).

extend from 18,000 ft to 45,000 ft AMSL. Above 45,000 ft there are no designated routes, and aircraft are handled on an individual basis. The numbering system for the jet routes is the same as that for the victor airways. Jet routes are designated on aeronautical charts as J-1, J-2, etc.

Since the VOR stations delineating a jet route are often more than 120 nmi apart, the jet route includes the airspace, as indicated in Fig. 4-4, for the case of two VOR stations 260 nmi apart.

Area navigation.   For many years all aircraft were required to fly on designated routes, airways, or jet routes. That is, all aircraft had to fly from one VOR station to the next since the VOR stations delineate the airways and jet routes. This required the funneling of all traffic on the designated routes which resulted in congestion on certain routes. Also the designated route was often not the shortest distance

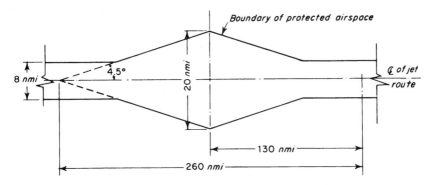

Figure 4-4   Delineation of a jet route airway.

between two points, resulting in additional fuel consumption, flight time, and cost. Furthermore, if the designated route penetrates an area of thunderstorms, aircraft have to be vectored around the storm by controllers on the ground. This imposes an extra workload on the controllers that is compensated for by the use of the Severe Weather Avoidance Program (SWAP).

Area navigation, referred to as RNAV, is a method of aircraft navigation that permits aircraft operation on any desired course within the coverage of station-referenced navigational signals or within the limits of a self-contained system capability. Random area navigation routes are direct routes, based upon the area navigation capability of aircraft, between way points defined in terms of latitude and longitude coordinates, degree and distance fixes, or offsets from established routes and airways. The major types of equipment used in area navigation are VORTAC, a VOR with distance measuring capability; or course line computer (CLC) systems; OMEGA/VLF, a long-range navigation system based upon very low-frequency radio signals transmitted from 17 stations worldwide; inertial navigation systems (INS); microwave landing systems (MLS) in the terminal area; and LORAN-C.

Area navigation provides a more flexible routing capability that allows for better utilization of the airspace. The greater utilization reduces delays in the airspace and results in more economical operation of the aircraft. For example, routes parallel to the designated routes from one VOR station to another can be established without requiring additional aids to navigation on the ground. Another example is the establishment of a more direct route from one point to another by establishing way points that provide for a shorter trip. Routing around a thunderstorm without continuous radar guidance from the ground is another example.

Area navigation is accomplished by the installation of special computers in the aircraft. These computers are tuned to VOR stations.

Each station provides information on the distance to the station and the azimuth of the aircraft relative to the station. First a route must be established relative to selected VOR stations from which information is fed to the computer. The computer then maintains the selected route by inputs of azimuth and distance from the selected VOR stations. In the aircraft the pilot selects a specified route, in terms of azimuth, and a pictorial display in the cockpit indicates whether the aircraft is following the selected route and any deviation from this route. The route is delineated by intersections called *way points*. An intersection is a point in space that is established by providing its latitude and longitude and by indicating its distance from the nearest VORTAC station. Airborne equipment uses the azimuth and distance from the intersections as inputs for the computers or the latitude and longitude of the way points for inertial navigation systems.

Area navigation is not limited to the horizontal plane but can also be utilized in the vertical plane, referred to as VNAV. It can also include a time-reference capability. A properly equipped aircraft could arrive at a specified point in space, called a *fix,* with no need for ground vectoring or directions and could also be at that point at a specified altitude and time. This is a four-dimensional capability giving latitude, longitude, altitude, and time (4D RNAV). Thus area navigation has the potential of increasing airspace capacity, enhancing safety, and reducing the workload of the pilot and the air traffic controller.

Area navigation routes that are frequently used by more than one user appear on published charts, much as victor airways and jet routes do. The numbering system is as follows. Series 700 is reserved for low altitudes, i.e., below 18,000 ft. Each route is labeled V700R, V701R, etc. The symbol V indicates victor airways as the reference, and R indicates area navigation. Above 18,000 ft, series 800 and series 900 are used, and the designation is J801R, J802R, etc. As an additional area navigation route is established, it is given the next available number regardless of whether the route is north-south or east-west. Each intersection is given a name similar to VORTAC stations. An area navigation route includes the same airspace as the jet route shown in Fig. 4-4, except that the angle from the VORTAC station may, e.g., be 3.25 instead of 4-5.

Major Components and Functions of the National Airspace System

The *national airspace system* (NAS) consists of a network of navigational aids and a number of air traffic control facilities. The specific purpose of the air traffic control service is twofold: to prevent collisions between aircraft and on the maneuvering area between aircraft and obstructions and to expedite and maintain an orderly flow of air traffic [3]. To properly manage the traffic in the system, the jurisdiction of control is divided into three parts: en route, terminal, and oceanic. Terminal air traffic control may be divided into terminal radar-approach control (TRACON) and air traffic control tower (ATCT) operations at an airport. Each part has a specific function.

Air route traffic control centers

Air route traffic control centers (ARTCC) have the responsibility of controlling the movement of en route aircraft along the airways and jet routes and in other parts of the airspace. Each of the 20 air traffic control centers within the continental United States has control of a definite geographical area which may be bigger than 100,000 mi^2. At the boundary point, which marks the limits of the control area of the center, control of aircraft may be transferred to an adjacent center or an approach control facility, or radar service may be terminated and aircraft using VFR are free to contact the next center. Air traffic control centers are normally not located at airports. Air traffic control centers can also provide approach control service to nontowered airports and to nonterminal radar-approach control airports. The ARTCC is concerned primarily with the control of aircraft operating under instrument flight rules (IFR).

Under IFR pilots are required to file a flight plan indicating the route and altitude they desire to fly. The ARTCC will then check to determine whether the flight plan, as filed, can be approved so that a safe separation between aircraft can be ensured. Changes in flight plans en route are permitted if approved by each ARTCC along the route of the flight.

Each ARTCC geographic area is divided into sectors. The configuration of each sector is based on equalizing the workload of the controllers. Control of aircraft is passed from one sector to another. The geographic area is sectored in both the horizontal plane and the vertical plane. Thus there can be a high-altitude sector above one or more low-altitude sectors. Each sector is staffed by one or more controllers, depending on the volume and complexity of traffic. The average number of aircraft that each sector can handle depends on the number of people assigned to the sector, the complexity of traffic, and the degree of automation provided.

Each sector is normally provided with one or more air route surveillance radar (ARSR) units which cover the entire sector and allow for monitoring of the separation between aircraft. In addition, each sector has data on the identification of the aircraft, destination, flight plan route, estimated speed, and flight altitude, which are posted on pieces of paper called *flight progress strips* or may be superimposed on the radarscope adjacent to the blips which specify the position and identity of the aircraft. The strips are continuously updated as the need arises.

At present, communication between the pilot and the controller is by voice. Therefore each ARTCC is assigned a number of very high and ultra high radio communication frequencies. The controller in turn assigns a specific frequency to the pilot.

Terminal approach control facility

The terminal approach control facility monitors the air traffic in the airspace surrounding airports with moderate to high-density traffic. It has jurisdiction in the control and separation of air traffic from the boundary area of the air traffic control tower at an airport to a distance of up to 50 mi from the airport and to an altitude ranging up to 17,000 ft. This is commonly referred to as the *terminal area*. Where there are several airports in an urban area, one facility may control traffic to all the airports. In essence, the facility receives aircraft from the ARTCC and guides them to one of several airports. In providing this guidance, the facility performs the important function of metering and sequencing aircraft to provide uniform and orderly flow to airports.

The radar-approach control facility is referred to as TRACON, an abbreviation for terminal radar-approach control. There are various degrees of automation in an approach control facility depending on the volume of traffic normally handled. Various abbreviations are used to designate the type of hardware in an approach control facility. As an example, ARTS III is an acronym for *automated radar terminal system*. The designation III denotes the highest level of automation, while I is the lowest level of automation. Thus one can have ARTS I, II, III automation capability in a TRACON facility. ARTS IIIA and ARTS IIIE are updated enhancements of the ARTS III system capability to accommodate automation data.

The organizational structure of an approach control facility is very similar to that of the ARTCC. Like the ARTCC, the geographic area of the facility is divided into sectors to equalize the workload of the controllers. The approach control facility transfers control of an arriving aircraft to the airport control tower when the aircraft is lined up with the runway about 5 mi from the airport. Likewise control of departing aircraft is transferred to the approach control facility by the airport control tower.

If the flow of aircraft is greater than the facility can handle, traffic management and the air traffic system command center (ATSCC), formerly called the central flow control facility (CFCF), manipulate aircraft on the ground and en route to adjust, or meter, the arrival flows to their destination airports. This may result in delays to departing aircraft or delays en route. In the past such aircraft were delayed by either reducing their speed en route or detaining them at specified radio fixes within the area of the destination facility. The latter method is referred to as *stacking*. In a stack, aircraft navigate around a fix in a racetrack pattern, a holding pattern, and are separated vertically by 1000-ft intervals. There may be as many as 10 aircraft in a stack, and each is directed in turn to a landing by the approach control facility. As a matter of procedure, stacking is no longer performed except when the arrival capacity at an airport is reduced due to unexpected events.

In 1990 there were more than 200 various types of terminal-area approach control facilities operated by the FAA in the United States.

Airport traffic control tower

The airport traffic control tower is the facility which supervises, directs, and monitors the arrival and departure traffic at the airport and in the immediate airspace within about 5 mi from the airport. The tower is responsible for issuing clearances to all departing aircraft; providing pilots with information on wind, temperature, barometric pressure, and operating conditions at the airport; and controlling all aircraft on the ground except in the maneuvering area immediately adjacent to the aircraft parking positions called the *ramp area*. In the United States in 1990, 400 air traffic control towers were operated by the FAA, and 25 air traffic control towers operated under contract to the FAA.

Flight service stations

The flight service stations (FSS) are located along the airways and at airports. Flight service stations are not air traffic control facilities but provide essential information to pilots. Their principal function is to accept and close flight plans and to brief pilots, before flight and in flight, on weather, navigational aids, airports and navaids that are out of commission, and changes in procedures and new facilities. A secondary function is to relay traffic control messages between aircraft and the appropriate control facility on the ground. The FAA operated 183 domestic and international flight service stations in 1990. Flight service stations are in the process of being converted to automated flight service stations (AFSS), and when the transition is completed, there will be 60 AFSS in the United States.

Air Traffic Separation Rules

Air traffic rules governing the minimum separation of aircraft in the vertical, horizontal or longitudinal, and lateral directions are established in each country by the appropriate government authority. The current rules described in this text are those prescribed by the FAA for use in the United States. The separation rules are prescribed for IFR operations, and these rules apply whether or not IMC conditions prevail. Minimum separations are a function of aircraft type, aircraft speed, availability of radar facilities, navigational aids, and other factors such as the severity of wake vortices [3].

Vertical separation in the airspace

The minimum vertical separation of aircraft outside the terminal area from the ground up to and including 29,000 ft above mean sea level (AMSL) is 1000 ft. Higher than 29,000 ft AMSL the minimum separation is 2000 ft. Within a terminal area a vertical separation of 500 ft is maintained between aircraft, except that a 1000-ft vertical separation is maintained below a heavy aircraft.

Use of VFR altitudes

VFR altitudes below 18,000 ft AMSL are designated as the odd 1000-ft altitudes plus 500 ft beginning at 3500 ft AMSL for course headings from 0° to 179° magnetic azimuth and the even 1000-ft altitudes plus 500 ft beginning at 4500 ft AMSL for course headings from 180° to 359° magnetic azimuth. The authorized VFR altitudes between 18,000 ft AMSL and 29,000 ft (flight level 290, or FL 290) are designated as the odd 1000-ft altitudes plus 500 ft beginning at FL 195 for course headings from 0° to 179° magnetic azimuth and the even 1000-ft altitudes plus 500 ft beginning at FL 185 for course headings from 180° to 359° magnetic azimuth. The authorized VFR altitudes above FL 290 are designated as the even 1000-ft altitudes at 4,000-ft intervals beginning at FL 300 for course headings from 0° to 179° magnetic azimuth and the even 1000-ft altitudes at 4000-ft intervals beginning at FL 320 from 180° to 359° magnetic azimuth.

These altitudes are in effect under VFR at altitudes of 3000 ft above the surface in both controlled and uncontrolled airspace.

Assigned IFR altitudes

The assigned IFR altitudes below 18,000 ft AMSL are designated as the odd 1000-ft altitudes for course headings from 0° to 179° magnetic azimuth and the even 1000-ft altitudes for course headings from 180° to 359° magnetic azimuth. The assigned IFR altitudes from 18,000 ft

AMSL up to but not including 29,000 ft (flight level 290, or FL 290) are designated as the odd 1000-ft altitudes for course headings from 0° to 179° magnetic azimuth and the even 1000-ft altitudes plus 500 ft for course headings from 180° to 359° magnetic azimuth. The assigned IFR altitudes at and above FL 290 are designated as the odd 1000-ft altitudes at 4000-ft intervals beginning at FL 290 for course headings from 0° to 179° magnetic azimuth and the odd 1000-ft altitudes at 4000-ft intervals beginning at FL 310 from 180° to 359° magnetic azimuth.

Longitudinal separation in the airspace

The minimum longitudinal separation depends on a number of factors; among the most important are aircraft size, aircraft speed, and availability of radar for the control of air traffic. For the purposes of maintaining aircraft separations, aircraft are classified by the FAA as heavy, large, or small based upon their maximum certified takeoff weight. Heavy aircraft have a maximum certificated takeoff weight of 300,000 lb or more. Large aircraft have a maximum certificated takeoff weight in excess of 12,500 lb but less than 300,000 lb. Small aircraft have a maximum certificated takeoff weight of 12,500 lb or less. Aircraft size is related to wake turbulence. Heavy aircraft create trailing wake vortices which are a hazard to lighter aircraft following them.

The minimum longitudinal separations en route are expressed in terms of time or distance as follows:

1. For en route aircraft following a preceding en route aircraft, if the leading aircraft maintains a speed at least 44 kn faster than the trailing aircraft, 5 mi between aircraft using distance-measuring equipment (DME) or area navigation (RNAV) and 3 min between all other aircraft

2. For en route aircraft following a preceding en route aircraft, if the leading aircraft maintains a speed at least 22 kn faster than the trailing aircraft, 10 mi between aircraft using DME or RNAV and 5 min for all other aircraft

3. For en route aircraft following a preceding en route aircraft, if both aircraft are at the same speed, 20 mi between aircraft using DME or RNAV and 10 min for all other aircraft

4. When an aircraft is climbing or descending through the altitude of another aircraft, 10 mi for aircraft using DME or RNAV if the descending aircraft is leading or the climbing aircraft is following and 5 min for all other aircraft

5. Between aircraft in which one aircraft is using DME or RNAV and the other is not, 30 mi

The minimum longitudinal separation over the oceans is normally 10 min for supersonic flights and 15 min for subsonic flights, but in some locations it can be slightly more or less than these values [3].

When the aircraft mix is such that wake turbulence is not a factor and radar coverage is available, the minimum longitudinal separation for two aircraft traveling in the same direction and at the same altitude is 5 nmi, except that when the aircraft are in the terminal environment within 40 nmi of the radar antenna, the separation can be reduced to 3 nmi. For this reason the minimum spacing in the terminal area is 3 nmi because the airport is almost always within 40 nmi of a radar antenna. Under certain specified conditions, a separation between aircraft on final approach within 10 nmi of the landing runway may be reduced to 2.5 nmi [3].

If wake turbulence is a factor, the minimum separation in the terminal area between a small or large aircraft and a preceding heavy aircraft is 5 nmi. The spacing between two heavy aircraft following each other is 4 nmi. The spacing between a heavy aircraft and a preceding large aircraft is 3 nmi.

For landing aircraft, when wake turbulence is a factor, the longitudinal separation is increased between a small aircraft and a preceding large aircraft to 4 mi and between a small aircraft and a preceding heavy aircraft to 6 mi.

The instrument flight separation rules for consecutive arrivals on the same runway which are used when wake vortices are a factor are shown in Table 4-1. This table also displays the observed average spacings maintained by pilots in VFR at Chicago O'Hare International Airport. Note that the VFR separations in Table 4-1 are not maintained by air traffic control but are the average observed separations maintained by pilots and, as such, are not valid measures upon which runway or airspace capacity studies can be based. They are simply presented for comparative purposes.

TABLE 4-1   Horizontal Separation in Landing for Arrival—Arrival Spacing of Aircraft on Same Runway Approaches in VFR and IFR Conditions, nmi

Leading aircraft type	VFR*			IFR (wake vortex)		
	Trailing aircraft type			Trailing aircraft type		
	Heavy	Large	Small	Heavy	Large	Small
Heavy	2.7	3.6	4.5	4.0	5.0	6.0
Large	1.9	1.9	2.7	3.0	3.0	4.0
Small	1.9	1.9	1.9	3.0	3.0	3.0

*These values are shown to appropriately represent these operations and are not regulatory in nature.

SOURCE: Federal Aviation Administration [18].

TABLE 4-2   Separation for Same-Runway Consecutive Departures in VFR and IFR Conditions, s

Leading aircraft type	VFR*			IFR		
	Trailing aircraft type			Trailing aircraft type		
	Heavy	Large	Small	Heavy	Large	Small
Heavy	90	120	120	120	120	120
Large	60	60	50	60	60	60
Small	50	45	35	60	60	60

*These values are shown to appropriately represent these operations and are not regulatory in nature.

SOURCE: Federal Aviation Administration [18].

The visual and instrument flight separation rules for consecutive departures from the same runway are expressed in terms of time and are shown in Table 4-2.

Lateral separation in the airspace

The minimum en route lateral separation below 18,000 ft AMSL is 8 nmi, and at and above 18,000 ft AMSL the minimum en route lateral separation is 20 nmi. Over the oceans the separation varies from 60 to 120 nmi depending on the location [3].

General considerations

The longitudinal-separation standards significantly influence the capacity of the airspace and the airport runways since separations reflect the size of headways between aircraft. The significant influence of radar in reducing headways can be illustrated as follows. With radar the minimum en route separation is 5 nmi for aircraft equipped with DME or RNAV. If radar is not available, the separation must be increased to 20 mi for aircraft equipped with distance-measuring or area navigation equipment and to about 30 mi if none of this equipment is installed in the aircraft. Over the oceans the longitudinal separation varies from 60 to 120 nmi. These large separations reduce capacity of the airspace and increase delays, and for this reason efforts are being made to reduce aircraft spacing. Figure 4-5 illustrates the impact of a variety of minimum-spacing rules for arriving aircraft on runway capacity.

Navigational Aids

Aids to aerial navigation can be broadly classified into two groups: those that are located on the ground, or *external* aids, and those

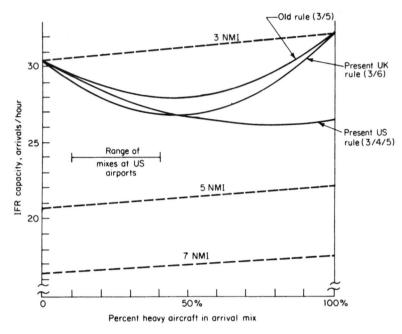

Figure 4-5 Effect of various wake vortex separation rules on IFR arrival capacity of a typical runway (*MITRE Corporation [14]*).

installed in the cockpit, or *internal* aids. Some aids are primarily for flying over the oceans, other aids are only applicable to flight over land masses, and still other aids can be used over either land or water. Some aids are used only during the en route portion of the flight, while other aids are necessary in terminal areas or near airports.

External overland en route aids

Very high-frequency omnirange radio. The advances in radio and electronics during and after World War II led to the installation of the very high-frequency omnirange radio (VOR) stations. These stations are located on the ground, and they send out radio signals in all directions. Each signal can be considered as a course or a route, referred to as a *radial,* that can be followed by an aircraft. In terms of 1° intervals, there are 360 courses or routes that are radiated from a VOR station, from 0° pointing toward magnetic north increasing to 359° in a clockwise direction. The VOR transmitter station is a small square building topped with what appears to be a white derby hat. It broadcasts on a frequency just above that of FM radio stations. The very high frequencies it uses are virtually free of static. The system of VOR stations establishes the network of airways and jet routes and is essential to area navigation. The range of a VOR station varies but is usually less than 200 nmi.

Aircraft equipped with a VOR receiver in the cockpit have a dial for tuning in the desired VOR frequency. A pilot can select the VOR radial or route to follow to the VOR station. In the cockpit a position deviation indicator (PDI) specifies the heading of the aircraft relative to the direction of the desired radial and whether the aircraft is to the right or left of the radial. Figure 4-6 shows schematically the type of information the PDI provides. At *A* the aircraft is on the selected radial, and the needle is pointed vertically and passes through the cross, which is a symbol for the aircraft. In other words, the aircraft is heading in the same direction as the desired radial. At *B* the aircraft is flying parallel to but to the right of the desired radial. At *C* the aircraft is to the right of the radial and is heading across the radial.

Distance-measuring equipment.   Distance-measuring equipment (DME) has been installed at nearly all VOR stations in the United States. Those so equipped are called VORTAC facilities. The DME shows to the pilot the slant distance between the aircraft and a particular VOR station. Since it is the air distance in nautical miles that is measured, the receiving equipment in an aircraft flying at 35,000 ft directly over the DME station will read 5.8 nmi.

An en route air navigation aid which best suited the tactical needs of the military was developed by the Navy in the early 1950s. This aid is known as TACAN, which stands for *tactical air navigation*. This aid combines azimuth and distance measuring into one unit instead of two and is operated in the ultra-high-frequency band. As a compro-

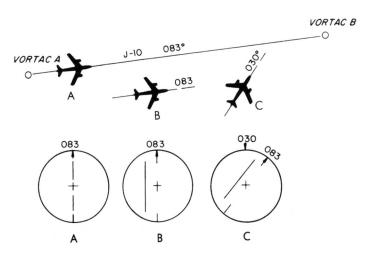

Figure 4-6   Illustration of position deviation indicator for aircraft utilizing a VOR radial.

mise between civilian and military requirements, the FAA replaced the DME portion of its VOR facilities with the distance-measuring components of TACAN. These stations are known as VOR-DMET. If a station has full TACAN equipment, both azimuth- and distance-measuring equipment, and VOR, it is designated as VORTAC.

Air route surveillance radar.   Long-range radar for tracking en route aircraft has been established throughout the continental United States and in other parts of the world. While in the United States there is complete radar coverage in the 48 contiguous states, this is not the case elsewhere in the world. These radars have a range of about 250 nmi. Strictly speaking, radar is not an aid to navigation. Its principal function is to provide air traffic controllers with a visual display of the position of each aircraft so they can monitor the spacings and intervene when necessary. However, radar can be and is used by air traffic controllers to guide aircraft whenever necessary. For this reason it has been included as an aid to navigation.

External overland terminal aids

The principal aids in the terminal area are used for landing aircraft, and these are described below.

Instrument landing system.   The most widely used method is the *instrument landing system* (ILS). It consists of two radio transmitters located at the airport. One radio beam is called the localizer, and the other is the glide slope. The localizer indicates to pilots whether they are left or right of the correct alignment for approach to the runway. The glide slope indicates the correct angle of descent to the runway. Glide slopes measure on the order of 2° to 3° minimum to 7.5° maximum.

To further help pilots on their ILS approach, two low-power fan markers, called ILS markers, are usually installed so that pilots will know just how far along the approach to the runway they have progressed. The first is called the *outer marker* (LOM) and is located about 3.5 to 5 mi from the end of the runway. The other is called the *middle marker* (MM), and it is located about 3000 ft from the end of the runway. For category II operations, when visibility is quite poor, an additional marker called the *inner marker* (IM) is located 1000 ft from the end of the runway. This inner marker is placed so as to alert pilots that they must have visual reference with the ground at that point and if they do not, abandon the approach. When the plane passes over a marker, a light goes on in the cockpit and a high-pitched tone sounds. The configuration of the ILS is shown in Fig. 4-7. At many locations the fan markers have been replaced with DME using the airport VORTAC or TACAN at the airport.

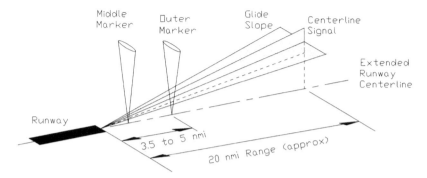

Figure 4-7 Schematic diagram of an instrument landing system.

The localizer consists of an antenna, which is located on the extension of the runway centerline approximately 1000 ft from the far end of the runway, and a localizer transmitter building located about 300 ft to one side of the runway at the same distance from the end of the runway as the antenna. The glide slope facility is placed 750 to 1250 ft down the runway from the threshold and is located to one side of the runway centerline at a distance which can vary from 400 to 650 ft. The functioning of the localizer and the glide slope facility is affected by the close proximity of moving objects such as vehicular and aircraft traffic. During inclement weather, the use of the ILS critical areas keeps aircraft and vehicles from entering areas that would impede an aircraft inside of the outer marker from receiving a clear signal. Stationary objects nearby can also cause a deterioration of the signals. Abrupt changes of slope in proximity to the antennas are not permitted, or else the signal will not be transmitted properly. Another limitation of the ILS is that the glide slope beam is not reliable below a height of about 200 ft above the runway.

Microwave landing system.   The ILS has a number of problems which have made the development of more sophisticated landing systems necessary. The ILS is based on signals reflecting from the surface of the ground. Thus the area adjacent to the antennas must be relatively smooth and must be kept clear of any obstructions such as buildings and taxiing aircraft; otherwise the beams are distorted. There have been improvements in the transmission of the localizer beam brought about by the installation of a waveguide antenna, which confines the beam spray and reduces the probability of reflections from buildings and other obstructions. But this has not solved all the problems associated with the ILS. The ILS provides only one path in space, which all aircraft must follow if they are using the system. Some aircraft, particularly the STOL (short takeoff and landing) type, can use a steeper approach angle, about 7°, than conventional aircraft, which use 2.5° to 3° approaches. Other aircraft may wish to make a two-segment approach to reduce noise beneath the flight path. The ILS is unable to provide for these types of operations. Finally, only a limited number of frequency channels are available for the ILS, and as the number of installations has increased, it is becoming difficult to provide the necessary discrete channels required.

To overcome these limitations, the *microwave landing system* (MLS) was developed. Instead of providing only one glide slope as the ILS does, the MLS provides for a number of slopes. In the horizontal plane, the MLS provides for any desired routes as long as they are within an area that is from 20° to 60° on each side of the runway centerline, whereas the ILS provides only one route to the runway. Distance-measuring capability can be incorporated into the MLS, providing the pilot with continuous information on the aircraft distance from the end of the runway and removing the need for establishing markers as in the ILS. The MLS is far less susceptible to interference from surrounding objects than the ILS. With the MLS a pilot can choose any desired route to the runway at any glide slope within the vertical coverage of the system. A microwave landing system is shown schematically in Fig. 4-8.

From the standpoint of airport planning, one of the most significant advantages of the MLS is the potential reduction of noise since air-

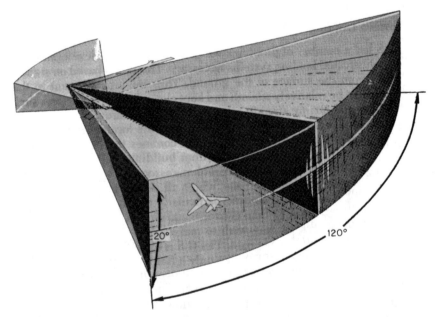

Figure 4-8 Schematic diagram of a microwave landing system. (*Federal Aviation Administration [5]*).

craft can be kept at higher altitudes before they make the descent to the airport or follow curved routes which do not affect as much land as the ILS routes do. The difference between an ILS and an MLS approach into Kennedy (JFK) and LaGuardia (LGA) airports in the New York area is shown in Fig. 4-9. Another advantage is the elimination of the requirement that all aircraft, large or small, follow a common approach route to the runway.

Precision-approach radar.   At a number of military airports, another landing aid known as *ground-controlled approach* (GCA) has been installed. The GCA operates either with the airport surveillance radar alone or with both the airport surveillance radar and *precision-approach radar* (PAR). The latter equipment was developed by the military during World War II in order to provide a mobile unit that is not dependent on airborne navigation equipment. The precision-approach radarscope gives controllers a picture of the descending aircraft in both plan and elevation; i.e., one-half the radarscope is in plan, and one-half is in elevation. Thus controllers can determine whether an aircraft is on the glide path and whether it is on the correct alignment. Instructions from controllers to pilots are given by voice communication, and thus no airborne navigation equipment is necessary. Commercial airline pilots use the ILS almost exclusively, because using PAR places too much dependence on the controller and does not provide any direct information to the pilot. At airports where there are both ILS and PAR facilities, commercial airline pilots use ILS but often request that they be monitored by PAR.

Airport surveillance radar.   To provide the TRACON and control tower operators with an overall picture of what is going on within the airspace surrounding the terminal, *airport surveillance radar* (ASR) has been installed at many of the major U.S. airports. The ASR rotates through 360°, and the information is received by an ARTS (automated radar terminal system) type of computer system in the TRACON and relayed to a bright radar tower equipment (BRITE) radarscope in the control tower. The primary range of ASR is from 30 to 60 mi. It shows the aircraft in their relative horizontal positions on the radarscope as blips. With nonmosaic radars, the blips of moving aircraft leave a luminous trail and indicate the direction in which the aircraft are moving. ARTS can also determine and show an indication of the aircraft speed. ASR does not indicate the altitude of aircraft since it simply responds to the reflection of the signal from the skin of the aircraft. This type of return radar is called *primary radar* or *skin paint*.

Approach lighting systems.   The most critical point of approach to landing comes when the aircraft breaks through the overcast and the

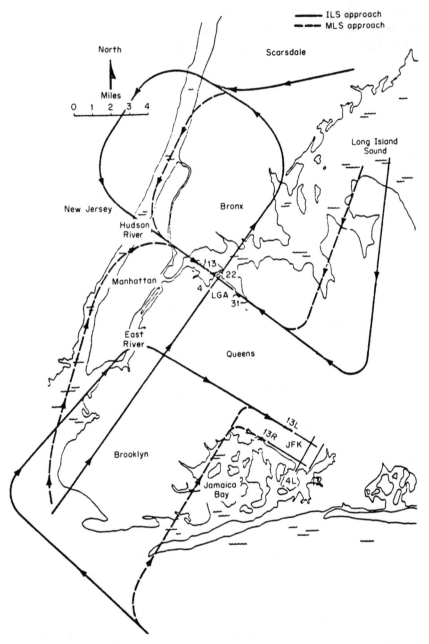

Figure 4-9 Differences between an ILS and an MLS approach to LaGuardia (LGA) and Kennedy (JFK) airports in the New York metropolitan area (*Federal Aviation Administration [5]*).

pilot must change from instrument to visual conditions. Sometimes, only a few seconds are available for the pilot to make the transition and complete the landing. To aid in making this transition, lights are installed on the approach to the runways and on the runways themselves. These are generally termed *approach lighting systems* (ALS). A number of types and configurations are used, and others are under experimental testing today. More details concerning these systems are contained in Chap. 13 on signing, marking, and lighting.

Airport surface detection.   At large high-density airports, controllers have difficulty in regulating taxiing aircraft because they cannot see the aircraft in poor-visibility conditions. A specially designed radar, called *airport surface detection equipment* (ASDE), often referred to as *ground radar,* has been developed to aid the controller. The system gives the air traffic controller in the control tower a pictorial display of the runways, taxiways, and terminal area, with radar indicating the positions of aircraft and other vehicles moving on the surface of the airport.

Visual-approach slope indicators.   These systems (VASI and PAPI) provide, through a system of lights, the proper approach slope to the runway much the same as the glide slope of an ILS system. VASI systems are intended for day or night use during good (VFR) weather conditions, and they cannot be used under very poor-visibility conditions. A refined version of the visual approach slope indicator—the *precision approach path indicator* (PAPI) system—is presently being installed at airports in the United States. The PAPI system gives a more definitive indication of approach slope to the pilot and uses only a single set of electronic devices at one point down the runway. A more detailed explanation of the VASI and PAPI systems is contained in Chap. 13 on signing, marking, and lighting.

Runway end identifier lights.   Runway end identifier lights (REIL) are installed to give the pilot positive visual identification of the approach end of the runway when there are no approach lights.

The relative location of terminal-area navigational aids is shown in Fig. 4-10. The siting requirements for visual aids and the ILS antennas can be found in the References [1].

External overwater en route aids

The principal overwater aid to navigation is LORAN, which consists of stations located on the ground. LORAN stands for *long-range aerial navigation*. The system was developed during World War II. LORAN stations are located in all parts of the world. The particular system in use today is designated LORAN-C. The principle of the

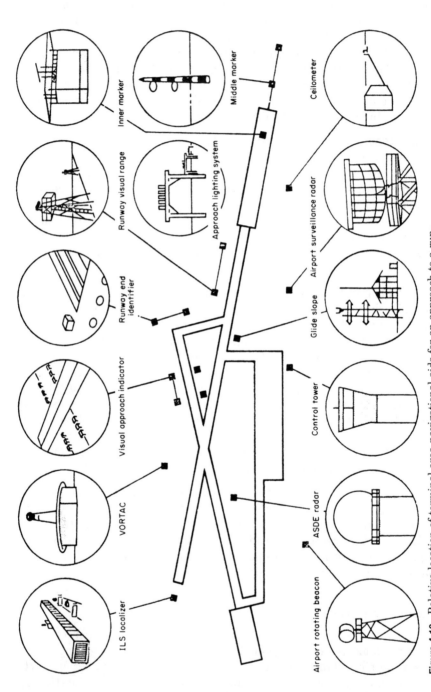

Figure 4-10 Relative location of terminal-area navigational aids for an approach to a runway system (*Federal Aviation Administration [1]*).

LORAN system is as follows. Each element consists of a master station and a slave station located some distance from the master. The master station sends radio signals into space, and at the same time one of the signals goes to the slave station, where it is delayed a specified amount of time and then sent into space. At any point in space, there is a difference in time between the original signal from the master and its intersection with the delayed signal from the slave. Thus a contour of equal time differences can be drawn in space. The same thing can be done from another master and slave station, resulting in another contour. The intersection of the two contours establishes a position in space. In the aircraft the LORAN receiver tunes in on two master and slave stations, establishing an intersection of two time-difference contours in space. The range of LORAN is affected by the time of day, being greater at night than during daylight. LORAN requires the use of a navigator in the cockpit.

Internal overwater en route aids

There are two principal aids used in overwater operation: the doppler navigation system and the *inertial navigation system* (INS). A third system which derives from use on ships at sea is celestial navigation. It was quite popular prior to the development of the doppler and inertial navigation systems. The advantages of these later systems are related to the economics of aircraft operation: They do not require the use of a navigator.

Doppler navigation system. This is a long-range-radar type of aid that provides the pilot with the ground speed, angle of the aircraft axis relative to the desired course (drift angle), distance of the aircraft right or left of the desired course, and distance to the destination or way point. Suppose an aircraft is to fly from point A to point B via a great circle route of length L. The length L is usually divided into several shorter lengths or segments. The ends of these segments are established in space by way points. A way point is an imaginary point in space. Inputs into the system are the latitudes and longitudes of points A and B and of all way points along the route. The number of way points depends on the length of the trip.

The doppler system is based on the following: The aircraft sends to the ground four beams of continuous wave energy (8800 Mc), two forward and two toward the rear. The change in frequency of the energy return from the ground is measured. This change in frequency is known as the *doppler shift* and is proportional to the aircraft speed in the directions the beam is pointing. By checking the speed in the four directions in which the beams are pointing, the system derives the ground speed and drift angle. The smoother the surface, the less chance there is for the radiated energy to reflect to the aircraft's antenna. This is a limitation of the system and is encountered over smooth bodies of water.

Inertial navigation system. The inertial navigation system is by far the most popular overwater long-range aid. It provides all the information that the doppler system provides as well as the wind speed and direction, latitude and longitude of the aircraft at any instant, and time to reach the next way point. As for the doppler system, the inputs are the latitudes and longitudes of the origin, destination, and way points. The inertial guidance system is a development of the space program. It is quite accurate and reliable. Both the inertial and doppler navigation systems provide azimuth information referenced to true north, not magnetic north.

Internal overland en route aids

Both the doppler and the inertial navigation systems can be used over land masses. There are also area navigation systems (RNAV) which can be used only over land. They use as inputs the distance and azimuth information provided by VORTAC stations. The desired course is referenced to way points, as for the other systems. The way points are established by distances and azimuth from the nearest adjacent VORTAC station. Information provided in the cockpit is similar to that for the doppler and inertial navigation systems.

Internal overland terminal aids

Area navigation systems used for en route navigation can also be used in the terminal area. In addition to guidance in the horizontal plane, these systems provide guidance in the vertical plane. This latter capability is particularly useful for guidance to runways.

Global positioning system

The *global positioning system* (GPS) is a space-based satellite radio positioning and navigation system. The system is designed to provide highly accurate position and velocity information on a continuous global basis to an unlimited number of properly equipped users. The system is unaffected by weather and provides a common worldwide grid reference system. The GPS concept is predicated upon accurate and continuous knowledge of the spatial position of each satellite in the system with respect to time and distance from the transmitting satellite to the user. It is expected that the full GPS will consist of 24 satellites in near-circular orbit about the earth. The GPS receiver automatically selects the appropriate signals from the satellite which is in view of the receiver and translates these signals to a three-dimensional position, velocity and time. It is expected that the GPS will have a horizontal accuracy on the order of 100 m. The global positioning system offers considerable assistance in navigation by providing precise position information to aircraft, and therefore it is likely to become an external navigational aid in the near future.

A variation of the system, called the *relative global positioning system,* has the potential to assist aircraft in conducting precision instrument approaches to airports. Relative GPS is based on the concept that the knowledge of the location of the GPS receiver in an aircraft relative to the GPS receiver at an airport is more accurate than the knowledge of the absolute position of each receiver. An aircraft on approach would receive GPS satellite position signals, and the beacon radar at the airport would relay its own position at the airport. By comparing the beacon radar position with the position of the aircraft, the aircraft can compute a vector to the beacon radar site. Knowledge of the range and bearing from the beacon radar site to the point of touchdown allows the aircraft system to compute a vector to touchdown. It is expected that the relative GPS will reach an accuracy on the order of 2 m, which would make it useful in precision instrument approaches to runways [20]. The FAA is currently conducting tests on another variation of the system, the differential global positioning system, for use in precision instrument approaches to runways.

The global positioning system has the capability of providing external overland and overwater precise navigational assistance to aircraft as well as assistance to aircraft in conducting terminal-area precision approaches to runways.

Aids for the Control of Air Traffic

The principal aids for the control of air traffic are voice communication and radar. The controller monitors the spacing between aircraft on the radarscope and instructs pilots by voice communication. There are two types of radar: primary and secondary. The primary radar returns appear on the radarscope as small blips. These are reflections from the aircraft body. The primary radar as it appears on a radarscope is shown in Fig. 4-11. Primary radar requires the installation of rotating antennas on the ground, and the range of the primary radar is a function of its frequency. Beacon radar, sometimes referred to as secondary radar, consists of a radar receiver and transmitter on the ground that transmit a coded signal to an aircraft if that aircraft has a transponder. A transponder is an airborne receiver and transmitter which receives the signal from the ground and responds by returning a coded reply to the interrogator on the ground. The coded reply normally contains information about the aircraft's identity and

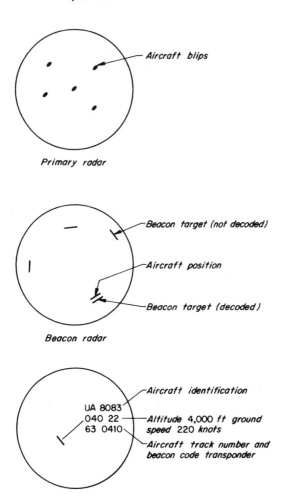

Figure 4-11 Various displays on radarscope used for air traffic control.

altitude. The interrogator (receiver and transmitter) is the beacon radar antenna. It is usually installed as an integral part of the primary radar antenna. Beacon radar returns are presented on the radarscope as two slashes if they are decoded and a single slash if they are not decoded. Prior to the development of ARTS, which provides alphanumeric information about the flight in a data block, controllers would decode only those aircraft that they were controlling. The slashes always appear at a right angle to the radial from the location of the antenna to the aircraft, as shown in Fig. 4-11. The center of the slash closest to the antenna is the location of the aircraft. Prior to the installation of ARTS, the primary and beacon radar presentations did not provide identity of the aircraft or its altitude. This was obtained by voice communication and was recorded on flight progress strips.

To overcome the deficiencies of the original beacon radar presentation, and to reduce the amount of communication, a video presentation which includes identity and altitude was developed. This is shown in Fig. 4-11 and is referred to as the alphanumeric display. The first line shows the identity of the aircraft; the second line shows its altitude and ground speed; and the third line gives the beacon code transponder number and the aircraft track number. To be able to have this information presented on the radarscope, the aircraft must carry a mode C transponder that has the capability of reporting altitude along with aircraft identity. All commercial airline aircraft carry a mode C transponder, which satisfies the requirement for reporting altitude. One problem with the original beacon radar system was its inability to selectively interrogate aircraft. A modified beacon radar system has been incorporated into ATC to alleviate this difficulty in congested areas through the use of a *discrete-address beacon system* (DABS) using a mode S transponder [15].

If all aircraft, including general aviation, were equipped with transponders, there would be no need for primary radar except possibly in a backup role.

Automation in Terminal and En Route Air Traffic Control

There are a number of reasons for updating and automating the air traffic control system. The more important are as follows [17]:

1. Having an operational system that is capable of being technically expanded in incremental steps to meet the needs of aviation as time requires

2. Accommodating increasing demands in a manner that allows users to operate in the airspace with minimal regulatory constraints and in a fuel-efficient way

3. Reducing the risks of midair and surface traffic collisions, landing and weather-related accidents, and collisions with the ground

4. Increasing the productivity of air traffic control personnel in terms of the amount of air traffic handled

5. Decreasing the technical staff required to maintain and operate the system

6. Maintaining the overall operating costs of the system at reasonable levels

7. Increasing the capacity of the airspace by reducing the spacing between aircraft through improved systems to monitor aircraft with more accurate airborne systems

Increasing the productivity of a controller is a very important objective for the following reasons. One is simply a matter of economics. In 1981 the number of air traffic and airway facility personnel employed by the FAA for the en route and terminal air traffic control system was nearly 26,000. It has been estimated that, with the implementation of automation as indicated in the National Airspace System Plan [17], the number of air traffic and airway facility personnel for the en route and terminal air traffic control system could be reduced to about 16,000 by the year 2000, a personnel reduction of about one-third. Without even a modest amount of automation, the requirements for controllers would quadruple from the 1980 level before the year 2000 [8, 13, 17, 21]. It has been questioned whether it would be possible to maintain such a workforce or even to manage it. The need for more controllers in the present system stems from the need to increase capacity by establishing more control sectors, but as the number of sectors is increased, more coordination between sectors is required. Coordination therefore imposes a severe constraint on the system.

There have been many discussions of the means of increasing controller productivity, but in a broad sense two general areas are being pursued. One automates the routine duties of controllers, such as preparing and updating flight strips, detecting conflicts in flight plans, etc., thus allowing the controller more time to monitor aircraft flow. The other reduces communication between pilots and controllers, again allowing the controller more time to monitor aircraft flow.

Some argue that one way to reduce the requirement for additional controllers is to place more responsibility on pilots. These responsibilities would include following a prescribed route and maintaining a time profile over a route. This would leave the controller as a flight monitor, intervening only when necessary.

The FAA is presently implementing a National Airspace System Plan [17] with the purpose of modernizing the air traffic control system of the United States through the year 2010. It is to be a highly automated system that would replace the existing system which has evolved over time. A discussion of the evolution of the existing system follows.

States of automation

In the United States, the several states of the air traffic control system have been referred to as second-generation, third-generation, etc.

The first-generation system can be considered the state of the system when it first evolved in the early 1940s. The second-generation system which followed was primarily a manual system with regard to control and separation of air traffic, although some functions such as flight data processing were gradually automated. This was largely the system until the late 1960s, when the transition to the third-generation system began. Both the first-and second-generation systems can be considered largely manual. The third-generation system provides for limited automation to assist the controller (flight plan processing, updating flight plans, improved video display, reduction in communication) and automated data acquisition and display, including altitude and position provided by the radar beacon system. In this system communication is still by voice but has been reduced by having alphanumeric displays on the radarscope. Sequencing of aircraft onto the final approach course to a runway is still done manually. It can be considered a system in which computers assist controllers in routine tasks and alert them to potential conflicts, but do not enter into the process of resolving conflicts and recommending solutions.

The beacon radar in the third-generation system lacks measurement accuracy and has limited aircraft capacity and reliability. Furthermore, the system relies heavily on voice communication. There is a limit to the number of voice channels available for air traffic control; consequently the demand could conceivably exceed the capacity of the communication system. An upgrade of the third-generation system was implemented in the 1970s. This upgrade included an improvement in the accuracy and capacity of beacon radars and the incorporation of a ground-to-air and air-to-ground data-link capability. The upgraded third-generation system is fairly automated, providing for flow control [22] and aircraft sequencing and spacing in the en route and terminal density airspace. It was estimated that the upgraded third-generation system had the capacity to handle traffic volumes projected into the early 1990s. Sometime thereafter the system would exhibit significant deficiencies. Consequently studies of alternative concepts for a fourth-generation system began in the early 1970s, referred to as *advanced air traffic management system study,* with a target date of operation around 1990 [15]. The several concepts studied included the use of satellites for surveillance, navigation, and communication for part of or all these functions. Data processing would be accomplished primarily with a highly integrated network of digital computers. Communication between the ground and aircraft would be largely by data link. This has already been implemented at some major airport terminal areas, but only for the issuance of clearances to airline pilots prior to departure.

An automated air traffic control facility

The automated en route system revises the role of controllers as active participants and allows them to discharge their primary responsibility, that of monitoring and controlling aircraft. Information that was manually acquired and processed is now automatically obtained, processed, and presented on the radar display with greater speed and accuracy. Transponder-equipped airplanes provide continuous altitude and identity information, reducing both pilot and communication workload. All information is fed into a computer complex that processes and updates flight data and presents them on a radar display in alphanumeric form with automatic target tracking. Subsequent stages of development provide for the automatic prediction of potential traffic conflicts, suggestions for resolving conflicts, and preplanning of traffic flow. Another method for reducing the communication workload will involve the use of automated ground-to-air and air-to-ground communication (data link). In the terminal area and en route environments, the ARTS (automated radar terminal system) provides for automatic tracking of both primary and beacon radar targets and displays on the radarscope alphanumeric information for each aircraft being tracked, as in the en route system. VORTAC continues to be the primary support of en route navigation, although it will be upgraded where required.

The National Airspace System Plan

The FAA is presently implementing a comprehensive plan for updating and modernizing the air traffic control system of the United States [17]. This plan spells out the specific improvements that will be made to the airspace systems and facilities to meet the projected demands of air traffic. The key elements of the plan include the replacement of the current air traffic control computer systems to permit the introduction of higher levels of automation, modernization of the flight service station network to improve the dissemination of flight and weather data to pilots, and the deployment of new radar, communications, and landing systems, to enhance safety and provide for more efficient traffic flow.

The level and projected growth of national airspace system activity from 1982 to 2000 are shown in Table 4-3. To meet the aviation demand to be imposed upon the system the following specific items are being implemented in the National Airspace System Plan:

1. Modernizing and enhancing of air traffic control equipment and facilities as well as the installation of new high-capacity, high-reliability computer systems known as the *advanced automation system* (AAS) that will replace the existing computer systems, radarscopes,

TABLE 4-3　National Airspace System Plan Activity, 1982–2000

	1982	1990	2000
NPIAS airports	3195	3638	4000
Airport operations (millions)			
Aircraft operations	127.6	185.8	255.3
Itinerant operations	68.0	98.2	136.1
Tower operations	61.6	83.4	105.6
Instrument operations	31.6	46.0	58.0
ARTCC operations			
IFR aircraft handled (millions)	27.8	37.4	47.8
FSS services (millions)	62.4	80.6	101.0
Domestic enplanements (millions)			
Air carrier	272.8	409.0	598.5
Commuter or regional	17.1	33.6	58.6
Aircraft fleet (thousands)			
Air carrier	2.5	2.9	3.4
Commuter	1.9	2.2	2.7
General aviation	213.2	251.3	321.9
Hours flown (millions)			
Air carrier	6.3	7.5	9.7
General aviation	37.3	49.6	65.4
Active pilots (thousands)			
Total	764.2	888.7	1070.7
Instrument-rated	252.5	327.3	422.8

and voice communications. This system will have the capability of automating many air traffic controller functions. The system will have modern controller workstations, called *sector suites,* which will be common to both the en route and terminal air traffic control functions. Sector suites will have multiple and greatly enhanced display monitors capable of displaying a plan view of air traffic and weather situations, alphanumeric flight and weather data, and other aeronautical information such as notices to pilots and traffic planning data.

2. The utilization of the advance automation system in both en route and terminal air traffic control facilities and the blending and integration of these facilities into area control facilities (ACFs) from the present level of over 200 en route and terminal area control facilities to about 20 area control facilities, 10 metroplex control facilities, and some 120 local control facilities by the year 2010. Metroplex control facilities will consist of several TRACONs consolidated into a single location, while the local control facilities are TRACONs that will retain their current identity.

3. The implementation of the automated en route air traffic control system (AERA) that will automatically probe aircraft routes; detect and resolve conflicts with other aircraft, weather, and terrain; and issue clearances to ensure the safe, metered, and efficient flow of traffic. These data can be provided directly to aircraft through a data link.

4. The consolidation and automation of the present system of flight service stations to provide for the rapid retrieval and transmission of flight and weather information to pilots. Pilots can now directly access the computer base by using direct user access terminals (DUAT) that are simply laptop computers.

5. The implementation of a new secondary beacon radar system to selectively address aircraft through the mode S transponder.

6. The expansion of the new secondary radar system to include the ability for automatic air-to-ground and ground-to-air data-link communications.

7. The phasing out of primary radar for en route air traffic control but its retention for terminal-area air traffic control.

8. The development of an enhanced interfacility communications network utilizing state-of-the-art telecommunications technology.

9. The implementation of a network of automated weather observation systems (AWOS) to provide greatly enhanced weather information that can be transferred by data link to pilots.

10. The installation of microwave landing systems as a replacement for instrument landing systems.

11. A modernization of flight inspection and facility maintenance programs.

References

1. *Airport Design,* Advisory Circular AC 150/5300-13, Federal Aviation Administration, Washington, 1989.
2. *Air Route Traffic Control,* Airway Planning Standard no. 2, Order 7031.3, Federal Aviation Administration, Washington, September 1977.
3. *Air Traffic Control Handbook,* Order 7110.65G, Federal Aviation Administration, Washington, March 1992.
4. *Air Traffic Management Plan (ATMP) Program, Development and Control Procedures,* Order 7000.3, Federal Aviation Administration, Washington, February 1988.
5. *An Analysis of the Requirements for, and the Benefits and Costs of, the National Microwave Landing System (MLS),* Office of Systems Engineering Management, Rep. FAA-EM-80-7, Federal Aviation Administration, Washington, June 1980.
6. *Aviation System Capacity Plan 1991-92,* Rep. DOT/FAA/ASC-91-1, Department of Transportation, Federal Aviation Administration, Washington, 1991.
7. *Chicago Delay Task Force Technical Report,* vol. 1: *Chicago Airport/Airspace Operating Environment,* Landrum and Brown, Inc., Chicago, April 1991.
8. *Civil Aviation Research and Development Policy Study,* Department of Transportation and National Aeronautics and Space Administration, Washington, March 1971.
9. *Designation of Federal Airways, Area Low Routes, Controlled Airspace, Reporting Points, Jet Routes and Area High Routes,* Part 71, Federal Aviation Regulations, Federal Aviation Administration, Washington, February 1992.
10. *Enroute High Altitude—U.S.,* Flight Information Publication, National Ocean Survey, National Oceanic and Atmospheric Administration, Department of Commerce, Washington, December 1975.
11. *Establishment of Jet Routes and Area High Routes,* Part 75, Federal Aviation Regulations, Federal Aviation Administration, Washington, December 1991.

12. *FAA Long-Range Aviation Projections, Fiscal Years 2004–2015,* Rep. FAA-APO-92-4, Office of Aviation Policy, Plans, and Management Analysis, Federal Aviation Administration, Washington, May 1992.
13. *FAA Report on Airport Capacity,* The MITRE Corporation, Rep. FAA-EM-74-5, Federal Aviation Administration, Washington, 1974.
14. "Future ATC Technology Improvements and the Impact on Airport Capacity," R. M. Harris, The MITRE Corporation, Air Transportation Systems Division, McLean, Va.
15. "Future System Concepts for Air Traffic Management," W. E. Simpson, Office of Systems Engineering, Department of Transportation, Presented at 19th Technical Conference of the International Air Transportation Association, Washington, October 1972.
16. *General Operating and Flight Rules,* Part 91, Federal Aviation Regulations, Federal Aviation Administration, Washington, February 1992.
17. *National Airspace System Plan: Facilities, Equipment and Associated Development,* Federal Aviation Administration, Washington, 1989.
18. *Parameters of Future ATC Systems Relating to Airport Capacity and Delay,* Rep. FAA-EM-78-8A, Federal Aviation Administration, Washington, 1978.
19. *Planning the Metropolitan Airport System,* Advisory Circular AC 150/5070-5, Federal Aviation Administration, Washington, May 1970.
20. "Relative Navigation Offers Alternatives to Differential GPS," *Aviation Week and Space Technology,* vol. 137, no. 22, November 1992.
21. *Report of Department of Transportation Air Traffic Control Advisory Committee,* Department of Transportation, Washington, December 1969.
22. *Summary Report 1972,* National Aviation System Planning Review Conference, Federal Aviation Administration, Washington.
23. "The Advanced Air Traffic Management System Study," R. L. Maxwell, Office of Systems Engineering, Department of Transportation, Presented at the 19th Technical Conference of the International Air Transportation Association, Washington, October 1972.
24. *Traffic Management System (TMS) Air Traffic Operation Requirements,* Order 7032.9, Federal Aviation Administration, Washington, September 1992.
25. *United States Aeronautical Information Publication,* 12th ed., Federal Aviation Administration, Washington, October 1992.
26. *United States Standard for Terminal Instrument Procedures (TERPS),* Order 8260.3B, Federal Aviation Administration, Washington, May 1992.

5

Airport Planning

The planning of an airport is such a complex process that the analysis of one activity without regard to its effect on other activities will not provide acceptable solutions. An airport encompasses a wide range of activities which have different and often conflicting requirements. Yet they are interdependent so that a single activity may limit the capacity of the entire complex. In the past, airport master plans were developed on the basis of local aviation needs. In more recent times, these plans have been integrated into an airport system plan which assessed not only the needs at a specific airport site but also the overall needs of the system of airports which service an area, region, state, or country. If future airport planning efforts are to be successful, they must be founded on guidelines established on the basis of comprehensive airport system and master plans.

The elements of a large airport are shown in Fig. 5-1. It is divided into two major components, the airside and the landside. The aircraft gates at the terminal buildings form the division between the two components. Within the system, the characteristics of the vehicles, both ground and air, have a large influence on planning. The passenger and shipper of goods are interested primarily in the overall door-to-door travel time, not just the duration of the air journey. For this reason, access to airports is an essential consideration in planning.

The problems resulting from the incorporation of airport operations into the web of metropolitan life are complex. In the early days of air transport, airports were located at a distance from the city, where inexpensive land and a limited number of obstructions permitted maximum flexibility in airport operations. Because of the nature of aircraft and the infrequency of flight, noise was not a problem to the community. In many cases this audible evidence of the arrival and departure of passenger and cargo planes was often a source of local pride. In addition, low population density in the vicinity of the airport

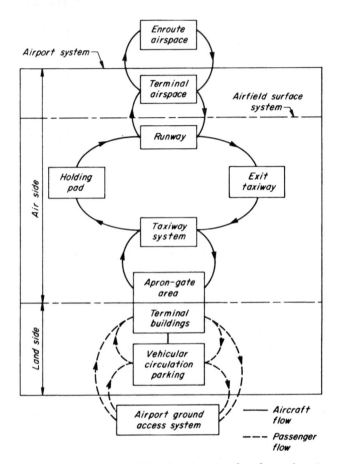

Figure 5-1 Components of the airport system for a large airport.

and light air traffic prevented occasional accidents from alarming the community. In spite of early lawsuits, the relationship between airport and community was relatively free of strife resulting from problems of nuisance or hazard.

Airport operations have been increasingly hampered by obstructions resulting from industrial development related to the airport, industry attracted by adjacent inexpensive land, and access to the transportation afforded by the airfield and its associated highways. While increasingly dense residential development has resulted from this economic stimulation, one must not overlook the effects of the unprecedented suburban spread during the post–World War II era, resulting from the backlog of housing needs and a period of economic prosperity.

Radical developments in the nature of air transport have created new problems. The phenomenal growth of air traffic has increased the probability of unfavorable community reaction, but developments in the aircraft themselves have had the most profound effect on airport-community relations. The greater size and speed of aircraft has resulted in increases in approach and runway requirements, while increases in the output of power plants have brought almost unavoidable increases in noise. Faced with these problems, the airport must cope with the difficulties of securing sufficient airspace for access to the airport, sufficient land for ground operations, and, at the same time, adequate access to the metropolitan area.

Types of Airport Planning Studies

Many different types of studies are performed in aviation and airport planning. These include studies related to facility planning, financial planning, traffic and markets, economics, and the environment. However, each of these studies can usually be classified as being performed at one of three levels: the system planning level, the master planning level, or the project planning level.

The airport system plan

An airport system plan is a representation of the aviation facilities required to meet the immediate and future needs of a metropolitan area, region, state, or country. The National Plan of Integrated Airport Systems (NPIAS) [35] is a system plan representing the airport development needs of the United States, the Michigan Aviation System Plan [33] is an example of a system plan representing the airport development needs of Michigan, and the Southeast Michigan Regional Aviation System Plan [46] is a system plan representing the airport development needs of a seven-county region comprising the Detroit metropolitan area.

The system plan presents the recommendations for the general location and characteristics of new airports and heliports and the nature of expansion for existing ones to meet forecasts of aggregate demand. The system plan identifies the aviation role of existing and recommended new airports and facilities. It includes the timing and estimated costs of development, and it relates airport system planning to the policy and objectives of the relevant jurisdiction. Its overall purpose is to determine the extent, type, nature, location, and timing of airport development needed to establish a viable, balanced, and integrated system of airports [14, 41, 42]. It also provides the basis for definitive and detailed airport planning such as that contained in the airport master plan.

The airport system plan provides both broad and specific policies, plans, and programs required to establish a viable and integrated system of airports to meet the needs of the region. The objectives of the system plan include

1. The orderly and timely development of a system of airports adequate to meet present and future aviation needs and to promote the desired pattern of regional growth relative to industrial, employment, social, environmental, and recreational goals

2. The development of aviation to meet its role in a balanced and multimodal transportation system and to foster the overall goals of the area as reflected in the transportation system plan and comprehensive development plan

3. The protection and enhancement of the environment through the location and expansion of aviation facilities in a manner which avoids ecological and environmental impairment

4. The provision of the framework within which specific airport programs may be developed consistent with the short- and long-range airport system requirements

5. The implementation of land-use and airspace plans which optimize these resources in an often constrained environment

6. The development of long-range fiscal plans and the establishment of priorities for airport financing within the government budgeting process

7. The establishment of the mechanism for the implementation of the system plan through the normal political framework, including the necessary coordination between governmental agencies, involvement of both public and private aviation and nonaviation interests, and compatibility with the content, standards, and criteria of existing legislation

A flowchart identifying a typical airport system planning process is shown in Fig. 5-2.

The airport master plan

An airport master plan is a concept of the ultimate development of a specific airport. The term *development* includes the entire airport area, for both aviation and nonaviation uses, and the use of land adjacent to the airport [12, 14]. The plan presents the development concept graphically and contains the data and rationale upon which the plan is based. Master plans and master plan updates are prepared to support modernization of existing airports and the creation of new airports.

The overall objective of the airport master plan is to provide guidelines for future development which will satisfy aviation demand in a financially feasible manner and will be compatible with the environ-

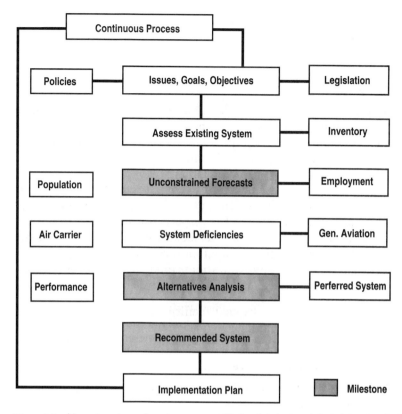

Figure 5-2 Airport system planning process (*Federal Aviation Administration [42]*).

ment, community development, and other modes of transportation. More specifically it is a guide for:

1. Developing the physical facilities of an airport
2. Developing land on and adjacent to the airport
3. Determining the environmental effects of airport construction and operations
4. Establishing access requirements
5. Establishing the technical, economic, and financial feasibility of proposed developments through a thorough investigation of alternative concepts
6. Establishing a schedule of priorities and phasing for the improvements proposed in the plan
7. Establishing an achievable financial plan to support the implementation schedule

8. Establishing a continuing planning process which will monitor conditions and adjust plan recommendations as circumstances warrant

The airport project plan

An airport project plan is a detailed plan of the specific development planned in the immediate future for an airport. The project plan focuses on specific elements of the airport master plan which are to be implemented in the short term and which may include such items as the addition of a new runway, modification of existing runways, provision of taxiways or taxiway exits, addition of gates, addition to or renovation of terminal building facilities, or modification of ground access facilities. The overall objective of the airport project plan is to provide the specific details of the development identified in the airport master plan that will satisfy immediate aviation needs and will be consistent with the objectives and constraints identified in the airport master plan. More specifically, the project plan is a detailed plan for

1. Developing the specific physical facilities at an airport including the architectural and engineering design for these facilities

2. Assessing the environmental effects of this development through the construction and operational phases

3. Determining the detailed costs and financially planning the development

4. Establishing a schedule for the construction and phasing of the specific items of development in the plan

Airport planning is based upon a multitude of procedures and criteria for evaluating needs, proposing alternative concepts, ranking and assigning priorities, and justifying the selected alternatives. In the early days of airport planning, the project plan evolved primarily as a technically viable solution geared almost entirely to aircraft operational requirements. Today, however, many complex technical, environmental, social, economic, financial, and political considerations influence the airport plan. As a result the plan adopted may not necessarily be the best technical plan, but it usually represents a compromise among many diverse physical and nonphysical planning requirements.

Elements of an Airport Planning Study

Coordination

Airport planning efforts draw widespread interest from citizens, community organizations, airport users, areawide planning agencies, and conservation groups. If these groups are not used in a coordinated, consultative, and advisory role at critical stages in the plan development, the plan will likely be unsuccessful when it is presented to the public. Interested parties should be brought in early in the plan development and should be given access to all relevant information. This process allows vital issues and concerns to be raised and resolved in the planning process and often assists in ultimate acceptance of the plan.

Content

The content of an airport plan varies in both level of detail and specific requirements depending upon the type of planning effort undertaken. However, the system plan and the master plan should include at least the following elements:

1. An inventory of existing airport facilities, current conditions, and issues related to the airport, and an identification of other planning studies which may impact the airport plan.

2. The demand forecast including aircraft operations, number of passengers, volume of cargo and mail, and vehicular traffic. The forecast must be made not only on an annual basis but also for the busiest hours of the day.

3. An analysis of the interaction between the various demand parameters and the capacity of the relevant facilities, including those which affect airfield, terminal, and ground access system operations.

4. The development of alternative concepts and solutions to reasonably satisfy forecasted demand by taking into account such factors as the functional role of the airport(s) under study and the impact on the environment, safety, economy, and fiscal resources of the area. An examination of alternative sites, including land-use and ground access plans, and development concepts is essential for the proper consideration and identification of viable alternatives.

5. The determination of the cost-effectiveness of alternative concepts and recommended solutions, including both tangible and intangible benefits and costs. Tangible benefits include such items as reduction in delays to aircraft, more efficient use of the runway and taxiway systems, and elimination of congestion in terminal buildings and on the ground access system. These benefits can be readily quantified in physical or monetary terms. Social costs, however, are much more difficult to quantify in physical or monetary terms and are often referred to as intangible. Many important benefits are significant in the decision-making process but which are also difficult to quantify. One of the more important is aircraft noise. Therefore, one must search for measures of effectiveness in other than monetary terms. For example, in the case of noise, it could be established that one run-

way layout was more effective than another because the number of people exposed to noise was reduced by half. Clearly a cost-effectiveness analysis is not an end in itself, but merely an aid to decision making.

6. A financial feasibility analysis which differs from economic feasibility, for there is no guarantee that if a proposed development is economically feasible, it is also feasible to finance that development. Investment priorities must be established among the various individual airport improvements. Frequently airport planning is separated from financial and management planning. The latter may be undertaken only after a physical plan is adopted. Because of financial constraints, financial planning should be undertaken concurrently with planning of the physical facilities.

7. The environmental impact of alternative concepts and recommended solutions must be considered and incorporated into the cost-effectiveness analysis. Although aircraft noise is the principal environmental problem faced by airport authorities, other factors must be considered. These are described in Chap. 15 discussing environmental impact.

8. The preparation of implementation schedules and costs and sources of revenue for the various phases of proposed development.

Inventory.   The initial step in the preparation of the plan is the collection of data on existing airport facilities, relevant aviation issues, and areawide planning efforts. State and regional transportation authorities should be consulted for they may be sources of valuable data. A key source of data, especially for airport traffic volumes, is the FAA. The planner should identify all the physical facilities at the airport, how they are used, and the volumes of traffic in and out of the airport. An identification of the uses of the airspace, availability of aids to air navigation, air traffic control, and communication facilities which serve the airport should be made. An inventory of land uses adjacent to the airport is also necessary to ascertain the environmental impact of improvements on the airport's neighbors. The collection of socioeconomic and demographic data, such as for population, employment, income levels, industrial and commercial activity, and land use, for the service area for the airport is valuable in demand forecasting and for determining the economic consequences of airport development [32]. A study of the financing mechanisms available to support airport improvements is necessary for the development of a financial plan.

Key issues which may bear on the preparation of the plan include aviation growth, changes in scheduled airline service including the impact of small regional or commuter airlines, the roles of existing airports in the area, capacity and delay considerations, ground access system congestion or capacity limitations, adverse environmental impact of adjacent land uses, and the presence of obstructions in the vicinity of the airport.

Forecasts. An airport plan must be developed on the basis of forecasts. From forecasts of demand, the performance and operational effectiveness of the various airport facilities can be evaluated. Forecasts are usually needed for the short, intermediate, and long range, or approximately 5, 10, and 20 years. A typical forecast associated with an airport master plan [25] is shown in Table 5-1. This table lists items normally forecast and the range of forecasts applicable to an airport master plan.

As the range of a forecast is increased, it becomes less precise and should be viewed as only an approximation. Earlier it was stated, depending upon the level of detail required in the planning effort, that for some activities, such as aircraft movements and number of passengers, both annual and busy-hour forecasts are necessary, while for air cargo and mail, an annual forecast may be sufficient. There are many ways in which to forecast future demand. Forecasting methods differ substantially, some are much more sophisticated than others, but all have a certain degree of uncertainty. Some methods are more suited for long-range forecasting. The simplest forecasting techniques simply project past trends of travel volumes into the future. The more complex techniques relate demand to a number of social, economic, and technological factors that affect air travel.

A relationship between social, technological, and economic variables, on one hand, and travel demand, on the other hand, is called a *demand model*. The development and use of demand models can be outlined by the following steps:

1. Observe past and current trends of air travel demand.

2. Inventory the variations in economic, social, and technological factors affecting air travel demand.

3. Establish a relationship between travel demand and those factors found to be of significance in altering travel demand.

4. Project the values of those factors affecting travel demand into the future.

5. Use the model and the forecasts to obtain forecasts of future air travel demand.

The forecast of demand should be performed in the context of determining the likely range in forecasts and the impact of variations in demand forecasts on the airport needs. A detailed discussion of the

TABLE 5-1   Miami International Airport Master Plan Update, Preferred Forecast Summary, and Extended Forecast

	Actual 1990	Forecast				
		1995	2000	2005	2010	2020
Passenger Forecast						
Domestic O&D passengers	9,919,660	10,820,000	11,890,000	12,800,000	13,620,000	15,350,000
Domestic connecting passengers	5,909,005	7,020,000	8,960,000	10,990,000	12,970,000	16,940,000
Total international passengers	10,008,780	14,950,000	19,400,000	24,080,000	28,650,000	37,760,000
Total passengers	25,837,445	32,790,000	40,250,000	47,870,000	55,240,000	70,050,000
Aircraft Operations Forecasts						
Air carrier operations	281,180	322,700	368,100	408,700	450,100	526,600
Air taxi operations	113,146	121,800	127,300	134,300	138,700	139,300
General aviation operations	79,415	74,700	76,700	79,700	82,000	86,900
Military operations	7,246	7,000	7,000	7,000	7,000	7,000
Total operations	480,987	526,200	579,100	629,700	677,800	759,800

Air Carrier Operational Fleet Mix

Commercial passenger aircraft

	Actual 1990	1995	2000	2005	2010	2020
60–120 seats	17.3%	11.5%	8.8%	7.6%	6.7%	5.4%
121–170 seats	49.0%	49.5%	48.5%	45.7%	45.0%	43.7%
171–240 seats	9.5%	10.7%	11.8%	13.6%	14.5%	16.4%
241–350 seats	12.8%	12.6%	11.7%	11.4%	10.9%	10.5%
351 + seats	3.4%	5.3%	7.7%	9.5%	10.0%	10.8%
Subtotal	91.9%	89.6%	88.5%	87.7%	87.1%	86.8%

Air Carrier Operational Fleet Mix (*Continued*)

Air cargo aircraft

Large piston/turboprop	2.0%	1.4%	0.0%	0.0%	0.0%	0.0%
Small jet	0.5%	1.2%	2.0%	2.8%	2.9%	3.0%
Medium jet	5.1%	7.0%	8.1%	7.7%	8.0%	8.2%
Large jet	0.4%	0.8%	1.4%	1.8%	1.9%	2.0%
Subtotal	8.1%	10.4%	11.5%	12.3%	12.9%	13.2%
Total	100.0%	100.0%	100.0%	100.0%	100.0%	100.0%

Air Taxi Operational Fleet Mix

Commercial passenger aircraft

Up to 19 seats	73.1%	67.0%	60.9%	54.2%	47.2%	29.4%
20–59 seats	24.7%	31.0%	38.1%	44.6%	51.5%	69.0%
Subtotal	97.9%	97.9%	99.0%	98.9%	98.7%	98.5%
Air cargo aircraft: small piston/turboprop	2.1%	2.1%	1.0%	1.1%	1.3%	1.5%
Total	100.0%	100.0%	100.0%	100.0%	100.0%	100.0%

Total Cargo Forecasts

International, tons	703,320	1,061,000	1,311,000	1,571,000	1,833,000	2,207,000
Domestic, tons	263,123	312,000	363,000	416,000	470,000	546,000
Total air cargo, tons	966,443	1,373,000	1,674,000	1,987,000	2,303,000	2,753,000
Total on-airport employees	19,070	27,000	30,000	33,000	36,000	42,000

SOURCE: Landrum and Brown, Inc. [25].

demand techniques available and their applicability to airport planning is found in Chap. 6.

Analysis of capacity and delay. The determination of the capacity, delay, and processing times associated with several viable alternative schemes for the improvement of existing airports or the development of new ones is an essential step in airport planning. The comparison of demand with capacity provides basic information for determining the extent of the required facilities [6, 28, 31, 39]. Delay is an essential input in a benefit-cost analysis. The money saved from reduced delays can be compared with the cost of airfield improvements, thus establishing a relationship between benefits and costs. The capacity analysis of airports includes not only the airfield but also the terminal airspace, aircraft gates, passenger terminals, vehicular circulation and parking within the airport boundary, and surface access to the airport. The proximity of airports to each other, the orientation of runways, and the type and mix of aircraft operations are all factors which can influence the capacity of an individual airport. As a first approximation, it is suggested that for instrument operations, a rectangular area 10 mi by 25 mi be reserved for the airspace for large aircraft with the airport runway approximately centered in the rectangle. For small airplanes, the size of the rectangle is reduced to 8 mi by 15 mi. If these requirements cannot be met, the planner should consult with the appropriate government authorities.

The capacity of roadways providing access to the airport can be obtained from references [1, 7, 28]. For large airports, access can be a constraint on the development of the entire airport complex [11].

The capacity of the airport system for an area must also be viewed in light of the functional roles of the airports in the system. A distinction is needed between facilities that accommodate commercial and general aviation operations. Based upon the projected needs ascertained in the demand forecast, sufficient facilities must be provided for the many diverse aviation uses consistent with overall policy guidelines.

Facility requirements. The general location of the various types of airport facilities and the requirements for runways, taxiways, aprons, terminals, cargo and servicing facilities, roadways, and parking at the various sites are developed from an analysis of the demand and capacity requirements and from geometric and other standards governing the design of airport components. This yields the number, length, and configuration of runways; the configuration of the taxiways; number of gates; and size of passenger terminal buildings; cargo buildings, and facilities for general aviation aircraft. This information enables the planner to obtain a first approximation of the overall size and shape of new airports or the expansion of existing sites and to analyze the impacts on land use, the environment, the infrastructure, and the airspace.

Evaluation criteria. The various development alternatives should be evaluated with respect to operational capability, capacity potential, ground access, development costs, environmental consequences, socio-economic factors, and consistency with areawide planning activities.

Airport site selection

The individual who undertakes the selection of a site suitable for new airports must first establish certain criteria which will guide the determination of its proper location and size. Most of these criteria, however, are also applicable to the expansion of existing airports. The location of an airport is influenced by the following factors:

1. Type of development of the surrounding area

2. Atmospheric conditions

3. Accessibility to ground transport

4. Availability of land for expansion

5. Presence of other airports and availability of airspace in the area

6. Surrounding obstructions

7. Economy of construction

8. Availability of utilities

9. Proximity to aeronautical demand

Type of development of surrounding area. This is an extremely important factor, since airport activities, particularly from the standpoint of noise, are often quite objectionable to the airport's neighbors. Thus a study of current and prospective uses of land adjacent to an airport site is essential. Sites which offer the closest compatibility with airport activities should be given priority [36].

Proximity to residential areas and schools should be avoided whenever possible. If the site is sparsely developed, enactment of zoning ordinances controlling the use of land adjacent to the airport should be considered in order to avoid future conflicts.

An airport is essential to the transportation complex of a community, yet it is also an integral part of the community. Hence it is subject to the same principles and policies that govern other elements of a community plan, and the airport must be coordinated with existing as well as planned features. It is desirable to provide a buffer zone between the runways, taxiways, aprons, etc., and the boundaries of the airport property, so that nuisances created by activities of the airport can be at least partially attenuated.

Noise is an extremely important factor where jet aircraft operations are anticipated. Since the introduction of jet aircraft, adverse community reaction to noise has grown rapidly. In many large urban areas, the FAA has prescribed specific flight patterns for aircraft arriving at and departing from the airport to minimize the adverse consequences of noise. Regulations on a national scale have been formulated to decrease aircraft noise. Aircraft and engine manufacturers are well aware of the noise problem, and every effort is being made to reduce noise consistent with the operating economics and safety of the aircraft.

Atmospheric and meteorological conditions.   The presence of fog, haze, and smoke reduces visibility and has the effect of lowering the traffic capacity of an airport, since the capacity when visibility is poor is less than that when visibility is good. Fog has a tendency to settle into areas where there is little wind. Lack of wind can be caused by surrounding topography. Likewise, smoke and haze are present in the vicinity of large industrial areas. An examination of specific site conditions as well as a detailed analysis of available weather records should be undertaken for all potential sites. Comparative site evaluations must be made to ensure that the selected site offers characteristics commensurate with aviation needs.

Accessibility to ground transportation.   Transit time from the point of origin to the ultimate destination for passengers and shippers of air freight is a matter of major concern. In many cases the ground time exceeds the air time by a considerable margin. With the introduction of jet transports, the margin has increased even more. For a journey of 400 mi between two large metropolitan areas, the ground time can be as much as twice the air travel time.

Surveys of ground transport to the airport indicate that in the United States the majority of passengers, visitors, and airport and airline employees travel by private automobile [47]. Indications are that this trend will continue in the future. Origins and destinations of travelers and shippers relative to airports are widely scattered throughout a metropolitan area. Out-of-town passengers' origins and destinations seem to concentrate more in centralized hotel locations near business centers. Still, many of these travelers are making use of rented automobiles.

Because of the lack of concentration of origins and destinations of air passengers in a metropolitan area and because of the popularity of the automobile as a personal means of transportation, the use of public transit up to now has not been large. However, as air transport continues to grow, the volume of passengers may be so large as to warrant special means of transportation to the airport. This may be especially true in large urban areas whenever the normal peak vehicular traffic periods coincide with the peak traffic periods at the airport. In some cities a train connects the airport with the downtown area [38]. Other cities are planning an expansion of existing urban-area rail service to airports. While such installations are undoubtedly quite expensive and probably cannot be economically justified on the basis of serving the airport alone, they may become useful in the future as part of a rapid transit facility for an entire metropolitan area. In any event, the private automobile will continue to be an important means of transportation to the airport. Consequently the planning of streets and highways to the airport and parking at the airport are important factors which must be given consideration.

Access to airports is required by both air passengers and other users of the airport, such as employees, visitors, trucks hauling freight and express mail, and businesses dealing with the airport's tenants. All modes of access should be considered. The need for satellite terminals in downtown areas should also be examined. Statistics show that the private automobile is the major form of access to the airport for both passengers and employees. It is expected that this trend will continue in the future despite the greater availability of mass transit. Although air cargo is growing rapidly, truck traffic is not a major contributor to airport traffic during peak periods because much air cargo is processed at night. On the other hand, the number of trips by airport employees can be larger than the number generated by the airline passengers. This depends on the size of the airport and the presence of commercial enterprises or large maintenance facilities at the airport.

The starting point for estimating the ground traffic generated by air passengers is a forecast of future air travel. It is very desirable to have a forecast of the daily distribution of passenger demand in terms of enplaning and deplaning passengers at least for the busy hours of the day. The next step is to estimate the modal split of ground transportation among the various alternatives including private automobiles, taxis, limousines, rental cars, and mass transit. After the modal split is estimated, it is necessary to estimate the vehicle occupancy for each mode. By using the modal split of airline passengers and the average occupancy by mode, the number of vehicles generated by airline passengers can be determined. From highway capacity standards the number of lanes required to serve passengers can be established [1, 11, 28]. In addition to an estimate of the number of airline passengers, an estimate must be made of the number of trips generated by visitors accompanying passengers. In some cases this may amount to

15 to 25 percent of the air passenger traffic. Another approach is to attempt to correlate air passenger activity hour by hour with corresponding ground activity by using multiple regression analysis. The regression models include the inherent assumption that the current ground-to-air traffic relationship will be maintained in the future, which may not be true if the modal split is altered significantly by providing rail rapid transit. For this reason, the regression approach is satisfactory for a first approximation, but more complete knowledge of the various factors generating ground traffic is necessary to arrive at more accurate future requirements of ground access. Data have been collected for various trip generators, including the various types of airport facilities, and these may be used as a guide for estimating highway system requirements [48].

As stated earlier, traffic generated by employees during peak periods can exceed that generated by passengers and visitors. Large airports may employ in excess of 25,000 people. This makes it imperative to consider employee access requirements separately. Employees usually have a different origin-destination pattern, which may have some influence on access requirements. Studies have shown that no consistent relationship exists between the number of airport employees and the annual number of air passengers.

Once the total volume of traffic entering the airport is known, an assignment must be made within the airport boundary. It is essential that the circulation of vehicular traffic in the terminal building area be well planned; otherwise congestion and delays are likely.

The circulation of traffic in an airport should generally be one-way, and the traffic flow should be directed first to the departure area on a one-level roadway system. The roads should be wide enough to permit passing. Direction information to locations of specific airlines, arrivals and departures, and public parking facilities should be adequate in number, size, and legibility to convey a precise message to airport users. To the extent possible, standard and consistent message signs should be used at airports [16, 29].

Pedestrian routes should be direct, well marked, and adequately lighted and should provide security to users. Covered walkways from public parking lots or to the entrance of the terminal building should be considered if bad weather occurs a substantial amount of the time and walking distances are long.

Provisions for automobile parking must be adequate to ensure that the airport functions effectively and efficiently. Despite the use of other means of ground transport to and from the airport, the use of the private automobile will continue to be substantial. Facilities required to satisfy automobile parking demand for major airports have grown to such magnitude that they have become a dominant consideration in the planning of an airport. A major goal for locating parking facilities for airline passengers is to minimize walking distances and therefore to bring the automobile as close as possible to the aircraft. The volume and characteristics of users play a major role in planning parking facilities. Each class of user has a different requirement depending on the reason for being at the airport. Adequate provisions must also be made for the elderly and handicapped to access airport facilities [2].

Parking at an airport must be provided for airline passengers, visitors accompanying passengers, other visitors, people employed at the airport, car rentals and limousines, and people having business with the airport tenants.

Separate parking areas should be provided for employees. These should be located as close as possible to the facilities in which employees work. Parking requirements for car rental should be determined by consultation with the rental concessionaires. Although it is often desirable to locate the car rental parking area as close as possible to the terminal building to minimize the passenger walking distance at large airports, there is a tendency to locate rental-car parking facilities remote from the terminal building complex and to shuttle passengers between terminal buildings and the rental-car parking lots. Close-in parking, when provided, need not be for the entire car-rental fleet but can be only for cars previously reserved by arriving passengers. Departing passengers can deposit rental cars in an outlying rental-car parking area within the airport. This should be conveniently located to the main airport internal access road. From there the car rental firm provides transportation to the passenger terminal. This is a common arrangement at large airports.

Public parking facilities are provided for airline passengers, visitors, and others. Surveys of a number of airline airports in the United States indicate that a large number of these users, on the order of 80 percent, park 3 h or less, and a very much smaller group park from 12 h to several days or longer. The short-term parkers, however, represent only about 15 to 20 percent of the maximum vehicle accumulation in the parking facility. Therefore, many airports designate the most convenient spaces to short-term parkers, who represent the highest number of users, and regulate these facilities by charging premium rates for parking. At large airports, additional parking is often provided outside the boundary of the airport by private concessionaires who provide transportation to the airport for patrons.

Projections of future public parking demand at an airport are generally made by some method of correlation with projected growth in air traffic, usually airline passengers. At an existing airport, information on the hourly distribution of automobile traffic entering and leaving the parking facility daily is required. The difference between entering and leaving traffic gives the accumulation of vehicles in the

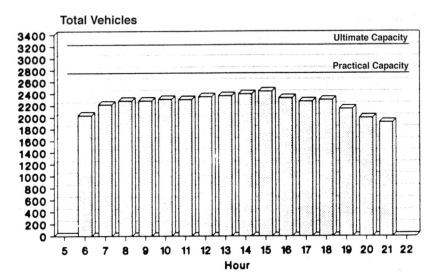

Figure 5-3 Total accumulation of vehicles in all parking facilities at Chicago Midway Airport (*Landrum and Brown, Inc. [22]*).

parking facility. A plot of the hourly distribution of accumulated vehicles will yield the peak accumulation and when it occurs. An example of such a plot for determining the parking requirements at Chicago Midway Airport is shown in Fig. 5-3 [22]. The peak accumulation represents current needs if the peak demand is to be accommodated.

To project into the future, a relationship between the number of cars entering and leaving a parking facility in a specific time, usually 1 h, and the total number of air passengers arriving and departing during the same period of time is necessary. For example, if there are 2000 arriving and departing passengers per hour and the number of cars entering and leaving the facility in the same time period is 1000, then the number of vehicles is one-half the number of passengers. Vehicle occupancy can be affected by the number of visitors accompanying passengers. The number can vary considerably depending on the country or area in which the airport is located. The forecast of passengers and vehicle occupancy can be used to determine the size of the entrance and exit to the parking facility and the number of spaces required. The busy-hour vehicle flow into and out of the facility can be determined by dividing the forecast of busy-hour passenger demand by vehicle occupancy. The number of spaces in the parking facility can be determined in several ways. One way is to obtain a projection of the daily distribution of passengers entering or leaving the airport and to convert the number of passengers to vehicles to find the peak accumulation of vehicles. Another way is to correlate the maximum accumulation of vehicles with busy-hour passenger demand during known years and to apply the correlation to projected future busy-hour demand. The most acceptable way is to perform a simulation of parking needs based upon the pattern of parking use at the airport.

It may not be feasible to satisfy demand during the busiest hours of the year. This is a matter of policy which rests with airport management. In part, it will be influenced by available space and the costs associated with providing parking.

Methods used for projecting parking requirements range from rules of thumb based on nationwide averages to trend analysis, mathematical models, and simulation. The choice of method depends on the type of planning being undertaken and the accuracy and sensitivity desired. Relatively simple methods based on parking experience at the airport may be sufficient for long-range master planning, but more sophisticated techniques may be required for more precise planning and design, especially when costly parking structures are involved and parking revenue projections are needed. The planner should always bear in mind that the accuracy of parking projections depends on the accuracy of passenger projections.

Availability of land for expansion.   In a field as dynamic as aviation, it is necessary to acquire land in advance or to be able to acquire sufficient real estate in the future for expanding the physical plant of the airport. Historically, as aircraft size and traffic volume increased, runways have had to be lengthened, terminal facilities expanded, and additional support facilities provided. Sufficient real estate must be available to accommodate new facilities. An investigation of policies related to "land banking" can lead to considerable economies and can minimize future environmental impact.

Factors influencing airport size.   The necessary size of an airport will depend on such primary factors as

1. The performance characteristics and size of aircraft expected to use the airport
2. The anticipated volume of traffic
3. Meteorological conditions
4. The elevation of the airport site

The performance characteristics of aircraft will have a direct bearing on the length of runways. Airport designers normally have access to information concerning the performance of various types of transport aircraft [8, 45]. This type of information is also available from airline engineering departments and aircraft manufacturers. However, most airport planners and designers are not sufficiently familiar with all the technical aspects concerning aircraft performance. To provide the airport designer with the information needed on runway length, agencies such as the FAA and the International Civil Aviation Organization (ICAO) have examined the performance characteristics of many types of aircraft and have documented runway lengths [3, 45]. Whenever possible, the lengths should be determined by consultation with airlines and other users of the airport.

The volume and character of traffic influence the number and configuration of runways required, configuration of the taxiways, and size of the ramp area. Important meteorological conditions which can affect the size of an airport are the wind and temperature. Temperature influences runway length, with high temperatures requiring longer lengths. The direction of the wind influences the number of runways and their configuration. Techniques for the analysis of the effects of wind are discussed in Chap. 7.

Presence of other airports and availability of airspace.   The operations at other airports in the area must be carefully analyzed when a site for a new airport is selected or when runways are added at an existing airport. Airports should be located a sufficient distance from each other to prevent aircraft which are maneuvering for a landing at one airport from interfering with the movements of aircraft at other airports. The minimum distance between airports depends entirely on the volume and type of traffic and on whether the airports are equipped to operate under poor-visibility conditions. Many metropolitan areas are experiencing severe air traffic congestion in the limited airspace available to serve the many airports in the region.

Maneuvering in the air is vastly more complicated during periods of poor visibility. Under instrument meteorological conditions (IMC), air traffic control segregates aircraft using the airways and maintains control until each in turn is cleared for an instrument approach to the airport.

The location of several airports in a metropolitan area can greatly influence their respective capacities. If the locations are too near each other, operations can be restricted to the extent that two airports will have no more capacity during IMC than a single airport.

Airport location must be consistent with the overhead airway traffic patterns if conflicts in traffic flow are to be avoided. It is imperative that the airport planner consult with the FAA on the adequacy of a airport site insofar as air traffic control is concerned. An illustration of the interaction of the airspace for arrivals and departures at Chicago O'Hare International and Midway airports is given in Fig. 5-4.

Surrounding obstructions.   Sites should be selected so that approaches necessary for the ultimate development of the airport are

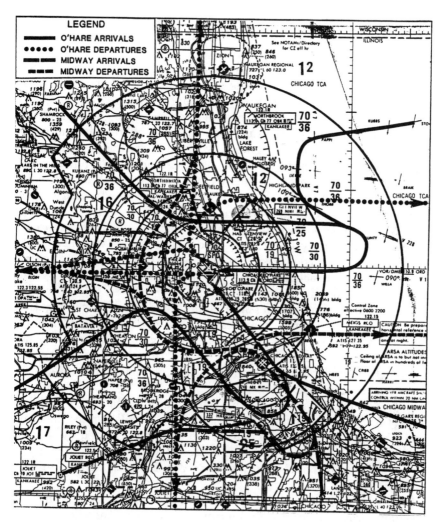

Figure 5-4 Airspace interaction between Midway and O'Hare International airports (*Landrum and Brown, Inc. [21]*).

free of obstructions or can be cleared if obstructions exist. The provision and protection of adequate approaches to an airport will necessitate height restrictions in the airport turning zones and in line with the runways. Steps should be taken in the planning stage to preclude the possibility of future erection of obstructions to aircraft using the airport. The purchase of the real estate necessary to protect approaches may not be economically feasible. Therefore, zoning for height restrictions should be initiated as soon as a site has been chosen.

The required clearances for the approaches to the runways and in the maneuver area directly above and adjacent to the airport are outlined in detail by the FAA in FAR part 77 [37] and by ICAO in Annex 14 [3, 17]. An object that protrudes above the imaginary surfaces specified in these two references is considered an obstruction to air navigation. If the planner finds that there is an obstruction to air navigation, the appropriate government agency must be contacted so that a special aeronautical study may be undertaken to determine the feasibility of a waiver being granted without degrading safety.

In FAR part 77, airport runways are classified as shown in Fig. 5-5. A visual runway is intended solely for the operation of aircraft using visual-approach procedures. A utility runway is constructed for and intended to be used by propeller-driven aircraft weighing 12,500 lb or less. A non-precision-instrument runway has an instrument-approach procedure with only horizontal guidance or area-type navigation equipment. A precision-instrument runway has an existing instrument-approach procedure utilizing an instrument landing system (ILS) or precision-approach radar (PAR).

To determine whether an object is an obstruction to air navigation, several imaginary surfaces are established with relation to the airport and to each runway. The size of the imaginary surfaces depends on the category of each runway, such as utility, and on the type of approach planned for that runway, such as visual, non-precision-instrument, or precision-instrument.

If the federal government is to participate in the financing of a new airport or in the expansion of an existing one, it will require that runway protection zones at the ends of runways be provided by the airport operator. Runway protection zones [8] are areas comprising the innermost portions of the runway approach areas, as defined in FAR part 77 [37].

The federal government requires that the airport owner have an adequate property interest in the runway protection zone area so that

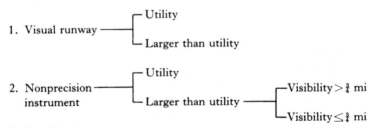

1. Visual runway
   - Utility
   - Larger than utility

2. Nonprecision instrument
   - Utility
   - Larger than utility
     - Visibility > ¾ mi
     - Visibility ≤ ¾ mi

3. Precision instrument

Figure 5-5 Classification of airport runways in FAR part 77 (*Federal Aviation Administration [37]*).

the requirements of FAR part 77 can be met and the area can be protected from future encroachments. Adequate property interest may be in the form of ownership in fee simple, long-term lease, or any other demonstration of legal ability to prevent future obstructions in the runway protection zone.

The federal government has long recognized that the provision of adequate approaches very near the ends of runways is essential to the interest of safety, yet often they are quite difficult to maintain unless positive control of the area can be exercised by the airport owner. Experience has also shown that the exercise of this control by zoning alone has not proved satisfactory, hence the necessity of obtaining some kind of interest in the land itself. Recognizing that such an interest imposes a financial burden on the airport owner, the federal government has made provision to participate in obtaining adequate property interests in runway protection zone areas.

The ICAO requirements are similar to FAR part 77 with several notable exceptions. The ICAO specifies dimensions for each category of runway by an aerodrome reference code numbers 1, 2, 3, and 4 based upon the reference field length, as shown in Table 5-2. The approach surface defined in FAR part 77 is for both arriving and departing aircraft. ICAO separates arrivals and departures and specifies the dimensions for approach surfaces for arrivals and takeoff climb surfaces for departures [3, 4]. The horizontal surface specified by ICAO is a circle whose center is at the *airport reference point* (ARP), whereas in FAR part 77 it is not a circle. A detailed description of these surfaces is presented in Chap. 7 on airport configuration.

Economy of construction.   It is clear that if alternative sites are available and equally adequate, the site which is more economical to construct should be given consideration. Sites lying on submerged lands are much more costly to develop than those on dry land. Rolling terrain requires much more grading than flat terrain. The availability of local construction materials, including aggregates which can be obtained on site or at nearby locations, can have a significant impact on construction cost.

TABLE 5-2   Codification of Airports by the ICAO for Obstruction Surfaces

Aerodrome code number	Reference field length, m
1	Less than 800
2	800–1200
3	1200–1800
4	1800 and over

SOURCE: International Civil Aviation Organization [2].

Availability of utilities.   An airport, particularly a large one, requires large quantities of water, natural gas or oil, electric power, and fuel for aircraft and surface vehicles. In site selection, the provision for these utilities must be given consideration. On a new site which is not near available sewers, a disposal plant may have to be constructed. In the case of electric power, most large airports must provide generating plants of their own to be used in emergencies in the event that a commercial source fails. In some cases, airports are providing generating facilities which meet a portion of the power needs on a continuing basis due to the efficiencies of new technology in energy generation.

Proximity to aeronautical demand.   In the selection of a new airport site, it is quite important that the location be such as to result in the shortest ground access time possible. While much has been said about the establishment of regional airports at considerable distances from the population centers, the facts remain that about half of the passengers in the United States travel no more than about 500 mi and that nearly half of all scheduled flights are in these short-haul operations. This trend is expected to continue.

In the short-haul operations, access time to the airport is quite important. The airline traveler is more interested in the overall door-to-door time than in just the time spent in the air. Locating an airport a considerable distance from the population center not only negates the effect of the increased speed provided by short-range turbojet transports but also can result in a loss of patronage.

Land-use planning

A land-use plan for property within the airport boundary and in areas adjacent to the airport is an essential part of an airport master plan. The plan for the use of land both on and off the airport is an integral part of an areawide comprehensive planning program, and therefore it must be coordinated with the objectives, policies, and programs for the area which the airport is to serve. Incompatibility of the airport with its neighbors stems primarily from the objections of people to aircraft noise. A land-use plan must therefore project the aircraft noise generated by airport operations in the future. Contours of equal-intensity noise can be drawn and overlaid on a land-use map, as shown in Fig. 5-6. From these contours an estimate can be made of the compatibility of existing land use with airport operations. If the land outside the airport is underdeveloped, the contours are the basis for establishing comprehensive land-use zoning requirements.

Although zoning is used as a method for controlling land use adjacent to an airport, it is not effective in areas which are already built up because it is usually not retroactive. Furthermore, jurisdictions

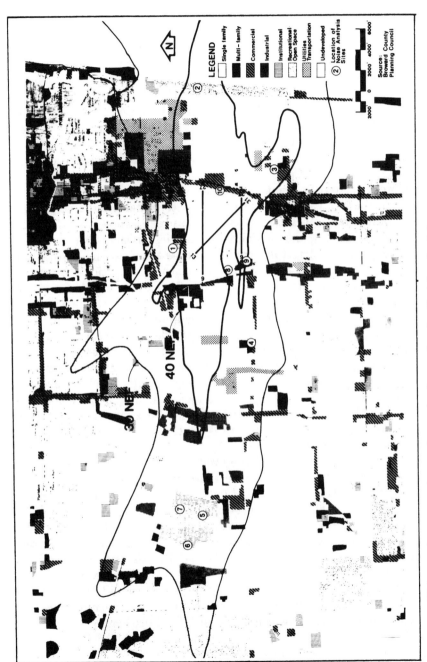

Figure 5-6 Noise contours superimposed on a land-use plan for Fort Lauderdale–Hollywood International Airport (*Environmental Sciences and Engineering, Inc.* [23]).

having zoning powers may not take effective zoning action. Aircraft operations into and out of the airport may be made unnecessarily complex to minimize noise encroachment on incompatible land uses. Despite these shortcomings, the planner should utilize zoning as a vehicle to achieve compatibility wherever this approach is feasible.

Airports become involved in two types of zoning. One type is height and hazard zoning, which is mainly to protect the approaches to the airport from obstructions [17, 20]. Both FAR part 77 [37] and ICAO in Annex 14 [3, 4] are the bases for height and hazard zoning. The other type is land-use zoning.

The extent of land use in the airport depends a great deal on the acreage available. Land uses can be classified as either closely or remotely related to aviation. Those uses closely related to aviation include the runways, taxiways, aprons, terminal buildings, parking, and maintenance facilities. Nonaviation uses include space for recreational, industrial, and commercial activities. In considering commercial or industrial activities, care should be taken to ensure that they do not interfere with aircraft operations, communications equipment, and aids to navigation on the ground. Smoke can obscure visibility. Recreational facilities such as golf courses may be suitable within the immediate proximity of the airport boundary. Certain agricultural uses are also appropriate as long as they do not attract birds. When there is acreage within the airport boundary in excess of aviation needs, it is sound fiscal planning to provide the greatest financial return from leases of the excess property. Thus the land-use plan within the airport is a very effective tool in helping airport managers make decisions concerning requests for land use by various interests, and often airports delineate areas on the airport property for the development of industrial parks.

The principal objective of the land-use plan for areas outside the airport boundary is to minimize the disturbing effects of noise. As stated earlier, the delineation of noise contours is the most promising approach for establishing noise-sensitive areas. The contours define the areas which are and are not suitable for residential use or other use and, likewise, those which are suitable for light industrial, commercial, or recreational activity. Although the responsibility for developing land uses adjacent to the airport lies with the governing bodies of adjacent communities, the land-use plan provided by the airport authority will greatly influence and assist the governing bodies in their task of establishing comprehensive land-use zoning.

Environmental impact assessment

Environmental factors must be considered carefully in the development of a new airport or the expansion of an existing one. This is a requirement of the Airport and Airway Improvement Act of 1982 (Public Law 97-248) and the Environmental Policy Act of 1969 (Public Law 91-190). Studies of the impact of the construction and operation of a new airport or the expansion of an existing one upon acceptable levels of air and water quality, noise levels, ecological processes, and demographic development of the region must be conducted to determine how the airport requirements can best be met with minimal adverse environmental and social consequences [10, 13].

Aircraft noise is the severest environmental problem to be considered in the development of airport facilities. Much has been done to quiet engines and to modify flight procedures, resulting in substantial reductions in noise. Another effective means for reducing noise is through proper planning of land use for areas adjacent to the airport. For an existing airport this may be difficult, for the land may have already been built up. Every effort should be made to orient air traffic away from noise-sensitive land development.

Other important environmental factors include air and water pollution, industrial wastes and domestic sewage originating at the airport, and disturbance of natural environmental values. In regard to air pollution, the federal government and industry have worked jointly toward alleviating the problem, and there is reason to believe that it will probably be eliminated in the near future as an environmental factor. An airport can be a major contributor to water pollution if suitable treatment facilities for airport wastes are not provided [30]. Chemicals used to deice aircraft are a major source of potential groundwater pollution, and provisions need to be made to safely dispose of this waste product. The environmental study must include a statement detailing the methods for handling sources of water pollution.

The construction of a new airport or the expansion of an existing one may have a major impact on the natural environment. This is particularly true for large developments where streams and major drainage courses may be changed, habitats of wildlife may be disrupted, and wilderness and recreational areas may be reshaped. The environmental study should indicate how these disruptions might be alleviated.

In the preparation of an environmental study or an environmental impact statement, the findings must cover the following items [43]:

1. The environmental impact of the proposed development

2. Any adverse environmental effects which cannot be avoided, should the development be implemented

3. Alternatives to the proposed development

4. The relationship between local short-term uses of the environment and the maintenance and enhancement of long-term productivity

5. Any irreversible environmental and irretrievable commitments of resources which would be involved in the proposed development, should it be implemented

6. Growth-inducing impact

7. Mitigation measures to minimize impact

In the application of these guidelines, attention must be directed to the following questions. Will the proposed development

1. Cause controversy?

2. Noticeably affect the ambient noise level for a significant number of people?

3. Displace a significant number of people?

4. Have a significant aesthetic or visual effect?

5. Divide or disrupt an established community or divide existing uses?

6. Have any effect on areas of unique interest or scenic beauty?

7. Destroy or derogate important recreational areas?

8. Substantially alter the pattern of behavior for a species?

9. Interfere with important wildlife breeding, nesting, or feeding grounds?

10. Significantly increase air or water pollution?

11. Adversely affect the water table of an area?

12. Cause excessive congestion on existing ground transportation facilities?

13. Adversely effect the land-use plan for the region?

The preparation of an environmental impact statement based upon an environmental assessment study is an extremely important part of the airport planning process. The statement should clearly identify the problems that will affect environmental quality and the proposed actions to alleviate them. Unless the statement is sufficiently comprehensive, the entire airport development may be in jeopardy. The details of the preparation of environmental assessments are discussed in Chap. 15.

Economic and financial feasibility

The economic and financial feasibility of alternative plans for a new airport or for expansion of an existing site must be clearly demonstrated by the planner. Even if the selected alternative is shown to be economically feasible, it is also necessary to show that the plan will generate sufficient revenues to cover the annual costs of capital investment, administration, operation, and maintenance. This must be determined for each stage or phase of development detailed in the airport master plan.

An evaluation of economic feasibility requires an analysis of benefits and costs. A comparison of benefits and costs of potential capital investment programs indicates the desirability of a project from an economic point of view. The economic criterion used in evaluating an aviation investment is the total cost of facilities, including quantifiable social costs, compared with the value of the increased effectiveness measured in terms of total benefits. The costs include capital investment, administration, operation, maintenance, and any other costs that can be quantified. The benefits include a reduction in aircraft and passenger delays and improved operating efficiency. The costs and benefits are usually determined on an annual basis.

There are a number of techniques for comparing benefits with costs. Most consider the time value of money based on an appropriate discount rate which reflects the opportunity cost of capital. The discount rate is a value by which a unit of money received in the future is multiplied to obtain its present value or present worth. In other words, a cost incurred in 1986 has a different economic value from the cost of the same item incurred in 1990.

If the time value of money is not considered, the ratio of benefits to costs is made for each year by merely dividing the benefits in a particular year by the cost of the project in that year. A project is considered economically feasible when the ratio of the benefits to costs is greater than unity, i.e., when the benefits exceed the costs. The larger the ratio, the more attractive the project from an economic standpoint. A ratio can also be obtained by comparing the present value of benefits with the present value of costs. This approach recognizes the time value of money. Another approach is to plot the net present value for each year versus time. The *net present value* (NPV) is the present value of benefits minus the present value of costs. An example of the comparison of three alternatives using the NPV method is shown in Fig. 5-7.

The financing of capital improvements for airports was discussed in Chap. 2. We stated that, in the early years of airport development, substantial capital improvement programs were financed at the local level by the sale of general obligation bonds backed by the taxing power of the community. As air transportation matured and the requirements of the community for capital spending programs increased, airports began to utilize revenue bonds as a source of

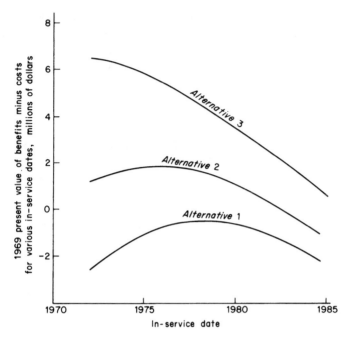

Figure 5-7 Comparison of alternatives by the net present value method.

financing. A financial feasibility study is therefore an analysis to determine whether bonds are marketable at reasonable interest rates. It also includes the feasibility of other forms of financing. The analysis requires a thorough evaluation of the revenues to be developed by a proposed improvement and the corresponding costs. Usually this is done in a traffic-and-earnings study performed over the planning horizon. In such a study, the forecast of demand is utilized, and rates and charges are established for the various revenue categories. This results in annual revenue projections. To make revenue bonds attractive to buyers, a typical airport revenue bond should show an expected coverage by net revenues (gross revenues minus costs) of at least 1.25 times the debt service requirements. If the analysis indicates that the revenues will be insufficient, then revisions in the scheduling or scope of the proposed development may have to be made or the rates and charges to the users of the airport may require adjustment.

Continuous planning process

A continuous airport system planning process is needed to respond to the needs of air transportation in a changing environment [42].

Changes in aviation demand, community policies, new technology, financial constraints, and other factors can alter the need for and the timing of facility improvements. Current data must be continually collected and assessed regarding airport needs, operations and utilization, environmental impact, and financial capabilities. The staging of airport improvements assists in the reevaluation of continuing needs at the points when implementation decisions are required.

In the airport system planning process, the overall objective of establishing and maintaining a *continuous* airport planning process is to insure that the airport system plan remains responsive to public needs. Specific objectives associated with the continuous airport system planning process include the following [42]:

1. Surveillance, maintenance, inventory, and update of the basic data, such as aviation activity, and socioeconomic and environmental factors relating to the existing airport system plan

2. Review and validation of data affecting the airport system plan

3. Reappraisal of the airport system plan in view of changing conditions

4. Modification of the airport system plan to retain its viability

5. Development of a continuous mechanism for ensuring the interchange of information between the system planning and master planning processes

6. Provision of a means for receiving and considering public comment in order to maintain and ensure public awareness of the role that airports play in the transportation system of an area

7. Redefinition of air transportation goals and policies

8. Integration of airport system planning into a multimodal planning process

9. Analysis of special issues

10. Publication of interim reports and formal plan updates

References

1. *Access to Commercial Service Airports, The Planning and Design of On-Airport Ground Access System Components,* Francis X. McKelvey, Final Report, Federal Aviation Administration, Washington, 1984.
2. *Access Travel: Airports, A Guide to Accessibility of Terminals,* Federal Aviation Administration, Washington, 1979.
3. *Aerodromes, Annex 14 to the Convention on International Civil Aviation,* vol. 1: *Aerodrome Design and Operations,* 1st ed., International Civil Aviation Organization, Montreal, Canada, July 1990.
4. *Aerodrome Design Manual, Part 1: Runways,* 2d ed., Document 9157-AN/901, International Civil Aviation Organization, Montreal, Canada, 1984.

5. *Aircraft Fuel Storage, Handling, and Dispensing on Airports,* Advisory Circular AC 150/5230-4, Federal Aviation Administration, Washington, 1982.
6. *Airport Capacity and Delay,* Advisory Circular AC 150/5060-5, Federal Aviation Administration, Washington, 1983.
7. *Airport Curbside Planning Manual,* Wilbur Smith and Associates, Transportation Systems Center, U.S. Department of Transportation, Cambridge, Mass., March 1980.
8. *Airport Design,* Advisory Circular AC 150/5300-13, Federal Aviation Administration, Washington, 1989.
9. *Airport Economics Manual,* 1st ed., Document 9562, International Civil Aviation Organization, Montreal, Canada, 1991.
10. *Airport Environmental Handbook,* Order 5050.4A, Federal Aviation Administration, Washington, October 1988.
11. *Airport Ground Access,* Rep. FAA-EM-79-4, Federal Aviation Administration, Washington, October 1978.
12. *Airport Master Plans,* Advisory Circular AC 150/5070-6A, Federal Aviation Administration, Washington, 1985.
13. *Airport Noise Compatibility Planning,* Part 150, Federal Aviation Regulations, Federal Aviation Administration, Washington, 1985.
14. *Airport Planning Manual,* Part 1: *Master Planning,* 2d ed., International Civil Aviation Organization, Montreal, Canada, 1987.
15. *Airport Planning Manual,* Part 2: *Land Use and Environmental Control,* 2d ed., International Civil Aviation Organization, Montreal, Canada, 1985.
16. *Airport Roadway Guide Signs,* Institute of Transportation Engineers, Washington, 1988.
17. *Airport Services Manual,* Part 6: *Control of Obstacles,* 2d ed., International Civil Aviation Organization, Montreal, Canada, 1983.
18. *Airport System Capacity, Strategic Choices,* Special Report 226, Transportation Research Board, Washington, 1990.
19. *Airport Systems Planning,* R. DeNeufville, MIT Press, Cambridge, Mass., 1976.
20. *A Model Zoning Ordinance to Limit Heights of Objects around Airports,* Advisory Circular AC 150/5190-4A, Federal Aviation Administration, Washington, 1987.
21. *Chicago Delay Task Force Technical Report,* vol. 1: *Chicago Airport/Airspace Operating Environment,* Landrum and Brown, Inc., Chicago, April 1991.
22. *Chicago Midway Airport Master Plan Study: Determination of Facility Requirements,* Working Paper no. 1, draft, Landrum and Brown, Inc., Chicago, September 1990.
23. *Environmental Impact Assessment for the Expansion of the Existing Terminal Complex at the Fort Lauderdale–Hollywood International Airport,* Environmental Sciences and Engineering, Inc., Tampa, Fla., April 1980.
24. *Environmental Protection, Annex 16 to the Convention on International Civil Aviation,* vol. 1: *Aircraft Noise,* 2d ed., International Civil Aviation Organization, Montreal, Canada, 1988.
25. *Forecasts of Aviation Demand, Miami International Airport Master Plan Update,* Draft Report, Landrum and Brown, Inc., Chicago, March 1992.
26. *Fort Lauderdale–Hollywood International Airport, Master Plan Report,* vol. 1: *Executive Summary,* Landrum and Brown, Inc., Cincinnati, Ohio, August 1977.
27. *Guide for the Planning of Small Airports,* Roads and Transportation Association of Canada, Ottawa, 1980.
28. *Highway Capacity Manual,* Special Report 209, Transportation Research Board, Washington, 1985.
29. *International Signs to Provide Guidance to Persons at Airports,* Document 9430, International Civil Aviation Organization, Montreal, Canada, 1984.
30. *Management of Airport Industrial Waste,* Advisory Circular AC 150/5320-15, Federal Aviation Administration, Washington, 1991.
31. *Measuring Airport Landside Capacity,* Special Report 215, Transportation Research Board, Washington, 1987.
32. *Measuring the Regional Economic Significance of Airports,* Rep. DOT/FAA/PP/87-1, Federal Aviation Administration, Washington, 1986.

33. *Michigan Aviation System Plan,* Technical Report, Michigan Department of Transportation, Lansing, 1989.
34. *Minimum Standards for Commercial Aeronautical Activities on Public Airports,* Advisory Circular, AC 150/5190-1A, Federal Aviation Administration, Washington, 1985.
35. *National Plan of Integrated Airport Systems (NPIAS), 1990–1999,* Federal Aviation Administration, Department of Transportation, Washington, 1991.
36. *Noise Control and Compatibility Planning for Airports,* Advisory Circular AC 150/5020-1, Federal Aviation Administration, Washington, 1983.
37. *Objects Affecting Navigable Airspace,* Part 77, Federal Aviation Regulations, Federal Aviation Administration, Washington, 1989.
38. "Opportunities for Fixed Rail Service to Airports," W. J. Sproule and S. Mandalapu, *International Air Transportation,* Proceedings of the 22d Conference on International Air Transportation, American Society of Civil Engineers, New York, 1992.
39. *Planning and Design Guidelines for Airport Terminal Facilities,* Advisory Circular AC 150/5360-13, Federal Aviation Administration, Washington, 1988.
40. *Planning the Airport Industrial Park,* Advisory Circular AC 150/5070-3, Federal Aviation Administration, Washington, September 1965.
41. *Planning the Metropolitan Airport System,* Advisory Circular AC 150/5070-5, Federal Aviation Administration, Washington, May 1970.
42. *Planning the State Airport System,* Advisory Circular AC 150/5050-3B, Federal Aviation Administration, Washington, 1989.
43. *Policies and Procedures for Considering Environmental Impacts,* Order 1050.1D, Federal Aviation Administration, Washington, December 1986.
44. *Recommended Methods for Computing Noise Contours around Airports,* Circular 205-AN/1/25, International Civil Aviation Organization, Montreal, Canada, 1988.
45. *Runway Length Requirements for Airport Design,* Advisory Circular AC 150/5325-4A, Federal Aviation Administration, Washington, 1990.
46. *Southeast Michigan Regional Aviation System Plan,* Tech. rep., Southeast Michigan Council of Governments, Detroit, April 1992.
47. *Survey of Airport Ground Access,* U.S. Aviation Industry Working Group, Washington, June 1981.
48. *Trip Generation,* 5th ed., Institute of Transportation Engineers, Washington, 1991.

6

Forecasting in Aviation and Airport Planning

Plans for the development of the various components of the airport system depend to a large extent upon the activity levels forecast for the future. Since the purpose of an airport is to process aircraft, passengers, freight, and ground transport vehicles in an efficient and safe manner, airport performance is judged on the basis of how well the demand placed upon the facilities in the system is handled. To adequately assess the causes of performance breakdowns in existing airport systems and to plan facilities to meet future needs, it is essential to predict the level and distribution of demand on the various components of the airport system. Without a reliable knowledge of the nature and expected variation in the loads placed upon a component, it is impossible to realistically assess its physical and operational requirements. For example, a forecast to project the mix of aircraft and the types of aviation activity at an airport site is necessary to identify the critical aircraft which dictate the elements of geometric and structural design, the type and extent of physical facilities, the navigational aid requirements, and any special or unique facility needs at the airport [20, 25, 26].

An understanding of future demand patterns allows the planner to assess future airport performance in light of existing and improved facilities; to evaluate the impact of various quality-of-service options on the airlines, travelers, shippers, and the community; to recommend development programs consistent with the overall objectives and policies of the airport operator; to estimate the costs associated with these facility plans; and to project the sources and level of revenues to support the capital improvement program.

It is essential in the planning and design of an airport to have realistic estimates of the future demand of airports. This is a basic requirement in developing either an airport master plan or an aviation system plan. These estimates determine the future needs for which the physical facilities are designed. A financial plan to achieve the recommended staged development along with required land-use zoning usually accompanies the plan. Clearly an airport is designed for a projected level and pattern of demand, and alterations in the magnitude or configuration of this demand require the implementation of facility modifications or airport operational measures to meet changing needs. Facility planning is necessary to provide adequate levels of service to airport users.

The development of accurate forecasts requires a considerable expenditure of time and other resources because of the complex methodologies that must be used and the extensive data acquisition that is often required. The usual justification for a demand forecast in an aviation plan is that the expected level of uncertainty associated with the estimation of essential variables will be reduced, thereby reducing the probability of errors in the planning process and enhancing the decision-making process. The implication, of course, is that the benefits gained from better knowledge of the magnitude and fluctuation in demand variables will outweigh the costs incurred in performing the forecast.

To assess the characteristics of future demand, the development of reliable predictions of airport activity is necessary. Pertinent data to be collected to estimate future volumes of aircraft, passenger, freight, mail, and express traffic include information concerning:

1. The region served by the airport, i.e., the airport service area
2. The origins and destinations of trips by both residents and nonresidents of the area, i.e., the air travel market area
3. The demographic and population growth characteristics of the area to be served by the airport
4. The economic character of the area, as indicated by such factors as
   a. Per capita disposable income levels
   b. The types and levels of business activity as well as the nature (professional, skilled, unskilled) and size of the labor force employed in local industries (agriculture, construction, manufacturing, tourism, mining, finance, insurance, real estate, etc.)
   c. Retail, commercial, and wholesale trade sales, and bank deposits
   d. Hotel and motel registrations
5. Trends in the existing transportation activities for passengers, freight, express, and mail by various modes
6. Trends in national traffic including estimates of the variables affecting future development
7. Distance, population, and industrial character of nearby areas having air service

8. Geographic factors influencing transportation requirements
9. The existence and degree of competition between airlines and among the several modes of transportation with respect to price, travel time, service frequency, and other quality-of-service parameters

Forecasts utilizing these types of data give the airport planner a solid foundation on which to initiate the development of plans to meet the future needs for airport facilities.

Over the years, certain techniques have evolved which enable airport designers to project future aeronautical demand. These are the principal items for which estimates are usually needed:

1. The volumes and peaking characteristics of passengers, aircraft, vehicles, freight, express, and mail

2. The number and types of aircraft needed to serve the above traffic

3. The number of based general aviation aircraft and the number of movements generated

4. The performance and operating characteristics of ground access systems

By using forecasting techniques, estimates of these parameters and a determination of the peak-period volumes of passengers and aircraft movements can be made. From these estimates concepts for the layout and sizing of terminal buildings, runways, taxiways, apron areas, and ground access facilities may be examined.

Forecasting demand in an industry as dynamic as aviation is an extremely difficult task, and if it could be avoided, it undoubtedly would. Nonetheless, estimates of traffic must be made as a prelude to the planning and design of facilities. It is important to remember that forecasting is not a precise science and that considerable subjective judgment must be applied to any analysis, no matter how sophisticated the mathematical techniques involved. By anticipating and planning for variations in predicted demand, the airport designer can correct projected service deficiencies before serious deficiencies in the system occur.

Levels of Forecasting

Demand estimates are prepared for a variety of reasons in aviation. Broad, large-scale aggregated macroforecasts are made by aircraft and equipment manufacturers, aviation trade organizations, government agencies, and others to estimate the market requirements for aviation equipment, trends in travel, personnel needs, air traffic control requirements, and other factors. Similarly, forecasts are made on a smaller scale to examine these needs in particular regions of an area and at specific airports.

In economics, forecasting is done on two levels, macroforecasting and microforecasting, and the same holds true in aviation. From the inception of the planning process for an airport, consideration is given to both levels. In airport planning, the designer must view the entire airport system as well as the airport under immediate consideration. Macroforecasts are forecasts of the total aviation activity in a large region such as a country, state, or metropolitan area. Typical macroforecasts are made for such variables as the total revenue passenger-miles, total enplaned passengers, and number of aircraft operations, aircraft in the fleet, and licensed pilots in the country. Microforecasts deal with the activity at individual airports or on individual routes. Microforecasts for airport planning address such variables as the number of originations, passenger origin-destination traffic, number of enplaned passengers, and number of aircraft operations by air carrier and general aviation aircraft at an airport. Separate forecasts are usually made, depending upon the need in a particular study, for cargo movements, commuter service, and ground access vehicular traffic. These forecasts are normally prepared to indicate annual levels of activity and are then disaggregated for airport planning purposes to provide forecasts of the peaking characteristics of traffic during the busy hours of the day, days of the week, and months of the year. As appropriate to the requirements of an airport planning study, such quantities as the number of general aviation aircraft based at an individual airport and the number of general aviation and military operations are also forecast.

In macroforecasting, the entire system of airports is examined relative to the geographic, economic, industrial, and growth characteristics of a region to determine the location and nature of airport needs in a region. The microforecast then examines the expected demand at local airports and identifies the necessary development of the airside, landside, and terminal facilities to provide adequate levels of service. Within the two levels of forecasting certain techniques enable the aviation planner to project such parameters as annual, daily, and hourly aircraft operations; passenger enplanements; freight and mail tonnage; and general aviation activity. In microforecasting, there are quite a few variables of significance. The forecast of each variable is quite important because it ultimately determines the size requirements of the facilities that will be necessary to accommodate demand. Often the forecasts of the different variables are linked by a series of steps, i.e., originating passengers are forecast first, and then this becomes a component in the forecast of enplaned passengers, which leads to a forecast of annual operations, and so on. These techniques are discussed further in the next section.

Forecasting Methodologies

There are a considerable variety of forecasting techniques available to airport planners, ranging from subjective judgment to sophisticated mathematical modeling. The selection of the particular methodology is a function of the use of the forecast, availability of a database, complexity and sophistication of the techniques, resources available, time frame in which the forecast is required and is to be used, and degree of precision desirable. A large number of variables affect a facility plan. It is important that each variable be considered in the context of its use in the plan and that forecasting techniques be utilized so as to minimize the uncertainty associated with the range of the forecast variable. For those variables which significantly affect the nature and extent of facilities, more sophisticated techniques are often warranted, and redundancy might be sought through the utilization of several techniques. Such a procedure tends to narrow the expected range of the forecast parameter over the planning horizon. Quite often forecasts are prepared under a varying set of assumptions or scenarios to further define the reasonable bounds of the design parameters.

Forecasting by judgment

An important element in any forecasting technique is the use of professional judgment. A forecast prepared through the use of mathematical relationships must ultimately withstand the test of rationality. In the preparation of the forecast of the annual enplaned passengers at Fort Lauderdale–Hollywood International Airport, the observed trends were modified through the use of judgment factors. The reason for selecting the projection shown in Fig. 6-1 was knowledge of the fact that, although the enplaned passenger growth rate at this airport was outpacing the national growth rate, this growth rate was also decreasing.

Frequently a group of professionals knowledgeable about aviation and the factors influencing aviation trends are assembled to examine forecasts from several different sources, and composite forecasts are prepared in accordance with the information in these sources and the collective judgment of the group. In some cases, judgment becomes the principal approach used with or without an evaluation of economic and other factors believed to affect aviation activity. A common approach being utilized more often today for preparing forecasts by judgment is known as the *Delphi method.* In this method a panel of experts on a particular subject are asked to rate or otherwise prioritize a series of questions or projections through a survey technique. The results of the survey are then distributed to the members of the panel, and an opportunity is given to each member to reevaluate the original rating based upon the collective ratings of the group.

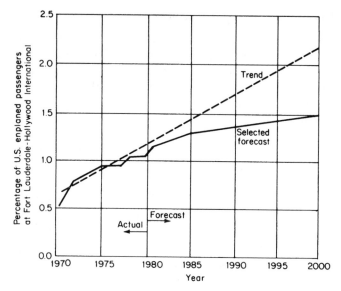

Figure 6-1 Percentage of total U.S. enplaned passengers at Fort Lauderdale–Hollywood International Airport (*Aviation Planning Associates, Inc. [32]*).

The reevaluation process is often sent through several iterations to arrive at a better result. In the Delphi method, the results of the technique do not have to represent a consensus of the panel; in fact, it is often quite useful to have a forecast that indicates the spread of the panel in reaching conclusions on a particular issue.

The preparation of judgment forecasts which reflect the collective wisdom of a broad range of professionals has proved very successful in many instances, principally due to the large number of factors which may be considered in such a process. Although there is often a lack of mathematical sophistication in the process, the knowledge and consideration of the many diverse factors influencing aviation forecasts usually improve the results. The disadvantages of this forecasting technique include the absence of statistical measures on which to base the results and the inability, except in the most obvious cases, to gain a significant consensus relative to the expected performance of the explanatory factors in the future.

Trend projection and extrapolation

Extrapolation is based upon an examination of the historical pattern of activity and assumes that those factors which determine the variation of traffic in the past will continue to exhibit similar relationships in the future. This procedure utilizes times-series type of data and seeks to analyze the growth and growth rates associated with a particular aviation activity. In practice, trends appear to develop in situations in which the growth rate of a variable is stable in either absolute or percentage terms, there is a gradual increase or decrease in growth rate, or there is a clear indication of a market saturation trend over time [16]. Statistical techniques are used in defining the reliability and expected range in the extrapolated trend. The analysis of the pattern of demand generally requires that upper and lower bounds be placed on the forecast, and statistics are used to define the confidence levels within which specific projections may be expected to be valid. From the variation in the trends and the upper and lower bounds placed on the forecast, a preferred forecast is usually developed. Quite often smoothing techniques are incorporated into the forecast to eliminate short-run, or seasonal, fluctuations in a pattern of activity which otherwise demonstrates a trend or cyclical pattern in the long run [16]. The simplest smoothing function might be a running average. Several types of extrapolation are commonly used, including linear and curvilinear trend extrapolation. In either case, variables which are suspected to demonstrate relationships to time periods are usually plotted on graph paper or are analyzed in spreadsheet computer programs, and a determination of the probable functional relationship between the variables is made. In practice, it may be necessary to combine several different types of trend extrapolation methods to prepare forecasts, particularly when these forecasts span relatively long times. The necessity of such an approach becomes intuitively obvious when one considers the likelihood of variables demonstrating consistent functional relationships throughout the diverse periods of aviation development.

Linear extrapolation. This technique is used for a pattern of demand that demonstrates an historical linear relationship with a time variable. The underlying relationship may be observed to be constant or to vary in a regular, seasonal, or cyclical pattern. Examples of these types of trends are shown in Fig. 6-2.

Exponential extrapolation. For a situation in which the dependent variable demonstrates a growth rate that is constant with time, an exponential extrapolation is usually indicated. This phenomenon occurs often in aviation for projections of levels of activity which have shown long-term trends to increase or decrease by an average annual percentage. Population statistics have been shown to demonstrate such a variation in the past. Examples of trends which demonstrate exponential variations are shown in Fig. 6-3. Observe that the exponential relationship shown plots as a straight line on log-linear scale graph paper.

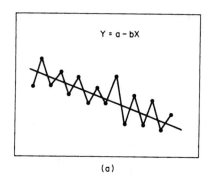

$$Y = a - bX$$

(a)

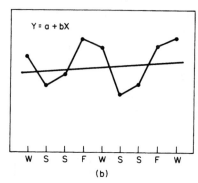

$$Y = a + bX$$

W   S   S   F   W   S   S   F   W

(b)

Figure 6-2 Typical linear trend data for linear extrapolation of aviation projections. (*a*) Declining linear trend for annual data; (*b*) increasing cyclical trend for seasonal data.

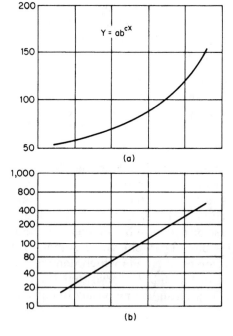

$$Y = ab^{cx}$$

(a)

(b)

Figure 6-3 Typical exponential-curve trend variation of data. (*a*) Trend on linear scale, graphical plot; (*b*) trend on log-linear scale graphical plot.

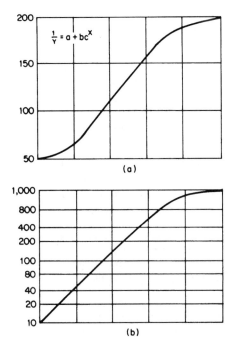

Figure 6-4 Typical logistics-curve trend variation of data. (a) Trend on linear scale, graphical plot; (b) trend on log-linear graphical plot.

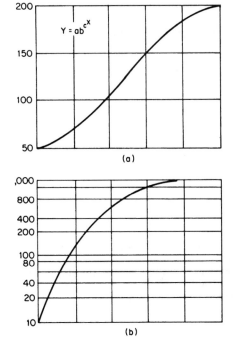

Figure 6-5 Typical Gompertz-curve trend variation of data. (a) Trend on linear scale, graphical plot; (b) trend on log-linear graphical plot.

Logistics curve extrapolation.   In situations where the average annual growth rate begins to gradually diminish with time, it may be appropriate to analyze trends through the use of a logistics curve. As aviation markets evolve, there is often an initial period of gradually increasing annual growth, an intermediate period of fairly constant growth, and a final period in which the growth rate diminishes to some point where market saturation has occurred. This type of phenomenon exhibits itself as an S-shaped curve, in which the dependent variable asymptotically approaches an upper limit. A curve quite similar to the logistics curve, but of different functional form, is the Gompertz curve. Both curves are shown in Figs. 6-4 and 6-5.

An illustration of the application of a trend line analysis to forecast annual enplanements at an airport is given in Example Problem 6-1.

Example Problem 6-1   The historical data shown in Table 6-1 have been collected for the annual passenger enplanements in a region and one of the commercial service airports in this region. It is necessary to prepare a forecast of the annual passenger enplanements at the study airport in the design years 1995 and 2000 by doing a trend line analysis.

In applying the trend line analysis to these data, a forecast of the annual enplanements at the study airport will be made by forecasting the historical trend to the design years. A plot of the trend in the annual passengers enplaned at the study airport is given in Fig. 6-6.

By extrapolating the trend into the future, an estimate of the annual enplaned passengers at the study airport in 1995 is found to be 2,100,000 and in the year 2000 is found to be 2,900,000 passengers.

The inability of trend projection techniques to show a causal relationship between the dependent and independent variables is a seri-

TABLE 6-1   Enplanement Data for Airport
Demand Forecast for Example Problems 6-1
through 6-3

| Year | Annual enplanements | |
	Regional	Airport
1983	13,060,000	468,900
1984	14,733,000	514,300
1985	16,937,000	637,600
1986	21,896,000	758,200
1987	24,350,000	935,200
1988	28,004,000	995,500
1989	31,658,000	1,139,700
1990	37,226,000	1,360,700
1991	40,753,000	1,488,900
1992	44,018,000	1,650,600

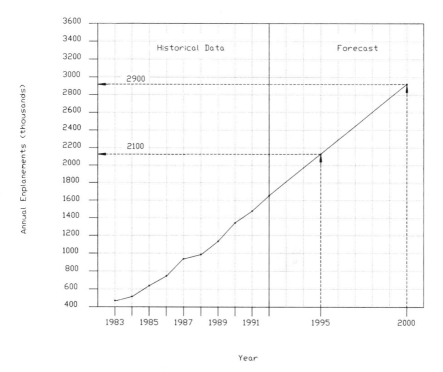

Figure 6-6   Trend line forecast of annual enplanements for Example Problem 6-1.

ous drawback. This is particularly true because, in the absence of such relationships, the degree of uncertainty in such forecasts increases with time. However, trend projections are useful for short-term forecasts in which the response to changes in those factors which stimulate the dependent variables is usually less dynamic. In those cases where cyclic variations may be expected, trend projections may also be quite beneficial.

Market analysis methods

Market share models. Forecasting techniques which are utilized to scale a large-scale aviation activity down to a local level are called *market share, ratio,* or *top-down models.* Inherent to the use of such a method is the demonstration that the proportion of the large-scale activity which can be assigned to the local level is a regular and predictable quantity. This method has been the dominant technique for aviation demand forecasting at the local level, and its most common use is in the determination of the share of total national traffic activi-

TABLE 6-2   Active General Aviation Aircraft in Florida, Historical
and Projected Data

Year	Active aircraft in fleet		Percentage of Florida aircraft to U.S. aircraft
	United States	Florida	
Historical			
1960	68,727	2,238	3.26
1961	75,655	2,561	3.39
1962	80,632	2,770	3.44
1963	84,121	2,882	3.43
1964	85,088	3,018	3.55
1965	88,492	3,280	3.71
1966	95,176	3,491	3.67
1967	104,378	4,035	3.87
1968	113,830	4,399	3.86
1969	123,791	4,831	3.90
1970	130,641	5,321	4.07
1971	131,743	5,531	4.20
1972	131,148	5,654	4.31
1973	145,010	6,514	4.49
1974	153,540	7,241	4.72
Forecast			
1980	186,000	9,430	5.07
1985	241,000	13,350	5.54
1990	308,000	18,510	6.01
1995	400,000	25,920	6.48

SOURCE: Landrum and Brown, Inc. [17].

ty that will be captured by a particular region, traffic hub, or airport. Historical data are examined to determine the ratio of local airport traffic to total national traffic, and the trends are ascertained. From exogenous sources the projected levels of national activity are determined, and these values are then proportioned to the local airport based upon the observed and projected trends. The ratio method is most commonly used in the development of microforecasts for regional airport system plans or for airport master plans.

An example of the data used to prepare a forecast of the number of active aircraft in Florida is given in Table 6-2. These data are plotted in Fig. 6-7 to extrapolate a market share or ratio which can be used to estimate the dependent variable in later years.

These methods are particularly useful in applications in which it can be demonstrated that the market share is a regular, stable, or predictable parameter. For example, the number of annual enplaned passengers at major air traffic hubs has been shown to be a consistent and relatively stable factor, and so this method is often used to predict this parameter [7]. An indication of this stability is shown in Table 6-3 for O'Hare International Airport.

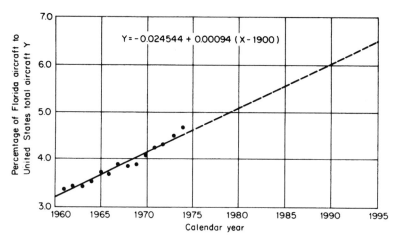

Figure 6-7 Historical trend and projection of trends for ratio of Florida aircraft to total U.S. aircraft (*Landrum and Brown, Inc. [17]*).

Quite often application of the market share technique is a two-step process in which a ratio is applied to disaggregate activity forecasts from a national to a regional level and then another ratio is applied to apportion the regional share among the airports in the region. Table 6-4 illustrates this procedure for the projection of the volume of air cargo at Fort Lauderdale–Hollywood International Airport.

The most compelling advantage of the market share method is its dependence upon existing data sources which minimizes forecasting

TABLE 6-3   Historic Domestic Scheduled Enplaned Passengers on Domestic and Local Service Carriers for the United States and Chicago, 1962–1974

Year	Domestic scheduled enplanements, thousands		Chicago market share, %
	United States	Chicago	
1962	59,087	5,753	9.74
1963	67,544	6,898	10.21
1964	76,985	7,898	10.26
1965	89,588	9,081	10.14
1966	102,718	10,254	9.98
1967	124,886	11,917	9.54
1968	142,199	13,336	9.38
1969	150,858	13,851	9.18
1970	149,338	13,330	8.91
1971	151,783	13,590	8.95
1972	167,091	14,885	8.91
1973	177,203	15,485	8.74
1974	183,193	16,183	8.83 ·

SOURCE: Landrum and Brown, Inc. [7].

TABLE 6-4   Domestic Air Cargo Forecast for Fort Lauderdale–Hollywood
International Airport (FLL), 1985–2000

Year	U.S. cargo, 10^6 lb	Regional share, %	Regional cargo, 1000 lb	FLL share of regional, %	FLL cargo, 1000 lb
Historical					
1974	6,854	2.19	149,895	8.06	12,081
1975	6,364	2.05	130,495	6.74	8,794
1976	6,758	2.01	135,869	9.33	12,681
1977	7,174	2.01	144,059	11.64	16,775
1978	7,686	2.09	160,816	15.53	24,967
1979	7,472	2.10	157,123	16.50	25,930
1980*	7,020	2.28	159,842	15.88	25,384
Forecast					
1985	9,494	2.08	197,500	19.00	37,500
1990	12,710	2.08	264,400	21.00	55,500
1995	16,580	2.08	344,900	22.00	75,900
2000	21,720	2.08	451,800	22.50	101,700

*Estimated.
SOURCE: Aviation Planning Associates, Inc. [32].

cost. However, its principal disadvantages lie in its dependence upon the stability and predictability of the ratios from which the forecasts are made and the uncertainty which may surround market shares in specific applications. An analysis of the factors which determine the scaling of shares may not be explicitly considered, and the introduction of a stimulus or restraint can significantly alter the relationships used to obtain the forecast. This type of situation arose in the early days following the deregulation of the airlines in the United States where significant changes in the service patterns of airlines occurred. Therefore, in such cases several forecasts may be required under different sets of assumptions which are deemed appropriate to the determination of market shares.

An illustration of the application of a market share analysis is given in Example Problem 6-2.

Example Problem 6-2   The historical data shown in Table 6-1 could also be used to prepare a forecast of the annual passenger enplanements at the study airport in the design years 1995 and 2000 by a market share method.

In applying the market share method to these data, a top-down forecast technique will be used. The implicit assumption in such a technique is that the same relationship between regional annual enplanements and annual enplanements at the study airport will be maintained in the future. To prepare such a forecast, a projection of the percentage of regional annual enplanements captured at the study airport is performed, and then a forecast is made of the regional annual enplanements. The study airport forecast percentage is applied to the regional forecast to arrive at the forecast of the study airport annual

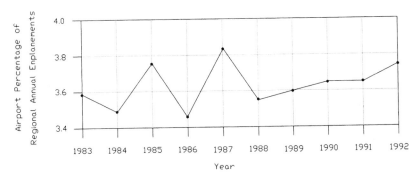

Figure 6-8 Trend line of airport percentage of regional annual enplanements for Example Problem 6-2.

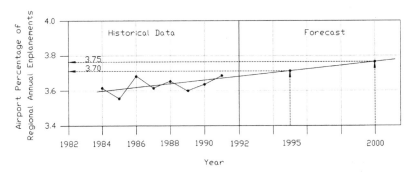

Figure 6-9 Trend line forecast of airport percentage of regional annual enplanements by applying smoothing function to trend data for Example Problem 6-2.

enplanements in the design years. A plot of the trend in the percentage of regional annual passengers enplaned at the study airport is shown in Fig. 6-8.

Because the variations shown in Fig. 6-8 often make it difficult to determine whether trends exist, a smoothing function is often applied to the data to assist in identifying trends which may not be obvious. In this case, a smoothing of the data was obtained by computing a running 3-year average of the data points. As shown in Fig. 6-9, this tends to smooth out the variations in the original data and more readily identifies trends in these data.

Although it may not be apparent in the original plot, the smoothing mechanism does indicate a very slight upward trend in the percentage of regional annual passengers captured by the study airport. This trend is shown by the dashed line on this figure. This trend line, when projected to the design years, forecasts 3.70 percent in the year 1995 and 3.75 percent in the year 2000 as the proportion of regional annual enplanements by the study airport.

To complete the forecast by the market share method, the regional annual enplanements must be determined. This is done by extrapolating the trend in regional annual enplanements as shown in Fig. 6-10. This extrapolation indicates regional annual enplanements in the year 1995 of 52,200,000 and in the year 2000 of 63,500,000.

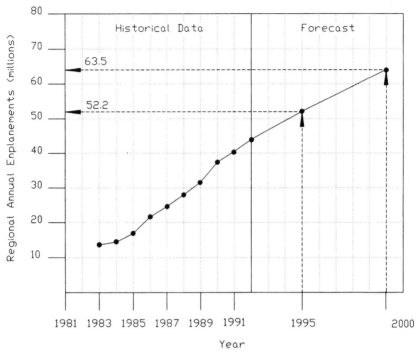

Figure 6-10 Trend line forecast of regional annual enplanements for Example Problem 6-2.

Therefore, the forecast for the annual enplanements at the study airport becomes $0.0370 \times 52,200,000 = 1,931,400$ passengers in the year 1995 and $0.0375 \times 63,500,000 = 2,381,250$ passengers in the year 2000.

Market definition models. This type of technique seeks to examine the behavioral characteristics of travelers and to separate travelers into distinct classifications based upon these characteristics. Inherent in the choice of such a method is the belief that certain socioeconomic characteristics influence the propensity for travel and that through the identification and classification of travelers according to such factors, it is possible to forecast travel patterns. In general, the population is grouped according to income, occupation, age, type and location of residence, education, and similar factors. From this a study of the travel characteristics of the individual groups is made. By utilizing the knowledge of these group travel patterns, it is possible to prepare forecasts by simply projecting the magnitude of the group in the future.

This technique does not depend significantly upon the assumed distribution of the underlying data, as is the case in statistically based techniques such as multiple regression analysis, but only requires sample sizes of significant magnitude to reasonably determine the travel preference of the group. In an early market study performed in New York [4], travel was divided into both personal travel and business travel. To analyze personal travel, the population was divided into 160 different groups, each characterized by a combination of income, occupation, education, and age. To analyze business travel, the labor force was divided into 130 groups, each characterized by a combination of income, occupation, and type of industry. In a more recent study, the travel habits and preferences of firms employing more than 100 persons were determined through surveys to determine air service requirements [21]. The market method does not require complex mathematical relationships and generally uses simple equations to generate a classification table or matrix.

One of the principal advantages of this method is that it allows for discrimination between discretionary and nondiscretionary travelers and those factors which influence travel behavior in these two distinct groups. The chief disadvantage is that it requires a fairly large population sample to identify the socioeconomic factors underlying travel choice.

Econometric modeling methods

The most sophisticated and complex technique normally used in aviation and airport demand forecasting is econometric models. Trend extrapolation methods do not explicitly examine the underlying relationships between the projected activity descriptor and the many variables which affect its change. A wide range of economic, social, market, and operational factors affect aviation. Therefore, to properly assess the impact of predicted changes in the other sectors of society upon aviation demand and to investigate the effect of alternative assumptions on aviation, it is often desirable to use mathematical techniques to study the correlations between dependent and independent variables. Econometric models that relate measures of aviation activity to economic and social factors are extremely valuable techniques in forecasting the future.

Multiple regression analysis. A great variety of techniques are used in econometric modeling for airport planning. Classical trip generation and gravity models are quite common in forecasting passenger and aircraft traffic. Simple and multiple regression analysis techniques, both linear and nonlinear, are often applied to a great variety of forecasting problems to ascertain the relationships between the dependent variables and such explanatory variables as economic and population growth, market factors, travel impedance, and intermodal competition. The form of the equations used in multiple linear regression analysis is given in Eq. (6-1):

$$Y_{est} = a_0 + a_1 X_1 + a_2 X_2 + a_3 X_3 + \cdots + a_n X_n \qquad (6\text{-}1)$$

where $\qquad Y_{est}$ = dependent variable or variable being forecast

$X_1, X_2, X_3, \ldots, X_n$ = dependent variables or variables being used to explain variation in dependent variable

$a_0, a_1, a_2, a_3, \ldots, a_n$ = regression coefficients or constants used to calibrate the equation

Statistical testing of models

Many statistical tests can be performed to determine the validity of econometric models in accurately portraying historical phenomena and to reliably project demand. Although the constants may be found to define the general equation of the model, it is possible that the range of error associated with the equation may be large or that the explanatory variables chosen do not directly determine the variation in the dependent variable.

There may be a tendency when one is performing sophisticated mathematical modeling to become disassociated from the significance of the results. It is incumbent upon the analyst to consider the reasonableness as well as the statistical significance of the model. Adequate consideration must be given to the rationality of the functional form and variables chosen for the analysis, and to the logic associated with calibrated constants.

One of the first statistical tests run on a model is the computation of the *coefficient of multiple correlation.* A high value indicates that there is a close correlation between the dependent variable and the independent variables, whereas a low value indicates a low correlation between the dependent variable and the independent variables. The *coefficient of multiple determination* measures the amount of variation in the dependent variable that is explained by the variation in the independent variables. A high value indicates that the high amount of the variation in the dependent variable is explained by the variation in the independent variables. The relationship used to determine the coefficient of multiple determination is given in Eq. (6-2):

$$R^2 = \frac{\Sigma(Y_{est} - Y_{av})^2}{\Sigma(Y - Y_{av})^2} \qquad (6\text{-}2)$$

where R^2 = coefficient of multiple determination that explains proportion of variation in dependent variable Y which is explained in the equation by variation in independent variables $X_1, X_2, X_3, ..., X_n$

Y = original values of dependent variable for each data point

Y_{est} = computed or estimated values of dependent variable for each data point

Y_{av} = average value of dependent variable over all data points

The coefficient of multiple correlation is simply the square root of the coefficient of multiple determination.

The *standard error of the estimate* is a measure of the dispersion of the data points about the regression line and may be used to establish the confidence limits within which projections made from the equation may be expected to lie. Similarly, tests may be performed to determine the statistical significance of calibrated constants in the equation. Knowledge of the range in these computed parametric quantities allows the degree of uncertainty in a demand forecast to be ascertained. The relationship used to determine the standard error of the estimate is given in Eq. (6-3):

$$\sigma_{y_{est}} = \left[\frac{\Sigma(Y - Y_{est})^2}{m - (n + 1)} \right]^{0.5} \tag{6-3)]}$$

where $\sigma_{y_{est}}$ = standard error of estimate

m = number of data points upon which equation is based

n = number of independent variables in equation

In many cases it is essential to determine the sensitivity of forecasts to changes in the explanatory variables. If a particular design parameter being forecast varies considerably with a change in a dependent variable, and if there is a significant degree of unreliability in this independent variable, then a great deal of confidence cannot be placed in the forecast and, more importantly, the design based upon the forecast. Tests are usually performed to determine the explanatory power of the independent variables and their interrelationships. The analyst should carefully investigate the sensitivity of projections within the expected variation of explanatory variables. It is also possible that certain explanatory variables do not significantly affect the modeling equation and that the need for collecting the data associated with these variables required for projections could be eliminated.

An illustration of the application of simple linear regression analysis is presented in Example Problem 6-3.

Example Problem 6-3 The historical data shown in Table 6-1 could also be used to prepare a forecast of the annual passenger enplanements at the study airport in the design years 1995 and 2000 by a simple regression analysis.

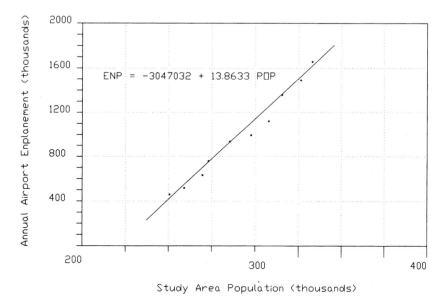

Figure 6-11   Regression line and data points for Example Problem 6-3.

In applying simple regression analysis to these data, let us assume that a relationship between the study airport annual enplanements (ENP) and the study area population (POP) is to be examined. Therefore, we assume that a linear relationship of the form shown in Eq. (6-1) exists between the variables.

$$\text{ENP} = a_0 + a_1(\text{POP})$$

By using a standard regression analysis computer program, the relationship is found to be

$$\text{ENP} = -3,047,032 + 13.8633(\text{POP})$$

where the coefficient of determination R^2 is 0.983815, the coefficient of correlation is 0.991874, and the standard error of the estimate $\sigma_{y_{est}}$ is 55,520.9.

The regression line and the data points upon which this regression line is based are shown in Fig. 6-11.

The coefficient of determination indicates that there is an extremely good relationship between the annual enplanements at the study airport and the study area population; that is, 98.4 percent of the variation in the study airport annual enplanements is explained by the variation in the study area population.

The standard error of the estimate, however, indicates that there is a large range of error associated with forecasting with this equation; i.e., there is a 68 percent probability that the forecast of annual enplanements at the study airport will have an error range of $\pm$ 55,520.9 annual enplanements. This may or may not be too high depending upon the level of annual operations forecast in the future and the sensitivity of various components of the airport system to such variations.

If the forecast of the area population in the year 1995 is found from a trend line extrapolation, as shown in Fig. 6-12, and is expected to be 363,000, then

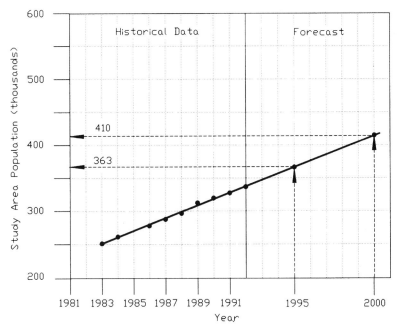

Figure 6-12 Trend line forecast of study-area population for Example Problem 6-3.

the forecast of the annual enplanements at the airport in the year 1995 can be found by substitution into the regression equation, yielding 1,985,300 annual enplanements. Similarly, if the forecast of the area population in the year 2000 is expected to be 410,000, then the forecast of the annual enplanements at the airport in the year 2000 is 2,636,900.

Given the range in the standard error of the estimate, it could be expected that in the year 1995 there is a probability of 68 percent that the forecast could range between 1,985,300 ± 55,500, or from 1,929,800 to 2,040,800 annual enplanements, about 68 percent of the time. Similarly, it could be expected that in the year 1995 the forecast could range between 2,636,900 ± 55,500, or from 2,581,900 to 2,692,400. It is likely that this range in forecast is acceptable since it represents about a 2 to 3 percent error.

It is interesting to compare the results found by the two different techniques used in Example Problems 6-1 through 6-3. It will be seen that the results compare very well. The determination of forecasts by two different means gives one some degree of confidence in the results when the two forecasts compare well. This is called *redundancy* in forecasting.

Based upon the results found in these example problems, a preferred forecast would be developed. If there is no reason to suspect that one technique is better than another, then a simple average might be used to develop the preferred forecast. If this is done in these examples, then the preferred forecast for the year 1995 is about 2 million and in the year 2000 is 2.6 million annual enplaned passengers.

TABLE 6-5   Explanatory Variables Used in Air Transportation Studies to Forecast Aviation and Airport Demand

Factors analyzed	Explanatory variables
Market	
Size and traffic potential	Population
	Industrial production
	Personal income:
	Total per capita
	Disposable
	Discretionary
	Personal expenditures:
	Total
	Travel and recreation
	Propensity for travel
	Leisure time:
	Attractions
	Availability
	Interregional linkages:
	Economic
	Social
Transportation	
Accessibility	Distance to airport
	Travel time to airport
Competition	Alternative airports
	Alternative modes of travel:
	Relative cost
	Relative travel time
	Schedule
	Reliability
Cost of air travel	Average fare
	Fare discounting policies
	Total travel cost
	Value of time
Schedule convenience	Service frequency
	Time of departure
	Necessary connections
Service reliability	On-time performance
	Cancellation history
Transport time	Airport-to-airport time
	Door-to-door time

One of the more common sensitivity tests conducted in aviation studies is the examination of fare and travel time elasticities of passenger and freight traffic. The *elasticity* is simply the percentage change in traffic for a 1 percent change in fare or travel time. Although it was an important factor in projecting demand fluctuations in aviation in the past, it has gained greater significance today due to vast changes in fares and service offerings in a deregulated industry.

A compilation of many of the variables used in aviation studies performed by the econometric modeling technique is given in Table 6-5. The FAA utilized an econometric model and multiple regression techniques to project the national level of annual air carrier enplanements from 1976 through 1987 [8]. This macroforecast model is given in Eq. (6-4):

$$ENP = -75.01 + 1.64CMP - 0.04APSU + 1.98PAT - 0.17REL - 5.59STR \qquad (6\text{-}4)$$

where
ENP = level of scheduled domestic revenue enplanements
CMP = number of civilians employed
APSU = annual purchase of automobiles
PAT = private investment in air transportation plants, equipment
REL = price of air transportation relative to other transport modes
STR = dummy variable to estimate impact of strikes on demand for air travel

An econometric model which has been used in several transportation planning studies, including the Northeast Corridor Project, was adapted for use in the Michigan State Airport System Plan [23]. In the series of equations which form the model to predict the total intercity passenger demand, projections are required for the population and user cost, travel time, and frequency of service by private automobile, rail, bus, and air transportation between the cities. The principal model equations are given in Eqs. (6-5) and (6-6).

$$V_{ij} = B_i B_j P_i P_j W_{ij}^{\ n} \qquad (6\text{-}5)$$

where
V_{ij} = total one-way annual passenger flow between city pair ij
B_i, B_j = calibrated constants related to measures of economic activity in cities
P_i, P_j = populations of cities
W_{ij} = measure of total transport impedance between cities, i.e., sum of travel impedance by each mode w_{ijk}
n = numerical constant

$$w_{ijk} = a_k t_{ijk}^{\ pk} c_{ijk}^{\ qk} (1 - e^{-mfijk})^{rk} \qquad (6\text{-}6)$$

where
w_{ijk} = measure of travel impedance between cities i and j on mode k
a_k, p_k, q_k, r_k = calibration constants for various modes of transport k
c_{ijk} = user travel cost on mode k between cities i and j
t_{ijk} = travel time on mode k between cities i and j

$$f_{ijk} = \text{frequency of service of mode } k \text{ between cities } i$$
$$\text{and } j$$
$$m, e = \text{numerical constants}$$

The success in applying mathematical modeling techniques to ascertain the level of future activity depends to a large extent upon the certainty associated with the independent variables and the relative influence of these variables on the dependent variable. Simple and multiple regression analysis methods are often applied to a great variety of forecasting problems to determine the relationships between transport-related variables and such explanatory factors as economic and population growth, market factors, travel impedance, and competitive forces. Table 6-6 lists many of the variables required for various purposes in aviation planning studies.

An excellent example of the ability of a mathematical equation to predict aviation traffic trends is shown in Fig. 6-13. In the study from which this figure is extracted, a nonlinear formulation of a regression model was constructed relating total miles flown per capita on domes-

TABLE 6-6   Typical Air Transportation Forecast Variables and Their Use in Aviation and Airport Studies

Application in planning studies	Forecast variables required
Macroforecast	Revenue passengers
National airport system needs	Revenue passenger-miles
State or regional airport system needs	Aircraft fleet
Airlines	Air carrier
Aircraft and equipment manufacturers	General aviation
Investment planning	Composition
Research and development needs	Size
Route planning	Capacity
Workforce requirements	Enplaned passengers
	Aircraft operations
Microforecast	Number of airline pilots and navigators
Airport facilities	
Airside	Aircraft operations
Runways	Air carrier:
Taxiways	Fleet mix
Apron area	Capacity
Navigational aids	Peak hour
General aviation needs	General aviation based aircraft
Landside:	
Gates	Passenger traffic:
Terminal facilities	Enplaned
Cargo processing	Origin, destination
Ground access:	Connections
Curb frontage	Freight and cargo tonnage
Parking	Airmail tonnage
Access highways	Peaking characteristics
Public transit	Originations
	Vehicular traffic

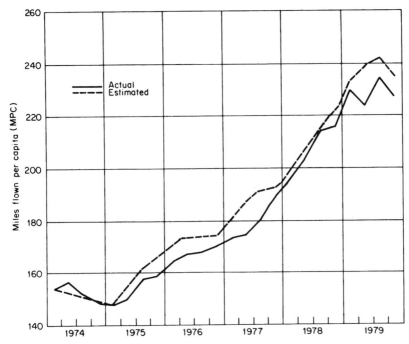

Figure 6-13 Comparison of actual and estimated total miles flown per capita (MPC) on domestic trunk and local service carriers (*Port Authority of New York and New Jersey [30]*).

tic trunk and local service carriers to the per-capita gross national product and the fare yield per passenger-mile [30]. Obviously the equation portrays quite well both the magnitude and the fluctuation of the known trends.

Forecasting Requirements and Applications

The specific forecasting needs depend upon the nature and scope of the study being undertaken. The requirements for a state aviation system plan are very different from those for an airport master plan. Facility planning requires projections of the parameters which determine physical design whereas financial planning requires projections of the cost elements and revenue sources associated with physical development. This section outlines the general forecasting requirements for various types of aviation and airport studies and discusses the more common methodologies used to arrive at these requirements.

The aviation system plan

The purpose of an aviation system plan is to identify the aviation development required to meet the immediate and future aviation needs of a region, state, or metropolitan area [25, 26]. The plan recommends the general location and characteristics of new and expanded airports; shows the timing, phasing, and estimated cost of development; and identifies revenue sources and legislation for implementation of the plan. The aviation system plan also provides a basis for the definitive, detailed development of individual airports in the system.

The primary forecasting requirement for the airport system plan is a projection of the level of aviation activity during the planning period. The forecasts usually are made on an annual basis for the planning entity as a whole and then are proportioned among the various individual airports within this entity. Specific projections are generally made for total aircraft operations, air carrier and general aviation operations, based aircraft, total air cargo and airmail tonnage, and passenger enplanements for the short-, medium-, and long-range time frames within the planning period. These demand projections are compared to the inventory of physical facilities to determine development needs. It should be emphasized that these projections are normally made in a very aggregate fashion and tend to examine overall determinates of regional activity rather than the specific factors affecting local activity.

The preparation of a forecast for a systems plan is initiated by the collection of historical data indicative of the various components of aviation activity. These data normally include broad measures of socioeconomic activity as well as aviation activity statistics. Because data collection is very expensive, most data collection in a system plan depends primarily upon secondary or existing sources, and very little survey work is performed. Many of the overall regional projections are made on the basis of trend projections or simple econometric models, and then these are apportioned to the individual airports within the region on the basis of ratio methods. In the analysis of observed trends and the preparation of future forecasts, broad indicators are generally used, including an examination of the consistency and realization of past trends and a comparison of growth rates and economic indicators.

The airport master plan

The purpose of the airport master plan is to provide the specific details for the future development of an individual airport to satisfy aviation needs consistent with community objectives. The airport master plan requires detailed projections of the level of demand on the various facilities associated with the airport. Various concepts and alternatives for development are examined and evaluated, and recommendations are made concerning the prioritizing, scheduling, and financing of the plan [2]. Although the basis for projections in a system plan is the aggregate level of annual demand, this is not sufficient for the master plan. Projections must be made for the magnitude, nature, and variation of demand on a monthly, daily, and hourly basis on the many facilities located at the airport.

Annual forecasts of airport traffic during the planning period are the basis for preparation of detailed forecasts in the master plan. These forecasts are made for each type of major airport user, including air carrier, commuter, and general aviation aircraft and are often expressed in terms of upper and lower bounds. The master plan forecasts are usually made under both constrained and unconstrained conditions. An *unconstrained* forecast is made relative to the potential aviation market. In this type of forecast, the basic factors tending to create aviation demand are utilized without any regard to constraining factors that could affect aviation growth at the location. A *constrained* forecast is made in the context of alternative factors which could limit growth at the specific airport. Constraining factors addressed in master plans include limitations on airport capacity due to land availability and noise restrictions, development of alternative reliever airports to attract general aviation demand, policies which alter access to airports by general aviation and commuter aircraft operations, and availability and cost of aviation fuel. The determination of the level of general aviation activity at an airport can be significantly changed when land availability is restricted, thereby placing limits on airside capacity. The available capacity may be utilized in the context of policies which favor commercial over general aviation growth [32]. The impacts of such constraints can be quite significant, as shown in Fig. 6-14.

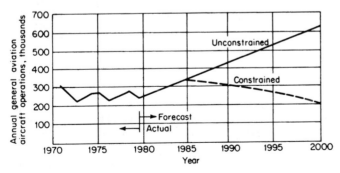

Figure 6-14 Constrained and unconstrained forecast of general aviation aircraft operations at Fort Lauderdale–Hollywood International Airport (*Aviation Planning Associates, Inc. [32]*).

For a master plan, forecasts are made for the annual, daily, and peak-hour operations by air carrier, commuter, air taxi, general aviation, cargo, and military aircraft; passenger originations, enplanements, and connections; annual tonnage and volumes of air freight and airmail; and daily and hourly ground access system and parking demand. Projections are also made for the mix or types of aircraft in each of the categories which will utilize the airport during the planning period. In the preparation of these forecasts, some variables are projected directly, and others are derived from these projections. For example, annual passenger enplanements might be forecasted from an econometric model, and then, based upon exogenous estimates of average air carrier fleet passenger capacity and boarding seat load factors, the annual air carrier operations could be derived from these data. An illustration of the derivation of projections of domestic air carrier operations from domestic passenger enplanements is given in Table 6-7. Similarly, by observing ratios which relate passenger enplanements to originating, connecting, and transfer passengers, the forecast of annual enplanements can be used to derive these other quantities.

It is essential for those responsible for preparing aviation forecasts to recognize that patterns of development exist for the various types and sizes of airports which are useful in the planning function. For example, there is an approximate relationship between the number of peak-hour enplanements on the peak-month average day and the level of annual enplanements for major air carrier airports serving the large hub areas in the United States. This relationship is shown in Fig. 6-15. Therefore, the analyst might wish to compare forecasts based upon modeling techniques to the relationship observed in this figure.

TABLE 6-7   Domestic Major and National Carrier Aircraft Operations at Fort Lauderdale–Hollywood International Airport, 1975–2000

Year	Domestic passenger enplanements	Average seats per departure	Boarding load factor	Domestic aircraft departures	Domestic aircraft operations
Historical					
1975	1,756,458	123.6	42.8	32,395	64,790
1976	1,944,402	126.4	41.2	36,283	72,566
1977	2,089,373	131.9	41.1	37,703	75,406
1978	2,664,211	142.8	46.8	39,424	78,848
1979	2,881,563	147.6	46.5	42,328	84,656
1980	2,810,525	144.1	46.9	41,612	83,224
Forecast					
1985	4,162,100	170.0	48.0	51,000	102,000
1990	5,436,200	190.0	50.0	57,200	114,400
1995	6,943,200	205.0	52.0	65,100	130,200
2000	8,735,600	210.0	53.0	78,500	157,000

SOURCE: Aviation Planning Associates, Inc. [32].

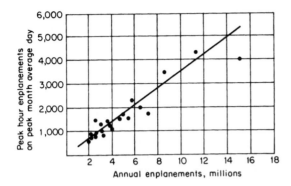

Figure 6-15 Relationship between peak-hour enplanements on the peak month average day and annual enplanements for large hub major air carrier airports (*Landrum and Brown, Inc. [7]*).

Similar characteristics have been shown to exist for monthly, daily, and hourly distribution of aircraft operations at air carrier and general aviation airports of differing levels of activity [24]. An illustration of the variation in peak-hour air carrier operations with annual enplaned passengers which is useful for order-of-magnitude estimates is given in Fig. 6-16.

Although these factors may not be precisely transferable between airports, they provide a general level of guidance to the airport planner in determining the reasonableness of forecasts in specific applications.

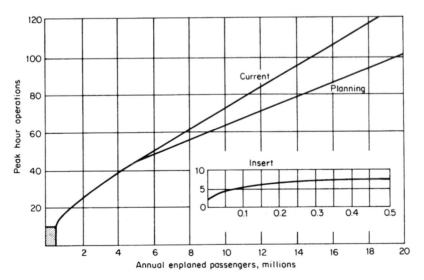

Figure 6-16 Estimating relation for peak-hour air carrier operations as a function of annual enplaned passengers (*Federal Aviation Administration [24]*).

For the most part, the methods used in the study of new airports are similar to those used for existing airports. However, the principal difference is the inability of the analyst to obtain a local historical database to generate extrapolation trends, market shares, or econometric models. To overcome this deficiency, an attempt is usually made to forecast by drawing an analogy between the subject airport and other existing airports which demonstrate similar traffic experience and which are located in areas with similar socioeconomic, demographic, and geographic characteristics. Forecasts are then made for the airport under consideration by using these airports as surrogates, and adjustments are performed to accommodate expected differences between the airports. In the past, the Air Transport Association and the FAA have collected and tabulated a significant amount of data for many airports [1, 29]. These data have included the number and distribution of commercial air carrier operations, fleet mix, and passengers on a peak and average monthly, daily, and hourly basis. It is apparent that one may expect a rather high degree of uncertainty associated with forecasting through such an analogy.

Future Aviation Forecasting Environment

Many of the forecasts made by the various aviation-related organizations become biased by the impact of recent events. Forecasts made in the early 1960s showed rather moderate growth whereas those made in the late 1960s showed fairly ambitious growth. These forecasts were made in the context of expectations which reflected the behavior of aviation at that time. In the 1970s and 1980s, the overall economic conditions, the availability and cost of petroleum-based fuels, and airline deregulation considerably affected aviation. Forecasts made in this era attempted to analyze the impact of these factors in projecting future demand for aviation. The effect of existing economic, social, and political factors on forecasts made in the 1960s and 1970s may be inferred from an examination of Fig. 6-17.

In a study of the major factors which will influence aviation over the period from 1980 through 2000, the following conclusions were drawn [5]:

1. Economic growth in most industrial nations will be moderate compared to historical trends.

2. Inflation rates will remain at or near the double-digit level.

3. Fuel prices will continue to rise at a pace which will exceed the inflation rate.

4. Concerns over jet fuel availability will dominate the demand for aviation services.

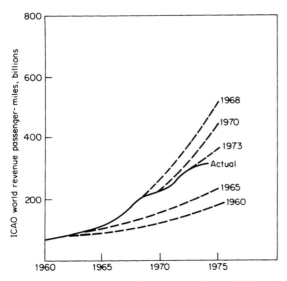

Figure 6-17 Variation in forecasts made at different points in times for ICAO world revenue passenger-miles (*Department of Transportation [27]*).

5. High to moderate unemployment rates will exist in the aviation industry.

6. The impact of recessions on aviation growth will rise.

7. Difficulties in attracting capital for investment in flight equipment and new markets will continue.

8. There will be continued and rising international tensions in the world.

On the basis of these assumptions, it is expected over this period that competition, high wage scales, and capital requirements will reduce the number of large carriers in long-haul markets and increase the number of specialized carriers providing low-fare and low-cost services in an increasingly competitive environment. Commuter and short-haul service will undergo considerable change which may be constrained by airspace and airside capacity limits, access to large hubs, and the uncertain role of small aircraft in the future. General aviation growth will be stimulated through the evolutionary effect of new technology, particularly avionics. Rising costs, however, will cause a shift away from personal and toward business general aviation flying.

As far as future trends in air travel are concerned, it is expected that there will be a greater growth of international air traffic which has been attributed to the globalization of the airline industry and to changing market forces in the United States. The factors which have contributed to the rise of the U.S. air transportation industry are changing. The steady decline in the real cost of air travel has reached a point where the unit costs in the 1990s will remain steady or slowly rise. The increased quality of service from improvements in the speed, comfort, convenience, and safety of air travel has been largely realized. Past demographic and cultural factors, such as the baby boom, are declining in importance. The rise in family discretionary income has peaked, and the use of discretionary income for air travel is meeting competition from mortgages, savings, and luxury goods. Foreign travel by U.S. citizens will be adversely affected by the decline in the exchange value of the dollar. Furthermore, this factor will encourage foreign nationals to travel to the United States [18].

A recent conference concerning aviation forecasting methodologies [6] concluded that financial concerns and financial forces are forcing the aviation industry to place more emphasis on short-term forecasting methods. Not only is this the case with the airlines, but also airports are shortening their planning horizons. Throughout the aviation industry there is a shift to simpler forecasting techniques requiring fewer variables and less detailed data. Wider use is being made of forecasting techniques that depend less on mathematical modeling and more on an analysis of different scenarios, judgment, and market segmentation. Scenarios are used to test basic assumptions and to explore alternatives. Although judgment has always played a significant role in demand forecasting, it is becoming more important as a subjective test of the reality associated with forecasting outcomes. There is a growing recognition that airline management strategies are important forces, shaping the future of aviation development.

Based upon the error inherent in forecasting and the differences in forecasts made by various organizations at the time, as may be observed in Fig. 6-18, recommendations were made concerning planning needs. These included the presentation of the range in key assumptions of the forecast, use of a variety of techniques to limit uncertainty, increased use of expert judgment to temper forecasts made from econometric models, and disaggregation of national forecasts to the regional, local, and market levels.

To adequately cope with the uncertainties associated with the traditional air transportation forecasting process and to react in a timely manner to inaccuracies found in estimates, the planning process is becoming a phase-oriented, continuing process. For example, the FAA prepares annual forecasts of aviation activity on a national and terminal-area basis which extend 10 years into the future [10, 11, 29]. Because of the high costs associated with the traditional planning process, the implementation of physical design changes, and the apparent inability to forecast with any degree of certainty, it is essen-

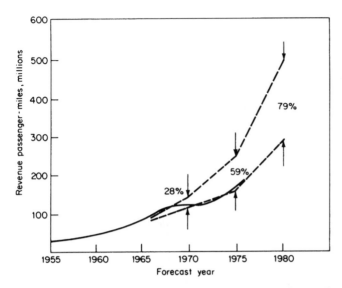

Figure 6-18 Percentage difference range between the highest and lowest forecasts made by different organizations at the same time for revenue passenger-miles in the U.S. domestic market (*Department of Transportation [27]*).

tial that planning techniques be developed which can respond to changes in the demand parameters prior to the investment decision. Perhaps the key to such a process lies in the recognition of the interaction of demand and supply parameters. Knowledge of the sensitivity of a physical facility component to a variation in demand can lead to more informed decisions and an understanding of the flexibility in facility design. A continued monitoring of the need for physical facilities in light of changing demand requirements provides a sound basis for the investment decision. Recognition of the uncertainties in the demand forecasting process can prevent a wasteful commitment of valuable resources. Explicit treatment of the variability of demand projections and facility modification recommendations through the use of sensitivity and tradeoff analyses is warranted.

References

1. *Aircraft Movement and Passenger Data for 100 U.S. Airports, Average Day in August 1978,* Air Transport Association of America, Washington, June 1979.
2. *Airport Master Plans,* Advisory Circular AC 150/5070-6A, Federal Aviation Administration, Washington, 1985.
3. *Airport System Capacity: Strategic Choices,* Special Report 226, Transportation Research Board, Washington, 1990.
4. "A Market Analysis Approach to Forecasting Domestic Air Travel," J. Legan, Port of New York Authority, New York, 1957.
5. *Assumptions and Issues Influencing the Future Growth of the Aviation Industry,* Circular 230, Transportation Research Board, Washington, August 1981.

6. *Aviation Forecasting Methodology: A Special Workshop,* Circular 348, Transportation Research Board, Washington, August 1989.
7. *Chicago O'Hare International Airport Master Plan Study,* vol. 3: *Aviation Demand Forecast,* Landrum and Brown, Inc., Cincinnati, Ohio, November 1979.
8. *FAA Aviation Forecasts, Fiscal Years 1976–1987,* Federal Aviation Administration, Washington, September 1975.
9. *FAA Aviation Forecasts, Fiscal Years 1981–1992,* Office of Aviation Policy, Federal Aviation Administration, Washington, September 1980.
10. *FAA Aviation Forecasts, Fiscal Years 1992–2003,* Federal Aviation Administration, Washington, February 1992.
11. *FAA Aviation Forecasts, Air Traffic Hub Series,* Forecast Branch, Office of Aviation Policy and Plans, Federal Aviation Administration, Washington, periodic.
12. *FAA Statistical Handbook of Civil Aviation,* Federal Aviation Administration, Washington, 1990.
13. *Forecasting Civil Aviation Activities,* Circular 372, Transportation Research Board, Washington, 1991.
14. "Forecasts of Aviation Demand, Miami International Airport Master Plan Update," draft report, Landrum and Brown, Inc., Chicago, March 1992.
15. *Forecasts of Commuter Airlines Activity,* Federal Aviation Administration, Washington, July 1977.
16. *Forecasting Methods for Management,* 5th ed., S. G. Makridakis, Wiley, New York, 1989.
17. *Fort Lauderdale–Hollywood International Airport Master Plan Report,* vol. 2: *Technical Supplement,* Landrum and Brown, Inc., Cincinnati, Ohio, August 1977.
18. *Future of Aviation,* Circular 329, Transportation Research Board, Washington, 1988.
19. *Future Development of the U.S. Airport Network,* preliminary report and recommended study plan, Transportation Research Board, Washington, 1988.
20. *Guide for the Planning of Small Airports,* Roads and Transportation Association of Canada, Ottawa, 1980.
21. *Lambert–St. Louis International Airport, 1980 Air Service Market Study,* Landrum and Brown, Inc., Cincinnati, Ohio, September 1980.
22. *Manual on Air Traffic Forecasting,* 2d ed., Document 8991-AT/722/2, International Civil Aviation Organization, Montreal, Canada, 1985.
23. *Michigan State Airport System Plan,* Technical Report, Michigan Aeronautics Commission, Department of State Highways and Transportation, Lansing, Mich., May 1975.
24. *Planning and Design Guidelines for Airport Terminal Facilities,* Advisory Circular AC 150/5360-13, Federal Aviation Administration, Washington, 1988.
25. *Planning the Metropolitan Airport System,* Advisory Circular AC 150/5070-5, Federal Aviation Administration, Washington, May 1970.
26. *Planning the State Airport System,* Advisory Circular AC 150/5050-3B, Federal Aviation Administration, Washington, 1989.
27. *Review of Aviation Forecasts and Forecasting Methodology,* Office of Transportation Planning Analysis, Department of Transportation, Washington, May 1975.
28. *Terminal Area Air Traffic Relationships, Peak Day-Busy Hour Fiscal Year 1980,* Office of Management Systems, Federal Aviation Administration, Washington, 1981.
29. *Terminal Area Forecasts, Fiscal Years 1991–2005,* Federal Aviation Administration, Washington, February 1991.
30. "Some Regression Models: U.S. Domestic Traffic," J. G. Augustinus, presentation to the Committee on Aviation Demand Forecasting of the Transportation Research Board, International Air Transportation Conference, Cincinnati, Ohio, May 1980.
31. *Trends and Issues in International Aviation,* Circular 393, Transportation Research Board, Washington, 1992.
32. *Traffic and Earnings Analysis, Fort Lauderdale–Hollywood International Airport,* Aviation Planning Associates, Inc., Cincinnati, Ohio, January 1982.

7

Airport Configuration

The *airport configuration* is the number and orientation of runways and the location of the terminal area relative to the runways. The number of runways provided at an airport depends on the volume of traffic. The orientation of these runways depends to a large extent on the direction of the prevailing wind patterns in the area, the size and shape of the area available for airport development, and land-use or airspace restrictions in the vicinity of the airport. The terminal buildings serving passengers should be located so as to provide easy and timely access to runways [12, 13].

Runways

In general, runways and connecting taxiways should be arranged so as to (1) provide adequate separations between aircraft in the air traffic pattern; (2) cause the least interference and delay in landing, taxiing, and takeoff operations; (3) provide the shortest taxi distance possible from the terminal area to the ends of the runways; and (4) provide adequate taxiways so landing aircraft can exit the runways as quickly as possible and follow the shortest possible routes to the terminal area.

At busy airports, holding or run-up aprons should be provided adjacent to the takeoff ends of the runways. These aprons should be designed to accommodate three or possibly four aircraft of the maximum size anticipated, with sufficient space for aircraft to bypass one another.

Taxiways

The principal function of taxiways is to provide access between runways and the terminal area and service hangars. Taxiways should be arranged so that aircraft which have just landed do not interfere with aircraft taxiing to take off. At busy airports where taxiing traffic is expected to move simultaneously in both directions, parallel one-way taxiways should be provided. Routes should be selected which will result in the shortest practicable distances from the terminal area to the ends of runways used for takeoff. Again, at busy airports taxiways should be located at various points along runways, so landing aircraft can leave the runways as quickly as possible to clear them for use by other aircraft; these are commonly referred to as *exit taxiways* or *turnoffs*. Whenever possible, taxiways should be routed so as to avoid crossing of active runways.

During peak traffic periods when a continuous supply of aircraft is available, the capacity of a runway is dependent to a large degree on how quickly landing aircraft can be vacated from a runway. An aircraft which has landed delays succeeding aircraft until it has cleared the runway. At many airports, exit taxiways are at right angles to the runways, so that aircraft must decelerate to very low speeds before they can safely turn off the runways. An exit taxiway with geometry designed to permit higher turnoff speeds reduces the time that a landing aircraft occupies the runway. This permits succeeding aircraft to be more closely spaced in terms of time, or it might allow a takeoff to be inserted between two successive landings.

Runway Configurations

Many runway configurations exist. Most are combinations of several basic configurations: single runways, parallel runways, intersecting runways, and open-V runways.

Single runway

This is the simplest of the runway configurations, and it is shown in Fig. 7-1a. It has been estimated that the hourly capacity of a single runway in visual flight rule (VFR) conditions is somewhere between 50 and 100 operations per hour, while in instrument flight rule (IFR) conditions this capacity is reduced to 50 to 70 operations per hour, depending on the composition of the aircraft mix and navigational aids available [4].

Parallel runways

The capacities of parallel-runway systems depend a great deal on the number of runways and on the spacing between them. Two and four parallel runways are common. There are airports with three sets of parallel runways. Airports with more than four parallel runways did not exist when this text was written, although plans to ultimately expand certain airports to five or more parallel runways did exist. It

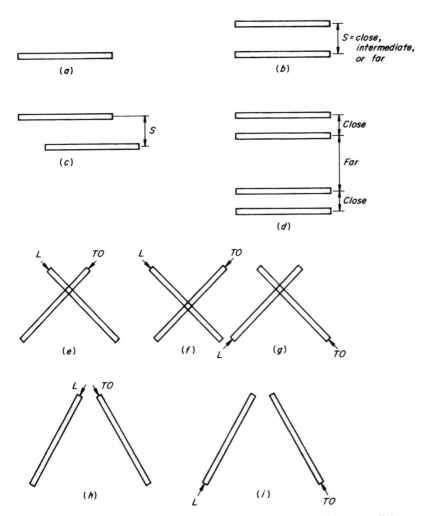

Figure 7-1 Typical runway configurations: (a) Single runway; (b) two parallel runways—even threshold; (c) two parallel runways—staggered threshold; (d) four parallel runways; (e) intersecting runways; (f) intersecting runways; (g) intersecting runways; (h) open-V runways; (i) open-V runways.

is unlikely that more than four parallel runways will exist except in rare circumstances, since few locations can generate the demand to match the capacity of five or more parallel runways. Furthermore, the ability of the air traffic control system to supply five or more runways at the same time becomes progressively more difficult, and the airspace requirement becomes very large. Finally, the availability of land to expand airports to this size is very restricted.

The spacing between parallel runways varies widely. For the purpose of this discussion, the spacing is classified as *close, intermediate,* and *far,* depending on the centerline separation between two parallel runways (Fig. 7-1*b*). Close parallel runways are spaced from a minimum of 700 ft (for air carrier airports) to less than 2500 ft [5]. In IFR conditions, operation of one runway is dependent upon the operation on the other runway. Intermediate parallel runways are spaced from 2500 to less than 4300 ft [5]. In IFR conditions, an arrival on one runway is independent of a departure on the other runway. Far parallel runways are spaced at least 4300 ft apart [5]. In IFR conditions, the two runways can be operated independently for both arrivals and departures. Therefore, as noted above, the centerline separation of parallel runways determines the degree of interdependence between operations on each of the parallel runways. In the future, the spacing requirements for simultaneous operations on parallel runways may be reduced. If this occurs, new spacings can be applied to the same classifications.

If the terminal buildings are placed between parallel runways, then runways are always spaced far enough apart to allow room for the buildings, adjoining apron, and appropriate taxiways. When there are four parallel runways, each pair is close, but the pairs are spaced far apart to allow space for terminal buildings (Fig. 7-1*d*).

In VFR conditions, close parallel runways allow simultaneous arrivals and departures; i.e., arrivals may occur on one runway while departures are occurring on the other runway. Aircraft operating on the runways must have wingspans less than 171 ft (airplane design groups I through IV, see Table 9-1) for centerline spacings at the minimum of 700 ft [5]. If larger-wingspan aircraft are operating on these runways (airplane design groups V and VI), the centerline spacing must be at least 1200 ft for such simultaneous operations [5]. In either case, wake vortex avoidance procedures must be used for *simultaneous* operations on closely spaced parallel runways. Furthermore, simultaneous arrivals to both runways or simultaneous departures from both runways are not allowed in VFR conditions for closely spaced parallel runways. In IFR conditions, close parallel runways cannot be used simultaneously but may be operated as dual-lane runways, as discussed later.

Intermediate parallel runways may be operated with simultaneous arrivals in VFR conditions. Intermediate parallel runways may be operated in IFR conditions with simultaneous departures if the centerline spacing is at least 3500 ft in a nonradar environment and at least 2500 ft in a radar environment [5]. Simultaneous arrivals and departures are also permitted if the centerline spacing is at least 2500 ft if the thresholds of the runways are not staggered [5]. At times it may be desirable to stagger the thresholds of parallel runways. The staggering may be necessary because of the shape of the acreage available for runway construction, or it may be desirable for reducing the taxiing distance of takeoff and landing aircraft. The reduction in taxiing distance, however, is based on the premise that one runway is to be used exclusively for takeoff and the other for landing. In this case the terminal buildings are located between the runways so that the taxiing distance for each type of operation (takeoff or landing) is minimized (Fig. 7-1c). If the runway thresholds are staggered, adjustment to the centerline spacing requirement is allowed for simultaneous arrivals and departures [5]. If the arrivals are on the near threshold, then the centerline spacing may be reduced by 100 ft for each 500 ft of threshold stagger down to a minimum centerline separation of 1000 ft for aircraft with wingspans up to 171 ft and a minimum of 1200 ft for larger-wingspan aircraft. If the arrivals are on the far threshold, the centerline spacing must be increased by 100 ft for each 500 ft of threshold stagger. Simultaneous arrivals in IFR conditions are not permitted on intermediate parallel runways but are permitted on far parallel runways with centerline spacings of at least 4300 ft [5].

The hourly capacity of a pair of parallel runways in VFR conditions varies greatly, from 60 to 200 operations depending on the aircraft mix and the manner in which arrivals and departures are processed on these runways [4]. Similarly, in IFR conditions, the hourly capacity of a pair of closely spaced parallel runways ranges from 50 to 60 operations, of a pair of intermediate parallel runways from 60 to 75 operations, and for a pair of far parallel runways from 100 to 125 operations [4].

Dual-lane runways

A dual-lane runway consists of two closely spaced parallel runways with appropriate exit taxiways. Although both runways can be used for mixed operations subject to the conditions noted above, the desirable mode of operation is to dedicate the runway farthest from the terminal building (outer) for arrivals and the runway closest to the terminal building (inner) for departures. It is estimated that a dual-lane runway can handle at least 70 percent more traffic than a single runway in VFR conditions and about 60 percent more traffic than a single runway in IFR conditions. Capacity has been found to be insensitive to the centerline spacings between the runways from 1000 to 2400 ft. It is therefore recommended that the two runways be spaced not less than 1000 ft apart (1200 ft where particularly larger-wingspan aircraft are involved). This spacing also provides sufficient runout distance for an arrival to stop between the two runways. A parallel taxiway between the runways will provide for a nominal increase in capacity, but it is not essential. The major benefit of a dual-lane runway is to provide an increase in IFR capacity with minimal acquisition of land [7, 14].

Intersecting runways

Many airports have two or more runways in different directions crossing each other. These are referred to as *intersecting runways.* Intersecting runways are necessary when relatively strong winds come from more than one direction, resulting in excessive crosswinds when only one runway is provided. When the winds are strong, only one runway of a pair of intersecting runways can be used, reducing the capacity of the airfield substantially. If the winds are relatively light, both runways can be used simultaneously. The capacity of two intersecting runways depends a great deal on the location of the intersection (i.e., midway or near the ends); the manner in which runways are operated for takeoffs and landings, referred to as the *runway-use strategy*; and the aircraft mix. This is illustrated in Fig. 7-1e to g. The farther the intersection is from the takeoff end of the runway and the landing threshold, the lower the capacity (Fig. 7-1g). The highest capacity is achieved when the intersection is close to the takeoff and landing threshold (Fig. 7-1e). For the strategy shown in Fig. 7-1e, the capacity ranges from 70 to 175 operations per hour in VFR conditions and from 60 to 70 operations per hour in IFR conditions [4]. For the strategy shown in Fig. 7-1f, the capacity ranges from 60 to 100 operations per hour in VFR and from 45 to 60 operations per hour in IFR [4]. For the strategy shown in Fig. 7-1g, the capacity ranges from 50 to 100 operations per hour in VFR and from 40 to 60 operations per hour in IFR [4].

Open-V runways

Runways in different directions which do not intersect are referred to as *open-V runways*. This configuration is shown in Fig. 7-1h and i. Like intersecting runways, open-V runways revert to a single runway when winds are strong from one direction. When the winds are light, both runways may be used simultaneously.

The strategy which yields the highest capacity occurs when operations are away from the V (Fig. 7-1h), referred to as a *diverging pattern*. In VFR conditions, the hourly capacity for this strategy ranges from 60 to 180 operations, and in IFR conditions the corresponding hourly capacity is from 50 to 80 operations [4]. When operations are toward the V (Fig. 7-1i), referred to as a *converging pattern,* the capacity is reduced to 50 to 100 operations per hour in VFR and between 50 and 60 operations per hour in IFR [4].

Combinations of runway configurations

From the standpoint of capacity and air traffic control, a single-direction runway configuration is most desirable. All other things being equal, this configuration will yield the highest capacity compared with other configurations. For air traffic control, the routing of aircraft in a single direction is less complex than routing in multiple directions. In comparing the divergent configurations, the open-V runway pattern is more desirable than an intersecting-runway configuration. In the open-V configuration, an operating strategy that routes aircraft away from the V will yield higher capacities than if the operations were reversed. If intersecting runways cannot be avoided, every effort should be made to place the intersections of both runways as close as possible to their thresholds and to operate the aircraft away from rather than toward the intersection.

Holding Bays

Holding bays, often referred to as *run-up* or *warm-up pads,* are necessary at or very near the ends of runways for piston aircraft to make final checks prior to takeoff and for all types of aircraft to await takeoff clearance. Sometimes due to adverse weather conditions on particular routes from the airport, certain aircraft will be delayed while others are allowed to proceed with takeoff. Holding bays are a useful way for aircraft to bypass one another in this situation. These holding bays are made large enough that if an aircraft is unable to take off because of some malfunction, another aircraft ready to take off can bypass it. If an aircraft queued up on the taxiway leading to the end of the runway were to have a malfunction, the aircraft would have to proceed down the runway and exit at the nearest exit taxiway if there were no holding bay. This would consume more time than if the aircraft were able to be bypassed on the holding bay.

A holding bay should be designed to accommodate two to four aircraft and to allow sufficient space for one aircraft to bypass another. The area allotted for a waiting aircraft will depend on its size and maneuverability.

When possible, holding bays should be located so as to permit departing aircraft to enter the runway at an angle of less than 90°. Aircraft should be permitted to enter the runway as close to the end of the runway as possible. Holding aircraft should be placed outside the bypass route so that the blast from the holding aircraft will not be directed toward the bypass route.

As an alternative to a holding bay, a bypass taxiway parallel to the taxiway leading to the runway might be provided. Often a dual taxiway serving the runway can be used as a holding bay or bypass taxiway. It is reasonable to provide a holding bay or bypass taxiway if the number of peak-hour operations on the runway to which they are adjacent exceeds 30. Peak traffic volumes at many airports exceed the capacity of holding bays, resulting in aircraft queues on the taxiway leading to the end of the runway. Despite this situation, holding bays are useful for allowing one aircraft to bypass another.

Holding Aprons

Holding aprons are relatively small aprons placed at a convenient location near the terminal area at the airport for the temporary storage of aircraft. At some airports the number of gates may be insufficient to handle the demand during a busy period. In this case, aircraft are routed by air traffic control to a holding apron and are held there until a gate becomes available. Holding aprons are not required if capacity matches demand. However, fluctuations in future demand are difficult to predict so that a temporary storage facility may be necessary.

Relation of Terminal Area to Runways

The key to a desirable airport layout is to provide the shortest taxiing distances from the terminal area to the takeoff ends of the runway and to shorten the taxiing distance for landing aircraft as much as is practicable. This is illustrated schematically by the sketches in Fig. 7-2. The sketches are an attempt to demonstrate principles governing airport configuration and are not to be construed as optimum layouts. The sketches are not complete insofar as taxiways are concerned. For example, two exits for landing aircraft are generally shown, whereas three exits might be more desirable at a specific location, depending upon the composition of aircraft and other factors.

Figure 7-2a shows a single-runway airport, it being assumed that takeoffs and landings will be about equal in each direction. Note that the taxiing distances are equal, regardless of which end of the runway is used for takeoff, and the terminal is also conveniently located for landing from either direction.

If the volume of traffic justifies a second parallel runway, then the desirable location of the terminal area with respect to the parallel runways is shown in Fig. 7-2b. This plan assumes that the wind conditions are such that landings and takeoffs can be made in either direction. At high-traffic-volume airports, it is very desirable to have one runway always available for either a landing or a takeoff.

If it is desired to have one runway used exclusively for landing and one exclusively for takeoffs, then the plan shown in Fig. 7-2c should

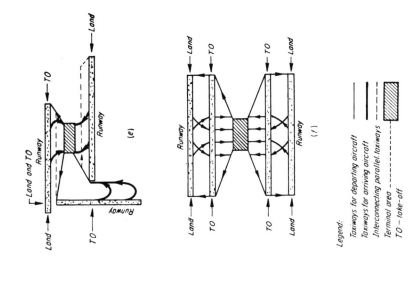

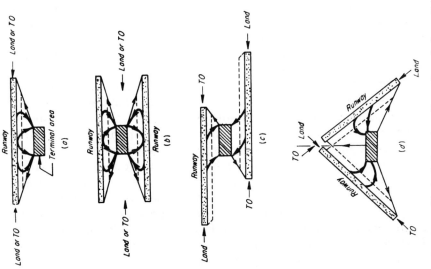

Figure 7-2 Typical airport configurations (schematic).

be given consideration. The principal advantage of this layout compared to the plan in Fig. 7-2b is the reduced taxiing distances for both takeoff and landing. The disadvantages are that the plan is based on the exclusive use of one runway for a specific function (landing or takeoff) and that staggering of the runways may require more land.

Close scrutiny of Figs. 7-2b and c makes it evident that it is not desirable to place the terminal area to one side of a parallel-runway configuration. Taxiing distances are longer, and aircraft traffic on the ground is required to cross active runways, thereby diminishing capacity.

If the winds at an airport require runways in more than one direction, it is desirable to locate the terminal area centrally, as shown in Fig. 7-2d. For this configuration it is assumed that when the winds are light, the air traffic controller is able to use both runways for landing or takeoff.

At some airport locations the winds come fairly regularly in the same general direction throughout the year, except for a small amount of time. If high volumes of traffic are anticipated, three runways may be required, with the terminal area located as shown in Fig. 7-2e.

At very high density airports serving transport aircraft, four parallel runways, as shown in Fig. 7-2f, are necessary. For this type of configuration it is desirable to reserve two runways exclusively for landing and two for takeoffs, in order to avoid interference from taxiing aircraft. Note that the runways adjacent to the terminal area, rather than the outer runways, have been designated for takeoff. This is done to prevent departing aircraft from crossing active runways for landing. Although landing aircraft must taxi across active takeoff runways, crossings of this kind are preferred by air traffic control to crossings by departing aircraft.

Whenever possible, the terminal area should be located so that aircraft taking off or landing do not pass directly over the area at very low altitudes, creating an aeronautical hazard as well as a nuisance.

The principles in laying out runways in relation to the terminal area can be illustrated by using several airports as examples. The air carrier runway on the South Field of Metropolitan Oakland International Airport (Fig. 7-3a) is an example of a single-runway layout shown schematically in Fig. 7-2a. Sky Harbor International Airport in Phoenix, Arizona (Fig. 7-3b), approximates the schematic arrangement shown in Fig. 7-2b. Orlando International Airport in Florida (Fig. 7-3c) illustrates the schematic layout shown in Fig. 7-2c. McCarron International Airport in Las Vegas, Nevada (Fig. 7-3d), is an example of the schematic layout shown in Fig. 7-2d. Dulles International Airport in Washington, D.C. (Fig. 7-3e), is an example of the schematic layout shown in Fig. 7-2e. Finally, the William B.

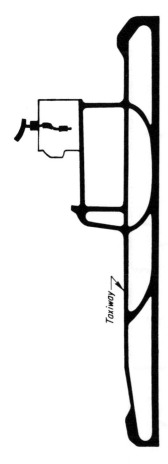

Figure 7-3   (a) Metropolitan
Oakland International Airport,
South Field.

Hartsfield International Airport in Atlanta, Georgia (Fig. 7-3*f*),
approximates the schematic layout in Fig. 7-2*f*.

Analysis of Wind

An analysis of wind is essential for planning runways. As a general
rule, the principal traffic runway at an airport should be oriented as
closely as practicable in the direction of the prevailing winds. When
landing and taking off, aircraft are able to maneuver on a runway as
long as the wind component at right angles to the direction of travel,
the *crosswind* component, is not excessive. The maximum allowable
crosswind depends not only on the size of the aircraft, but also on the
wing configuration and the condition of the pavement surface.
Transport-category aircraft can maneuver in crosswinds as high as 30
kn (35 mi/h), but it is quite difficult to do so, and thus lower values
are used for airport planning.

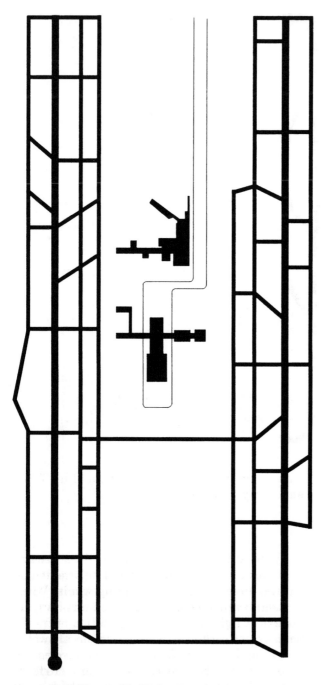

Figure 7-3 (*b*) Phoenix Sky Harbor International Airport.

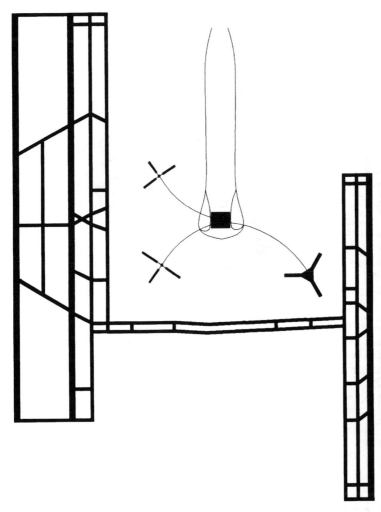

Figure 7-3   (*c*) Orlando International Airport.

The Federal Aviation Administration (FAA) recommends that runways be oriented so that aircraft may be landed at least 95 percent of the time with allowable crosswind components not exceeding specified limits based upon the airport reference codes (see Chap. 9). When the wind coverage is less than 95 percent, a crosswind runway is recommended. The allowable crosswind is 10.5 kn (12 mi/h) for airport reference codes A-I and B-I; 13 kn (15 mi/h) for airport reference codes A-II and B-II; 16 kn (18.5 mi/h) for airport reference codes A-III, B-III, C-I, C-II, C-III, and C-IV; and 20 kn (23 mi/h) for airport reference codes A-IV through D-VI [5]. In some cases, two runways may be

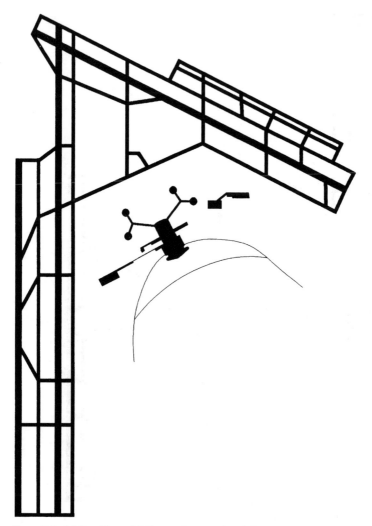

Figure 7-3 (*d*) Las Vegas McCarron International Airport.

required at an airport to achieve the 95 percent wind coverage criterion. The International Civil Aeronautics Organization (ICAO) also specifies that runways should be oriented so that aircraft may be landed at least 95 percent of the time with crosswind components of 20 kn (23 mi/h) for runways of 1500 m or more, 13 kn (15 mi/h) for runways between 1200 and 1500 m, and 10 kn (11.5 mi/h) for runways less than 1200 m long [1, 2].

Once the maximum permissible crosswind component is selected, the most desirable direction of runways for wind coverage can be

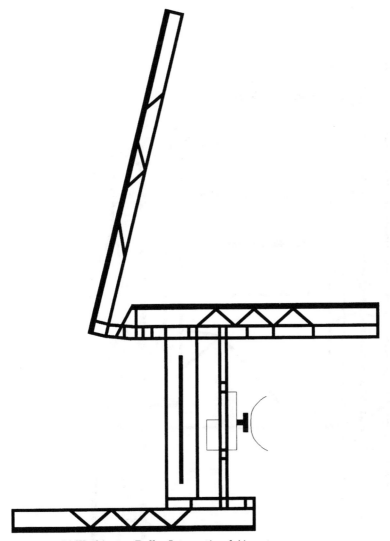

Figure 7-3 (e) Washington Dulles International Airport.

determined by examining the wind characteristics for the following conditions:

1. The entire wind coverage regardless of visibility or cloud ceiling

2. Wind conditions when the ceiling is at least 1000 ft and the visibility is at least 3 mi

3. Wind conditions when the ceiling is between 200 and 1000 ft and/or the visibility is between ½ and 3 mi

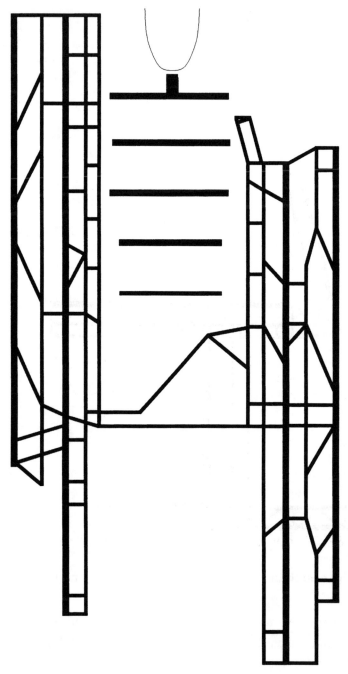

Figure 7-3   (*f*) Atlanta William B. Hartsfield International Airport.

The first condition represents the entire range of visibility, from excellent to very poor, and is termed the *all-weather* condition. The next condition represents the range of good-visibility conditions not requiring the use of instruments for landing, termed the *visual meteorological condition* (VMC). The last condition represents various degrees of poor visibility requiring the use of instruments for landing, termed the *instrument meteorological condition* (IMC). It is useful to know the strength of the winds under each of these weather conditions, but particularly when visibility is restricted. Normally when visibility approaches ½ mile and the ceiling is 200 ft, there is very little wind present, the visibility being reduced by fog, haze, smoke, or smog. Sometimes the visibility may be extremely poor, yet there is no distinct cloud ceiling; for that matter, no clouds need to be present at all. Examples of this condition are fog, smoke, smog, haze, and so on.

The 95 percent criterion suggested by the FAA and ICAO is applicable to all conditions of weather; nevertheless, it is still useful to examine the data in parts whenever this is possible.

In the United States, weather records can be obtained from the Environmental Data and Information Service of the National Climatic Center at the National Oceanic and Atmospheric Administration, located in Ashville, North Carolina. Weather data are collected from weather stations throughout the United States on an hourly basis and are recorded on computer tapes for analysis. The data collected include the ceiling, visibility, wind speed, wind direction, storms, barometric pressure, amount and type of liquid and frozen precipitation, temperature, and relative humidity. A report illustrating the tabulation and representation of some of the data used in airport studies was prepared for the FAA [15]. In the past, wind velocities were divided into velocity ranges in 22.5° increments (16 points on the compass), but now these data are obtainable in 10° increments (36 points on a compass). The weather records contain the percentage of time that certain combinations of ceiling and visibility occur (e.g., ceiling, 500 to 900 ft; visibility, 3 to 6 mi) and the percentage of time that winds of specified velocity ranges come from different directions (e.g., from NNE, 4 to 7 mi/h). The directions are referenced to true north.

The wind rose

The directions of the runway or runways at an airport can be determined through graphical vector analysis on a *wind rose*. A standard wind rose consists of a series of concentric circles cut by radial lines on polar-coordinate graph paper. The radial lines are drawn to the

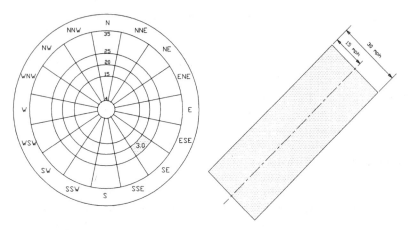

Figure 7-4 Wind rose coordinate system and template.

scale of the wind magnitude such that the area between each pair of successive lines is centered on the wind direction. A typical wind rose polar-coordinate system is shown on the left side of Fig. 7-4. The shaded area indicates that the wind comes from the southeast (SE) and has a magnitude between 20 and 25 mi/h 3.0 percent of the time. A template representing the crosswind component limits is also drawn to the same radial scale. A template drawn with crosswind component limits of 15 mi/h is shown on the right side of Fig. 7-4. On this template three equally spaced parallel lines have been plotted. The middle line represents the runway centerline, and the distance between the middle line and each outside line is, to scale, the allowable crosswind component (in this case, 15 mi/h). The template is placed over the wind rose so that the centerline on the template passes through the center of the wind rose.

By overlaying the template on the wind rose and rotating the centerline of the template through the origin of the wind rose, one may determine the percentage of time that a runway in the direction of the centerline of the template can be used such that the crosswind component does not exceed 15 mi/h. Optimum runway directions can be determined from this wind rose by the use of the template, typically made on a transparent strip of material. With the center of the wind rose as a pivot point, the template is rotated until the sum of the percentages included between the outer lines is a maximum. The reader should recognize that if a wind vector from a segment lies outside either outer line on the template for the given direction of the runway, that wind vector must have a crosswind component which exceeds the allowable crosswind component plotted on the template. When one of the outer lines on the template divides a segment of wind direction, the fractional part is estimated visually to the nearest 0.1 percent. This procedure is consistent with the accuracy of the wind data and assumes that the wind percentage within the sector is uniformly distributed within that sector. In practice, it is usually easier to add the percentages contained in the sectors outside the two outer parallel lines and subtract these from 100 percent to find the percentage of wind coverage.

Example Problem 7-1. As an example, assume that the wind data for all conditions of visibility are those shown in Table 7-1. These wind data are plotted to scale, as indicated above, to obtain a wind rose, shown in Fig. 7-5.

The percentage of time that the winds correspond to a given direction and velocity range is marked in the proper sector of the wind rose on a polar-coordinate scale for both wind direction and wind magnitude. The template is rotated about the center of the wind rose, as explained earlier, until the direction of the centerline yields the maximum percentage of wind between the parallel lines.

Once the optimum runway direction has been found in this manner, the next

TABLE 7-1   **Percentage of Time that Winds Come from Particular Directions at Various Velocities in All-Weather Conditions**

Sector	True azimuth	Wind speed range, mi/h				Total
		4–15	15–20	20–25	25–35	
		Percentage of time				
N	0.0	2.4	0.4	0.1	0.0	2.9
NNE	22.5	3.0	1.2	1.0	0.5	5.7
NE	45.0	5.3	1.6	1.0	0.4	8.3
ENE	67.5	6.8	3.1	1.7	0.1	11.7
E	90.0	7.1	2.3	1.9	0.2	11.5
ESE	112.5	6.4	3.5	1.9	0.1	11.9
SE	135.0	5.8	1.9	1.1	0.0	8.8
SSE	157.5	3.8	1.0	0.1	0.0	4.9
S	180.0	1.8	0.4	0.1	0.0	2.3
SSW	202.5	1.7	0.8	0.4	0.3	3.2
SW	225.0	1.5	0.6	0.2	0.0	2.3
WSW	247.5	2.7	0.4	0.1	0.0	3.2
W	270.0	4.9	0.4	0.1	0.0	5.4
WNW	292.5	3.8	0.6	0.2	0.0	4.6
NW	315.0	1.7	0.6	0.2	0.0	2.5
NNW	337.5	1.7	0.9	0.1	0.0	2.7
Subtotal		60.4	19.7	10.2	1.6	91.9
Calms						8.1
Total						100.0

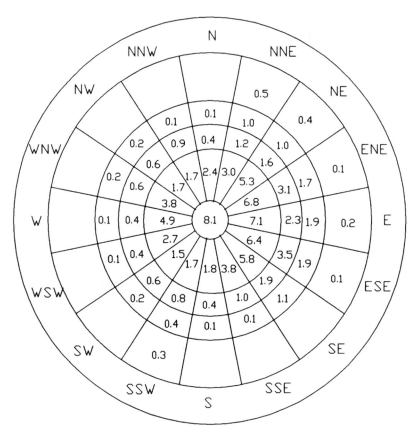

Figure 7-5 Wind data plot for Example Problem 7-1.

step is to read the bearing of the runway on the outer scale of the wind rose, where the centerline on the template crosses the wind direction scale. Because true north is used for published wind data, this bearing usually is different from that used in numbering runways since runway designations are based on the magnetic bearing. In Fig. 7-6, it will be found that a runway oriented on an azimuth to true north of 90° to 270° (N90°E to S90°W true bearing) will permit operations 90.8 percent of the time with the crosswind components not exceeding 15 mi/h. If the magnetic declination at this airport is 0°, then the runway will be called runway 9-27.

Should the wind analysis not give the desired wind coverage, the template may then be used to determine the direction of a second runway, a crosswind runway, which will increase the wind coverage to 95 percent. This is done by blocking out the area between the two outer parallel lines for the direction of the primary runway (since this has already been counted in the wind coverage for the primary runway) and rotating the template until the percentage between the outer parallel lines for the remaining area for another direction is

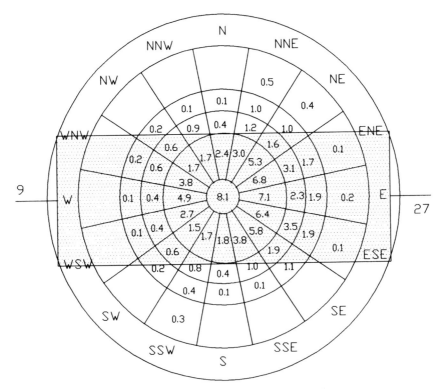

Figure 7-6 Wind coverage for runway 9-27 for Example Problem 7-1.

maximized. If this is done in this problem, it is found that the crosswind runway should be located in an orientation of 12° to 192° (N12°E to S12°W true
bearing). This will permit an additional wind coverage of 6.2 percent above that
provided by the runway oriented 90° to 270°, for a total wind coverage for both
runways of 97.0 percent.

Let us say that because of noise-sensitive land uses in the direction of the
optimal crosswind runway, a crosswind runway will be located in the orientation of 30° to 210° which results in an additional wind coverage of 5.8 percent.
This runway orientation, called runway 3-21, is shown in Fig. 7-7. The total
wind coverage for both runways is then 96.6 percent. The total wind coverage
for a runway in the orientation of 30° to 210° direction is found to be 84.8 percent from Fig. 7-7. The combined wind coverage of 96.6 percent for the use of
either runway is shown in Fig. 7-8.

The wind coverage for the simultaneous use of the various runways can also
be obtained from the wind rose. The simultaneous use of two runways requires
that the crosswind component not exceed 15 mi/h on either runway. Therefore
Fig. 7-9 shows the simultaneous use of the runways oriented in the 30° and 90°
directions, and Fig. 7-10 shows the simultaneous use of runways oriented in the
210° and 270° directions. It is assumed that any runway may be used during
periods when the wind magnitude is equal to or less than 4 mi/h.

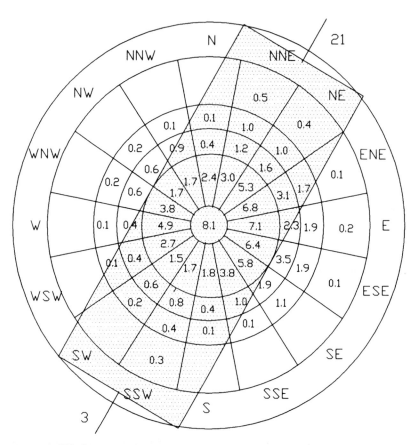

Figure 7-7 Wind coverage for runway 3-21 for Example Problem 7-1.

Wind data can be displayed in a variety of ways for various purposes. One commonly used method is to indicate the percentage of time that the winds come from various directions under the weather conditions specified above. Figure 7-11 shows graphically the percentage of time that winds come from particular directions based upon the data contained in Example Problem 7-1. Figure 7-12 shows the average wind speed of winds coming from each direction.

In practice, wind analyses are usually conducted by using computer programs. A typical output from a computer program based upon the same wind data in Example Problem 7-1 to determine the primary runway orientation is shown in Table 7-2. This table indicates the wind coverage available for runways oriented in various directions in 2° increments of azimuth. As may be observed, the same primary runway orientation of 90° is found by computer analysis. Note in Table 7-

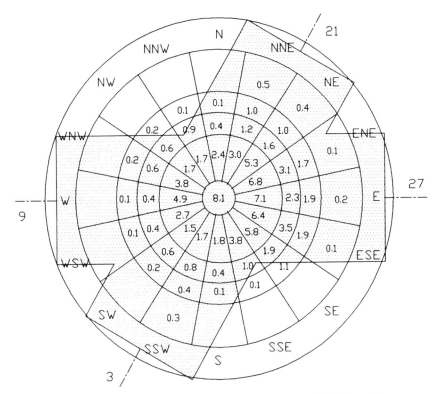

Figure 7-8 Wind coverage for runways 9-27 and 3-21 for Example Problem 7-1.

2 that several directions (88° to 94°) seem to have the same wind coverage, but the computer is printing out the results only to the nearest 0.1 percent while it retains more precise numbers in determining the maximum wind coverage orientation. Table 7-3 shows the analysis for the optimal crosswind runway based upon wind coverage, and it indicates that a runway oriented at 12° is the optimal crosswind runway for the primary runway at an orientation of 90°.

Most computer programs will also indicate the directional use of the primary and crosswind runways. Table 7-4 indicates that a primary runway oriented in the 90° to 270° direction can be used 68.0 percent of the time in the 90° direction and 30.9 percent of the time in the 270° direction with crosswind components not exceeding 15 mi/h. Similarly, this table indicates that a crosswind runway oriented in the 30° to 210° direction can be used 53.3 percent of the time in the 30° direction and 39.6 percent of the time in the 210° direction with crosswind components not exceeding 15 mi/h.

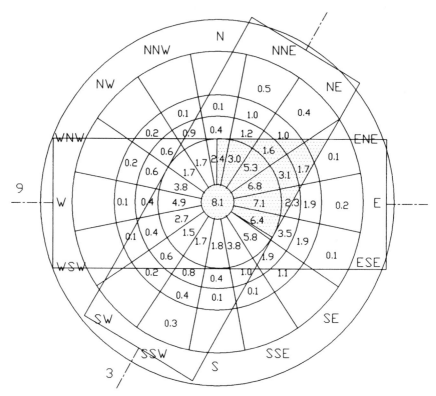

Figure 7-9 Wind coverage for simultaneous use of runways 3 and 9 for Example Problem 7-1.

Thus far the procedure has been illustrated as it applies to wind data records with a wind velocity break at 15 mi/h. However, it can also be used to obtain estimates of wind coverage for any other wind velocity break. The concentric circles on the wind rose are drawn to scale and represent breaks in the wind velocity data. Suppose the break in the wind data is at 12 mi/h instead of 15 mi/h. Then the two outer parallel lines representing the 15 mi/h maximum allowable crosswind component would not correspond to or be tangent to the 12 mi/h circle, but would lie outside it. An estimate must then be made of the fractional percentage in each segment between the 12 mi/h circle on the wind rose and the 15 mi/h outer parallel lines, and this percentage is added to the percentage within the 12 mi/h circle. Clearly the given wind rose could also be used for other maximum allowable crosswind components by choosing a template scaled to represent the maximum allowable crosswind component.

The next step is to examine wind data during both unrestricted and restricted ceiling and visibility conditions cited earlier (ceiling

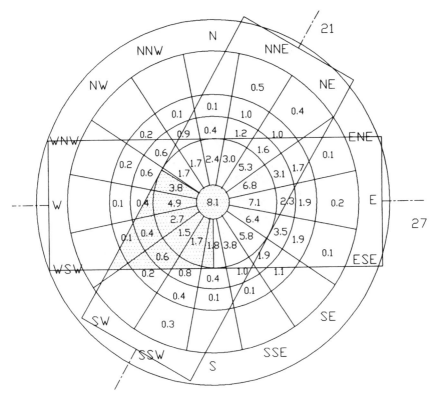

Figure 7-10 Wind coverage for simultaneous use of runways 21 and 27 for Example Problem 7-1.

between 200 and 1000 ft and visibility between ½ and 3 mi) and to plot wind roses for this condition. From these analyses it can be ascertained whether the runways are capable of accepting aircraft at least 95 percent of the time when restricted-visibility conditions prevail. The analysis will also yield information on the percentage of the total time that each of the conditions prevails. Often wind data for an entirely new location have not been recorded. In that case, records of nearby weather-measuring stations should be consulted. If the surrounding area is fairly level, the records of these stations should indicate the winds at the site of the proposed airport. However, if the terrain is hilly, the wind pattern often is dictated by the topography, and it is dangerous to utilize records of stations at some distance from the site. In this event, a study of the topography of the region and consultation with long-time residents may prove useful. In critical situations it is probably prudent to collect at least 1 year's worth of weather data at the site, from which the runway orientation can be determined.

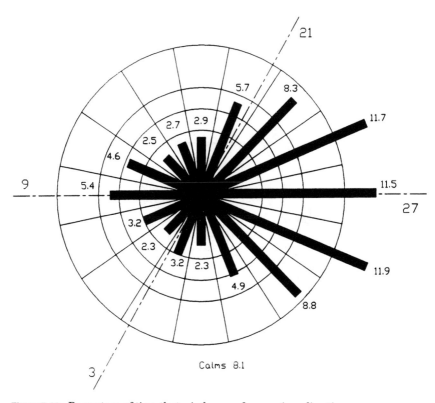

Figure 7-11   Percentage of time that wind comes from various directions.

Many computer programs have been developed to analyze wind based upon various forms of wind data. These programs typically determine primary and crosswind runway orientation, wind coverage for each runway, and wind coverage for the directional use of each runway. These results were shown in Tables 7-2 through 7-4 for a typical wind analysis program. Often some programs will plot the results in a useful manner. One such program, airport design 3-2, which has been developed by the FAA [5], tabulates the weather data in a spreadsheet format for computer analysis and graphically displays the results of the wind analysis.

Obstructions in the Vicinity of Airports

In the United States, the FAA requires that protection zones be provided at the ends of runways. *Runway protection zones,* formerly called *runway clear zones,* are trapezoidal areas comprising the innermost portions of the runway approach surfaces, as defined in FAR part 77 [5, 8]. The runway protection zone is the area on the ground

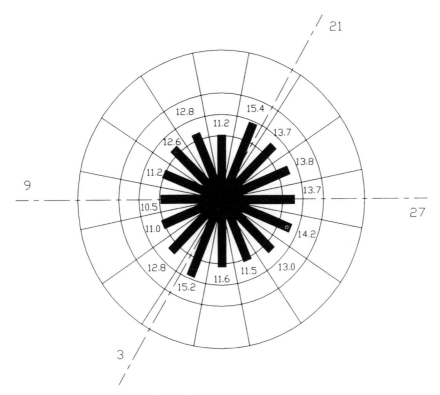

Figure 7-12 Wind speed in miles per hour from various directions.

beneath the approach surface to a runway from the end of the primary surface to the point where the approach surface is 50 ft above the primary surface, as shown in Fig. 7-13. The dimensions of the runway protection zone can be derived from the approach surface requirements for a runway. These dimensions are shown in Table 7-5. Note that when runways are authorized for different types of approaches from each end, the inner width of the runway protection zone W_1 is the width of the primary surface for the runway, determined by the larger of the primary surfaces for each approach authorized (see Table 7-6).

When the runway protection zone begins at a location other than 200 ft beyond the end of the runway due to the application of the declared-distance concept, discussed in Chap. 3, two runway protection zones are usually required, an approach runway protection zone and a departure runway protection zone. The dimensions of the approach runway protection zone are given in Table 7-6; but the departure runway protection zone begins 200 ft beyond the far end of the takeoff run available, and the portion of the runway between the takeoff run available and the end of the runway is declared unavail-

TABLE 7-2   Computer Analysis for Primary Runway Direction for Wind Data of Example Problem 7-1

Runway azimuth coverage	Wind coverage	Runway azimuth	Wind coverage	Runway azimuth	Wind coverage, %
2	82.3	62	89.1	122	87.8
4	82.4	64	89.4	124	87.4
6	82.5	66	89.6	126	87.1
8	82.6	68	89.8	128	86.7
10	82.7	70	89.9	130	86.4
12	82.8	72	90.1	132	86.0
14	83.0	74	90.3	134	85.7
16	83.2	76	90.4	136	85.4
18	83.4	78	90.5	138	85.1
20	83.6	80	90.6	140	84.8
22	83.8	82	90.6	142	84.6
24	84.1	84	90.7	144	84.3
26	84.3	86	90.7	146	84.1
28	84.5	88	90.8	148	83.9
30	84.8	90	90.8	150	83.6
32	85.0	92	90.8	152	83.5
34	85.3	94	90.8	154	83.3
36	85.5	96	90.7	156	83.1
38	85.8	98	90.7	158	83.0
40	86.1	100	90.6	160	82.8
42	86.4	102	90.5	162	82.7
44	86.7	104	90.3	164	82.6
46	87.0	106	90.1	166	82.5
48	87.2	108	89.9	168	82.5
50	87.5	110	89.7	170	82.4
52	87.8	112	89.4	172	82.4
54	88.1	114	89.1	174	82.3

The true azimuth of the least-crosswind runway is 90°; crosswinds are under 15.0 mi/h 90.8 percent of the time.

able and unsuitable for the takeoff run. The dimensions of the departure runway protection zone are as follows:

1. For runways serving only small aircraft in aircraft approach categories A and B, the length is 1000 ft, the inner width is 250 ft, and the outer width is 450 ft.

2. For runways serving large aircraft in aircraft approach categories A and B, the length is 1000 ft, the inner width is 500 ft, and the outer width is 700 ft.

3. For runways serving aircraft in aircraft approach category C, D, or E, the length is 1700 ft, the inner width is 500 ft, and the outer width is 1010 ft.

The runway protection zone actually consists of two subareas: the runway *object-free area* (OFA) and the controlled-activity area. The

TABLE 7-3 Computer Analysis for Crosswind Runway Direction for Primary Runway Oriented at 90° for Wind Data of Example Problem 7-1

Runway azimuth	Additional wind coverage	Runway azimuth	Additional wind coverage	Runway azimuth	Additional wind coverage, %
2	6.1	62	2.9	122	1.9
4	6.1	64	2.6	124	2.0
6	6.1	66	2.4	126	2.1
8	6.1	68	2.1	128	2.2
10	6.2	70	1.9	130	2.3
12	6.2	72	1.6	132	2.4
14	6.2	74	1.3	134	2.5
16	6.1	76	1.1	136	2.7
18	6.1	78	0.9	138	2.8
20	6.1	80	0.6	140	2.9
22	6.1	82	0.4	142	3.1
24	6.0	84	0.3	144	3.2
26	6.0	86	0.1	146	3.3
28	5.9	88	0.0	148	3.5
30	5.8	90	0.0	150	3.7
32	5.7	92	0.0	152	3.8
34	5.6	94	0.1	154	4.0
36	5.5	96	0.2	156	4.2
38	5.4	98	0.4	158	4.3
40	5.2	100	0.5	160	4.5
42	5.1	102	0.7	162	4.7
44	4.9	104	0.8	164	4.9
46	4.7	106	1.0	166	5.1
48	4.6	108	1.1	168	5.3
50	4.4	110	1.2	170	5.5
52	4.1	112	1.3	172	5.6
54	3.9	114	1.4	174	5.8
56	3.7	116	1.5	176	5.9
58	3.4	118	1.6	178	6.0
60	3.2	120	1.8	180	6.0

Locate crosswind runway at true azimuth of 12°. Crosswinds are under 15.0 mi/h 6.2 percent of the time. Total wind coverage for use of both runways is 97.0 percent.

TABLE 7-4 Computer Analysis Results for Primary and Crosswind Runway Coverage in All Directions for Wind Data of Example Problem 7-1

All-Weather Condition Summary of Wind Analysis for Directional Coverage	
Runway azimuth	Wind coverage, %
90 or 270	90.8
30 or 210	84.8
Both	96.6
90	68.0
270	30.9
30	53.3
210	39.6
Calms	8.1

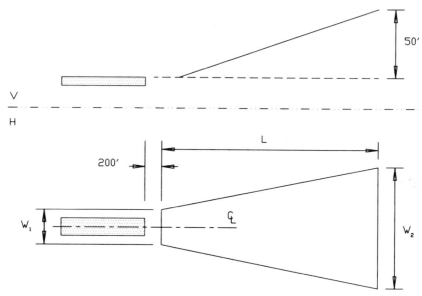

Figure 7-13 Dimensions of runway protection zone.

runway OFA is a two-dimensional ground area surrounding the runway. It has a width varying between 250 and 800 ft and a length beyond the end of the runway varying between 300 and 1000 ft, depending upon the type of approach made to the runway and the approach category and airplane design group of the aircraft for which the runway is designed. The specific dimensions of the runway OFA are given in Tables 9-6 and 9-7. The runway OFA clearing standard precludes parked aircraft and objects, except those objects which are aids to air navigation whose location is fixed by function. The controlled-activity area is the portion of the runway protection zone beyond and to the sides of the runway OFA.

FAR part 77

Among other things, part 77 of the Federal Aviation Regulations establishes standards for determining obstructions in navigable airspace, sets forth the requirements for notice to the FAA due to certain proposed construction or alteration activities, and provides for aeronautical studies of obstructions to air navigation to determine the effect of these obstructions on the safe and efficient use of airspace [8, 9]. The airport operator has the responsibility to ensure that the aerial approaches to the airport will be adequately cleared and protected and that the land adjacent to or in the immediate vicinity of the airport is reasonably restricted to the extent possible through the use of such measures as the adoption of zoning ordinances. A model zoning

TABLE 7-5   Dimensions of Runway Protection Zone

Aircraft served	Runway approach* Approach end	Opposite end	Length L, ft	Width Inner W_1, ft	Outer W_2, ft	Area, acres
Small	V	V	1000	250	450	8.035
		NP	1000	500	650	13.200
		NP+	1000	1000	1050	23.542
		P	1000	1000	1050	23.542
	NP	V	1000	500	800	14.922
		NP	1000	500	800	14.922
		NP+	1000	1000	1200	25.252
		P	1000	1000	1200	25.252
	NP+	V	1700	1000	1510	48.978
		NP	1700	1000	1510	48.978
		NP+	1700	1000	1510	48.978
		P	1700	1000	1510	48.978
	P	V	2500	1000	1750	78.914
		NP	2500	1000	1750	78.914
		NP+	2500	1000	1750	78.914
		P	2500	1000	1750	78.914
Large	V	V	1000	500	700	13.770
		NP	1000	500	700	13.770
		NP+	1000	1000	1100	24.105
		P	1000	1000	1100	24.105
	NP	V	1700	500	1010	29.465
		NP	1700	500	1010	29.465
		NP+	1700	1000	1425	47.320
		P	1700	1000	1425	47.320
	NP+	V	1700	1000	1510	48.978
		NP	1700	1000	1510	48.978
		NP+	1700	1000	1510	48.978
		P	1700	1000	1510	48.978
	P	V	2500	1000	1750	78.914
		NP	2500	1000	1750	78.914
		NP+	2500	1000	1750	78.914
		P	2500	1000	1750	78.914

*V = visual approach; NP = non-precision-instrument approach with visibility minimums more than $\frac{3}{4}$ mi; NP + = non-precision-instrument approach with visibility minimums as low as $\frac{3}{4}$ mi; and P = precision- instrument approach.

SOURCE: Federal Aviation Administration [5].

ordinance to limit the height of objects around airports is published by the FAA [6].

Subpart C of FAR part 77 establishes standards for determining obstructions to air navigation. The standards apply to existing and manufactured objects, objects of natural growth, and terrain.

To determine whether an object is an obstruction to air navigation, several imaginary surfaces are established with relation to the airport and to each end of a runway. The size of the imaginary surfaces depends on the category of each runway (e.g., utility or transport) and

TABLE 7-6   Imaginary Surface Dimensions, FAR Part 77, ft

Item	Visual runway* A	Visual runway* B	Non-precision-instrument runway† A	Non-precision-instrument runway† B C	Non-precision-instrument runway† B D	Precision-instrument runway
Width of primary surface and approach surface at inner end	250	500	500	500	1,000	1,000
Radius of horizontal surface	5,000	5,000	5,000	10,000	10,000	10,000
Approach surface width at outer end	1,250	1,500	2,000	3,500	4,000	16,000
Approach surface length	5,000	5,000	5,000	10,000	10,000	50,000‡
Approach slope	20:1	20:1	20:1	34:1	34:1	50:1§

*A = utility runways; B = runways larger than utility.

†C = visibility minimum greater than ¾ mi; D = visibility minimum as low as ¾ mi.

‡Inner length 10,000 ft; outer length 40,000 ft.

§Inner length 50:1; outer length 40:1.

SOURCE: Federal Aviation Administration [8].

on the type of approach planned for that end of the runway (e.g., visual, non-precision-instrument, or precision-instrument).

The principal imaginary surfaces are shown in Fig. 7-14 and are described as follows:

1. *Primary surface.* The *primary surface* is a surface longitudinally centered on a runway. When the runway is paved, the primary surface extends 200 ft beyond each end of the runway. When the runway is unpaved, the primary surface coincides with each end of the runway. The elevation of the primary surface is the same as the elevation of the nearest point on the runway centerline.

2. *Horizontal surface.* The *horizontal surface* is a horizontal plane 150 ft above the established airport elevation, the perimeter of which is constructed by swinging arcs of specified radii from the center of each end of the primary surface of each runway and connecting each arc by lines tangent to those arcs.

3. *Conical surface.* The *conical surface* is a surface extending outward and upward from the periphery of the horizontal surface at a slope of 20 horizontal to 1 vertical for a horizontal distance of 4000 ft.

4. *Approach surface.* The *approach surface* is a surface that is longitudinally centered on the extended runway centerline and extends

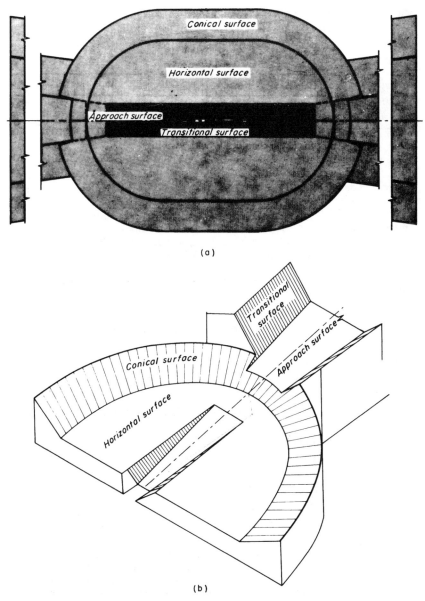

Figure 7-14 FAR part 77 imaginary surfaces: (*a*) plan view, (*b*) isometric view. (*Federal Aviation Administration [8]*).

outward and upward from each end of a runway at a designated slope based on the type of available or planned runway approach.

5. *Transitional surface.* *Transitional surfaces* extend outward and upward at right angles to the runway centerline plus the runway

centerline extended at a slope of 7 to 1 from the sides of the primary surface up to the horizontal surface and from the sides of the approach surfaces. The width of the transitional surface provided from each edge of the approach surface is 5000 ft.

Dimensions of the several imaginary surfaces are shown in Table 7-6. The imaginary surfaces for military airports differ from those specified above, and the dimensions of these are also contained in FAR part 77.

Example Problem 7-2. An analysis of the obstruction surfaces defined in FAR part 77 is to be made. The obstruction surface is to be specified for approaches to both ends of a 6500-ft-long by 100-ft-wide runway designed for use by utility aircraft for non-precision-instrument approaches. The mean elevation of the runway is 1257 ft above mean sea level. Assume a coordinate system with the origin at the midpoint X of the centerline of the runway, Y along the runway centerline, and Z at mean sea level. The problem is to define the X, Y, and Z coordinates of the critical points on the FAR part 77 imaginary surfaces for this runway.

The FAR part 77 specifications for all runway approaches are found from the above data:

Length of the primary surface: 200 ft from end of runway.

Horizontal surface elevation: 150 ft above runway elevation.

Conical surface width: 4000 ft.

Conical surface slope: 20 to 1.

Approach surface transitional surface width: 5000 ft.

Transitional surface slope: 7 to 1.

The specific FAR part 77 specifications in Table 7-6 for this type of runway require the following:

Width of primary and approach surface at inner end: 500 ft.

Radius of horizontal surface: 5000 ft.

Approach surface width at outer end: 2000 ft.

Approach surface length: 5000 ft.

Approach surface slope: 20 to 1.

The various components of the FAR part 77 imaginary surfaces can then be determined.

Figure 7-15 indicates critical points on the paved runway, as well as the primary surface, horizontal surface, and conical surface surrounding the runway. The center of the runway is denoted by point a, which has coordinates of $X = 0$, $Y = 0$, and $Z = 1257$ ft. The physical end of the runway at the edge of the pavement width is denoted by point b. It has coordinates of $X = 50$, $Y = 3250$, and $Z = 1257$ ft. Point c is the end and edge of the primary surface surrounding the runway. The primary surface is 200 ft beyond the end of the runway and is 500 ft wide centered on the runway centerline. Therefore, the coordinates of point c are $X = 250$, $Y = 3450$, and $Z = 1257$ ft.

The horizontal surface is 150 ft above the runway elevation. Therefore its elevation is $1257 + 150 = 1407$ ft above mean sea level (AMSL). It has a radius of 5000 ft centered at the intersection of the extended runway centerline and the outer edge of the primary surface. Points d and e are at the far edge of the hori-

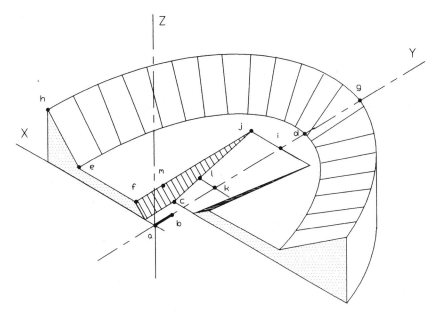

Figure 7-15 Critical obstruction surfaces for Example Problem 7-2.

zontal surface. The coordinates of point d are $X = 0$, $Y = 8450$, and $Z = 1407$ ft; and the coordinates of point e are $X = 5000$, $Y = 0$, and $Z = 1407$ ft.

A transitional surface is provided from the edge of the primary surface parallel to the runway centerline, extending upward at a slope of 7 to 1 until it reaches the elevation of the horizontal surface. To reach an elevation of 150 ft, this surface extends horizontally a distance of $150 \times 7 = 1050$ ft from the edge of the primary surface to intersect the horizontal surface. Point f is the point where the transitional surface from the primary surface surrounding the runway intersects the horizontal surface. The coordinates of point f are $X = 1300$, $Y = 0$, and $Z = 1407$ ft.

The conical surface extends upward from the outer perimeter of the horizontal surface at a slope of 20 to 1 for a horizontal distance of 4000 ft. Its elevation at its outer edge is $150 + 4000/20 = 350$ ft above the elevation of the runway. Therefore, it has an elevation at the outer edge of $1257 + 350 = 1607$ ft AMSL. Points g and h are at the far edge of the conical surface. The coordinates of point g are $X = 0$, $Y = 12{,}450$, $Z = 1607$ ft; and the coordinates of point h are $X = 9000$, $Y = 0$, and $Z = 1607$ ft.

Figure 7-15 also illustrates critical points along the required approach surface along the extended runway centerline as it cuts through the horizontal surface and the conical surface as it rises from the primary surface near the end of the runway. The required approach surface begins at the outer edge of the primary surface and extends outward for a distance of up to 5000 ft and upward at a slope of 20 to 1. Its required elevation at the far end is therefore $5000/20 = 250$ ft above the elevation of the runway, or $1257 + 250 = 1507$ ft AMSL. Its required width at the far end is 2000 ft and at the inner end is 500 ft. Points i and j are on the far edge of the approach surface. However, the approach surface reaches the elevation of the horizontal surface at a distance of $150 \times 20 = 3000$ ft from the beginning of the approach surface. This means that both the horizontal surface and the conical surface are at an elevation below the required approach surface elevation beyond points i and j. Therefore, the actual elevations of points i and j can be no greater than the elevation of the horizontal surface. This means that for this application the approach surface does not exist beyond points i and j and that the elevations of either the horizontal surface or the conical surface govern. Therefore, for both points i and j, the value of the coordinate $Z = 1407$ ft, that of the horizontal surface. The coordinates of point i are $X = 0$, $Y = 6450$, and $Z = 1407$ ft; and the coordinates of point j are $X = 700$, $Y = 6450$, and $Z = 1407$ ft.

The runway protection zone is an area on the ground directly under the approach surface. It begins at the intersection of the primary surface and the approach surface. Therefore its inner width is 500 ft. It ends at the point where the approach surface is 50 ft above the runway elevation. Since the approach surface is at a 20 to 1 slope, the approach surface is at an elevation of 50 ft above the runway elevation at a distance $50 \times 20 = 1000$ ft from the end of the primary surface. At this point the approach surface can be computed to be 800 ft wide, since it varies from 500 ft wide at the primary surface to 2000 ft wide at its far end, at a distance of 5000 ft from the end of the primary surface. The dimensions of the runway protection zone can also be verified by reference to Table 7-5. Points k and l are on the far end of the runway protection zone. Point k has coordinates of $X = 0$, $Y = 4450$, and $Z = 1257$ ft; and point l has coordinates of $X = 400$, $Y = 4450$, and $Z = 1257$ ft. Note that the actual elevations of the points in the runway protection zone are the elevations of the ground points under that portion of the approach slope defining the runway protection zone.

Another transitional surface extends upward from the approach surface at a slope of 7 to 1 for a horizontal distance of up to 5000 ft, intersecting both the horizontal surface and the conical surface. However, this transitional surface is provided for this type of an approach only in the area which lies within the area bounded by the horizontal surface, since the radius of the horizontal surface and the length of the approach surface are both 5000 ft. Since the approach surface attains a height of 150 ft, the height of the horizontal surface, at a distance of 3000 ft from the end of the primary surface, as noted earlier, this transitional surface exists only from the beginning of the approach surface to this point. Beyond this point the elevations of the horizontal surface and the conical surface are below that of the approach surface, and the lower elevation governs. The point at which the transitional surface from the approach surface ends is also designated by point j, noted earlier. There is no transitional surface from the approach surface beyond this point because the approach surface does not exist beyond this point.

At its inner end, point m, the transitional surface intersects the horizontal surface at an elevation of 150 ft above the mean elevation of the runway at an X coordinate of $250 + 150 \times 7 = 1300$ ft and a Y coordinate of 3450 ft. Therefore, the coordinates of point m are $X = 1300$, $Y = 3450$, and $Z = 1407$ ft. The transitional surface continues at the same slope parallel to the runway centerline along the edges of the primary surface parallel to the runway centerline, as noted earlier.

The various coordinates and dimensions of the imaginary surfaces computed above are summarized in Table 7-7; the letter codes refer to Fig. 7-15.

In addition to the surfaces defined above, other standards for determining obstructions to air navigation are contained in FAR part 77. Existing and future objects, whether stationary or mobile, are consid-

TABLE 7-7   Coordinates for FAR Part 77 Imaginary Surfaces in Example Problem 7-2

Point	Description	Coordinates		
		X	Y	Z
a	Center of runway	0	0	1,257
b	Edge of runway	50	3,250	1,257
c	Edge of primary surface	250	3,450	1,257
d	Outer edge of horizontal surface	0	8,450	1,407
e	Edge of horizontal surface	5,000	0	1,407
f	Transitional and horizontal surfaces	1,300	0	1,407
g	Outer edge of conical surface	0	12,450	1,607
h	Edge of conical surface	9,000	0	1,607
i	Outer end of approach surface	0	6,450	1,407
j	Outer edge of approach surface	700	6,450	1,407
k	End of runway protection zone	0	4,450	1,257
l	Outer edge of runway protection zone	400	4,450	1,257
m	End of transitional surface	1,300	3,450	1,407

ered to be obstructions to air navigation if their heights are greater than any of the following heights or surfaces:

1. A height of 500 ft above ground level at the site of the object.

2. A height 200 ft above ground level or 200 ft above the established airport elevation, whichever is greater, within 3 nmi of the established reference point at an airport with its longest runway more than 3200 ft in actual length. This height increases in the ratio of 100 ft for each additional 1 nmi of distance from the reference point up to a maximum of 500 ft.

3. A height within a terminal obstacle clearance area, including an initial approach segment, a departure area, and a circling approach area, which would result in the vertical distance between any point on the object and an established minimum instrument flight altitude within that area or segment to be less than the required obstacle clearance.

4. A height within an en route obstacle clearance area, including turn and termination areas, of a federal airway or approved off-airway route that would increase the minimum obstacle clearance altitude.

5. The surface of a takeoff and landing area of an airport or any of the imaginary surfaces defined above.

6. Except for traverse ways on or near an airport with an operative ground traffic control service furnished by the air traffic control tower or by airport management and coordinated with the air traffic control service, the heights of traverse ways must be increased by 17 ft for interstate highways; 15 ft for any other public roadway; 10 ft or the height of the highest mobile object that would normally traverse the

road, whichever is greater, for a private road; 23 ft or an amount equal to the height of the highest mobile object that would normally traverse it for railroads, waterways, or any other thoroughfare not previously mentioned.

Subpart B of FAR part 77 identifies circumstances in which notice is required to be given to the FAA when certain construction or alteration activities are proposed. These include the circumstances associated with the standards given above and any construction or alteration of greater height than an imaginary surface extending outward and upward at one of the following slopes [9]:

1. A slope of 100 horizontal to 1 vertical for a horizontal distance of 20,000 ft from the nearest point of the nearest runway at an airport or seaplane base with at least one runway more than 3200 ft in actual length

2. A slope of 50 horizontal to 1 vertical for a horizontal distance of 10,000 ft from the nearest point of the nearest runway at an airport or seaplane base with its longest runway no more than 3200 ft in actual length

3. A slope of 25 horizontal to 1 vertical for a horizontal distance of 5000 ft from the nearest point of the nearest takeoff and landing area for a heliport

FAR part 77 imposes strict requirements on both airport sponsors and others associated with construction activities in the vicinity of airports which should be referenced prior to initiation of construction activities.

Example Problem 7-3. At present a 9000-ft paved runway exists at an airport and the approach threshold is marked at the physical end of the runway. The approach end of the runway is to be authorized for visual approaches by small aircraft. A railroad is located 1046 ft from the physical end of the runway along the extended centerline of the runway. The elevation of the end of the runway is 845 ft AMSL. The elevation of the centerline of the railroad is 899 ft AMSL. We wish to determine whether the threshold is properly located or should be displaced based upon the requirements of FAR part 77. If it must be displaced, find the amount of displacement required and the effective length of runway available from the displaced threshold.

A sketch of the physical layout in both horizontal and vertical planes is shown in Fig. 7-16.

From the above data, FAR part 77 specifies that the approach surface must have a vertical clearance above the railroad of 23 ft, and the runway approach is to be designed for an approach surface of 20 to 1.

Since the approach slope required is 20 to 1 and it must clear the railroad by 23 ft, the approach slope must be at an elevation of $899 + 23 = 922$ ft AMSL on the extended centerline of the runway at the railroad. The runway elevation is 845 ft AMSL, and therefore the approach slope must be $922 - 845 = 77$ ft

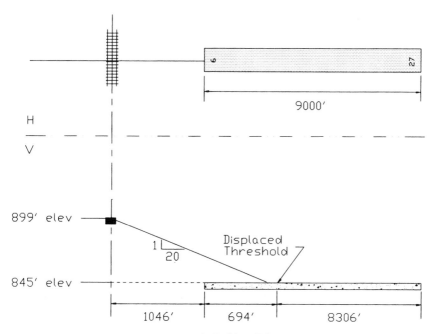

Figure 7-16 Approach surface for Example Problem 7-3.

above the runway at the railroad. Therefore, the approach slope intersects the extended runway centerline at the outer end of the primary surface, which is 200 ft beyond the end of the runway, at a distance of $20 \times 77 = 1540$ ft from the railroad centerline. The runway threshold must be located at least 1540 + 200 = 1740 ft from the railroad. The runway threshold must be displaced 1740 − 1,046 = 694 ft, and the available runway length for approaches to runway 9 is 8306 ft.

The available landing distance is 8306 ft, and a displaced threshold would be marked 1540 ft from the physical end of the runway. The available takeoff distance is still 9000 ft.

Before the runway threshold is displaced, consideration should be given to the impact of the railroad on the runway threshold approach surface, discussed in Chap. 9. The FAR part 77 restriction simply indicates that the railroad is a potential hazard to aviation and that a further study should be conducted to determine if the threshold should be displaced.

ICAO Annex 14

The ICAO requirements are similar to FAR part 77 with the following exceptions. The approach surface defined in FAR part 77 is for both arriving and departing aircraft. The ICAO separates arrivals and departures and specifies dimensions for approach surfaces and take-off climb surfaces for departures. The horizontal surface specified by ICAO is a circle whose center is at the *airport reference point,* where-

TABLE 7-8   ICAO-Recommended Dimensions and Slopes of Obstacle Limitation Surfaces for Approach Runways, m

	Runway Classification									
	Noninstrument				Non-precision-instrument approach			Precision approach		
								Category I		Category II or III
Reference code:	1	2	3	4	1, 2	3	4	1, 2	3, 4	3, 4
Conical										
Slope	20:1	20:1	20:1	20:1	20:1	20:1	20:1	20:1	20:1	20:1
Height	35	55	75	100	60	75	100	60	100	100
Inner Horizontal										
Height	45	45	45	45	45	45	45	45	45	45
Radius	2,000	2,500	4,000	4,000	3,500	4,000	4,000	3,500	4,000	4,000
Inner Approach										
Width								90	120	120
Distance from threshold								60	60	60
Length								900	900	900
Slope								40:1	50:1	50:1

	1	2	3	4	5	6	7	8	9	10
Approach										
Length of inner edge	60	80	150	150	150	300	300	150	300	300
Distance from threshold	30	60	60	60	60	60	60	60	60	60
Divergence (each side)	10:1	10:1	10:1	10:1	67:1	67:1	67:1	67:1	67:1	67:1
First section										
Length	1,600	2,500	3,000	3,000	2,500	3,000	3,000	3,000	3,000	3,000
Slope	20:1	25:1	30:1	40:1	30:1	50:1	50:1	40:1	50:1	50:1
Second slope										
Length						3,600*	3,600*	12,000	3,600*	3,600*
Slope						40:1	40:1	34:1	40:1	40:1
Horizontal section										
Length						8,400*	8,400*		8,400*	8,400*
Total length						15,000	15,000	15,000	15,000	15,000
Transitional										
Slope	5:1	5:1	7:1	7:1	5:1	7:1	7:1	7:1	7:1	7:1
Inner Transitional										
Slope								2.5:1	3:1	3:1
Balked Landing Surface										
Length of inner edge								90	120	120
Distance from threshold								†	1,800‡	1,800‡
Divergence (each side)								10:1	10:1	10:1
Slope								40:1	30:1	30:1

*Variable length.

†Distance to end of strip.

‡Or end of runway, whichever is less.

SOURCE: International Civil Aviation Organization [1].

TABLE 7-9   ICAO-Recommended Dimensions and Slopes of Obstacle Surfaces for Takeoff Runways, m

	Reference code		
Surface and dimensions	1	2	3 or 4
Takeoff Climb			
Length of inner edge	60	80	180
Distance from runway end*	30	60	60
Divergence (each side)	10:1	10:1	8:1
Final width	380	580	1,200
			1,800 †
Length	1,600	2,500	15,000
Slope	20:1	25:1	50:1

*The takeoff climb surface starts at the end of the clearway if the clearway length exceeds the specified distance.

†1,800 m when the intended track includes changes of heading greater than 15° for operations conducted at night.

SOURCE: International Civil Aviation Organization [1].

as in FAR part 77 it is not a circle, nor is the airport reference point used to determine the horizontal surface. The airport reference point is the geometric centroid of the runway system at the airport based upon the lengths of the runways. The height of this surface is 150 ft above the airport elevation, as in part 77. In FAR part 77 the conical surface extends horizontally 4000 ft at a slope of 20 to 1 irrespective of the type of runway and visibility. In ICAO Annex 14 [1, 2, 3], the slope of the conical surface is the same, but the horizontal distance varies depending upon the aerodrome reference code (see Table 9-2).

In FAR part 77 the slope of the transitional surface is a constant 7 to 1, whereas in ICAO Annex 14 this slope is specified for runway reference codes 3 and 4. For other runways the slope is 5 to 1.

The dimensions of the imaginary surfaces specified by the ICAO in Annex 14 are summarized in Tables 7-8 and 7-9.

References

1. *Aerodromes, Annex 14 to the Convention on International Civil Aviation,* vol. 1: *Aerodrome Design and Operations,* 1st ed., International Civil Aviation Organization, Montreal, Canada, July 1990.
2. *Aerodrome Design Manual,* pt. 1: *Runways,* 2d ed., International Civil Aviation Organization, Montreal, Canada, 1984.
3. *Aerodrome Design Manual,* pt. 2: *Taxiways, Aprons and Holding Bays,* 2d ed., International Civil Aviation Organization, Montreal, Canada, 1983.
4. *Airport Capacity and Delay,* Advisory Circular AC 150/5060-5 with Change 1, Federal Aviation Administration, Washington, 1983.
5. *Airport Design,* Advisory Circular AC 150/5300-13, Federal Aviation Administration, Washington, 1989.
6. *A Model Zoning Ordinance to Limit Height of Objects around Airports,* Advisory Circular AC 150/5190-4A, Federal Aviation Administration, Washington, 1987.

7. *Dual Lane Runway Study,* Rep. FAA-RD-73-60, Federal Aviation Administration, Washington, May 1973.

8. *Objects Affecting Navigable Airspace,* pt. 77, Federal Aviation Regulations (FAR), Federal Aviation Administration, Washington, 1989.

9. *Proposed Construction or Alteration of Objects That May Affect the Navigable Airspace,* Advisory Circular AC 70/7460-2I, Federal Aviation Administration, Washington, November 1988.

10. *Runway Capacity Criteria for Airport Planning Purposes,* 5th ed., Air Transportation Association of America, Washington.

11. *Techniques for Determining Airport Airside Capacity and Delay,* Rep. FAA-RD-74-124, Federal Aviation Administration, Washington, June 1976.

12. *The Apron and Terminal Building Planning Report,* Rep. FAA-RD-75-191, Federal Aviation Administration, Washington, 1975.

13. *The Apron-Terminal Complex,* Ralph M. Parsons Company, Federal Aviation Administration, Washington, September 1973.

14. *The Dual Lane Runway Concept,* R. M. Harris, Rep. M72-156, MITRE Corporation, McLean, Va., August 1972.

15. *Wind-Ceiling-Visibility Data at Selected Airports,* Report, Contract DOT-FA79WAI-057, National Climatic Center, Ashville, N.C., January 1981.

8

Airport Airside Capacity and Delay

The effectiveness of a transportation system is commonly measured in terms of its ability to efficiently process the transported unit. Since the system performance is dependent upon the individual components of that system, it is usually necessary to evaluate these components to determine overall system capabilities. In cases where use of the system requires the sequential utilization of a group of processors, the overall efficiency of the system is usually limited by the characteristics of the least efficient component.

In air transportation, particular concern is focused upon the movement of aircraft, passengers, ground access vehicles, and cargo through both the airport and the aviation system. The experienced air traveler has grown accustomed to delayed flights, overbooking, missed connections, ground congestion, parking shortages, and long lines in the terminal building during peak travel periods. For many air transportation trips, the relative advantage of the speed characteristics of aircraft is considerably diminished by ground access, terminal system, and airside delays.

In a more general sense, the unprecedented growth in the demand for air transportation services over the past 30 years has, in many situations, outpaced the ability to provide facilities to adequately accommodate this growth. To a greater and greater extent, elements of the air transport system are being stressed beyond their design capabilities, resulting in significant service deterioration at major airports in this country [5, 6, 9, 10, 18, 21, 27]. It is understandable, then, that considerable emphasis has been placed upon research to analyze the level and causes of capacity deficiencies. It is now possible to accurately determine the capability of airport and aviation system components to process demand and to pinpoint the causes of deficiencies in these systems. This knowledge allows one to propose solutions to the problems identified.

Information on airport capacity and delay is important to the airport planner. There is a strong belief within the aviation community that significant gains in air transportation efficiency can be realized through an understanding of the factors causing delays and by the application of technological innovation to alleviate delay.

Planners can compare capacity with the existing and forecast demand and ascertain whether improvements to increase capacity will be needed. Comparing the capacity of different configurations at airfields helps determine which are most efficient. Inadequate capacity leads to increasing delays at airports. Delay is an important factor in a benefit-cost analysis. If an economic value can be placed on delay, the delay reduction savings resulting from an improvement become benefits, which can be used to justify the cost of that improvement [19].

Capacity and Delay

The term *capacity* means the processing capability of a service facility over some period. However, for a service facility to realize its maximum or ultimate capacity, there must be a continuous demand for service. In aviation, it is virtually impossible to have a continuous demand throughout the operating period of the system. Even if a continuous demand were artificially created by causing a backlog at the service facilities by limiting operating periods or providing reduced operating staff, the delays at these facilities would result in such a deterioration in service quality as to make the situation undesirable. Therefore, the facility designer is faced with the problem of providing sufficient capacity to accommodate fluctuating demand with an acceptable level or quality of service. Typically, the design specifications at an airport require that sufficient capacity be provided that a relatively high percentage of the demand will be subjected to some minimal amount of delay.

This discussion leads to an awareness of the relationship between capacity and delay. To provide sufficient capacity to service a varying demand without delay will normally require facilities that are difficult to justify economically. Therefore, in design a level of delay acceptable from the perspectives of both the user and the operator is usually established, and system components of sufficient capacity are chosen to ensure that these delay criteria are met.

Capacity and delay in airfield planning

Although capacity is an important measure of the effectiveness of an airport, it should not be used as the sole criterion. In preliminary planning, several alternative airfield configurations are usually considered. Capacity estimates are useful for the initial screening of alternatives and for selecting those alternatives which should be subject to further analysis. When demand approaches capacity, delays to aircraft build up very rapidly. Congestion is usually associated with increasing delay, particularly when demand approaches capacity for more than very short periods. Because of economic factors, estimating the magnitude of delays often is more important for the justification and establishment of requirements for airfield improvements than the determination of capacity [26].

A primary objective of capacity and delay studies is to determine effective and efficient means to increase capacity and to reduce delay at airports. In practice, analyses are conducted to examine the implications of the changes in the nature of the demand, operating configurations of the airfield, and impact of facility modifications on the quality of service afforded this demand. Some of the typical applications of these analyses might include

1. The effect of alternative runway exit locations and geometry on runway system capacity

2. The impact of airfield restrictions due to noise abatement procedures, limited runway capacity, or inadequate airport navigational aids on aircraft processing rates

3. The consequences of introducing heavy aircraft into the aircraft mix at an airport, and an examination of alternative mechanisms for servicing the mix

4. The investigation of alternative runway use configurations on the ability to process aircraft

5. The generation of alternatives for new runway or taxiway construction to facilitate aircraft processing

6. The gains which might be realized in system capacity or in delay reduction by the diversion of general aviation aircraft to reliever facilities in large air traffic hub areas

As shown in Fig. 8-1, the airfield layout plan for O'Hare International Airport consists of three sets of parallel runways and a shorter runway which can be used for departures by small aircraft. In Fig. 8-2 the airfield layout plan for Detroit Metropolitan Airport consists of three parallel runways and a crosswind runway. The runway-use strategy employed at these airports at any point is largely a function of the magnitude and nature of aircraft demand, wind pattern and weather conditions, and air traffic operating rules. However, each runway-use strategy has a specific hourly capacity and results in different levels of delay.

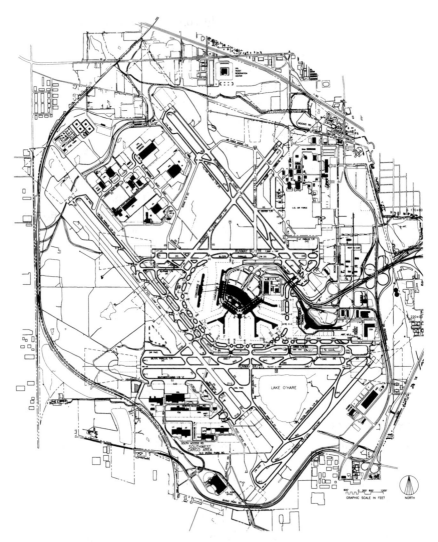

Figure 8-1 Airfield layout plan for O'Hare International Airport (*Landrum and Brown Aviation Consultants and Department of Aviation, City of Chicago*).

The FAA has indicated that in 1990 there were 23 commercial service airports in the United States which had a total annual delay in excess of 20,000 h. It has forecast that without airside capacity improvements there are likely to be 40 commercial service airports in the United States which will experience that level of annual delay in the year 2000 [9]. Although accurate estimates of the cost of delay are difficult to determine, based upon 1989 estimates of aircraft operating costs [2], an annual delay of 20,000 h costs in excess of $20 million at a typical commercial service airport.

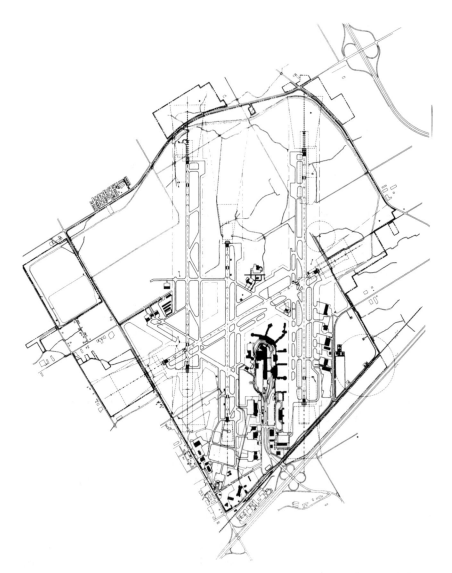

Figure 8-2 Airfield layout plan for Detroit Metropolitan Airport (*Landrum and Brown Aviation Consultants and Wayne County Department of Aviation*).

The distribution of the causes of delays greater than 15 min is given in Table 8-1 for the years 1985 and 1990. Observe that weather causes more than half of the delay, but the percentage of flights delayed due to weather is decreasing. The second most significant factor causing delays is terminal air traffic volume, and this factor has increased dramatically as a cause of delay from 1985 to 1990. In 1990, more than 10 percent of all U.S. flights were delayed for 15 min or more [9].

TABLE 8-1   Percentage Distribution of Causes of Flight
Delays Greater than 15 Min, 1985 and 1990

Cause	1985	1990
Weather	68	53
Terminal volume	12	36
Center volume	11	2
Runway-taxiway closures	6	4
National airspace system equipment	2	2
Other	1	3
Total delayed operations	334,000	404,000

SOURCE: Federal Aviation Administration [9].

TABLE 8-2   Average Aircraft Delay by Phase of Flight, 1987
to 1990

| Flight phase | Average delay per flight, min | | | |
	1987	1988	1989	1990
Gate hold	1.0	1.0	1.0	1.0
Taxi out	6.6	6.8	7.0	7.2
Airborne	3.9	4.0	4.3	4.3
Taxi in	2.1	2.1	2.2	2.3
Total	13.7	14.0	14.6	14.9

SOURCE: Federal Aviation Administration [9].

The average delay by the phase of each flight for 1987 through 1990 is given in Table 8-2. As may be observed, the most significant component of aircraft delay is in the taxi-out portion of flight, and this is due to the lack of availability of a takeoff runway, typically caused by congestion in the departure airspace. Airborne delay is the second most significant component of aircraft delay due principally to arrival traffic congestion in the terminal airspace [9].

The operational and economic implications of delay to aircraft increasingly dictate that delay analyses be included in airfield planning studies and that these analyses be conducted well before demand is expected to reach capacity levels. As an illustration of the magnitude of the impacts of airside delays, reference is made to the delay study conducted at O'Hare International Airport [19]. The computed variation of average delay per operation in both VFR and IFR conditions with the variation in peak-hour demand for this airport is plotted in Fig. 8-3.

Approaches to the Analysis of Capacity and Delay

In this chapter, analysis of capacity and delay is confined to the airfield, or aircraft operations area, which is composed of the runways,

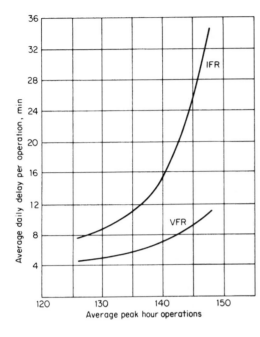

Figure 8-3 Variation of average daily aircraft delay at O'Hare International Airport (*Landrum and Brown Aviation Consultants [19]*).

taxiways, and apron areas. Capacity and delay have been evaluated by the use of analytical and computer simulation models. The focus first is on analytical models, often referred to as mathematical models.

Mathematical models of airport operations are tools for understanding the important parameters that influence the operation of systems and for investigating specific interactions in systems of particular interest. Depending upon the complexity of the system, a large number of conditions may be studied, perhaps more cheaply and quickly than by other methods. To make the mathematics tractable for a complex system, many simplifying assumptions must often be made, which may give unrealistic results. In such a case, one can resort to a computer simulation model or some other technique. Thus it is necessary, when one is contemplating the formulation and application of a mathematical model, to examine critically the correspondence between the real world being studied and the abstract world of the model and to determine the effect of their differences on the decisions to be made.

Computer simulation models are extremely useful for studying complex systems which cannot be represented by equations. These have been used successfully for solving many problems in air transport including airport planning [2, 10, 17, 22, 23, 30]. An important point to remember is that the prime justification for using computer simulation is to reduce the differences between the real world and the abstract world of the model. If the input data required for the model are not very detailed or are inaccurate, the results may not be any better than the results obtained from an analytical model of lesser complexity.

Definitions of capacity

For airport planning, the airfield capacity has been defined in two ways. One definition has been used extensively in the United States in the past: *Capacity* is the number of aircraft operations during a specified time corresponding to a tolerable level of average delay. This is shown in Fig. 8-4 and is referred to as *practical capacity.* Another definition is gaining favor: *Capacity* is the maximum number of aircraft operations that an airfield can accommodate during a specified time when there is a continuous demand for service [26]. The continuous demand for service means that there are always aircraft ready to take off or land. This definition has been referred to in several ways, namely, as the *ultimate capacity, saturation capacity,* and *maximum throughput rate,* and it is also illustrated in Fig. 8-4.

An important difference in these two measures of capacity is that one is defined in terms of delay and the other is not. There are several reasons for considering two definitions of capacity. There has been a general lack of agreement on the specification of acceptable levels of delay applicable to all airports and their airfield components. Because policies, expectations, and constraints differ from airport to airport, the amount of acceptable delay differs from airport to airport. The definition of ultimate capacity does not include delay and reflects the

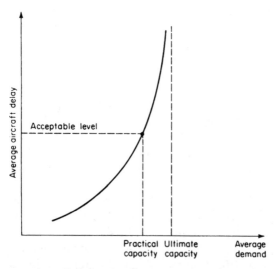

Figure 8-4 Relationship between delay-related and ultimate capacity.

capability of the airfield to accommodate aircraft during peak periods of activity. However, for this definition one does not have an explicit measure of the magnitude of congestion and delay. The magnitude of delay is greatly influenced by the pattern of demand. As an example, when several aircraft wish to use the airfield at the same time, the delay will naturally be larger than if these aircraft were spaced some interval of time apart. Since the fluctuation of demand within any hour can vary widely, there may be large variations in average delay for the same level of hourly aircraft demand. The shape of the curve in Fig. 8-4 is therefore influenced by the pattern of demand.

Experience has shown that the definition related to ultimate capacity yields values that are slightly larger than those from the definition which includes delay, but the difference is not large. Mathematically, the analysis of ultimate capacity is less complex than that for practical capacity since the determination of practical capacity implies a definition of the acceptable level of delay.

The airfield and its components

The *airfield* is the system of components upon which aircraft operate. A simplified diagram of the airfield and its relationship to the adjacent airspace is shown in Fig. 8-5. Air traffic control procedures, including those reflecting the effects of wake vortices, are major factors that influence runway component capacity and delay. Therefore, the runway component also encompasses the common approach and departure paths to and from the runways. The capacity of the taxiway component usually is much greater than the capacities of the runway or apron-gate components. The principal exception occurs when a taxiway crosses an active runway. General aviation aircraft do not operate on fixed schedules, and therefore the time spent by these aircraft in the general aviation apron areas at airports fluctuates widely. Techniques analyzing delay in the apron area generally consider only the capacity and delay on the air carrier aircraft parking apron or ramp. For determining capacity and delay, the operations on the runways, taxiways, and gates at most airports can be considered independent of each other and may be analyzed separately [25, 26]. For planning purposes, it is sufficiently accurate to assume that the capacity of the runways is not affected by operations on either the gates or the taxiways. Because operations on one airfield component generally do not affect the capacity of another component, the capacity of the entire airfield is governed by the capacity of each component which is most restrictive. Additionally, because operations on one component have little influence on the delay to aircraft on another component, the total delay to aircraft on the entire airfield can be estimated by adding the delay to aircraft on each airfield component.

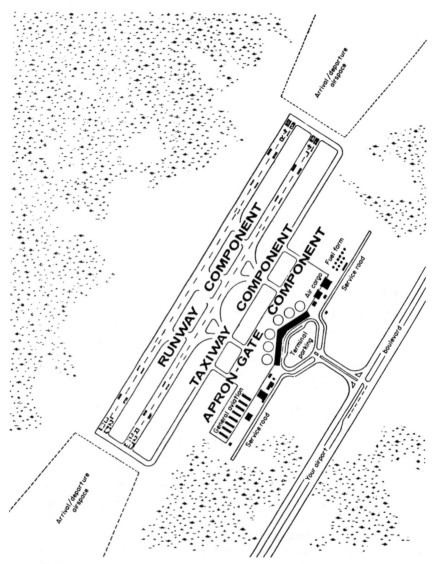

Figure 8-5 Components of the airfield system (*Federal Aviation Administration [26]*).

Factors that affect hourly capacity

Many factors influence the capacity of an airfield, and some are more significant than others. In general, the capacity depends on the configuration of the airfield, environment in which aircraft operate, availability and sophistication of aids to navigation, and air traffic control facilities and procedures. The most important factors include

1. The configuration, number, spacing, and orientation of the runway system

2. The configuration, number, and location of taxiways and runway exits

3. The arrangement, size, and number of gates in the apron area

4. The runway occupancy time for arriving and departing aircraft

5. The size and mix of aircraft using the facilities

6. Weather, particularly visibility and ceiling, since air traffic rules in good weather are different from those in poor weather

7. Wind conditions which may preclude the use of all available runways by all aircraft

8. Noise abatement procedures which may limit the type and timing of operations on available runways

9. Within the constraints of wind and noise abatement, the strategy which the controllers choose to operate the runway system

10. The number of arrivals relative to the number of departures

11. The number and frequency of touch-and-go operations by general aviation aircraft

12. The existence and frequency of wake vortices which require greater separations when a light aircraft follows a heavy aircraft than when a heavy follows a light

13. The existence and nature of navigational aids

14. The availability and structure of airspace for establishing arrival and departure routes

15. The nature and extent of the air traffic control facilities

The most significant factor which affects runway capacity is the spacing between successive aircraft. This spacing is dependent upon the appropriate air traffic rules, which are to a large extent functions of weather conditions and aircraft size and mix.

Formulation of Runway Capacity through Queueing Theory

In 1960 the FAA contracted with Airborne Instruments Laboratory to develop mathematical models for estimating runway capacity [3]. These models relied on steady-state queueing theory. Essentially there were two models, one for runways serving either arrivals or departures and the other for runways serving mixed operations. For runways used exclusively for arrivals or departures, the model was a simple Poisson-type queue with a first-come, first-served service discipline. The demand process for arrivals or departures is characterized as a Poisson distribution with a specified mean arrival or departure rate. The runway service process is a general service distribution specified by the mean service time and the standard deviation of the mean service time. For mixed operations, when runways are used for both takeoffs and landings, the process is more complicated, and a preemptive spaced-arrivals model was developed. In this model arrivals have priority over departures for use of the runways. The takeoff demand process is assumed to follow a Poisson distribution; however, the landing process encountered at the end of the runway is not Poisson, but is more like the output of an airborne queueing system.

It was recognized that steady-state conditions are rarely achieved at airports; however, it was argued that time-dependent solutions, although possible, were quite complex and were out of the question for the large number of situations required for the preparation of a runway capacity handbook to be used by airport planning and design professionals. Additional support for the use of steady-state solutions came from observations which showed that the average delay times yielded by the models were in general agreement with measured delays under a wide variety of operating conditions.

Mathematical formulation of delay

The delay for runways used exclusively for arrivals was computed from the following equation:

$$W_a = \frac{\lambda_a(\sigma_a^2 + 1/\mu_a^2)}{2(1 - \lambda_a/\mu_a)} \qquad (8\text{-}1)$$

where W_a = mean delay to arriving aircraft
λ_a = mean arrival rate of aircraft
μ_a = mean service rate for arrivals, or reciprocal of mean service time
σ_a = standard deviation of mean service time of arriving aircraft

The mean service time may be the runway occupancy time or the time separation in the air immediately adjacent to the runway threshold, whichever value is the larger.

The model for departures is identical to that for arrivals except for a change in subscripts. The following equation is therefore used for the departure delay:

$$W_d = \frac{\lambda_d(\sigma_d^2 + 1/\mu_d^2)}{2(1 - \lambda_d/\mu_d)} \tag{8-2}$$

where W_d = mean delay to departing aircraft
λ_d = mean departure rate of aircraft
μ_d = mean service rate for departures, or reciprocal of mean service time for departures
σ_d = standard deviation of mean service time of departing aircraft

For mixed operations, arriving aircraft are normally given priority, and the delay to these aircraft is given by the arrivals equation, Eq. (8-1). However, the average delay to departures in this situation can be found from the following equation:

$$W_d = \frac{\lambda_d(\sigma_j^2 + j^2)}{2(1 - \lambda_d j)} + \frac{g(\sigma_f^2 + f^2)}{2(1 - \lambda_a f)} \tag{8-3}$$

where W_d = mean delay to departing aircraft
λ_a = mean arrival rate of aircraft
λ_d = mean departure rate of aircraft
j = mean interval of time between two successive departures
σ_j = standard deviation of mean interval of time between successive departures
g = mean rate at which gaps between successive arrivals occur
f = mean value of interval of time within which no departure can be released
σ_f = standard deviation of interval of time in which no departure may be released

During busy periods the second term in Eq. (8-3) is expected to be zero if it is assumed that aircraft are in a queue at the end of the runway and are always ready to go when permission is granted. It must be emphasized that the above equations are valid only when the mean arrival or departure rate is less than the mean service rate, which is the condition for which the equations have been derived. The use of the model for the arrivals-only case is illustrated in Example Problem 8-1.

Example Problem 8-1 It is necessary to compute the average delay to arriving aircraft on a runway system which services only arrivals if the mean service time is 60 s per aircraft with a standard deviation in the mean service time of 12 s and an average rate of arrivals of 45 aircraft per hour.

The mean service rate for arrivals μ_a is the reciprocal of the mean service time, yielding 1 aircraft per minute, or 60 aircraft per hour. Substitution into Eq. (8-1) yields

$$W_a = \frac{45[(^{12}/_{3600})^2 + 1/60^2]}{2(1 - ^{45}/_{60})} = 0.026 \text{ h} = 1.6 \text{ min}$$

Therefore, the average aircraft delay is about 1.6 min per arrival.

The relationship between delay and capacity can be shown by determining the runway service rate which corresponds to a delay of 4 min by using the above equation. Assuming that the standard deviation of the mean service time is the same, we have

$$\frac{4}{60} = \frac{45[(^{12}/_{3600})^2 + 1/\mu_a^2]}{2(1 - 45/\mu_a)}$$

or μ_a is equal to 52 arrivals per hour. If the delay criterion were that arrival delays could not exceed 4 min, then the runway capacity related to delay would be 52 arrivals per hour.

Observe that a 15 percent increase in capacity from 52 to 60 arrivals per hour results in a 60 percent delay reduction of 2.4 min. This is typical at airports nearing saturation. Small increases in capacity can result in significant decreases in delay.

Formulation of Runway Capacity through the Space-Time Concept

The various intervals of time included in the above models are shown on the space-time diagram in Fig. 8-6. The space-time diagram is a very useful device for understanding the sequencing of aircraft opera-

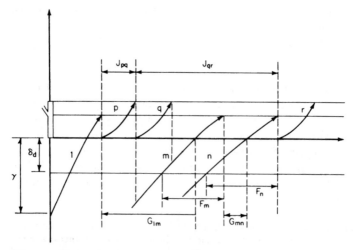

Figure 8-6 Space-time diagram concepts for mixed operations on runway system.

tions on a runway system and in the airspace adjacent to this system. In the case illustrated in Fig. 8-6, three arrivals and three departures are serviced as indicated. The basic sequencing rules being used to service these aircraft are as follows:

1. Two aircraft may not conduct an operation on the runway at the same time.

2. Arriving aircraft have a priority in the use of the runway over departing aircraft.

3. Departures may be released if the runway is clear and the subsequent arrival is at least a certain distance from the runway threshold.

Examination of the space-time diagram in Fig. 8-6 shows that the mean departure interval j is the average of the time interval between successive departures J_{pq} and J_{qr}. Also the mean time interval between arrivals—the gap between arrivals I_g during which it may be possible to release g departures—is the average of the quantities G_{lm} and G_{mn}. Finally, the value of the time interval in which departures cannot be released f is equal to the average of the quantities F_m and F_n.

Several other observations may be made about the sequence of operations shown on this space-time diagram. The initial departure p could have been released, if it was available, before the first arrival l reached the distance δ_d from the runway threshold, since the runway was clear. The second departure q was released when the previous departure p cleared the runway since the next arrival m was more than the distance δ_d from the threshold at that time. However, the third departure r was not released when that departure cleared the runway because the approaching aircraft m was closer than distance δ_d from the threshold at that time. For the same reason, this departure was not released until after the last arrival n cleared the runway. In this figure, then, the delays which aircraft would incur are due to the required separations between different types of operational sequences.

The use of the air traffic separation rules in accommodating a series of arrivals and departures may be best understood through a numerical example illustrating the space-time concept for processing aircraft on a runway system.

Example Problem 8-2   A runway is to service arrivals and departures. The common approach path is 7 mi long for all aircraft. During a particular time interval the runway is serving only two types of aircraft: type A with an approach speed of 120 mi/h and type B with an approach speed of 90 mi/h. Each arriving aircraft will be on the runway for 40 s before exiting the runway. The air traffic separation rules in effect are given in Table 8-3.

TABLE 8-3   Air Traffic Separation Rules for Example Problem 8-2

Operational sequence	Air traffic rules
Arrival-departure	Clear runway
Departure-arrival	Arrival at least 2 mi from arrival threshold
Departure-departure	120 s
Arrival-arrival	Miles:

$$\text{Trailing} \begin{array}{c} A \\ B \end{array} \begin{array}{cc} & \text{Leading} \\ & \begin{array}{cc} A & B \end{array} \\ \left[\begin{array}{cc} 4 & 3 \\ 5 & 3 \end{array} \right] \end{array}$$

During the period to be analyzed, five aircraft in an ordered arrival queue of B, A, A, B, A aircraft approach the runway. An identical ordered departure queue of aircraft is awaiting clearance to take off.

A space-time diagram to service these aircraft is drawn by assuming that the first arrival is at the entry gate at time 0 and that arrivals are given priority over departures.

The space-time diagram for arrivals is drawn first, since these aircraft normally have priority over departures. This is shown in Fig. 8-7. The dashed lines indicate points where the interarrival separation rules are enforced to ensure that the minimum interarrival spacing is maintained. The numbers in parentheses indicate the time that each aircraft is at the point indicated.

Since the first aircraft, a type B aircraft, is at the entry gate at time 0, and since it takes 280 s to travel the common approach path from the entry gate to the runway threshold, this aircraft passes over the runway threshold at time 280 s. Immediately behind this aircraft is a type A aircraft that is approaching the runway at a speed of 120 mi/h. In this case, the trailing aircraft A is flying faster than the leading aircraft B, and so it is closing in on the leading aircraft.

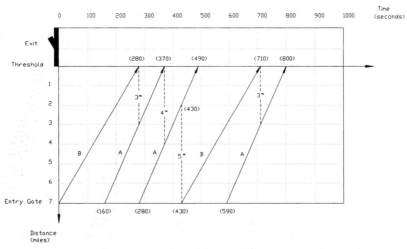

Figure 8-7 Space-time diagram for scheduling arrivals in Example Problem 8-2.

These two aircraft are closest when the leading aircraft passes over the arrival threshold, or at time 280 s. At this time the trailing aircraft can be scheduled no closer than 3 mi behind the leading aircraft, or the trailing aircraft is scheduled to pass over the 3-mi point at time 280 s. Since this aircraft is approaching the runway at a rate of 30 s/mi, it passes over the runway threshold 90 s later, or at time 370 s. It passes over the entry gate 210 s earlier, or at time 160 s.

This process is continued until all aircraft have been scheduled. Observe that when a type B aircraft is trailing a type A aircraft, since the type B aircraft is traveling more slowly than the type A, these two aircraft are closest when the trailing aircraft passes over the entry gate, and the required separation is maintained at that point.

Once all the aircraft are scheduled, as shown in Fig. 8-7, it is determined that it will take 800 s to service these five arriving aircraft. The time span at the runway threshold for serving these 5 arrivals is $800 - 280 = 520$ s. In this time span there are 4 pairs of arrivals. Therefore, the average time between arrivals, the *interarrival time,* is $520/4 = 130$ s per arrival. The capacity of the runway to service arrivals will be shown later to be

$$C_a = \frac{3600}{130} = 28 \text{ aircraft per hour}$$

The space-time diagram in Fig. 8-8 is then constructed from Fig. 8-7 and is used to determine whether a departure may be released in the time gaps between arrivals. Each arrival spends 40 s on the runway prior to exiting the runway. Therefore, the time when each arriving aircraft exits the runway is determined first. The results are shown in Fig. 8-8. At any time, if the runway is clear, a departure may be cleared for takeoff if (1) the incoming arrival is at least 2 mi from the arrival threshold and (2) it has been at least 120 s since the last departure was cleared for takeoff.

Again, in Fig. 8-8, the dashed lines indicate points where the separation rules are enforced, and the numbers in parentheses indicate the time each aircraft is at the point indicated. However, these comparisons are now made to ensure

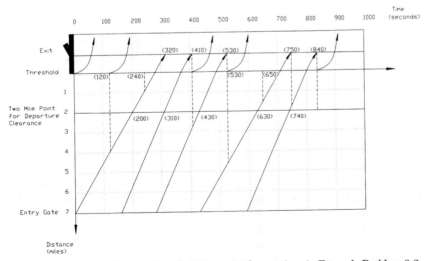

Figure 8-8 Space-time diagram for scheduling mixed operations in Example Problem 8-2.

that the departure-departure, arrival-departure, and departure-arrival spacings are each maintained.

It will take 840 s, measured at the runway threshold, to service all the arrivals and all the departures. It is also observed that departures can be inserted between a pair of arrivals on only two occasions. Therefore, the probability of inserting a departure between the 4 pairs of arrivals is 2 out of 4, or 0.50. The capacity to service mixed operations will be shown later to be

$$C_m = \frac{3600}{130}(1.0 + 0.50) = 42 \text{ aircraft per hour}$$

where the 1.0 represents the probability of an arrival at the threshold every 130 s and 0.50 represents the probability of inserting a departure in an interarrival time of 130 s.

The capacity of the runway to service only departures will be shown later to be

$$C_d = \frac{3600}{120} = 30 \text{ aircraft per hour}$$

Formulation of Ultimate Capacity

Capacity as defined here expresses the maximum physical capability of a runway system to process aircraft. It is the *ultimate capacity* or maximum aircraft operations rate for a set of specified conditions, and it is independent of the level of average aircraft delay. In fact, it has been shown that when traffic volumes reach hourly capacity levels, average aircraft delays may range from 2 to 10 min.

Delay is dependent upon the capacity as well as the magnitude, nature, and pattern of demand. Delays can occur even when the demand averaged over 1 h is less than the hourly capacity. Such delays occur because demand fluctuates within an hour so that, during some smaller intervals of time, demand is greater than capacity.

If the magnitude, nature, and pattern of demand are fixed, then delay can be reduced only by increasing capacity. But if demand can be manipulated to produce more uniform patterns, then delay can be reduced without increasing capacity. Thus, estimating capacity is an integral step in determining the delay to aircraft.

Mathematical formulation of ultimate capacity

These types of models determine the maximum number of aircraft operations that a runway system can accommodate in a specified time when there is continuous demand for service [26]. In these models, capacity is equal to the inverse of a weighted-average service time of all aircraft being served. For example, if the weighted-average service time is 90 s, the capacity of the runway is 1 operation every 90 s, or 40 operations per hour. The models treat the common approach path to the runway together with the runway as the runway system. The runway service time is defined as either the separation in the air between arrivals in terms of time, the *interarrival time*, or the *runway occupancy time*, whichever is larger. The material presented in this section is taken largely from several references [16, 25, 26].

Development of models for arrivals only. The capacity of a runway system used only for arriving aircraft is affected by the following factors:

1. The aircraft mix, which is usually characterized by segregating aircraft into several classes according to their approach speeds

2. The approach speeds of various classes of aircraft

3. The length of the common approach path from the entry gate to the runway threshold

4. The minimum air traffic separation rules or the practical observed separations if no rules apply

5. The magnitude of errors in arrival time at the entry point to the common approach path, entry gate, and speed variation of aircraft on the common approach path

6. The specified probability of violation of the minimum air traffic separations considered acceptable or attainable

7. The mean runway occupancy times of various classes of aircraft in the mix and the magnitude of the variation in these times

The error-free case. With little loss in accuracy and to make the computations simpler, aircraft are grouped into several discrete speed classes V_k. To obtain the weighted service time for arrivals, it is necessary to formulate a matrix of the times between aircraft arrivals at the runway threshold. Given this matrix and the percentage of various classes in the aircraft mix, the weighted service time can be computed. The inverse of the weighted service time is the capacity of the runway.

Let the error-free matrix, comprised of elements m_{ij}, be designated as $[M_{ij}]$, the minimum error-free time interval at the runway threshold for aircraft of speed class i followed by aircraft of class j. Let p_i be the percentage of aircraft of class i in the mix and p_j the percentage of aircraft of class j in the mix. Then

$$\Delta T_{ij} = T_j - T_i = m_{ij} \tag{8-4}$$

where ΔT_{ij} = actual time separation at runway threshold for two successive arrivals, an aircraft of speed class i followed by an aircraft of speed class j

T_i = time that leading aircraft i passes over runway threshold

T_j = time that trailing aircraft j passes over runway threshold

m_{ij} = minimum error-free interarrival separation at runway threshold, which is same as ΔT_{ij} in error-free case

$$E(\Delta T_{ij}) = \Sigma p_{ij} m_{ij} = \Sigma [p_{ij}][M_{ij}] \tag{8-5}$$

where $E(\Delta T_{ij})$ = expected value of service time, or interarrival time, at runway threshold for arrival aircraft mix

p_{ij} = probability that leading arriving aircraft i will be followed by trailing arriving aircraft j

$[p_{ij}]$ = matrix of these probabilities

$[M_{ij}]$ = matrix of minimum interarrival separations m_{ij}

The capacity for arrivals is given by

$$C_a = \frac{1}{E(\Delta T_{ij})} \tag{8-6}$$

where C_a is the capacity of the runway to process this mix of arrivals.

To obtain the interarrival time at the runway threshold, it is necessary to know whether the speed of the leading aircraft V_i is greater or less than that of the trailing aircraft V_j, since the separation at the runway threshold will differ in each case. This can be illustrated by drawing space-time diagrams representative of these conditions, as shown in Figs. 8-9, 8-10, and 8-11. In these diagrams the following notation is used:

γ length of common approach path

δ_{ij} minimum permissible separation distance between two arriving aircraft, a leading aircraft i and a trailing aircraft j, anywhere along common approach path

V_i approach speed of leading aircraft of class k

V_j approach speed of trailing aircraft j of class k

R_i runway occupancy time of leading aircraft

The closing case ($V_i \leq V_j$).   First let us consider the case where the leading aircraft's approach speed is less than that of the trailing aircraft, as shown in Fig. 8-9. The minimum time separation at the threshold may be written in terms of the minimum distance separation δ_{ij} and the speed of the trailing aircraft V_j. However, if the runway occupancy time of the arrival R_i is greater than the airborne separation, then it would be the minimum separation at the threshold. The equation for this case is

$$\Delta T_{ij} = T_j - T_i = \frac{\delta_{ij}}{V_j} \tag{8-7}$$

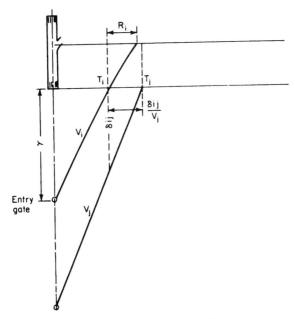

Figure 8-9 Space-time diagram for error-free interarrival spacing for the closing case when $V_i \leq V_j$.

The opening case ($V_i > V_j$). Next let us consider the case where the leading aircraft's approach speed V_i is greater than that of the trailing aircraft V_j, as shown in Figs. 8-10 and 8-11, the minimum time separation at the threshold is written in terms of the distance δ_{ij}, the length of the common approach path γ, and the approach speeds of the leading and trailing aircraft V_i and V_j. This corresponds to the minimum separation distance, δ_{ij} along the common approach path which now occurs at the entry gate instead of the threshold. The equation for the case shown in Fig. 8-10, when control is exercised only from the entry gate to the arrival threshold, is

$$\Delta T_{ij} = T_j - T_i = \frac{\delta_{ij}}{V_i} + \gamma \left(\frac{1}{V_j} - \frac{1}{V_i} \right) \tag{8-8}$$

When control is exercised to maintain the separations between both aircraft as the leading aircraft passes over the entry gate, as shown in Fig. 8-11, the equation is

$$\Delta T_{ij} = T_j - T_i = \frac{\delta_{ij}}{V_j} + \gamma \left(\frac{1}{V_j} - \frac{1}{V_i} \right) \tag{8-9}$$

Carefully note that the only difference between Eqs. (8-8) and (8-9) is in the first term of the equation, where V_i and V_j are interchanged.

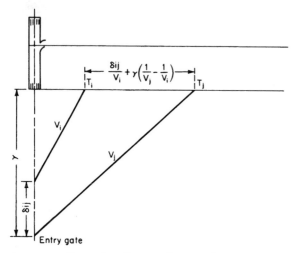

Figure 8-10 Space-time diagram for error-free interarrival spacing for the opening case when $V_i > V_j$ for aircraft control from entry gate to arrival threshold.

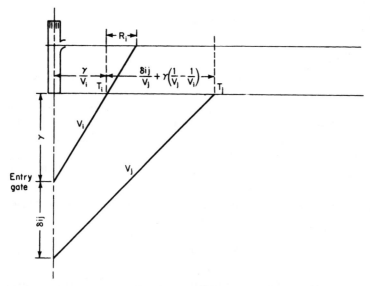

Figure 8-11 Space-time diagram for error-free interarrival spacing for the opening case when $V_i > V_j$ for both aircraft separated in vicinity of entry gate.

Example Problem 8-3 Now we will solve Example Problem 8-2, using the error-free analytical equations developed above. It is necessary to determine the arrival capacity of the runway in an error-free context where aircraft separations are maintained in the airspace along the common approach path between the entry gate and the arrival threshold.

There are four possible interarrival cases: leading A and trailing A, leading B and trailing B, leading A and trailing B, and leading B and trailing A. These cases are governed by Eqs. (8-7) and (8-8). Equation (8-7) gives the minimum time between arrivals at the runway threshold when the leading aircraft is approaching the runway at an approach speed less than or equal to that of the trailing aircraft.

If we have leading A and trailing A, or leading B and trailing B, both aircraft are traveling at the same speed. Therefore, Eq. (8-7) applies, and we have for type A following type A

$$\Delta T_{ij} = \frac{4(3600)}{120} = 120 \text{ s}$$

and for type B following type B

$$\Delta T_{ij} = \frac{3(3600)}{90} = 120 \text{ s}$$

When the leading aircraft is type B and the trailing aircraft is type A, Eq. (8-7) also applies, and we have

$$\Delta T_{ij} = \frac{3(3600)}{120} = 90 \text{ s}$$

When the leading aircraft is approaching the runway at an approach speed greater than that of the trailing aircraft, then the minimum time between arrivals at the runway threshold is given by Eq. (8-8). This is the case when type B aircraft follows type A aircraft. Therefore, we have

$$\Delta T_{ij} = \frac{5(3600)}{120} + 7\left(\frac{1}{90} - \frac{1}{120}\right)(3600)$$

$$= 220 \text{ s}$$

The ordered queue consists of pairs of arrivals B-A, A-A, A-B, and B-A. Therefore, we have the following interarrival matrix and probability matrix based upon the actual queue of arriving aircraft given:

$[T_{ij}]$:

Leading

		A	B
Trailing	A	120	90
	B	220	120

$[p_{ij}]$:

Leading

		A	B
Trailing	A	0.25	0.50
	B	0.25	0.00

From Eq. (8-5), since in the error-free case $[T_{ij}]$ is equal to $[M_{ij}]$, we know that the expected value of the interarrival time is

$$E(\Delta T_{ij}) = 0.25(120) + 0.50(90) + 0.25(220) + 0.00(120)$$

$$= 130 \text{ s}$$

From Eq. (8-6) for the arrival capacity of the runway, we then have

$$C_a = \frac{3600}{130} = 28 \text{ operations per hour}$$

This agrees exactly with the results of this problem done by the space-time diagram method in Example Problem 8-2.

Consideration of position error.   The above models represent the situation of a perfect system with no errors. To take care of position errors, a *buffer time* is added to the minimum separation time to ensure that the minimum interarrival separations are maintained. The size of the buffer depends upon what probability of violation of the minimum separation rules is acceptable. Figure 8-12 shows the position of the trailing aircraft as it approaches the runway threshold. In the top portion of this illustration, the trailing aircraft is sequenced so that its mean position is exactly determined by the minimum separation between the leading and trailing aircraft. However, if aircraft position is a random variable, there is an equal probability that it can be either ahead or behind schedule. Naturally if it is ahead of schedule, the minimum separation criterion will be violated. If the position error is normally distributed, then the shaded area of the bell-shaped curve will correspond to a probability of violation of the minimum separation rule of 50 percent. Therefore, to lower this probability of violation, the aircraft may be scheduled to arrive at this

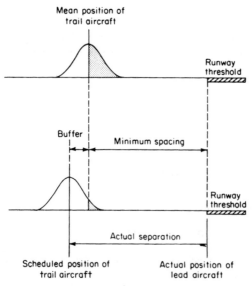

Figure 8-12   Illustration of buffer spacing on actual separation between aircraft when position error is considered.

position later, by building in a buffer to the minimum separation criterion, as shown in the bottom portion of the illustration. In this case, only when the aircraft is so far ahead of schedule as to encroach upon the smaller shaded area of the bell-shaped curve would a separation violation occur. There is, of course, a lower probability of this occurring. In practice, air traffic controllers schedule aircraft with a buffer so that the probability of violation of minimum separation rules is at an acceptable level.

As will be shown, in the closing case the buffer is a constant value. However, in the opening case the buffer need not be a constant value and will normally be less than the buffer for the closing case. Given the models for the buffer, a matrix of buffer times $[B_{ij}]$ for aircraft of speed class i followed by aircraft of speed class j is developed. This matrix is added to the error-free matrix to determine the actual interarrival time matrix from which the capacity may be found. The relationship is given in Eq. (8-10).

$$E(\Delta T_{ij}) = \Sigma[p_{ij}][M_{ij} + B_{ij}] \tag{8-10}$$

The closing case. In this case the leading aircraft's approach speed is less than that of the trailing aircraft, and the separations are as shown in Fig. 8-9. Let us call ΔT_{ij} the actual minimum time between aircraft of class i and class j, and let us assume that the runway occupancy is less than ΔT_{ij}. Designate the mean or expected value of ΔT_{ij} as $E(\Delta T_{ij})$ and e_0 as a zero-mean normally distributed random error with a standard deviation of σ_0. Then for each pair of arrivals $\Delta T_{ij} = E(\Delta T_{ij}) + e_0$. In order not to violate the minimum separation distance, the value of ΔT_{ij} must be increased by a buffer amount b_{ij}. Therefore, we have

$$\Delta T_{ij} = m_{ij} + b_{ij}$$

and

$$\Delta T_{ij} = m_{ij} + b_{ij} + e_0$$

For this case the minimum separation distance at the runway threshold is given by Eq. (8-7). The objective is to find for a specified probability of violation p_v the required amount of buffer. Thus

$$p_v = \text{Prob}\left(\Delta T_{ij} < \frac{\delta_{ij}}{V_j}\right)$$

or

$$p_v = \text{Prob}\left(\frac{\delta_{ij}}{V_j} + b_{ij} + e_0 < \frac{\delta_{ij}}{V_j}\right)$$

which simplifies to

$$p_v = \text{Prob}(b_{ij} < -e_o)$$

By using the assumption that errors are normally distributed with standard deviation σ_0, the value of the buffer can be derived as [16]

$$b_{ij} = q_v \sigma_0 \qquad (8\text{-}11)$$

where q_v is the value for which the cumulative standard normal distribution has the value $1 - p_v$. Stated differently, this simply means the number of standard deviations from the mean in which a certain percentage of the area under the normal curve would be found. For example, if $p_v = 0.05$, then q_v is the 95th percentile of the distribution and equals 1.65. Therefore, in the closing case, the buffer time is a constant that depends on the magnitude of the dispersion of the position error and the acceptable probability of violation p_v.

The opening case. Next consider the case when the leading aircraft is approaching the runway threshold at a speed greater than that of the trailing aircraft. In this case, the separation between the aircraft increases from the entry gate. The model is based on the supposition that the trailing aircraft should be scheduled not less than a distance δ_{ij} behind the leading aircraft when the latter is at the entry gate; but it is assumed that strict separation is enforced by air traffic control only when the trailing aircraft reaches the entry gate. This assumption was shown in Fig. 8-10.

For this case the probability of violation is simply the probability that the trailing aircraft will arrive at the entry gate before the leading aircraft is a specified distance inside the entry gate. This may be expressed mathematically as

$$p_v = \text{Prob}\left(T_j - \frac{\delta_{ij} + \gamma}{V_j} < T_i - \frac{\gamma}{V_i}\right)$$

or

$$p_v = \text{Prob}\left[T_j - T_i < \frac{\delta_{ij}}{V_j} + \left(\frac{\gamma}{V_j} - \frac{\gamma}{V_i}\right)\right]$$

Using Eq. (8-9) with this equation to compute the actual spacing at the arrival threshold and simplifying, we get

$$b_{ij} = \sigma_0 q_v - \delta_{ij}\left(\frac{1}{V_j} - \frac{1}{V_i}\right) \qquad (8\text{-}12)$$

Therefore, for the opening case, the amount of buffer is reduced from that required in the closing case, as shown in Eq. (8-12). Negative values of buffer are not allowed, so the buffer is some finite positive number with a minimum of zero. The matrix of the buffer $[B_{ij}]$ for each pair of aircraft with an interarrival buffer of b_{ij} can then be found. The application of position error to the arrivals-only runway capacity problem is illustrated in Example Problem 8-4.

Example Problem 8-4   Assume that the aircraft approaching the runway in Example Problem 8-3 have a position error of 20 s which is normally distributed. In this environment the probability of violating the minimum separation rule for arrival spacing is allowed to be 5 percent.

It is necessary to determine the hourly capacity of the runway to service arrivals.

The error-free interarrival matrix $[M_{ij}]$ was found earlier, and now it is only necessary to find the buffer matrix $[B_{ij}]$ and to solve Eq. (8-10) for the expected value of the interarrival time.

In the closing case where the leading aircraft is slower than the trailing aircraft, Eq. (8-11) gives the buffer. For a 5 percent probability of violation, q_v can be found from statistics tables to be 1.65. For each of these cases, the buffer is independent of speed, and therefore

$$b_{ij} = 20(1.65) = 33 \text{ s}$$

In the opening case where the leading aircraft is faster than the trailing aircraft, Eq. (8-12) gives the buffer. Therefore, for $V_i = 120$ and $V_j = 90$ we have

$$b_{ij} = 20(1.65) - 5(\tfrac{1}{90} - \tfrac{1}{120})(3600) = -17 \text{ s}$$

However, the minimum value of the buffer is always zero. Summarizing the values of b_{ij} found for the buffer in a matrix and recalling the $[M_{ij}]$ matrix for the error-free case, we have

$[M_{ij}]$:   Leading

		A	B
Trailing	A	120	90
	B	220	120

$[B_{ij}]$:   Leading

		A	B
Trailing	A	33	33
	B	0	33

$[M_{ij}] + [B_{ij}]$:   Leading

		A	B
Trailing	A	153	123
	B	220	153

Substitution into Eq. (8-10) then gives the expected value of the interarrival time.

$$E(\Delta T_{ij}) = 0.25(153) + 0.50(123) + 0.25(220) = 155 \text{ s}$$

and from Eq. (8-6) we have

$$C_a = \frac{3600}{155} = 23 \text{ operations per hour}$$

which is a reduction from the arrival capacity found in Example Problem 8-3. Therefore, we conclude that the existence of position error reduces the arrival capacity of a runway.

Development of a model for departures only. Since departures are normally cleared for takeoff based upon maintaining a minimum time interval between successive departures, the interdeparture time t_d, the departure-only capacity of a runway C_d is given by

$$C_d = \frac{3600}{E(t_d)} \tag{8-13}$$

and

$$E(t_d) = \Sigma[p_{ij}][t_d] \tag{8-14}$$

where $E(t_d)$ = expected value of service time, or interdeparture time, at runway threshold for departure aircraft mix
   $[p_{ij}]$ = matrix of probabilities that leading departing aircraft i will be followed by trailing departing aircraft j
   $[t_d]$ = matrix of interdeparture times

Development of models for mixed operations. This model is based on the same four operating rules as the model developed by Airborne Instruments Laboratory [3]:

1. Arrivals have priority over departures.

2. Only one aircraft can occupy the runway at any time.

3. A departure may not be released if the subsequent arrival is less than a specified distance from the runway threshold, usually 2 nmi in IFR conditions.

4. Successive departures are spaced at a minimum time separation equal to the departure service time.

A space-time diagram may be drawn to show the sequencing of mixed operations under the rules stated above, and this is done in Fig. 8-13. In this figure, T_i and T_j are the times that the leading aircraft i and the trailing aircraft j, respectively, pass over the arrival threshold, δ_{ij} is the minimum separation between arrivals, T_1 is the time when the arriving aircraft clears the runway, T_d is the time when the departing aircraft begins its takeoff roll, δ_d is the minimum distance that an arriving aircraft must be from the threshold to release a departure, T_2 is the time which corresponds to the last instant when a departure can be released, R_i is the runway occupancy time for an arrival, G is the time gap in which a departure may be released, and t_d is the required service time for a departure.

Since arrivals are given priority over departures, the arriving aircraft are sequenced at the minimum interarrival separation and a departure cannot be released unless there is a gap G between arrivals. Therefore, we may write

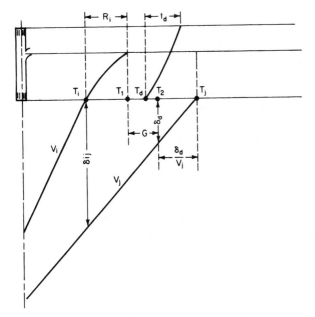

Figure 8-13 Space-time diagram for error-free interarrival spacing for mixed operations on a runway system.

$$G = T_2 - T_1 \geq 0$$

but we know that

$$T_1 = T_i + R_i$$

and

$$T_2 = T_j - \frac{\delta_d}{V_j}$$

Therefore, we may write

$$T_2 - T_1 \geq \left(T_j - \frac{\delta_d}{V_j}\right) - (T_i + R_i) \geq 0$$

or to release one departure between a pair of arrivals, we have

$$T_j - T_i \geq R_i + \frac{\delta_d}{V_j}$$

Through a simple extension of this equation, it is apparent that the required mean interarrival time $E(\Delta T_{ij})$ to release n_d departures between a pair of arrivals is given by

$$E(\Delta T_{ij}) \geq E\ (R_i) + E\left(\frac{\delta_d}{V_j}\right) + (n_d - 1)E(t_d) \qquad (8\text{-}15)$$

Note that the last term of this equation is equal to zero when only one departure is to be inserted between a pair of arrivals. An error term $\sigma_G q_v$ may be added to the above equation to account for the violation of the gap spacing. The use of Eq. (8-15) with gap error will be illustrated in Example Problem 8-5.

The capacity for mixed operations is given by

$$C_m = \frac{1}{E(\Delta T_{ij})}(1 + \Sigma n_d p_{nd}) \qquad (8\text{-}16)$$

where
C_m = capacity of runway to process mixed operations
$E(\Delta T_{ij})$ = expected value of interarrival time
n_d = number of departures which can be released each gap between arrivals
p_{nd} = probability of releasing n_d departures in each gap

The application of the equations for mixed operations on a runway is shown now.

Example Problem 8-5 Assume that the aircraft approaching the runway in Example Problem 8-4 have an error in the gap between arrivals of 30 s which is normally distributed. In this environment, the probability of violating the minimum gap in which departures can be released is 10 percent. The runway occupancy time for type A aircraft is 50 s and for type B aircraft is 40 s. A departure can be released if the arriving aircraft is at least 2 mi from the arrival threshold. The minimum time between successive departures is 60 s. The arrival mix and the departure mix are identical.

It is necessary to determine the minimum separation between arrivals to ensure that one departure can be released between each pair of arrivals.

The required interarrival time to release a departure between every pair of arrivals is given by Eq. (8-15).

There are three type A aircraft and two type B aircraft in both the arrival and departure queues. Therefore, there is a 60 percent chance of a type A aircraft and a 40 percent chance of a type B aircraft. Substitution into Eq. (8-15) yields

$$E(\Delta T_{ij}) \geq 0.6(50) + 0.4(40) + \left[\frac{0.6(2)}{120} + \frac{0.4(2)}{90}\right]3600 + (1 - 1)(60)$$
$$\geq 114 \text{ s}$$

Therefore, to release a departure between a pair of arrivals in an error-free context, there must be a minimum gap between successive arrivals of 114 s. In a position error context, the error term must be computed and added to this value. For a probability of violation of 10 percent, the value of q_v is found to be 1.28. The error in the gap between arrivals is $\sigma_g q_v = 30(1.28) = 38$ s. Therefore, in a position error context, there must be a gap between arrivals of at least $114 + 38 = 152$ s to release at least one departure between each pair of arrivals, and at least 212 s to release at least two departures between a pair of arrivals.

If the actual interarrival time matrix in Example Problem 8-4 is examined, it can be seen that a departure cannot be released only when an arrival of a type B aircraft is followed by an arrival of a type A aircraft since this is the only case where the required interarrival time of 152 s is not attained. This occurs 50 percent of the time. Also, anytime a type A aircraft is followed by a type B aircraft, two departures may be released. Therefore, the capacity of the runway to service mixed operations in a position error context is

$$C_m = \frac{3600}{155}[1.0 + 1(0.25) + (2)(0.25)] = 41 \text{ operations per hour}$$

A comprehensive example problem using the analytical equations developed for determining the hourly capacity of a runway is presented to summarize this section.

Example Problem 8-6 A runway is to service arrivals and departures. The common approach path is 6 mi long for all aircraft. During a particular interval the runway is serving three types of aircraft with the mix and operating characteristics shown in Table 8-4. The air traffic separation rules in effect are given in Table 8-5.

Assume that the standard deviation of the position of airborne aircraft and the error in the gaps between arrivals are known to be 20 s. And assume that the minimum separation rules may be violated 10 percent of the time.

First let us find the capacity of the runway system to service only arrivals.

The error-free interarrival time equations are given in Eqs. (8-7) and (8-8). From these the interarrival matrix can be computed. For example, for a leading B followed by a trailing A, we have

TABLE 8-4 Aircraft Mix and Operating Characteristics for Example Problem 8-6

Aircraft type	Approach speed, mi/h	Runway occupancy time, s	Mix, % Arrival	Mix, % Departure
A	135	50	20	15
B	110	40	45	55
C	90	30	35	30

TABLE 8-5 Air Traffic Separation Rules for Example Problem 8-6

Operational sequence	Air traffic separation rules
Arrival-departure	Clear runway
Departure-arrival	2 mi
Departure-departure	Seconds
Arrival-arrival	Miles

Departure-departure, Seconds, Leading:

$$\text{Trailing} \begin{array}{c} \text{A} \\ \text{B} \\ \text{C} \end{array} \begin{bmatrix} \begin{array}{ccc} \text{A} & \text{B} & \text{C} \\ 90 & 90 & 60 \\ 90 & 90 & 60 \\ 120 & 60 & 60 \end{array} \end{bmatrix}$$

Arrival-arrival, Miles, Leading:

$$\text{Trailing} \begin{array}{c} \text{A} \\ \text{B} \\ \text{C} \end{array} \begin{bmatrix} \begin{array}{ccc} \text{A} & \text{B} & \text{C} \\ 4 & 3 & 3 \\ 5 & 4 & 3 \\ 6 & 4 & 3 \end{array} \end{bmatrix}$$

$$\Delta T_{BA} = \frac{3(3600)}{135} = 80 \text{ s}$$

and for a leading A followed by a trailing B we have

$$\Delta T_{AB} = \frac{5(3600)}{135} + 6\left(\frac{1}{110} - \frac{1}{135} \right)(3600) = 170 \text{ s}$$

Continuing these computations for all combinations of leading and trailing aircraft and computing the arrival mix probabilities result in the matrices below. The probability matrix is computed based upon the mix percentages as opposed to the actual ordering of aircraft. This is because in long-term planning applications, the actual order of aircraft is unknown.

$[M_{ij}]$:

		Leading		
		A	B	C
Trailing	A	107	80	80
	B	170	131	98
	C	240	175	120

$[P_{ij}]$:

		Leading		
		A	B	C
Trailing	A	0.04	0.09	0.07
	B	0.09	0.20	0.16
	C	0.07	0.16	0.12

The interarrival buffer time which must be added to the error-free case when position error is present is given by Eqs. (8-11) and (8-12). If the probability of violation is 10 percent, then $q_v = 1.28$. Using Eqs. (8-11) and (8-12) to solve for the buffer, we have for a leading type B aircraft and a trailing type A aircraft

$$b_{BA} = 20(1.28) = 26 \text{ s}$$

and for a leading type A aircraft followed by a trailing type B aircraft

$$b_{AB} = 20(1.28) - 5(\tfrac{1}{110} - \tfrac{1}{135})(3600) = -5 \text{ s}$$

But the minimum value of the buffer is always zero.

Continuing this for all combinations of leading and trailing aircraft gives the buffer and interarrival time matrices:

$[B_{ij}]$:

		Leading		
		A	B	C
Trailing	A	26	26	26
	B	0	26	26
	C	0	0	26

$[M_{ij} + B_{ij}]$: Leading

		A	B	C
	A	133	106	106
Trailing	B	170	157	124
	C	240	175	146

The expected value of the interarrival time becomes from Eq. (8-10)

$$E(\Delta T_{ij}) = 0.04(133) + 0.09(106) + \cdots + 0.12(146)$$
$$= 151 \text{ s}$$

The arrival capacity is then, from Eq. (8-6),

$$C_a = \frac{3600}{151} = 24 \text{ operations per hour}$$

Next let us find the capacity of the runway system to service only departures. The expected value of the departure time is computed from Eq. (8-14) by using the departure-departure time matrix given and the departure mix probability matrix below. This matrix is based on actual departure-departure times and always considers error.

$[p_{ij}]$: Leading

		A	B	C
	A	0.0225	0.0825	0.0450
Trailing	B	0.0825	0.3025	0.1650
	C	0.0450	0.1650	0.0900

The expected value of the departure time is then

$$E(t_d) = 0.0225(90) + 0.0825(90) + \cdots + 0.0900(60)$$
$$= 77 \text{ s}$$

The departure capacity of the runway is given by Eq. (8-13)

$$C_d = \frac{3600}{77} = 47 \text{ operations per hour}$$

Next let us find the probability of releasing a departure after each arrival and the capacity of the runway system to service mixed operations in the case where arrivals are given priority over departures.

To release n_d departures, the required interarrival time is given by Eq. (8-15) with a buffer term added. Solving for each term in this equation, we have

$$E(R_i) = 0.20(50) + 0.45(40) + 0.35(30) = 38 \text{ s}$$

$$E\left(\frac{\delta_d}{V_j}\right) = \left[0.20\left(\frac{2}{135}\right) + 0.45\left(\frac{2}{110}\right) + 0.35\left(\frac{2}{90}\right)\right](3600)$$

$$= 68 \text{ s}$$

$$E(t_d) = 77 \text{ s}$$

$$E(B_{ij}) = 26(0.68) + 0(0.32) = 18 \text{ s}$$

and therefore

$$E(\Delta T_{ij}) \geq 38 + 68 + 18 + 77(n_d - 1)$$

$$\geq 124 + 77(n_d - 1)$$

For one departure we then have a required interarrival time of 124 s, for two successive departures we have a required interarrival time of 201 s, and for three successive departures we have a required interarrival time of 278 s.

Therefore, anytime the interarrival time is greater than or equal to 124 and less than 201 s, one departure may be released between a pair of arrivals. Anytime the interarrival time is greater than or equal to 201 and less than 278 s, two departures may be released between a pair of arrivals. Anytime the interarrival time is greater than or equal to 278 s, at least three or more departures may be released between a pair of arrivals.

Examination of the interarrival time matrix gives the probability of releasing departures between arrivals. Therefore, for one departure the probability is 77 percent, for two successive departures the probability is 7 percent, and we cannot release more than two successive departures between a pair of arrivals while maintaining minimum interarrival separations.

The mixed-operation hourly capacity is then from Eq. (8-16)

$$C_m = \frac{3600}{151}[1 + 0.77(1) + 0.07(2)] = 46 \text{ operations}$$

Now let us find (1) the required interarrival time if at least one departure is to be released after each arrival and (2) the resulting capacity of the runway system to service mixed operations under this condition. For this to occur, all values of the interarrival matrix must be at least 124 s. Therefore, all values of the interarrival time less than 124 s must be increased to 124 s to release at least one departure between every pair of arrivals. Therefore, the new required interarrival time matrix becomes

$[T_{ij}]$:   Leading

		A	B	C
	A	133	124	124
Trailing	B	170	157	124
	C	240	175	146

which results in

$$E(\Delta T_{ij}) = 154 \text{ s}$$

and the hourly capacity for mixed operations becomes

$$C_m = \frac{3600}{154}[1 + 0.93(1) + 0.07(2)] = 48 \text{ operations}$$

Therefore, by increasing the interarrival separations anytime a type A aircraft follows a type B or type C aircraft, or anytime a type B aircraft follows a type C aircraft, the runway capacity can be increased from 46 to 48 operations per hour. Small increases in capacity can result in significant decreases in delay.

Application of Techniques for Ultimate Hourly Capacity

The hourly capacity of the runway system is the maximum number of aircraft operations that can take place on the runway system in an hour. The maximum number of aircraft operations depends on a number of conditions including, but not limited to, the following:

1. Ceiling and visibility conditions

2. Physical configuration of the runway system

3. Air traffic control system separation rules

4. Runway-use strategy

5. Mix of aircraft using the runway system

6. Ratio of arrivals to departures

7. Number of touch-and-go operations by general aviation aircraft

8. Number and location of exits from the runway system

It is important to point out that the definition of the hourly capacity of runways in this section differs from that which is delay-related, since the definition of capacity here contains no assumptions regarding acceptable levels of delay.

The determination of the runway system hourly capacity is normally made through the use of computer programs developed for that purpose [16, 17, 20, 28, 29]. These programs are capable of accommodating virtually any runway-use configuration at an airport and allow for the variation in all the parameters which might affect runway capacity. Based upon the use of these programs and by constraining many of the variables to conform to present operating scenarios, an airport capacity handbook has been developed which will allow for the computation of realistic estimates of the runway capacity [4, 26]. The material which follows is based upon the airport capacity handbook techniques for determining runway hourly capacity [4, 26].

Parameters required for runway capacity

As noted above, to determine the hourly capacity of the runway system, it is necessary to ascertain the parameters that will affect capacity. Because aircraft separation rules differ in VMC and IMC weather, it is first necessary to determine the ceiling and visibility conditions, or more appropriately, the separation rules applicable to flying conditions, when the ceiling at the airport is at least 1000 ft and visibility is at least 3 mi. This condition results in visual flight rules for arriving and departing aircraft. If either of or both these criteria are not met, then IFR are in effect. Of course, all airports have a period when conditions are such that IFR apply. Therefore, the hourly capacity of runways is normally specified for each of these conditions.

The physical runway surfaces at an airport can be used in several ways. For example, two parallel runways can be used with arrivals on one runway and departures on the other runway at some time. Also they could be used with arrivals and departures on one surface and only arrivals on the other surface. These runway-use configurations are called the *runway-use strategies* which are dependent upon weather conditions, aircraft types, and the spacing between runways. It is necessary to specify the runway use strategies and the percentage of time each strategy is used. It is also necessary to specify the types of aircraft which can utilize a given runway since quite often shorter pavements are constructed for use by general aviation aircraft only. The aircraft which may use a runway surface are defined in terms of a *mix index*. This index is simply an indication of the level of operations on the runway by large and heavy aircraft. For this procedure aircraft are classified as in Table 8-6. The mix index MI is given by Eq. (8-17).

$$MI = C + 3D \qquad (8\text{-}17)$$

where MI = mix index
      C = percentage of type C aircraft in mix of aircraft using runway
      D = percentage of type D aircraft in mix of aircraft using runway

The percentage of arrival operations which occur on the runway is also necessary. This is because the spacing rules for arrivals and departures differ. There are three types of operations: arrivals, departures, and *touch-and-go* operations. A touch-and-go operation is most

TABLE 8-6   Aircraft Classification for Ultimate-Capacity Techniques

Aircraft mix class	Aircraft wake turbulence class	Number of engines	Maximum certified takeoff weight, lb
A	Small	Single	12,500 or less
B	Small	Multiple	12,500 or less
C	Large	Multiple	12,500–300,000
D	Heavy	Multiple	300,000 or more

SOURCE: Federal Aviation Administration [26].

commonly used by general aviation pilots practicing approaches, landings, and takeoffs. These operations are seldom conducted in poor weather. For the purpose of determining capacity, the parameter called *percent arrivals* is used to define the proportion of each type of operation which occurs on the runway. In VFR conditions it is also necessary to find the percentage of touch-and-go operations. At times small general aviation airports may have touch-and-go operations which can approach 30 percent of all operations.

The location of runway exits for arriving aircraft must also be known since this affects the runway occupancy time. Depending upon the nature of the aircraft using a runway exits should be located at positions which will allow minimum runway occupancy times. If this is not the case, the capacity will be reduced because of excessive runway occupancy times.

As a result of extensive research conducted to determine the capacity of runway systems, the FAA has published a series of charts to determine the runway capacity [4, 26]. These charts are used to determine the runway capacity through Eq. (8-18).

$$C = C_b ET \tag{8-18}$$

where C = hourly capacity of runway-use configuration in operations per hour

C_b = ideal or base capacity of runway-use configuration

E = exit adjustment factor for number and location of runway exits

T = touch-and-go adjustment factor

The use of this equation and the charts is illustrated in Example Problem 8-7.

Example Problem 8-7 It is required to find the VFR and IFR hourly capacity of the runway system shown in Fig. 8-14. The runway-use strategy is as shown. In VFR weather, the traffic consists of 3 single-engine, 20 light twin-engine, 25 large transport-type, and 2 wide-bodied aircraft. Arrivals constitute 40 percent of the operations, and there are approximately 3 touch-and-go operations. In IFR, the small aircraft population count drops to 2 single-engine and 5 light twin-engine aircraft, the arrival rate increases to 50 percent, and there are no touch-and-go operations.

The capacity of intersecting runways is a function of the location of the intersection from both the arrival and the departure threshold. The closer the intersection to these thresholds, the greater the capacity. The charts used for this configuration in VFR and IFR are taken from the FAA [4, 26] and are given in Figs. 8-15 and 8-16. The aircraft are grouped into various classes in VFR and IFR conditions in Table 8-7. From the tabular data, the mix index MI can be found for VFR from Eq. (8-17) as

$$MI = C + 3D = 50.0 + 3(4.0) = 62.0$$

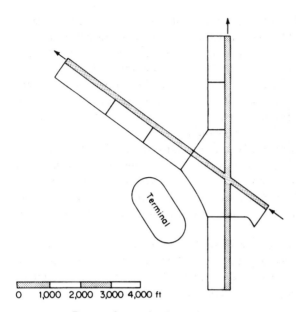

Figure 8-14   Runway layout for Example Problem 8-7.

and for IFR

$$\text{MI} = C + 3D = 73.5 + 3(5.9) = 91.2$$

By using the VFR mix index of 62.0 and the percent arrivals PA equal to 40, the base capacity C_b is found from the left side of Fig. 8-15 as about 95 operations per hour. This base value is then adjusted for touch-and-go operations and the location of exits using the right side of this figure. From the given data, the percentage of touch-and-go operations in VFR is equal to 6 percent. Therefore, the touch-and-go adjustment factor T is equal to 1.03. For the mix index of 62.0, only those exits located between 3500 and 6500 ft from the arrival threshold can be counted. There are two such exits, one at 4500 ft and the other at 6000 ft. The table then gives an exit factor E of 0.97 for 40 percent arrivals.

Therefore, the hourly capacity of the runway system in VFR is, from Eq. (8-18),

$$C = 95(1.03)(0.97) = 95 \text{ operations per hour}$$

The IFR capacity is determined similarly from Fig. 8-16. This will yield, for an IFR mix index of 91.2, a percent arrivals PA of 50 percent, base capacity, touch-and-go factor, and exit factor

$$C = 58(1.00)(0.97) = 56 \text{ operations per hour}$$

Computation of Delay on Runway Systems

Delay to aircraft is defined as the difference between the actual time it takes an aircraft to maneuver on the runway and the time it would take the aircraft to maneuver without interference from other air-

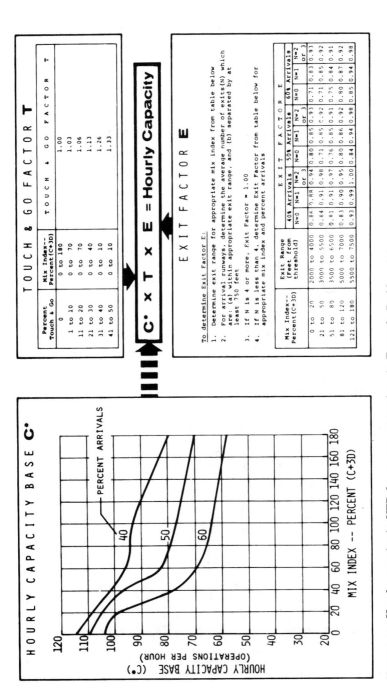

Figure 8-15 Hourly capacity in VFR for runway operations in Example Problem 8-7 (*Federal Aviation Administration [26]*).

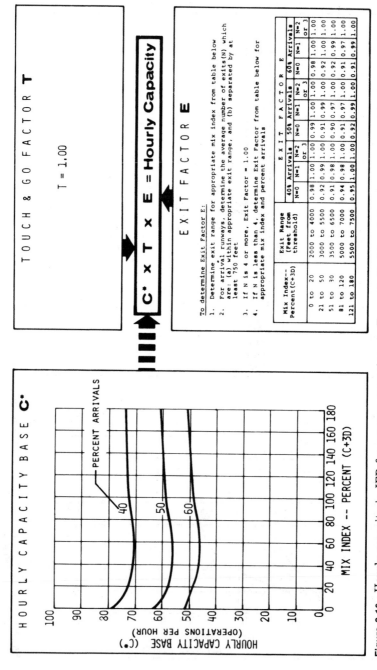

Figure 8-16 Hourly capacity in IFR for runway operations in Example Problem 8-7 (*Federal Aviation Administration [26]*).

TABLE 8-7   Tabulation of Aircraft Mix Index for Example Problem 8-7

Aircraft	Class	VFR mix		IFR mix	
		No.	%	No.	%
Single-engine	A	3	6.0	2	5.9
Twin-engine	B	20	40.0	5	14.7
Transport	C	25	50.0	25	73.5
Wide-bodied	D	2	4.0	2	5.9
Total		50	100.0	34	100.0

craft. The runway is defined as the entire runway system including approach and departure airspace [26]. To compute runway system delay, it is necessary to analyze each runway-use configuration for the demand placed upon it. To compute annual runway delay, it is necessary to determine the percentage of time each runway-use configuration is employed throughout the year. Normally this will require knowledge of the following factors:

1. The hourly capacity of the runway-use strategy in VFR and IFR conditions
2. The pattern of hourly, daily, and monthly aircraft demand during the design year
3. The peaking of demand during the design hour
4. The frequency of occurrence of runway-use strategies, ceiling, and visibility conditions

The techniques are outlined in detail by the FAA [4, 26], but it is sufficient to note that the computation of annual delay is a very tedious and time-consuming process, now generally performed on computers [17, 28, 29]. The elements of the process are exhibited in Example Problem 8-8, which uses charts from the FAA [4, 26].

Example Problem 8-8   The hourly delay to aircraft operating on the runway system in Example Problem 8-7 is to be found for both VFR and IFR conditions. The peak 15-min demand in the peak hour is 20 operations in VFR conditions and 10 operations in IFR conditions.

The hourly capacity of the runway system was found earlier to be 95 operations per hour in VFR conditions and 56 operations per hour in IFR conditions. The hourly demand was 50 operations in VFR and 34 operations in IFR. Therefore, the ratio of hourly demand to hourly capacity is computed for VFR conditions as

$$\frac{D}{C} = \frac{50}{95} = 0.53$$

and for IFR conditions as

$$\frac{D}{C} = \frac{34}{56} = 0.61$$

VFR Conditions

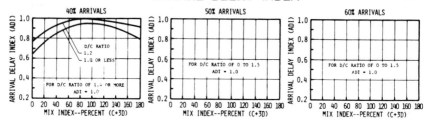

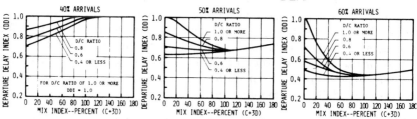

IFR Conditions

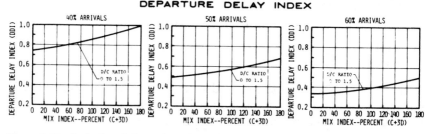

Figure 8-17 Arrival and departure delay indices for Example Problem 8-8 (*Federal Aviation Administration [26]*).

Figure 8-17, which is taken from references for this runway-use strategy [4, 26], gives the variation of the *arrival delay index* (ADI) and the *departure delay index* (DDI) for VFR and IFR conditions for this runway use for 50 percent arrivals.

Based upon the respective mix indices for VFR and IFR conditions, this chart gives for VFR

$$\text{ADI} = 1.00 \qquad \text{and} \qquad \text{DDI} = 0.65$$

and for IFR

$$\text{ADI} = 1.00 \qquad \text{and} \qquad \text{DDI} = 0.57$$

These indices are combined with the respective ratios of demand to capacity to arrive at the *arrival delay factor* (ADF) and the *departure delay factor* (DDF) as follows:

$$ADF = (ADI)\,\frac{D}{C}$$

and

$$DDF = (DDI)\,\frac{D}{C}$$

Therefore, for arrivals we have in VFR conditions

$$ADF = (1.00)(0.53) = 0.53$$
$$DDF = (0.65)(0.53) = 0.34$$

and in IFR conditions

$$ADF = (1.00)(0.61) = 0.61$$
$$DDF = (0.57)(0.61) = 0.35$$

The average delay for each aircraft is then found from Fig. 8-18 by using the above delay factors and the *demand profile factor* (DPF).

The demand profile factor is simply a measure of the peaking of demand in the hour; it is defined as the peak 15-min demand divided by the hourly demand. Therefore, the demand profile factors (DPFs) are in VFR conditions

$$DPF = \frac{20}{50}(100) = 40$$

and in IFR

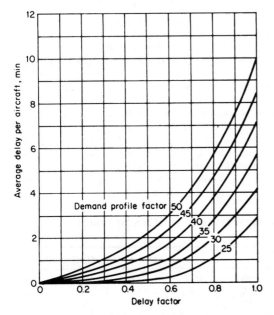

Figure 8-18 Variation of average aircraft delay with delay factor (*Federal Aviation Administration [26]*).

$$\text{DPF} = \frac{10}{34}(100) = 29$$

From Fig. 8-18, the average delays are found in VFR as

$$\text{Arrival delay} = 1.4 \text{ min per aircraft}$$

$$\text{Departure delay} = 0.7 \text{ min per aircraft}$$

and in IFR

$$\text{Arrival delay} = 0.8 \text{ min per aircraft}$$

$$\text{Departure delay} = 0.2 \text{ min per aircraft}$$

Since there are equal arrivals and departures, the total delay to all aircraft in this hour in both VFR and IFR conditions can be found:

$$\text{Total VFR delay} = 50[1.4(0.5) + 0.7(0.5)] = 52.5 \text{ min}$$

$$\text{Total IFR delay} = 34[0.8(0.5) + 0.2(0.5)] = 17.0 \text{ min}$$

The computation of delay at an airport over various periods requires that this procedure be repeated for each hour, day, and month for each runway-use strategy and weather condition and be summed over the period of interest. The process is shown in Example Problem 8-9 for the computation of delay on a daily basis.

Example Problem 8-9 To illustrate the airport capacity handbook method [4, 26] of computing the delay throughout a typical day, let us determine the daily delay to aircraft if the runway-use strategy shown in Fig. 8-14 is used throughout the day. The hourly capacity of the runway-use strategy is 80 operations per hour, and the number of arrivals equals the number of departures in each hour. The hourly demand throughout the day is given in Table 8-8.

Let us assume that the mix index MI is equal to 70 and that the demand profile factor DPF is equal to 25. Both the mix index and the demand profile factor are the same in each hour. Let us also assume that VFR conditions exist throughout the day.

To solve this problem, it is necessary to compute the arrival delay index, departure delay index, arrival delay factor, and departure delay factor for each hour during the day. The computational procedure differs depending upon whether when the hourly demand is less than or equal to the hourly capacity or the hourly demand is greater than the hourly capacity.

For the condition when hourly demand is less than or equal to the hourly capacity, the procedure is the same as that in Example Problem 8-8.

TABLE 8-8   Hourly Aircraft Runway Demand on Typical Day for Example Problem 8-9

Hour	Demand	Hour	Demand	Hour	Demand	Hour	Demand
0000	10	0600	20	1200	50	1800	100
0100	10	0700	40	1300	50	1900	70
0200	10	0800	60	1400	40	2000	40
0300	10	0900	50	1500	70	2100	30
0400	10	1000	60	1600	110	2200	20
0500	10	1100	50	1700	120	2300	10

For example, in hour 1000 the aircraft demand is 60 operations per hour, and the runway capacity is 80 operations per hour. Therefore, the ratio of hourly demand to hourly capacity is 60/80 = 0.75.

From Fig. 8-17, the arrival delay index (ADI) is 1.0, and the departure delay index (DDI) is 0.70. The arrival delay factor is then

$$\text{ADF} = \text{ADI}\,\frac{D}{C} = 1.0(0.75) = 0.75$$

and the departure delay factor is

$$\text{DDF} = \text{DDI}\,\frac{D}{C} = 0.70(0.75) = 0.53$$

From Fig. 8-18, by using these delay factors and a demand profile factor of 25, an average arrival delay of 1.2 min and an average departure delay of 0.2 min are found.

Since the number of arrivals is equal to the number of departures, the total delay in hour 1000 is then equal to

$$\text{Delay} = 1.2(0.5)(60) + 0.2(0.5)(60) = 42 \text{ aircraft-minutes}$$

The procedure is the same for hours 0000 through 1500.

However, beginning at hour 1600, the hourly demand exceeds the hourly capacity for 3 h. These are called *overloaded hours*. The cumulative demand for these 3 h, which is 330 operations, exceeds the cumulative capacity available for these 3 h, which is 240 operations. Therefore, some of these aircraft are not serviced in these 3 h and they spill over into later hours. The later hours service this backlog of demand until the backlog is cleared up. This is shown in Table 8-9. The time from hour 1600 to hour 2000 is called the *saturated period*.

For the overloaded period, hours 1600 through 1800, the total demand is divided by the total capacity to arrive at the average demand-to-capacity ratio during the overloaded hours. Therefore

$$\frac{D}{C} = \frac{110 + 120 + 100}{80 + 80 + 80} = 1.38$$

From Fig. 8-17, the arrival delay index is 1.00, and the departure delay index is 0.75 during the overloaded hours. This results in an arrival delay factor of 1.00 × 1.38 = 1.38 and a departure delay factor of 0.75 × 1.38 = 1.04.

Figure 8-19, which is taken from the FAA [4, 26], gives the aircraft delay in the saturated period when the period of overload is 3 h. From this figure, by using a demand profile factor of 25, an average arrival delay over this period is

TABLE 8-9   **Overloaded and Saturated Hours for Example Problem 8-9**

Hour	Demand	Capacity	Overload	Cumulative overload
1500	70	80	0	0
1600	110	80	+ 30	+ 30
1700	120	80	+ 40	+ 70
1800	100	80	+ 20	+ 90
1900	70	80	− 10	+ 80
2000	40	80	− 40	+ 40
2100	30	80	− 50	0

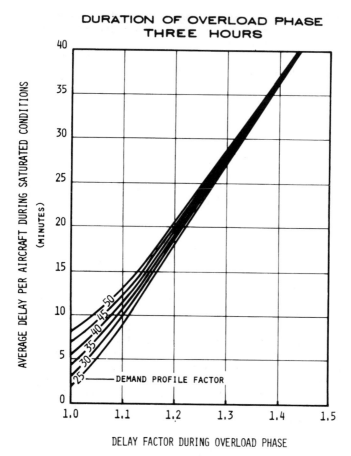

Figure 8-19 Average aircraft delay during saturated conditions for an overload period of 3 h (*Federal Aviation Administration [26]*).

found to be 35 min per arrival, and an average departure delay over this period is found to be 4 min per departure.

The delay in the saturated period is found by adding the total demand in the saturated period and multiplying by the average delay per operation. The total demand is then the demand from hours 1600 through 2000, or 440 operations.

Since the number of arrivals is equal to the number of departures in each hour, this yields a total delay in the saturated period of

$$\text{Delay} = 440(0.5)(35) + 440(0.5)(4) = 8580 \text{ aircraft-minutes}$$

In hours 2100 through 2300, the demand is once again less than the capacity, and the backlog has been cleared up. Therefore, the procedure is the same as that used for hours 0000 through 1500.

The results are displayed in tabular format in Table 8-10. The result is that the total delay on this day is equal to 8799 aircraft-minutes. The average delay to aircraft on this day is 8799/1050 = 8.4 min.

TABLE 8-10   Tabulation of Hourly Delay for Example Problem 8-9

Hour	Demand	D/C	Arrivals ADI	Arrivals ADF	Departures DDI	Departures DDF	Delay, min Per arrival	Delay, min Per departure	Delay, min Hourly total
0000	10	0.13	1.0	0.13	0.65	0.08	0	0	0
0100	10	0.13	1.0	0.13	0.65	0.08	0	0	0
0200	10	0.13	1.0	0.13	0.65	0.08	0	0	0
0300	10	0.13	1.0	0.13	0.65	0.08	0	0	0
0400	10	0.13	1.0	0.13	0.65	0.08	0	0	0
0500	10	0.13	1.0	0.13	0.65	0.08	0	0	0
0600	20	0.25	1.0	0.25	0.65	0.16	0	0	0
0700	40	0.50	1.0	0.25	0.67	0.34	0	0	0
0800	60	0.75	1.0	0.75	0.70	0.53	1.2	0.2	42
0900	50	0.63	1.0	0.63	0.68	0.43	0.4	0.1	13
1000	60	0.75	1.0	0.75	0.70	0.53	1.2	0.2	42
1100	50	0.63	1.0	0.63	0.68	0.43	0.4	0.1	13
1200	50	0.63	1.0	0.63	0.68	0.43	0.4	0.1	13
1300	50	0.63	1.0	0.63	0.68	0.43	0.4	0.1	13
1400	40	0.50	1.0	0.50	0.67	0.34	0.2	0	4
1500	70	0.88	1.0	0.88	0.73	0.64	1.8	0.4	77
1600	110								
1700	120								
1800	100	1.38	1.0	1.38	0.75	1.04	35	4	8580
1900	70								
2000	40								
2100	30	0.38	1.0	0.38	0.65	0.25	0.1	0	2
2200	20	0.25	1.0	0.25	0.65	0.16	0	0	0
2300	10	0.13	1.0	0.13	0.65	0.08	0	0	0
Daily	1050								8799

Graphical methods for approximating delay

A relatively simple technique for estimating delays when demand exceeds capacity has been used in aviation studies [19]. This method employs a *deterministic model.* In this method, a time scale is established on the X axis to represent the time period being analyzed. On the Y axis, a scale is established for the cumulative number of aircraft which have arrived by some time. Therefore, a point on the plot represents the total number of aircraft which have arrived at that time. The curve which results from plotting the succession of points is actually a representation of the demand $D(t)$. A line of constant, or for that matter variable, slope can be drawn on the same graph to show the service capabilities, or capacity, of a facility. This is the service function $S(t)$. An illustration of such a graph is given in Fig. 8-20.

In this figure, point A represents the time when the demand rate begins to exceed the service rate or capacity. Therefore, delays and queues begin to develop. At point B, the delays and queues which have built since time t_1 will have dissipated, and the demand rate is

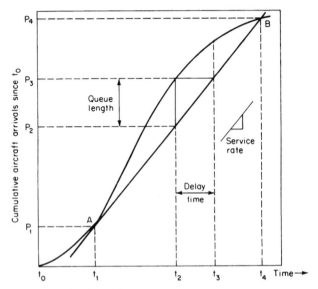

Figure 8-20 Graphical determination of aircraft delay when the demand rate exceeds the service rate. (a) Departures only; (b) arrivals only and mixed operations.

now less than the service rate. A review of the results displayed on this figure shows the following:

1. Delays occur from time t_1 to time t_4.

2. The total number of aircraft delayed is the difference between P_4 and P_1.

3. From time t_1 to time t_3 the demand rate exceeds the service rate, and delays and queues increase during this period.

4. The maximum delay and maximum queue length occur at time t_3 since the demand rate becomes less than the service rate then.

5. The delay to any aircraft is given by the magnitude of a horizontal line drawn between the two curves.

6. The length of a queue at any time is given by the magnitude of a vertical line drawn between the two curves at that point.

7. The area between the two curves represents the total delay to all aircraft which are delayed from time t_1 to time t_4.

This type of analysis is useful in airport planning to estimate the magnitude of delays, number of aircraft delayed, and cost of delay under assumed operating conditions. It does not, however, indicate the delays that occur when average demand is less than capacity. These are normally calculated by the equations or methods discussed earlier.

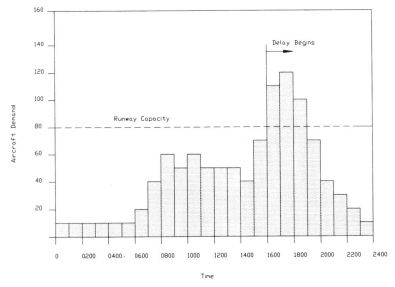

Figure 8-21   Pattern of hourly aircraft demand for Example Problem 8-10.

This next example problem illustrates the application of this technique to a runway system.

Example Problem 8-10   The hourly aircraft demand during a typical day at an airport is given in Table 8-8. The runway system has a capacity to service 80 aircraft per hour without delay. The delay when demand exceeds capacity is to be analyzed.

For discussion purposes, the hourly pattern of demand and the capacity are plotted in Fig. 8-21. The deterministic model makes the assumption that delay occurs only when demand exceeds capacity. This figure shows that up until hour 1600 the aircraft demand in any hour is less than the runway capacity, and therefore delays do not occur. However, beginning at hour 1600, the aircraft demand begins to exceed the runway capacity, and so delays begin to accrue from this point.

The data in Table 8-8 are plotted in Fig. 8-22, where the cumulative hourly demand $D(t)$ and cumulative service rate $S(t)$ are plotted versus time. This figure again shows that at hour 1600 the demand rate begins to exceed the capacity, and so delay will begin in hour 1600.

The shaded area between the curves represents the period when aircraft delays occur. However, due to the scale on this figure, it is difficult to determine the values of the greatest delay to any aircraft, the greatest X value within the shaded area, the greatest number (queue) of aircraft delayed, the largest Y value within the shaded area, or the total aircraft-hours of delay, the shaded area. Since only the period after hour 1600 contains delay and since the Y value within the shaded area represents the difference between demand and capacity, a plot of the cumulative difference between demand and capacity versus time from hour 1600, tabulated in Table 8-11, is shown in Fig. 8-23. This figure effectively expands the scale of the shaded area in Fig. 8-22 and is much easier to use to determine the values of the delay and queue length parameters.

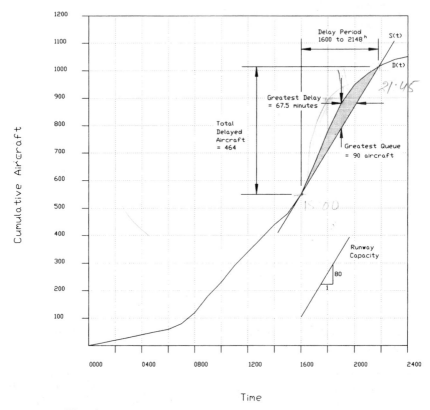

Figure 8-22 Plot of cumulative aircraft arrivals versus time for Example Problem 8-10.

TABLE 8-11 Cumulative Demand Minus Capacity during Delay Period for Example Problem 8-10

End hour	Demand	Demand − Capacity	Cumulative demand − Capacity
1500			0
1600	110	+ 30	+ 30
1700	120	+ 40	+ 70
1800	100	+ 20	+ 90
1900	70	− 10	+ 80
2000	40	− 40	+ 40
2100	30	− 50	− 10
2200	20	− 60	− 70

The greatest Y value in Fig. 8-23 represents the greatest number of aircraft delayed at any time, the period when the curve is above the X axis is the period during which delay occurs, and the area of the curve above the X axis is the total aircraft-hours of delay.

The largest number of aircraft delayed at any time is found to be 90. The service time for an aircraft is the reciprocal of the runway capacity, or the runway will service 1 aircraft in 0.75 min. The delay time for the aircraft which is

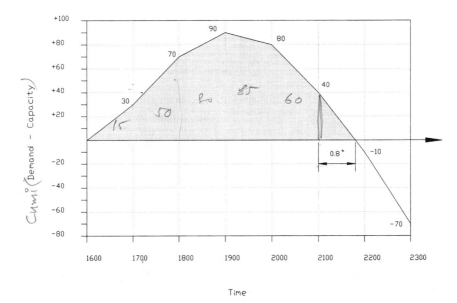

Figure 8-23 Plot of demand minus capacity for delay period for Example Problem 8-10.

delayed the longest is therefore 90(0.75) = 67.5 min. The total number of aircraft hours of delay is the area under this curve, or 306 aircraft-hours.

All the aircraft in hours 1600 to 2000 and 80 percent of the aircraft in hour 2100 are delayed. This amounts to 464 aircraft. Therefore, the average delay to delayed aircraft on this day is 306/464 = 0.66 h = 39.6 min. The average delay to all aircraft using the runway on this day is 306/1050 = 0.29 h = 17.5 min.

An analysis of the information contained in Figs. 8-22 and 8-23 should lead to exactly the same results, but it is easier to work with Fig. 8-23 to obtain the results.

Application of Techniques for Annual Service Volume

The *annual service volume* is a level of annual aircraft operations that may be used as a reference in preliminary planning. As annual aircraft operations approach the annual service volume, the average delay to each aircraft throughout the year may increase rapidly with relatively small increases in aircraft operations, thereby causing levels of service on the airfield to deteriorate.

When annual aircraft operations on the airfield are equal to the annual service volume, the average delay to each aircraft throughout the year is on the order of 1 to 4 min. A more precise estimate of actual average delay to aircraft at a particular airport can be obtained by using these procedures if this is required in the planning application.

If the number of annual operations exceeds the annual service volume, moderate or severe congestion may occur, similar to that presently being experienced at several major airports in the United States, such as O'Hare International Airport, LaGuardia Airport, and Atlanta International Airport.

For analyses of airfield improvements, aircraft delays also can be important at levels of annual aircraft operations less than the annual service volume. Therefore, delays to aircraft should also be considered in planning and evaluating airfield improvement at levels of annual operations less than the annual service volume. In some instances, when the annual demand is expected to approach one-half of the annual service volume within the planning horizon, the normal construction costs of airfield improvements may be balanced by savings in aircraft delay costs.

To determine the annual service volume that can be accommodated at an airport, it is necessary to find the weighted-hourly capacity of each runway-use configuration at the airport. The weighted hourly capacity C_w is given by the relationship

$$C_w = \frac{\Sigma P_n W_n C_n}{\Sigma P_n W_n} \tag{8-19}$$

where C_w = weighted average hourly capacity of runway system in year

C_n = hourly capacity of runway-use configuration n

P_n = percentage of time runway-use configuration n is used in year

W_n = weight assigned to runway-use configuration n to account for fact that different delay levels occur on various runway-use configurations

TABLE 8-12   ASV Weighting Factors for Weighted Hourly Capacity

Percentage of dominant capacity	Mix index in VFR, 0–180	Mix index in IFR		
		0–20	21–50	51–180
91 or more	1	1	1	1
81–90	5	1	3	5
66–80	15	2	8	15
51–65	20	3	12	20
0–50	25	4	16	25

SOURCE: Federal Aviation Administration [26].

Through empirical studies at airports, the weight factors W_n have been found to approximate the values in Table 8-12 [4, 26].

Once the weighted-hourly capacity of the runway system has been found, the annual service volume ASV is computed from

$$\text{ASV} = C_w HD \qquad (8\text{-}20)$$

where C_w = weighted-hourly capacity of runway system in year, as found above

H = *hourly ratio* or ratio of average daily operations in peak month to peak-hour operations in peak month

D = *daily ratio* or ratio of annual operations to average daily operations in peak month

The hourly ratio H and the daily ratio D defined above are used to simulate the fluctuation in the pattern of demand at the airport and can usually be ascertained from existing records or through demand forecasting. However, approximate values of these ratios have been determined for airports in the United States and can be used for estimating purposes [4, 24, 25, 26]. These approximate values are given in Table 8-13.

The computation of the annual service volume for an airport is illustrated in Example Problem 8-11.

Example Problem 8-11 The runway system at a large commercial service airport is shown in Fig. 8-24. The hourly capacity, percentage of annual use, and runway-use configurations are given in Table 8-14. The mix index at this airport is 100. The annual demand on this airport is 294,000 operations, the average daily traffic in the peak month is 877 operations, and the peak-hour traffic is 62 operations.

It is required to find the weighted-hourly capacity and the annual service volume at this airport if VFR conditions occur 82 percent of the time.

TABLE 8-13   Typical Hourly and Daily Ratios for Estimating Annual Service Volume

Mix index	Hourly ratio H	Daily ratio D
0–20	7–11	280–310
21–50	10–13	300–320
51–180	11–15	310–350

SOURCE: Federal Aviation Administration [26].

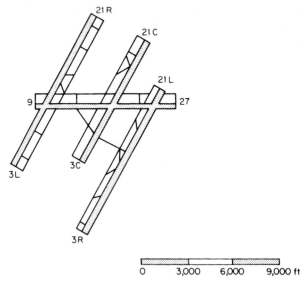

Figure 8-24 Runway layout for Example Problem 8-11.

From Table 8-14, the hourly capacity and percentage use of each configuration are found. The most frequently used runway configuration has a capacity of 122 operations per hour in VFR weather conditions. This is the dominant capacity. Therefore, the weighting factors for each runway-use configuration may be found from Table 8-12, and all the required values are tabulated in Table 8-15.

By using the data in Table 8-15, the weighted-hourly capacity is found from Eq. (8-19), which yields

$$C_w = \frac{0.52(1)(122) + 0.17(25)(61) + \cdots + 0.02(25)(48)}{0.52(1) + 0.17(25) + \cdots + 0.02(25)}$$

$$= \frac{720.19}{12.52} = 58 \text{ operations per hour}$$

The annual service volume ASV is then computed by using the above weighted-hourly capacity and the daily and hourly ratios from Table 8-13.

$$\text{ASV} = 58\left(\frac{877}{62}\right)\left(\frac{294,000}{877}\right) = 275,000 \text{ operations}$$

Estimating procedures for annual service volume and delay

Simplified estimating procedures exist for airport planning purposes to determine the annual service volume and average aircraft delay for runway configurations at airports. These procedures should be used

TABLE 8-14   Runway-Use Configurations and Percentage of Annual Use and Runway Hourly Capacities for Example Problem 8-11

Runway-use configuration	Percentage of annual use		Hourly capacity	
	VFR weather	IFR weather	VFR weather	IFR weather
A	63	0	122	—
B	21	78	61	48
C	9	12	52	52
D	7	10	59	48

TABLE 8-15   Tabulation of Factors for Weighted Hourly Capacity for Example Problem 8-11

Weather	Runway use	Percentage use	Hourly capacity	Percentage of dominant capacity	Weight
VFR					
	A	52	122	100	1
	B	17	61	50	25
	C	7	52	43	25
	D	6	59	48	25
IFR					
	A	0	—	—	—
	B	14	48	39	25
	C	2	52	43	25
	D	2	48	39	25

only for preliminary estimating purposes. These procedures allow for approximations of

1. The hourly capacity of runways in VFR and IFR conditions

2. The annual service volume of runways

3. The average annual delay to aircraft on runways

Hourly capacities and annual service volumes for a number of runway configurations are presented in Table 8-16. An approximate esti-

TABLE 8-16   Preliminary Estimates of Hourly and Annual Ultimate Capacities

Runway configuration	Mix index, % $(C + 3D)$	Hourly capacity, operations per hour		Annual service volume, operations per year
		VFR	IFR	
A	0–20	98	59	230,000
	21–50	74	57	195,000
	51–80	63	56	205,000
	81–120	55	53	210,000
	121–180	51	50	240,000
B 700' to 2,499'	0–20	197	59	355,000
	21–50	145	57	275,000
	51–80	121	56	260,000
	81–120	105	59	285,000
	121–180	94	60	340,000
C 4,300' or more	0–20	197	119	370,000
	21–50	149	114	320,000
	51–80	126	111	305,000
	81–120	111	105	315,000
	121–180	103	99	370,000
D 700' to 2,499' 2,500' to 3,499'	0–20	295	62	385,000
	21–50	219	63	310,000
	51–80	184	65	290,000
	81–120	161	70	315,000
	121–180	146	75	385,000
E 700' to 2,499' 3,500' or more 700' to 2,499'	0–20	394	119	715,000
	21–50	290	114	550,000
	51–80	242	111	515,000
	81–120	210	117	565,000
	121–180	189	120	675,000
F	0–20	98	59	230,000
	21–50	77	57	200,000
	51–80	77	56	215,000
	81–120	76	59	225,000
	121–180	72	60	265,000
G	0–20	150	59	270,000
	21–50	108	57	225,000
	51–80	85	56	220,000
	81–120	77	59	225,000
	121–180	73	60	265,000

TABLE 8-16   Preliminary Estimates of Hourly and Annual Ultimate Capacities (*Continued*)

Runway configuration	Mix index, % (C + 3D)	Hourly capacity, operations per hour		Annual service volume, operations per year
		VFR	IFR	
H	0–20	132	59	260,000
	21–50	99	57	220,000
	51–80	82	56	215,000
	81–120	77	59	225,000
	121–180	73	60	265,000
I	0–20	150	59	270,000
	21–50	108	57	225,000
	51–80	85	56	220,000
	81–120	77	59	225,000
	121–180	73	60	265,000
J	0–20	132	59	260,000
	21–50	99	57	220,000
	51–80	82	56	215,000
	81–120	77	59	225,000
	121–180	73	60	265,000
K (700' to 2,499')	0–20	197	59	355,000
	21–50	145	57	275,000
	51–80	121	56	260,000
	81–120	105	59	285,000
	121–180	94	60	340,000
L (4,300' or more)	0–20	197	119	370,000
	21–50	149	114	320,000
	51–80	126	111	305,000
	81–120	111	105	315,000
	121–180	103	99	370,000
M (700' to 2,499')	0–20	295	59	385,000
	21–50	210	57	305,000
	51–80	164	56	275,000
	81–120	146	59	300,000
	121–180	129	60	355,000

TABLE 8-16   Preliminary Estimates of Hourly and Annual Ultimate Capacities (*Continued*)

Runway configuration	Mix index, % $(C + 3D)$	Hourly capacity, operations per hour		Annual service volume, operations per year
		VFR	IFR	
N 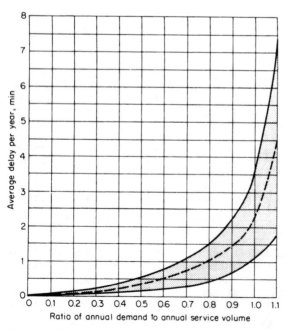	0–20	295	59	385,000
	21–50	210	57	305,000
	51–80	164	56	275,000
	81–120	146	59	300,000
	121–180	129	60	355,000
O	0–20	197	59	355,000
	21–50	147	57	275,000
	51–80	145	56	270,000
	81–120	138	59	295,000
	121–180	125	60	350,000

SOURCE: Federal Aviation Administration [26].

Figure 8-25   Relationship between average aircraft delay and ratio of annual demand to annual service volume (*Federal Aviation Administration [26]*).

mate of average aircraft delay per year for any runway configuration can be obtained from Fig. 8-25. The data shown in Table 8-16 and Fig. 8-25 are based on a number of assumptions, including these:

1. A representative range of mix indices is sufficient for estimating purposes.

2. The hourly capacities are those which correspond to the runway utilization that produces the largest capacity consistent with current air traffic control procedures and practices, and this configuration is used 80 percent of the time.

3. One-half of the demand for the use of the runways is by arriving aircraft, and thus the numbers of arriving and departing aircraft in a specified period of time are equal.

4. The percentage of touch-and-go operations is a function of the mix index of the airport.

5. Sufficient taxiways exist to permit the capacity of the runways to be fully realized.

6. The impact on capacity of a taxiway crossing an active runway is assumed to be negligible.

7. There is sufficient airspace to accommodate all aircraft wishing to use the runways, and aircraft operations are conducted in a radar environment with at least one runway equipped with an instrument landing system.

8. IFR conditions occur 10 percent of the time.

9. Representative hourly and daily ratios are a function of the mix index.

The order-of-magnitude relationship between the average annual delay per aircraft and the annual service volume depicted in Fig. 8-25 was derived from historical traffic records and a range of assumptions on likely operating conditions, as itemized above. Typically, the upper portion of the shaded band on this figure is representative of airports primarily serving air carrier operations. Airports serving primarily general aviation operations may typically fall anywhere within the entire shaded band. The dotted curve is the average of the upper and lower limits of the band indicated. Example Problem 8-12 shows the use of these approximate procedures.

Example Problem 8-12 An airport has a single runway available to service arrivals and departures. The projected annual demand in the year 2000 is 220,000 operations. The aircraft mix is estimated to consist of 21 percent small single-engine aircraft, 20 percent small multiengine aircraft, 55 percent large commercial aircraft, and 4 percent heavy aircraft. Air carrier operations predominate at the airport, and very few touch-and-go operations occur.

It is necessary to determine the annual service volume and average delay to aircraft for the single runway and for closely spaced parallel runways which may be constructed.

The mix index at the airport is

$$MI = 55 + 3(4) = 67$$

From Table 8-16, the annual service volume for each runway is found from runway configurations A and B with the mix index range of 51 to 80:

$$\text{Single-runway ASV} = 205{,}000 \text{ operations}$$

$$\text{Close parallel-runway ASV} = 260{,}000 \text{ operations}$$

The ratios of annual demand to annual service volume are then computed for both situations as for a single runway

$$\frac{\text{Demand}}{\text{ASV}} = \frac{220{,}000}{205{,}000} = 1.07$$

and for close parallel runways

$$\frac{\text{Demand}}{\text{ASV}} = \frac{220{,}000}{260{,}000} = 0.84$$

By using the upper half of the graph in Fig. 8-25, since this is predominantly an air carrier airport, yields the average annual delay per aircraft. These values are for a single runway between 4.0 and 6.0 min per operation and for close parallel runways between 1.2 and 1.7 min per operation.

It is clear that the construction of close parallel runways will represent a benefit in terms of decreased delays. However, a detailed analysis should be performed prior to making a decision on construction modifications, since this procedure is approximate and is based upon the assumptions noted above.

The determination of realistic estimates of aircraft delay is a tedious and time-consuming process, as shown in Example Problem 8-9. The airport capacity handbook outlines in detail the procedures which are required [4, 26]. Basically, to compute aircraft delay, it is necessary to have estimates of the hourly demand on the runway system, hourly capacity of each runway-use configuration, and percentage of time each runway-use configuration is employed in each weather condition. The process can be performed for a typical day, for several days, or on a monthly or annual basis.

Fortunately, the FAA has developed a computer program for the determination of the annual delay at an airport [17, 28, 29]. This program compiles the average aircraft delay by runway-use configuration and by weather condition on a daily, weekly, monthly, and annual basis. It presents an annual distribution of the magnitude of delay at the airport.

The annual service volume may be related to specific criteria for the level of average aircraft delay. To do this, the annual demand on the airfield is varied over a range of values which represent upper and lower bounds on the specified average-delay criteria. By plotting the average aircraft delay versus the annual demand, a curve is developed that shows the variation in average aircraft delay as a function of annual demand. By finding the annual demand which corresponds to the delay criteria, the annual service volume is defined. Note that the same process may be used on an hourly basis to find the practical hourly capacity of a runway-use configuration. The computer model cited above is very useful for this purpose.

Perhaps the most significant development concerning the determination of delay at an airport in the last few years has been the simulation model SIMMOD [22]. This model is a very powerful airside discrete event simulation model which allows for the simulation of various operating scenarios at an airport. The model is capable of simulating the airspace around an airport, the airfield, and the gate area.

A detailed discussion of the model is beyond the scope of this book, but basically the airspace and airfield are configured as a link-and-node diagram representing arrival and departure paths in the airspace and on the airfield up to the gate area. An airline schedule is developed which is representative of the period to be analyzed. Statistical data related to the lateness distribution historically found at a particular airport can be used in the model. Normally several runs of the model are made, and the results are averaged to arrive at indications of airport performance.

SIMMOD generates a series of standard reports detailing the flights simulated and the aggregated flight delay statistics. It will also prepare extended reports which compile delay statistics by hour for the various runway-use configurations for arrivals, departures, and total operations. These reports can be output in both tabular and graphical formats.

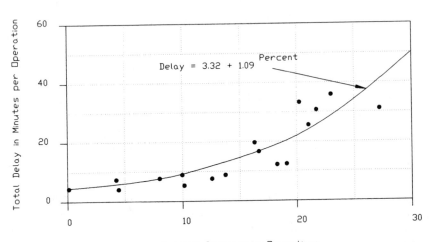

Figure 8-26 Combined ground and air delay for changes in airfield demand at Lindbergh Field (*Federal Aviation Administration [2]*).

Experience with the use of SIMMOD has shown that it realistically portrays the airfield operating environment and is a valuable aid in ascertaining the impact of both demand and facility changes at airports [30]. Figure 8-26 shows the results of a simulation performed at Lindbergh Field in San Diego to determine the impact of increased demand on the performance of the existing airfield [2].

Apron Gate Capacity

The *gate capacity* is the maximum number of aircraft that a fixed number of gates can accommodate during a specified interval when there is a continuous demand for service. The gate capacity can be calculated as the inverse of a weighted-average gate occupancy time of all aircraft being served. For example, if aircraft occupy a gate for an average of 30 min, the capacity of the gate equals 2 aircraft per hour.

The factors that affect gate capacity are as follows:

1. The number and type of gates available to aircraft

2. The mix of aircraft demanding apron gates and the gate occupancy time for various aircraft

3. The percentage of time that gates may be used, which reflects the facts that time is required to maneuver aircraft into and out of gate positions and that the delay often restricts the time actually available for aircraft gate occupancy

4. Restrictions in the use of any or all gates

The *type* of gate refers to its ability to accommodate a large, medium, or small aircraft. Normally gates at an airport are designated as

TABLE 8-17   Typical Aircraft Gate Occupancy Time Required for Aircraft Servicing

	Gate occupancy time, min	
Aircraft	Turnaround station	En route station
A-300-600	30	20
B-737	28	22
B-747-200	60	30
B-757-100	30	20
B-767-200	30	20
B-777	45	25
DC-9-51	30	20
DC-10-10	30	20
MD-11	52	24
MD-87	25	14

SOURCE: Aircraft manufacturers.

wide-bodied aircraft gates, narrow-bodied aircraft gates, and commuter aircraft gates. The *mix* of aircraft refers primarily to size but also to the required gate occupancy time. Very large aircraft require certain types of gates to process passengers. Time is spent maneuvering at a gate, and therefore the gate may not be used 100 percent of the time. If the gate occupancy time includes the time to maneuver at gate positions, as well as the normal processing times to load and unload passengers, to fuel and inspect the aircraft, and to perform cabin service and other routine service, then the utilization will approach 100 percent. Occupancy times vary depending on the size of the aircraft and whether it is an originating, turnaround, or through flight. Table 8-17 gives typical values of aircraft gate occupancy times. These times, however, can be considered minimal. The gate occupancy times expected by aircraft manufacturers are usually given in publications; however, this will vary with each airline and its operating procedures at different stations or airports. Often schedules or lateness of arrivals will result in much greater occupancy times than normally required.

Analytical models for gate capacity

The basis of gate capacity analysis is that the gate time demanded by aircraft should be less than or equal to the gate time available for them. Two analytical models have been developed for determining the capacity of gates at an airport. One model assumes that all aircraft can use all the gates available at an airport. This is termed an *unrestricted* gate-use strategy. The other model assumes that aircraft of a certain size or airline can use only gates specifically designed for these aircraft or airlines. This is called a *restricted* gate-use strategy. Both models are described, and most situations encountered in practice may be approached through one of the two models.

When there are no restrictions on the use of gates, i.e., all aircraft can use all the gates, the capacity of the gates C_g can be derived as

Gate time supplied $\geq$ gate time demanded

$$\mu_k N_k \geq E(T_g)C_g \tag{8-21}$$

where μ_k = gate utilization factor, or percentage of time in 1 hour that gates of type k may be used by aircraft of type i
   N_k = number of type k gates available to aircraft of type i
   $E(T_g)$ = expected value of gate occupancy time demanded by aircraft which can use gate type k
   C_g = capacity of type k gates, aircraft per hour

The expected value of the gate occupancy time $E(T_g)$ is

$$E(T_g) = \sum m_i T_{gi} \qquad (8\text{-}22)$$

where m_i = percentage of type i aircraft in fleet mix using gates at airport

T_{gi} = gate occupancy time required for type i aircraft at airport

The use of these equations is illustrated for unrestricted gate use in Example Problem 8-13.

Example Problem 8-13 An airport has four gates available to all aircraft. The aircraft mix at the airport in the peak hour consists of 30 percent type A, 50 percent type B, and 20 percent type C aircraft. Type A aircraft require a gate occupancy time of 60 min, type B require 45 min, and type C require 30 min. Normally, due to the distribution of demand, the maximum gate utilization which can be expected is 70 percent. It is required to find the capacity of the gates at this airport to process aircraft.

From Eq. (8-21) we have

$$0.70(4)(60) \geq [0.3(60) + 0.50(45) + 0.20(30)]C_g$$

which reduces to

$$C_g = 3.6 \text{ aircraft per hour}$$

Observe that since every aircraft at a gate entails two operations, an arrival and a departure, the hourly capacity of the gates could also be expressed as $2(3.6) = 7.2$ operations per hour. Also if the gate utilization factor is equal to 1, then the ultimate capacity of the gates equals 5.2 aircraft per hour or 10.4 operations per hour.

For restricted gate use, the mix of gates and the mix of aircraft using the airport may not be the same. Therefore, it is necessary to find the capacity of each type of gate and then to determine the overall capacity of the airport based upon gate capabilities as the minimum capacity gate capacity of any type gate. Mathematically, this becomes

$$C_g = \min(C_{gk}) \qquad (8\text{-}23)$$

The use of the above analysis for restricted gate use is shown in Example Problem 8-14.

Example Problem 8-14 An airport has 10 gates available for aircraft. These gates are restricted in the types of aircraft which can be accommodated. The five type I gates can accommodate any type of aircraft, the three type II gates cannot accommodate a type A aircraft, and the two type III gates can accommodate only a type C aircraft.

The mix and the gate occupancy times of the aircraft using the airport in the peak hour are the same as in Example Problem 8-13. The gate utilization factor is 1.0.

Determine the capacity of the gates to process aircraft at this airport.

The relationship shown in Eq. (8-21) must be solved for each type of gate. There are 5 gates available to type A aircraft, 8 gates for type B aircraft, and 10 gates for type C aircraft. Solving Eq. (8-21) for each gate type yields

$$1.0(5)(60) \geq 0.3(60)C_{gI}$$

$$C_{gI} = 16.67 \text{ aircraft per hour}$$

$$1.0(8)(60) \geq [0.3(60) + 0.5(45)]C_{gII}$$

$$C_{gII} = 11.85 \text{ aircraft per hour}$$

$$1.0(10)(60) \geq [0.3(60) + 0.5(45) + 0.2(30)]C_{gIII}$$

$$C_{gIII} = 12.90 \text{ aircraft per hour}$$

Therefore, the type II gates restrict the aircraft capacity of the airport from Eq. (8-23)

$$C_g = \min(16.67, 11.85, 12.90) = 11.9$$

Therefore, with this mix of aircraft demanding gates at the airport, the gate mix restricts the airport capacity to 11.9 aircraft per hour, or 23.8 operations per hour.

Note that only with this capacity is the gate time supplied greater than or equal to the gate time demanded. This is shown as

Gate time supplied $\geq$ gate time demanded

$$1.0(10)(60) \geq [0.3(60) + 0.5(45) + 0.2(30)](11.9)$$

$$600 \geq 554 \quad \text{as required}$$

Graphical technique for gate capacity

A graphical technique to determine the gate capacity has also been developed by the FAA [4, 26]. This method assumes a gate utilization factor of 1.0, and the results must be adjusted accordingly if this utilization is not obtainable. The method assumes that all gates which can accommodate a wide-bodied aircraft can accommodate any other aircraft, but non-wide-bodied gates will not accommodate the wide-bodied aircraft. The parameters required to use this technique include the number of gates, percentage of wide-bodied and non-wide-bodied aircraft and gates, and the gate occupancy times for wide-bodied and non-wide-bodied aircraft.

The chart used for gate capacity with this technique is shown in Fig. 8-27 and its use is shown in Example Problem 8-15.

Example Problem 8-15 Assume that in Example Problem 8-14 the only wide-bodied aircraft are type A. Using the data in this problem, determine the gate capacity of the airport by the FAA chart method.

The percentage of non-wide-bodied aircraft is 70 percent. The percentage of gates that accommodate wide-bodied aircraft is 50 percent. The number of

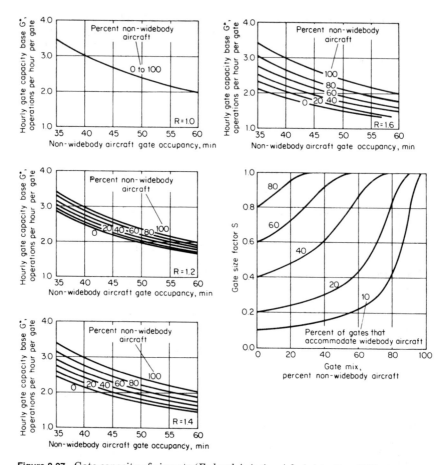

Figure 8-27 Gate capacity of airports (*Federal Aviation Administration [26]*).

gates is 10. The ratio of the gate occupancy time of wide-bodied aircraft to that of non-wide-bodied aircraft is

$$R = \frac{60}{(0.5/0.7)(45) + (0.2/0.7)(30)} = \frac{60}{40.7} = 1.47$$

Entering Fig. 8-27 in the upper right-hand corner with a gate mix of 70 and intersecting the interpolated curve for 50 percent wide-bodied gates, we see that the gate size factor S is 1.0, the chart limit. Using the non-wide-bodied gate time of 40.7 and interpolating between the results for the values of R of 1.4 and 1.6 on the left-hand curves, we see that the hourly base gate capacity is about 2.57 operations per hour per gate. Therefore

$$C_g = 2.57(1.0)(10) = 25.7 \text{ operations per hour}$$

This agrees well with the previous results.

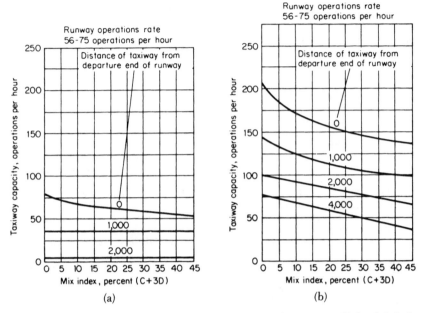

Figure 8-28 Hourly capacity of a taxiway crossing an active runway (*Federal Aviation Administration [26]*).

Although the hourly gate capacity can be determined by the techniques outlined above, in most airport studies the gate requirements at an airport are found by using gate simulation models. These models expand present-day schedules into future design day schedules to determine airport gate requirements by constructing ramp charts. The use of such a simulation model is discussed in Chap. 10.

Taxiway Capacity

In connection with the work described above on ultimate runway capacity, a model has been developed to determine the capacity of a taxiway crossing an active runway [26]. This capacity is a function of the mix of aircraft using the active runway, the distance of the taxiway from the departure end of the runway, and the nature of operations on the active runway, departures, arrivals, or mixed operations. The capacity of taxiways crossing an active runway for a runway operations rate of from 56 to 75 operations per hour is shown in Fig. 8-28.

These models of taxiway capacity are important at airports where the demand rate is high and the configurations of the taxiways and runways are such that crossings must occur during peak periods. A complete discussion of the methodology can be found in Refs. 20, 25, and 26. The evaluation of the throughput capacity of a taxiway was not evaluated as this capacity is virtually always greater than that of the runway and gate system.

References

1. "A Flexible Model for Runway Capacity Analysis," W. E. Weiss, Master's thesis, Massachusetts Institute of Technology, Cambridge, June 1980.
2. *Air and Ground Delay Study: Impact of Increased Operations at Lindbergh Field*, Rep. AOR-89-01, Operations Research Office, Federal Aviation Administration, Washington, September 1989.
3. *Airport Capacity*, Rep. BRD-136, Airborne Instruments Laboratory, Federal Aviation Agency, Office of Technical Services, Department of Commerce, Washington, June 1963.
4. *Airport Capacity and Delay*, Advisory Circular AC 150/5060-5, Federal Aviation Administration, Washington, 1983.
5. *Airport Ground Access*, Rep. FAA-EM-79-4, Federal Aviation Administration, Washington, October 1978.
6. *Airport System Capacity, Strategic Choices*, Special Report 226, Transportation Research Board, Washington, 1990.
7. *Air Traffic Control Handbook*, Order 7110.65G, Federal Aviation Administration, Washington, March 1992.
8. "A Methodology for Airport Capacity Analysis," S. L. M. Hockaday and A. Kanafani, *Transportation Research*, vol. 8, 1974.
9. *Aviation System Capacity Plan 1991–92*, Rep. DOT/FAA/ASC-91-1, Department of Transportation, Federal Aviation Administration, Washington, 1991.
10. *Chicago Delay Task Force Technical Report*, vol. 2: *Model Validation and Existing Airspace/Airfield Performance*, Landrum and Brown Aviation Consultants, Chicago, April 1991.
11. *Critique of the Aircraft Delay Curves in Techniques for Determining Airport Airside Capacity and Delay*, C. T. Ball, Graduate Rep. UCB-ITS-GR-78-1, Institute of Transportation Studies, University of California, Berkeley, August 1978.
12. *FAA Report on Airport Capacity*, Rep. FAA-EM-74-5, Federal Aviation Administration, Washington, 1974.
13. *Future Development of the U.S. Airport Network*, Preliminary Report and Recommended Study Plan, Transportation Research Board, Washington, 1988.
14. *Measuring Airport Landside Capacity*, Special Report 215, Transportation Research Board, Washington, 1987.
15. *Models for Estimating Runway Landing Capacity with Microwave Landing Systems (MLS)*, V. Tosic and R. Horonjeff, Special Report, National Aeronautics and Space Administration, Ames Research Center, Moffett Field, Calif., September 1975.
16. *Models for Runway Capacity Analysis*, R. M. Harris, Rep. FAA-EM-73-5, MITRE Corporation, Federal Aviation Administration, Washington, May 1974.
17. *Model Users' Manual for Airfield Capacity and Delay Models*, C. T. Ball, Rep. FAA-RD-76-128, Federal Aviation Administration, Washington, November 1976.
18. *National Plan of Integrated Airport Systems (NPIAS) 1990–1999*, Federal Aviation Administration, Washington, 1991.
19. *O'Hare Delay Task Force Study*, Rep. FAA-AGL-76-1, Executive Summary and Technical Report, Chicago O'Hare International Airport, prepared by Landrum and Brown, Inc., Cincinnati, Ohio, Federal Aviation Administration, Washington, July 1976.
20. *Procedures for Determination of Airport Capacity*, Rep. FAA-RD-73-111, vols. 1 and 2, Interim Report, Federal Aviation Administration, Washington, April 1973.
21. *Reauthorizing Programs of the Federal Aviation Administration, Future Capacity Needs and Proposals to Meet Those Needs*, Subcommittee on Aviation, Committee on Public Works and Transportation, House of Representative Rep. 101-37, House of Representatives, Washington, 1990.

22. *SIMMOD: The Airport and Airspace Simulation Model,* Reference Manual, Release 1.1, Federal Aviation Administration, Washington, October 1990.
23. *South Florida Supplemental Airport Study,* H. True, S. Wolf, and D. Winer, Operations Research Service, Federal Aviation Administration, Washington, February 1991.
24. *Supporting Documentation for Technical Report on Airport Capacity and Delay Studies,* Rep. FAA-RD-76-153, Federal Aviation Administration, Washington, June 1976.
25. *Technical Report on Airport Capacity and Delay Studies,* Rep. FAA-RD-76-153, Final Report, Federal Aviation Administration, Washington, June 1976.
26. *Techniques for Determining Airport Airside Capacity and Delay,* Rep. FAA-RD-74-124, Federal Aviation Administration, Washington, June 1976.
27. *Terminal Area Forecasts, Fiscal Years 1991–2005,* Rep. FAA-APO-91-5, Federal Aviation Administration, Washington, July 1991.
28. *Upgraded FAA Airfield Capacity Model,* vol. 1: *Supplemental User's Guide,* W. J. Swedish, Rep. FAA-EM-81-1, The MITRE Corporation, Federal Aviation Administration, Washington, February 1981.
29. *Upgraded FAA Airfield Capacity Model,* vol. 2: *Technical Descriptions of Revisions,* W. J. Swedish, Rep. FAA-EM-81-1, The MITRE Corporation, Federal Aviation Administration, Washington, February 1981.
30. *Validation of the SIMMOD Model,* J. C. Bobick, Final Report, ATAC Corporation, Mountain View, Calif., December 1988.

9

Geometric Design of the Airfield

Airport Design Standards

To provide assistance to airport designers and a reasonable amount of uniformity in the design of airport facilities for aircraft operations, design guidelines have been prepared by the FAA [6] and the International Civil Aviation Organization (ICAO) [2, 3, 4]. Any design criteria involving the widths, gradients, separations of runways, taxiways, and other features of the aircraft operations area must necessarily incorporate wide variations in aircraft performance, pilot technique, and weather conditions.

The FAA design criteria provide for uniformity at airport facilities in the United States and serve as a guide to aircraft manufacturers and operators with regard to the facilities expected to be available in the future. The FAA design standards are published in advisory circulars which are revised periodically as the need arises [1]. The ICAO strives toward uniformity and safety on an international level. Its standards, which are very similar to the FAA standards, apply to all member nations of the Convention on International Civil Aviation and are published as Annex 14 to that convention [2]. Requirements for military services are so specialized that they are not included in this chapter.

The design standards prepared by the FAA and the ICAO are presented in the text which follows under the general headings of airport classification, runways, taxiways, and aprons. The material is organized so that the various criteria may be readily compared. It is incumbent upon airport planners to review the latest specifications for airport design at the time studies are undertaken because changes are incorporated as conditions dictate.

Airport Classification

For the purpose of stipulating geometric design standards for various types of airports and the functions which they serve, alphabetic and numeric codes and other descriptors have been adopted to classify airports.

Federal Aviation Administration

The FAA classifies airports for geometric design purposes based upon the *airport reference code*. The airport reference code is a coding system used to relate the airport design criteria to the operational and physical characteristics of the aircraft intended to operate at the airport. It is based upon the *aircraft approach category* and the *airplane design group* to which the aircraft is assigned. The aircraft approach category, as shown in Table 9-1, is determined by the aircraft approach speed, which is defined as 1.3 times the stall speed in the landing configuration of aircraft at the maximum certified landing weight [6]. The airplane design group is a grouping of aircraft based upon wingspan, shown also in Table 9-1. The airport reference code is a two-designator code referring to the aircraft approach category and the airplane design group for which the airport has been designed. For example, an airport reference code of B-III is an airport designed to accommodate aircraft with approach speeds from 91 and to less than 121 kn (aircraft approach category B) with wingspans from 79 to less than 118 ft (airplane design group III). The FAA publishes a list of the airport reference codes for various aircraft [6]. An airport designed to accommodate the Boeing 767-200 which has an approach speed of 130 kn (aircraft approach category C) and a wingspan of 156 ft 1 in (airplane design group IV) is classified with an airport reference code C-IV.

Utility airports. A *utility airport* is one which has been designed, constructed, and maintained to accommodate aircraft from approach

TABLE 9-1   FAA Airport Reference Code

Aircraft approach category	Aircraft approach speed, kn	Airplane design group	Aircraft wingspan, ft
A	<91	I	<49
B	91–<121	II	49–<79
C	121–<141	III	79–<118
D	141–<166	IV	118–<171
E	≥166	V	171–<214
		VI	214–<262

SOURCE: Federal Aviation Administration [6].

categories A and B [6]. The specifications for utility airports are grouped for small aircraft (those with a maximum certified takeoff weight of 12,500 lb or less) and *large aircraft* (those with a maximum certified takeoff weight in excess of 12,500 lb).

Design specifications for utility airports are governed by the airplane design group and the types of approaches authorized for the airport runway, i.e., visual, non-precision-instrument, or precision-instrument approaches.

Utility airports for small aircraft are called *basic utility stage I, basic utility stage II,* and *general utility stage I.* Utility airports for large aircraft are called *general utility stage II.* Utility airports are further grouped for either visual and non-precision-instrument operations or precision-instrument operations. The visual and non-precision-instrument operation utility airports are the basic utility stage I, basic utility stage II, or general utility stage I airports. The precision-instrument operation utility airport is the general utility stage II airport.

A basic utility stage I airport can accommodate about 75 percent of the single-engine and small twin-engine aircraft used for personal and business purposes. This generally means an aircraft weighing on the order of 3000 lb or less, and it is given the airport reference code B-I, which indicates that it accommodates aircraft in aircraft approach categories A and B and aircraft in airplane design group I. A basic utility stage II airport can accommodate all the airplanes of a basic utility stage I airport plus some small business and air taxi type of airplanes. This generally means an aircraft weighing on the order of 8000 lb or less, and it is also given the airport reference code B-I. A general utility stage I airport accommodates all small aircraft. It is assigned the airport reference code B-II. A general utility stage II airport serves large airplanes in aircraft approach categories A and B and usually has the capability for precision-instrument operations. It is assigned the airport reference code B-III.

Transport airports. A *transport airport* is designed, constructed, and maintained to accommodate aircraft in approach categories C, D, and E [6]. The design specifications of transport airports are based upon the airplane design group.

The International Civil Aviation Organization

The ICAO uses a two-element code, the *aerodrome reference code,* to classify the geometric design standards at an airport [2, 3]. The code elements consist of a numeric designator and an alphabetic designator. Aerodrome code numbers 1 through 4 classify the length of runway available, or the *reference field length,* which includes the runway length and, if present, the stopway and clearway. The reference field length is the actual runway takeoff length converted to an equivalent length at mean sea level, 15°C, and 0 percent gradient. Aerodrome code letters A through E classify the wingspan and outer main gearwheel-span for the aircraft for which the airport has been designed. These aerodrome reference codes are given in Table 9-2. For example, an airport designed to accommodate a Boeing 767-200 with an outer main gearwheel-span 34 ft 3 in (10.44 m) wide, a wingspan of 156 ft 1 in (48 m), at a maximum takeoff weight of 317,000 lb, requiring a runway length of about 6000 ft (1830 m) at sea level on a standard day would be classified by the ICAO with aerodrome reference code 4-D. Note that this classification system does not explicitly include the function of the airport, the service it renders, or the type of aircraft accommodated.

There is an approximate correspondence between the airport reference code of the FAA and the aerodrome reference code of the ICAO [2, 3]. The FAA's aircraft approach categories of A, B, C, and D are approximately the same as the ICAO's aerodrome codes 1, 2, 3, and 4, respectively. Similarly the FAA's airplane design groups of I, II, III, IV, and V approximately correspond to the ICAO's aerodrome codes A, B, C, D, and E.

Runways

A runway is a rectangular area on the airport surface prepared for the takeoff and landing of aircraft. An airport may have one runway or several runways which are sited, oriented, and configured so as to provide for the safe and efficient use of the airport under a variety of conditions. Several factors which affect the siting, orientation, and number of runways at an airport include local weather conditions, particularly wind distribution and visibility, topography of the airport and surrounding area, type and amount of air traffic to be serviced at the airport, aircraft performance requirements, and aircraft noise [2].

TABLE 9-2   ICAO Aerodrome Reference Codes

Aerodrome code number	Reference field length, m	Aerodrome code letter	Wingspan, m	Outer main gearwheel span,* m
1	<800	A	<15	<4.5
2	800–< 1200	B	15–<24	4.5–<6
3	1200–< 1800	C	24–<36	6–<9
4	≥1800	D	36–<52	9–<14
		E	52–<65	9–<14

*Distance between outside edges of tires on the main wheel gears.

SOURCE: International Civil Aviation Organization [2].

Length

It is very difficult to specify runway lengths for different classes of airports because runway length depends on many factors. These factors are discussed in detail in Chap. 3. Some of the more important are the performance characteristics of a particular type of aircraft, trip length, and altitude and temperature at the airport. As a guide to airport planners, the FAA has published the runway length requirements for air carrier and general aviation aircraft [17]. This information is in the form of performance curves and tables which relate runway length to takeoff or landing weight, temperature, and altitude. The distance that an aircraft can normally travel for a specific takeoff weight is also given. The performance curves are an abstraction of information contained in an aircraft flight operations manual. They are not complete, and thus if the planner has access to aircraft FAA-approved flight operations manuals, these are the best sources of information for determining runway length. The aircraft manufacturers also prepare publications for specific use in airport planning which detail the characteristics of aircraft of importance in airport planning and design. In addition to the temperature and altitude considerations contained in the advisory circulars, performance curves in the flight operations manuals also include the effects of longitudinal gradient and wind. For the purpose of planning, it is assumed that there is no wind acting on the runway. Since the particular performance characteristics of similar aircraft operated by the airlines may differ, the airport planner should always consult with the airlines before planning facilities for aircraft operations.

If one can determine the runway length required at sea level in standard atmospheric conditions, ICAO Annex 14 provides corrections to this length for altitude, temperature, and runway gradient. The runway length correction for altitude is 7 percent for each 1000-ft elevation above mean sea level. This length is further corrected for temperature at the rate of 1 percent for every 1°C that the airport reference temperature exceeds the temperature of the standard atmosphere for that elevation. The temperature of the standard atmosphere is 15°C at sea level and decreases at approximately a rate of 1.981° for each 1000-ft increase in elevation. The airport reference temperature T_0 is given by Eq. (9-1).

$$T_0 = T_1 + \frac{T_2 - T_1}{3} \qquad (9\text{-}1)$$

where T_1 is the mean of the mean daily temperatures for the hottest month of the year, the hottest month being that which has the highest mean daily temperature, and T_2 is the mean of the maximum daily temperatures for the same month.

Both T_1 and T_2 should be averaged for a period of years. The length corrected for altitude and temperature is further corrected for slope at the rate of 10 percent for each 1 percent of effective runway gradient. It must be emphasized that these corrections are approximate and that the best source of information is the flight operations manual.

In planning airports the runways should be long enough to accommodate the aircraft which requires the greatest length. The approximate runway length requirements for various categories of aircraft are given in Table 9-3. These values have been determined through the use of airport design 3.2, a computer program prepared by the FAA for use in airport design [6].

For airports serving air carrier aircraft, runways of 12,000 ft will accommodate long-range flights involving trip lengths on the order of 6000 (statute) mi. Lengths of 9000 to 10,000 ft can readily serve flights on the order of 3000 mi. However, one should be aware of the fact that aircraft airframe, wing, and engine technology is making it possible to fly greater distances with greater weights at reduced runway lengths.

Parallel-runway system spacing

The spacing of parallel runways depends on a number of factors such as whether the operations are in VMC or IMC and, if in IMC, whether it is desired to accommodate simultaneous arrivals or simultaneous arrivals and departures. At those airports serving both heavy and

TABLE 9-3   Approximate Runway Length Requirements for Various Aircraft

Aircraft	Runway length,* ft
Small airplanes	
With less than 10 passenger seats	
75% of fleet	2,450
95% of fleet	3,010
100% of fleet	3,560
With 10 or more passenger seats	4,170
Large airplanes of 60,000 lb or less	
75% of fleet at 60% useful load	5,300
75% of fleet at 90% useful load	7,000
100% of fleet at 60% useful load	5,500
100% of fleet at 90% useful load	7,800
Large airplanes of more than 60,000 lb	
Stage length of 1000 mi	5,950
Stage length of 2000 mi	7,600
Stage length of 3000 mi	8,950
Stage length of 6000 mi	11,200

*At sea level, mean daily maximum temperature in hottest month of 85°F and no runway gradient.

SOURCE: Federal Aviation Administration, Airport Design computer program, Version 3.2 [6].

light aircraft, simultaneous use of runways even in VMC conditions may be dictated by separation requirements to safeguard against wake vortices.

Under VMC, the FAA requires parallel-runway centerline separations of 300 ft for single-engine propeller aircraft, 500 ft for twin-engine propeller aircraft, and 700 ft for all other aircraft, when the operations are in the same direction and wake vortices are not prevalent. It also recommends increasing the separation to 1200 ft for airplane design groups V and VI. During the daylight hours, 1400 ft is required for opposite-direction operations, and 2800 ft is required at all other times [6]. If wake vortices are generated by heavy jets and if it is desired to operate on two runways simultaneously in VMC when little or no crosswind is present, then the minimum distance specified by the FAA is 2500 ft. However, the designer should contact the nearest government aviation body to determine for the particular location whether it is necessary to have such a large separation.

For operations under VMC, the ICAO recommends that the minimum separations between the centerlines of parallel runways for simultaneous use, disregarding wake vortices, be 120 m (400 ft) for aerodrome code 1, 150 m (500 ft) for aerodrome code 2, and 210 m (700 ft) for aerodrome code 3 or code 4 runways.

In IMC, the FAA specifies 4300 ft and the ICAO specifies 1525 m (5000 ft) as the minimum separation between centerlines of parallel runways for simultaneous instrument approaches. However, there is evidence that these distances are conservative, and steps are being taken to reduce them. The ultimate goal is to reduce this distance by about one-half. For instrument-dependent approaches, both the FAA and the ICAO recommend centerline separations of 3000 ft (915 m).

Both the FAA and the ICAO specify that two parallel runways may be used simultaneously for radar departures in IMC if the centerlines are separated by at least 2500 ft (760 m). The FAA requires a 3500-ft centerline separation for simultaneous nonradar departures. If two parallel runways are to be operated independently of each other in IMC under radar control, one for arrivals and the other for departures, then both the FAA and the ICAO specify a minimum separation between the centerlines of 2500 ft (760 m) when the thresholds are even. If the thresholds are staggered, the runways can be brought closer together or must be separated farther, depending on the amount of stagger and which runways are used for arrivals and departures. If approaches are to the nearest runway, then the spacing may be reduced by 100 ft (30 m) for each 500 ft (150 m) of stagger down to a minimum of 1200 ft (360 m) for airplane design groups V and VI and 1000 ft (300 m) for all other aircraft. However, if the approaches are to the farthest runway, then the runway spacing must be increased by 100 ft (30 m) for each 500 ft (150 m) of stagger.

TABLE 9-4   Parallel-Runway Centerline Separation Criteria, ft (m)

Simultaneous VFR operations	
All aircraft	700*,† (210)
Simultaneous IFR operations	
Segregated operations	2500‡ (760)
Radar departures	2500 (760)
Nonradar departures	3500 (1000)
Dependent arrivals	3000 (915)
Independent arrivals	4300 (1300)

*Operations in same direction during daylight hours when wake vortices not present; for single-engine propeller aircraft 300 ft (100 m); for twin-engine propeller aircraft 500 ft (150 m); for airplane design groups V and VI, a separation of 1200 ft (360 m) is recommended; air traffic control practices may require a greater separation to hold aircraft between runways; centerline separations under 2500 ft (760 m) require the observance of wake turbulence avoidance practices.

†For opposite-direction operations during daylight hours, increase to 1400 ft (425 m); for nighttime operations increase to 2800 ft (850 m).

‡If the runway thresholds are staggered and the approach is to the near threshold, reduce by 100 ft (30 m) for each 500 ft (150 m) of stagger to a minimum separation of 1000 ft (300 m) or 1200 ft (360 m) for airplane design groups V and VI; if the runway thresholds are staggered and the approach is to the far threshold, increase by 100 ft (30 m) for each 500 ft (150 m) of stagger.

SOURCE: Federal Aviation Administration [6].

TABLE 9-5   Parallel-Runway Centerline Separation Criteria, m (ft)

Simultaneous VFR operations		
Aerodrome code 1	120	(400)
Aerodrome code 2	150	(500)
Aerodrome code 3 or 4	210	(700)
Simultaneous IFR operations		
Segregated operations	760	(2500)*
Independent departures	760	(2500)
Dependent arrivals	915	(3000)
Independent arrivals	1525	(5000)

*If the runway thresholds are staggered and the approach is to the near threshold, reduce by 30 m (100 ft) for each 150 m (500 ft) of stagger to a minimum separation of 300 m (1000 ft); if the runway thresholds are staggered and the approach is to the far threshold, increase by 30 m (100 ft) for each 150 m (500 ft) of stagger.

SOURCE: International Civil Aviation Organization [2, 3].

These parallel-runway separation criteria are summarized in Tables 9-4 and 9-5.

Geometric specifications of runway system

The runway system at an airport consists of the structural pavement, shoulders, blast pad, runway safety area, various obstruction-free

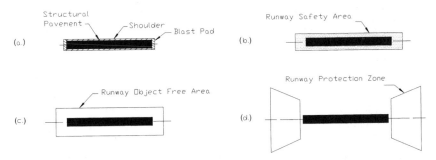

Figure 9-1 Runway system components: (*a*) Paved areas; (*b*) runway safety area; (*c*) runway object-free area; (*d*) runway protection zone.

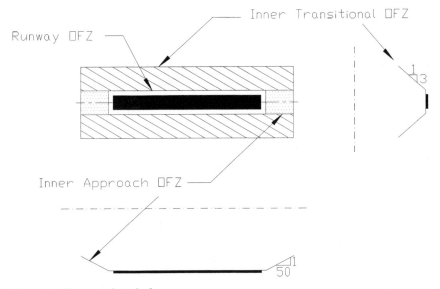

Figure 9-2 Runway obstacle-free zones.

surfaces, and the runway protection zone, as shown in Figs. 9-1 and 9-2.

1. The runway *structural pavement* supports the aircraft with respect to structural load, maneuverability, control, stability, and other operational and dimensional criteria.

2. The *shoulder* adjacent to the edges of the structural pavement resists jet blast erosion and accommodates maintenance and emergency equipment.

3. The *blast pad* is an area designed to prevent erosion of the surfaces adjacent to the ends of runways due to jet blast or propeller wash.

4. The *runway safety area* (RSA) is an area surrounding the runway that is prepared or suitable for reducing the risk of damage to aircraft in the event of an undershoot, overshoot, or excursion from the runway. ICAO refers to an area similar to the runway safety area as the *runway strip* and the *runway end safety area.* The runway safety area includes the structural pavement, shoulders, blast pad, and stopway, if provided. This area should be capable of supporting emergency and maintenance equipment as well as providing support for aircraft. The runway safety area is cleared, drained, and graded, and it should have no potentially hazardous ruts, humps, depressions or other surface variations. It should be free of objects except for objects that are required to be located in the runway safety area because of their function. These objects are required to be constructed on frangible mounted structures at the lowest possible height with the frangible point no higher than 3 in above grade.

5. The runway *object-free area* (OFA) is defined by the FAA as a two-dimensional ground area surrounding the runway which must be clear of parked aircraft and objects other than those whose location is fixed by function.

6. The runway *obstacle-free zone* (OFZ) is a defined volume of airspace centered above the runway which supports the transition between ground and airborne operations. The FAA specifies this as the airspace above a surface whose elevation is the same as that of the nearest point on the runway centerline and extending 200 ft beyond each end of the runway.

7. The *inner-approach obstacle-free zone,* which applies only to runways with approach lighting systems, is the airspace above a surface centered on the extended runway centerline beginning 200 ft beyond the runway threshold at the same elevation as the runway threshold and extending 200 ft beyond the last light unit on the approach lighting system. Its width is the same as that of the runway obstacle-free zone, and it slopes upward at the rate of 50 horizontal to 1 vertical.

8. The *inner transitional obstacle-free zone,* which applies only to precision-instrument runways, is defined by the FAA as the volume of airspace along the sides of the runway and the inner-approach obstacle-free zone. The surface slopes at the rate of 3 horizontal to 1 vertical out from the edge of the runway obstacle-free zone and the inner-approach obstacle-free zone until it reaches a height of 150 ft above the established airport elevation.

9. The *runway protection zone* (RPZ) is an area on the ground used to enhance the protection of people and objects near the runway approach. It is the inner portion of the FAR part 77 approach surface to the runway, as discussed in Chap. 7.

10. The runway *threshold approach surface* is a sloping surface of specified dimensions beginning either at the end of the runway or 200 ft from the end of the runway. No objects may penetrate this surface. If objects do penetrate this surface, either the objects must be removed or the approach threshold to the runway must be displaced. The threshold approach surface is not the same as the approach surface defined in FAR part 77, which is discussed in Chap. 7.

The FAA runway standards related to the pavement and shoulder width, safety area, blast pad, and obstacle-free surfaces are given in Tables 9-6 and 9-7. Similar data for the ICAO are given in Table 9-8. The FAA requirements for the runway threshold approach surface are given in Table 9-9 and are shown in Fig. 9-3.

As noted earlier, the FAA has prepared a computer program, airport design 3.2, to assist airport planners in determining the dimensions of these critical runway areas and surfaces for runways serving various types of aircraft under specific operating conditions [6]. The results obtained from using this program are illustrated in Example Problem 9-1.

Example Problem 9-1 It is necessary to determine the runway design dimensional criteria for a runway designed to accommodate a Boeing 767-200 aircraft as the critical design aircraft. The runway is to be used under precision-instrument operations at an airport with an elevation of 1000 ft above mean sea level and an average high temperature of 86°F.

The Boeing 767-200 aircraft has a wingspan of 156.1 ft, which places it in airplane design group IV, which accommodates aircraft with wingspans up to but not including 171 ft. The use of the specifications presented in Table 9-7, or the computer program airport design 3.2, yields the results shown in Table 9-10. The design criteria presented in this table are developed based upon the specifications for a runway designed to accommodate all airplanes in the aircraft approach category and airplane design group to which the Boeing 767-200 is assigned, i.e., FAA airport reference code C-IV.

Figures 9-4 through 9-7 illustrate the dimensions of the various runway surfaces shown in Table 9-10 for this example problem.

Sight distance and longitudinal profile. In addition to the information in Tables 9-6, 9-7, 9-8, and 9-9, other factors must be considered when one is establishing the longitudinal profile of a runway. Some of these factors are the sight distance, required length of vertical transition curves, required distance between vertical transition curves, and allowable longitudinal gradient.

The FAA requirement for sight distance on individual runways requires that the runway profile permit any two points 5 ft above the runway centerline to be mutually visible for the entire runway length. If, however, the runway has a full-length parallel taxiway, the runway profile may be such that an unobstructed line of sight will exist from any point 5 ft above the runway centerline to any other

TABLE 9-6   Runway Dimensional Standards for Aircraft of Approach Categories A and B, ft

	Approach type									
	Visual and non-precision-instrument					Precision-instrument				
	Airplane design group									
	I*	I	II	III	IV	I*	I	II	III	IV
Runway width	60	60	75	100	150	75	100	100	100	150
Shoulder width	10	10	10	20	25	10	10	10	20	25
Blast pad										
Width	80	80	95	140	200	95	120	120	140	200
Length	60	100	150	200	200	60	100	150	200	200
Safety area										
Width	120	120	150	300	500	300	300	300	400	500
Length†	240	240	300	600	1000	600	600	600	800	1000
Object-free area										
Width	250	400	500	800	800	800	800	800	800	800
Length†	300	500	600	1000	1000	1000	1000	1000	1000	1000
Obstacle-free zone										
Width‡	120§	250	250	250	250	300	300	300	300	300
Length¶	200	200	200	200	200	200	200	200	200	200

*Facilities for small airplanes only.

†From end of runway; with the declared-distance concept, these lengths begin at the stop end of each ASDA (accelerate-stop distance available) and at both ends of the LDA (landing distance available), whichever is greater.

‡For runways serving small aircraft only; for large aircraft the greater of 400 ft or 180 ft plus the wingspan of the most demanding aircraft plus 20 ft for each 1000 ft of airport elevation.

§For runways serving small aircraft with approach speeds of less than 50 kn; increase to 250 ft for runways serving small aircraft with approach speeds greater than 50 kn.

¶Beyond the end of each runway.

TABLE 9-7   Runway Dimensional Standards for Aircraft in Approach Categories C, D, and E, ft

	Airplane design group					
	I	II	III	IV	V	VI
Runway width	100	100	100[a]	150	150	200
Shoulder[b] width	10	10	20[a]	25	35	40
Blast Pad						
Width	120	120	140[a]	200	220	280
Length	100	150	200	200	400	400
Safety area						
Width[c]	500	500	500	500	500	500
Length[d]	1000	1000	1000	1000	1000	1000
Object-free area						
Width	800	800	800	800	800	800
Length[d]	1000	1000	1000	1000	1000	1000
Obstacle-free zone						
Width[e]	400	400	400	400	400	400
Length[f]	200	200	200	200	200	200

[a]For airplane design group III serving aircraft with maximum certified takeoff weight greater than 150,000 lb, the standard runway width is 150 ft, the shoulder width is 25 ft, and the blast pad width is 200 ft.

[b]Airplane design groups V and VI normally require stabilized or paved shoulder surfaces.

[c]For airport reference code C-I and C-II, a runway safety area width of 400 ft is permissible. For runways designed after Feb. 28, 1983, to serve aircraft approach category D aircraft, the runway safety area width increases 20 ft for each 1000 ft of airport elevation above mean sea level.

[d]From end of runway; with the declared-distance concept, these lengths begin at the stop end of each ASDA and at both ends of the LDA, whichever is greater.

[e]For large aircraft, the greater of 400 ft or 180 ft plus the wingspan of the most demanding aircraft plus 20 ft for each 1000 ft of airport elevation; for small aircraft, 300 ft for precision-instrument runways, 250 ft for all other runways serving small aircraft with approach speeds of 50 kn or more, and 120 ft for all other runways serving small aircraft with approach speeds less than 50 kn.

[f]Beyond the end of each runway.

point 5 ft above the runway centerline for one-half the runway length. The FAA recommends a clear line of sight between the ends of intersecting runways. The terrain must be graded and permanent objects designed and sited so that there will be an unobstructed line of sight from any point 5 ft above one runway centerline to any point 5 ft above an intersecting runway centerline within the runway visibility zone. The runway visibility zone is the area formed by imaginary lines connecting the visibility points of the two intersecting runways. The runway visibility zone for intersecting runways is shown in Fig. 9-8. The visibility points are defined as follows:

TABLE 9-8   ICAO Runway and Runway Strip Dimensional Standards, m

	Aerodrome code				
	A	B	C	D	E
Pavement width					
Aerodrome code					
1*	18	18	23		
2*	23	23	30		
3	30	30	30	45	
4			45	45	45
Pavement and shoulder width[†,‡]			60	60	60

	Aerodrome code			
	1	2	3	4
Runway strip width‡				
Precision approach	150	150	300	300
Non-precision-instrument approach	150	150	300	300
Visual approach	60	80	150	150
Clear and graded-area width‡				
Instrument approach	80	80	150§	150§
Visual approach	60	80	150	150

*The width of a precision-instrument approach runway should not be less than 30 m where the aerodrome code number is 1 or 2.

†Minimum width of pavement and shoulders when pavement width is less than 60 m.

‡Symmetric about the runway centerline.

§It is recommended that this be provided for the first 150 m from each end of the runway and that it be increased linearly from this point to a width of 210 m at a point 300 m from each end of the runway and remain at this width for the remainder of the runway.

TABLE 9-9   Dimensions of Runway Threshold Approach Surface

	Small airplanes		Large airplanes		
	Approach speed, kn		Visibility limits,* mi		
	<50	≥50	>1	≤1	$\leq \frac{3}{4}$
Trapezoidal section					
Inner portion					
Distance from runway end D	0	0	0	200	200
Width W_1	120	250	400	1,000	1,000
Outer portion					
Length L_1	500	2,250	1,500	10,000	10,000
Width W_2	300	700	1,000	4,000	4,000
Rectangular section					
Length L_2	2,500	2,750	8,500	0	0
Slope	15:1	20:1	20:1	20:1	34:1

*For category II operations, see FAA Advisory Circular AC 120-29 [9]; for STOL runways see FAA Order 8260.3B [21].

SOURCE: Federal Aviation Administration [6].

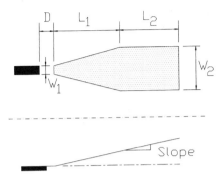

Figure 9-3 Runway threshold approach surface.
Dimensions are keyed to Table 9-9.

1. If the distance from the intersection of the two runway centerlines is 750 ft or less, the visibility point is on the centerline at the runway end designated by point a in Fig. 9-8.

2. If the distance from the intersection of the two runway centerlines is greater than 750 ft but less than 1500 ft, the visibility point is on the centerline 750 ft from the intersection of the centerlines designated by point b in Fig. 9-8.

3. If the distance from the intersection of the two runway centerlines is equal to or greater than 1500 ft, the visibility point is on the centerline equidistant from the runway end and the intersection of the centerlines designated by points c and d in Fig. 9-8.

The ICAO requirement for sight distance on individual runways requires that the runway profile permit an unobstructed view between any two points at a specified height above the runway centerline to be mutually visible for a distance equal to at least one-half the runway length. The ICAO specifies that the height of these two points should be 1.5 m (5 ft) above the runway for aerodrome code A runways, 2 m (7 ft) above the runway for aerodrome code B runways, and 3 m (10 ft) above the runway for aerodrome code C, D, or E runways.

In regard to the runway longitudinal profile, it is desirable to minimize longitudinal grade changes as much as possible. However, this may not be possible for reasons of economy. Therefore, both the ICAO and the FAA allow changes in grade but limit the number and size. The maximum longitudinal grade changes permitted by the FAA are listed in Table 9-11 and illustrated in Fig. 9-9. The maximum longitudinal grade changes permitted by the ICAO are listed in Table 9-12. Tables 9-11 and 9-12 also list the maximum longitudinal grade. The FAA limits both longitudinal gradient and longitudinal grade changes

TABLE 9-10 Runway Design Criteria Based upon Airport Reference Code from Airport Design Computer Program for Example Problem 9-1

Design of Boeing 767-200	
Fuselage length	159.0 ft
Wingspan	156.1 ft
Undercarriage width	34.25 ft
Tail height	52.90 ft
Aircraft approach category	C
Airplane design group	IV

Runway pavement	
Length (approximate)	8130 ft
Width	150 ft
Shoulder width	25 ft
Runway blast pad	
Width	200 ft
Length	200 ft
Runway safety area (RSA)	
Width	500 ft
Length beyond each runway end	1000 ft
Runway object-free area (OFA)	
Width	800 ft
Length beyond each runway end	1000 ft
Runway obstacle-free zone (OFZ)	
Width	400 ft
Length beyond each runway end	200 ft
Inner-approach obstacle-free zone	
Width	400 ft
Length beyond approach lighting system	200 ft
Slope from 200 ft beyond runway threshold	50:1
Inner transitional obstacle-free zone	
Slope	3:1
Runway protection zone	
Length	2500 ft
Width 200 ft from runway end	1000 ft
Width 2700 ft from runway end	1750 ft
Threshold approach surface	
Distance from threshold to start of surface	200 ft
Width at start of trapezoidal section	1000 ft
Width at end of trapezoidal section	4000 ft
Length of trapezoidal section	10000 ft
Length of rectangular section	0 ft
Slope	34:1

SOURCE: Federal Aviation Administration [6].

to 2.0 percent for runways serving aircraft in approach categories A and B aircraft and 1.5 percent for runways serving aircraft in approach categories C, D, and E. The ICAO limits both longitudinal gradient and longitudinal grade changes to 2.0 percent for aerodrome code 1 and 2 runways and 1.5 percent for aerodrome code 3 runways. For aerodrome code 4 runways, the maximum longitudinal gradient

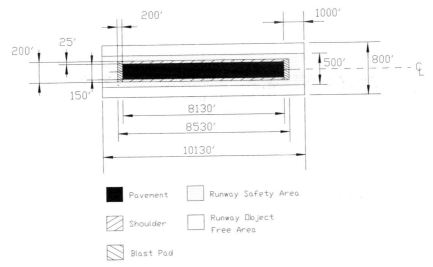

Figure 9-4 Runway pavement, shoulders, blast pad, runway safety area, and runway object-free area for Example Problem 9-1.

is 1.25 percent and the maximum change in longitudinal gradient is 1.5 percent. In addition, for runways equipped to be used in bad weather, the gradient of the first quarter and the last quarter of the length of the runway must be very flat for reasons of safety. Both the ICAO and the FAA require that this gradient not exceed 0.8 percent. In all cases it is desirable to minimize both longitudinal grades and grade changes.

Longitudinal slope changes are accomplished by means of vertical curves. The length of a vertical curve is determined by the magnitude of the changes in slope and the maximum allowable change in slope of the runway. Both these values are also listed in Tables 9-11 and 9-12.

The number of slope changes along the runway is also limited. The FAA requires that the distance between the points of intersection of two successive curves not be less than the sum of the absolute percentage values of change in slope multiplied by 250 ft for airports serving aircraft approach categories A and B and 1000 ft for airports serving approach categories C, D, and E. The ICAO requires that the distance between the points of intersection of two successive curves not be less than the sum of the absolute percentage values of change in slope multiplied by 50 m (165 ft) for aerodrome code 1 and 2 runways, 150 m (500 ft) for aerodrome code 3 runways, and 300 m (1000 ft) for aerodrome code 4 runways. The ICAO also specifies the minimum distance in all cases to be 45 m (150 ft).

For example, for an FAA runway serving transport aircraft, i.e., aircraft in approach category C, D, or E, if the change in slope were

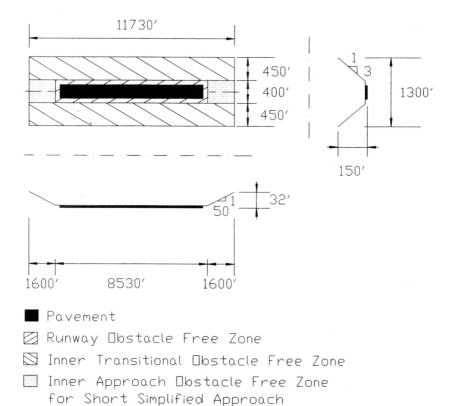

■ Pavement

▨ Runway Obstacle Free Zone

◨ Inner Transitional Obstacle Free Zone

▢ Inner Approach Obstacle Free Zone
  for Short Simplified Approach
  Lighting System (1400')

Figure 9-5 Runway obstacle-free zones for Example Problem 9-1.

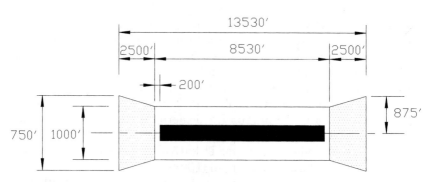

Figure 9-6 Runway protection zone for Example Problem 9-1.

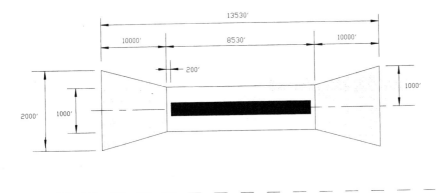

Figure 9-7 Runway threshold approach surface for Example Problem 9-1.

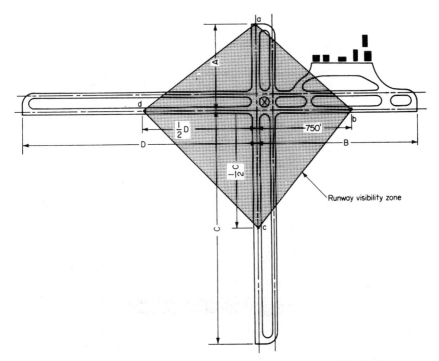

Figure 9-8 Runway visibility zone for intersecting runways (*Federal Aviation Administration [6]*).

TABLE 9-11 **Runway Surface Gradient Standards**

	Aircraft approach category				
	A	B	C	D	E
Gradient, %					
Pavement longitudinal[a]					
Maximum	2.0	2.0	1.5[b]	1.5[b]	1.5[b]
Maximum change	2.0	2.0	1.5	1.5	1.5
Pavement transverse					
Maximum	2.0	2.0	1.5	1.5	1.5
Shoulder transverse					
Minimum	3.0	3.0	1.5[c]	1.5[c]	1.5[c]
Maximum[d]	5.0	5.0	5.0	5.0	5.0
Runway end safety area					
Maximum longitudinal[e]	3.0	3.0	3.0	3.0	3.0
Maximum longitudinal grade change	2.0	2.0	2.0	2.0	2.0
Minimum transverse	1.5	1.5	1.5	1.5	1.5
Maximum transverse[d]	5.0	5.0	3.0	3.0	3.0
Vertical curve, ft					
Minimum length[a, f]	300[g]	300[g]	1000	1000	1000
Minimum distance between points of intersection[a, h]	250	250	1000	1000	1000

[a]Applies also to runway safety area adjacent to sides of runway.

[b]May not exceed 0.8 percent in the first and last quarters of runway.

[c]A minimum of 3 percent for turf.

[d]A slope of 5 percent is recommended for a 10-ft width adjacent to the pavement areas to promote drainage.

[e]For the first 200 ft from the end of the runway, and if it slopes, it must be downward. For the remainder of the runway safety area, the slope must be such that any upward slope does not penetrate the approach surface or clearway plane and any downward slope does not exceed 5 percent.

[f]For each 1 percent change in grade.

[g]No vertical curve is required if the grade change is less than 0.4 percent.

[h]Distance is multiplied by the sum of the absolute grade changes in percent.

SOURCE: Federal Aviation Administration [6].

1.5 percent, the required length of vertical curve would be 1500 ft. Vertical curves are normally not necessary if the change in slope is not more than 0.4 percent. The FAA specifies a minimum length of vertical transition curve of 300 ft for each 1 percent change in grade for runways serving approach categories A and B and 1000 ft for each 1 percent change in grade for runways serving approach categories C, D, and E. The ICAO specifies a minimum length of vertical transition curve of 75 m for each 1 percent change in grade for aerodrome code 1 runways, 150 m for each one percent change in grade for aerodrome code 2 runways, and 300 m for each 1 percent change in grade for aerodrome code 4 runways.

Transverse gradient. A typical cross section of a runway is shown in Fig. 9-10. The FAA and ICAO specifications for the transverse slope on runways are given in Tables 9-11 and 9-12, respectively. It is recommended that a 5 percent transverse slope be provided for the first 10 ft of shoulder adjacent to a pavement edge to ensure proper drainage.

Airfield separation requirements related to runways

The minimum distances from the runway centerline to parallel taxiways, taxilanes, aircraft holding lines, helicopter touchdown pads, and aircraft parking areas are also specified. These distances are given in Tables 9-13 and 9-14 for the FAA and Tables 9-15 and 9-16 for the ICAO.

Taxiways and Taxilanes

Widths and slopes

Since the speeds of aircraft on taxiways are considerably less than on runways, the criteria governing longitudinal slopes, vertical curves, and sight distance are not as stringent as for runways. Also the lower speeds permit the taxiway to be narrower than the runway. The principal geometric design features of interest are listed in Tables 9-17 and 9-18 for the FAA and Tables 9-19 and 9-20 for the ICAO.

Taxiway and taxilane separation requirements

Taxiways are defined paths on the airfield surface which are established for the taxiing of aircraft and are intended to provide a linkage between one part of the airfield and another. The term *dual parallel taxiways* refers to two taxiways parallel to each other on which airplanes can taxi in opposite directions. An *apron taxiway* is a taxiway located usually on the periphery of an apron, intended to provide a through taxi route across the apron. A *taxilane* is a portion of the aircraft parking area used for access between the taxiways and the aircraft parking positions. The ICAO defines an *aircraft stand taxilane* as a portion of the apron intended to provide access to only the aircraft stands.

To provide a margin of safety in the airport operating areas, the trafficways must be separated sufficiently from each other and from adjacent obstructions. Minimum separations between the centerlines of taxiways, between the centerlines of taxiways and taxilanes, and between taxiways and taxilanes and objects are specified so that aircraft may safely maneuver on the airfield.

(a.)

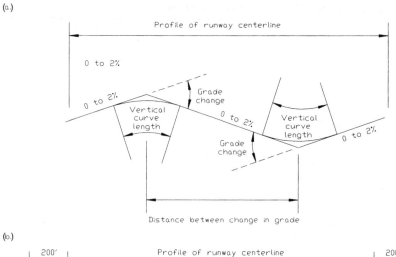

(b.)

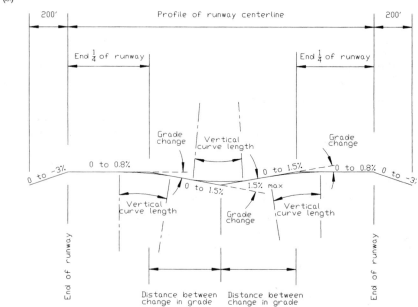

Figure 9-9 Runway longitudinal profile: (*a*) Utility airports; (*b*) transport airports.

FAA separation criteria. The separation criteria adopted by the FAA are predicated upon the wing tips of the aircraft for which the taxiway and taxilane system have been designed and provision of a minimum wing-tip clearance on these facilities. The required separation between taxiways, between a taxiway and a taxilane, or between a taxiway and a fixed or movable object requires a minimum wing-tip clearance of 0.2 times the wingspan of the most demanding aircraft in the airplane design group plus 10 ft. This clearance provides a mini-

TABLE 9-12   Runway Surface Gradient Standards

	Aerodrome code			
	1	2	3	4
Runway longitudinal				
Gradient, %				
Maximum	2.0	2.0	1.5 *	1.25 *
Maximum change	2.0	2.0	1.5	1.5
Maximum effective†	2.0	2.0	1.0	1.0
Vertical curve, m				
Minimum length of curve‡	75	150	300	300
Minimum distance between				
points of intersection§	50	50	150	300
Runway strips				
Gradient, %				
Maximum longitudinal	2.0	2.0	1.75	1.5
Maximum transverse	3.0	3.0	2.5	2.5

	Aerodrome code				
	A	B	C	D	E
Runway transverse gradient, %					
Maximum	2.0	2.0	1.5	1.5	1.5
Minimum	1.0	1.0	1.0	1.0	1.0
Shoulder transverse gradient, %					
Maximum	2.5	2.5	2.5	2.5	2.5

*May not exceed 0.8 percent in the first and last quarters of runway for aerodrome code 4 or for a category II or III precision-instrument runway for aerodrome code 3.

†Difference in elevation between high and low point divided by runway length.

‡For each 1 percent change in grade.

§Distance is multiplied by sum of absolute grade changes in percent; minimum length is 45 m.

SOURCE: International Civil Aviation Organization [3].

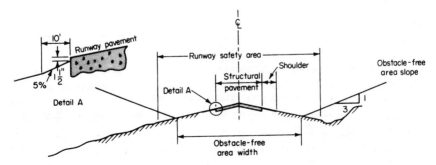

Figure 9-10 Runway cross section.

TABLE 9-13   Airfield Separation Criteria for Aircraft in Approach Categories A and B

	Airplane design group				
	I*	I	II	III	IV
Visual or non-precision-instrument runway centerline to:					
Taxiway or taxilane centerline†	150	225	240	300	400
Hold line†	125	200	200	200	250
Helicopter touchdown pad	400	400	400	400	400
Aircraft parking area	125	200	250	400	500
Precision-instrument runway centerline to:					
Taxiway or taxilane centerline†	200	250	300	350	400
Hold line†	175	250	250	250‡	250‡
Helicopter touchdown pad	400	400	400	400	400
Aircraft parking area	400	400	400	400	500

*For facilities for small aircraft only.

†Satisfies the requirement that no part of an aircraft at a holding location or on a taxiway centerline be within the runway safety area or penetrate the obstacle-free zone. Accordingly, at higher elevations, an increase in these separations may be needed to achieve this result.

‡For sea level up to 6000 ft. Increase by 1 ft for each 100 ft of airport elevation above 6000 ft.

SOURCE: Federal Aviation Administration [6].

TABLE 9-14   Airfield Separation Criteria for Aircraft in Approach Categories C and D

	Airplane design group					
	I	II	III	IV	V	VI
Visual or non-precision-instrument runway centerline to:						
Taxiway or taxilane centerline*	300	300	400	400	400 †	600
Hold line*	250	250	250	250	250	250
Helicopter touchdown pad	400	400	400	400	400	400
Aircraft parking area	400	400	500	500	500	500
Precision-instrument runway centerline to:						
Taxiway or taxilane centerline*	400	400	400	400	400 †	600
Hold line*,‡	250	250	250§	250§	280 ¶	325
Helicopter touchdown pad	400	400	400	400	400	400
Aircraft parking area	500	500	500	500	500	500

*Satisfies the requirement that no part of an aircraft at a holding location or on a taxiway centerline be within the runway safety area or penetrate the obstacle-free zone. Accordingly, at higher elevations, an increase in these separations may be needed to achieve this result.

†For airports at or below an elevation of 1345 ft; increase to 450 ft for airports at elevations between 1345 and 6560 ft and to 500 ft for airports at an elevation above 6560 ft.

‡For aircraft in aircraft approach category D, increase by 1 ft for each 100 ft of airport elevation above mean sea level.

§For aircraft in aircraft approach category C and airplane design groups III and IV, increase by 1 ft for each 100 ft of airport elevation greater than 3200 ft.

¶For aircraft in aircraft approach category C and airplane design group V, increase by 1 ft for each 100 ft of airport elevation above mean sea level.

SOURCE: Federal Aviation Administration [6].

TABLE 9-15   Runway-to-Taxiway Separation Criteria on the Airfield

	Aerodrome code				
	A	B	C	D	E
Runway centerline to:					
Parallel-taxiway centerline					
Non-precision-instrument runways					
Aerodrome code 1	37.5	42			
Aerodrome code 2	47.5	52			
Aerodrome code 3			93	101	
Aerodrome code 4				101	107.5
Precision-instrument runways					
Aerodrome code 1	82.5	87			
Aerodrome code 2	82.5	87			
Aerodrome code 3			168	176	
Aerodrome code 4				176	182.5

SOURCE: International Civil Aviation Organization [2, 3, 4].

TABLE 9-16   Runway-to-Holding-Line Separation Criteria on the Airfield

	Type of runway				
	Non-precision-instrument	Non-precision-approach	Precision-instrument approach category		Takeoff
			I	II & III	
Aerodrome code 1	30	40	60*	—	30
Aerodrome code 2	40	40	60*	—	40
Aerodrome code 3	75	75	90*,†	90*,†	75
Aerodrome code 4	75	75	90*,†	90*,†	75

*This distance may have to be increased to avoid interference with radio aids; for a precision-instrument category III runway, this increase may be on the order of 50 m.

†If a holding bay or a taxiway holding position is at a lower elevation compared to the runway threshold, the distance may be decreased by 5 m for every 1 m that the holding bay or holding position is lower than the threshold, contingent upon not interfering with the inner transitional surface; if a holding bay or a taxiway holding position is at a higher elevation compared to the runway threshold, the distance should be increased by 5 m for every 1 m that the holding bay or holding position is higher than the threshold.

SOURCE: International Civil Aviation Organization [2, 3, 4].

mum taxiway centerline to a parallel-taxiway centerline or taxilane centerline separation of 1.2 times the wingspan of the most demanding aircraft plus 10 ft, and between a taxiway centerline and a fixed or movable object of 0.7 times the wingspan of the most demanding aircraft plus 10 ft. This separation is also applicable to aircraft traversing through a taxiway on an apron or ramp. This separation may have to be increased to accommodate pavement widening on taxiway curves. It is recommended that a separation of at least 2.6 times the wheelbase of the most demanding aircraft be provided to accommodate a 180° turn when the pavement width is designed for tracking the nosewheel on the centerline.

TABLE 9-17   Taxiway Dimensional Standards, ft

	Airplane design group					
	I	II	III	IV	V	VI
Width	25	35	50^a	75	75	100
Edge safety margin^b	5	7.5	10^c	15	15	20
Shoulder width	10	10	20	25	35^d	40^d
Safety area width^e	49	79	118	171	214	262
Object-free area width						
Taxiway^f	89	131	186	259	320	386
Taxilane^g	79	115	162	225	276	334
Separations						
Taxiway centerline to:						
Taxiway centerline^h	69	105	152	215	267	324
Fixed or movable object^i	44.5	62.5	93	129.5	160	193
Taxilane centerline to:						
Taxilane centerline^j	64	97	140	198	245	298
Fixed or movable object^k	39.5	57.5	81	112.5	138	167

^{a}For airplanes in airplane design group III with a wheelbase equal to or greater than 60 ft, the standard taxiway width is 60 ft.

^{b}The taxiway edge safety margin is the minimum acceptable distance between the outside of the airplane wheels and the pavement edge.

^{c}For airplanes in airplane design group III with a wheelbase equal to or greater than 60 ft, the taxiway edge safety margin is 15 ft.

^{d}Airplanes in airplane design groups V and VI normally require stabilized or paved taxiway shoulder surfaces.

^{e}May use aircraft wingspan in lieu of these values.

^{f}May use 1.4 times the wingspan plus 20 ft in lieu of these values.

^{g}May use 1.2 times the wingspan plus 20 ft in lieu of these values.

^{h}May use 1.2 times the wingspan plus 10 ft in lieu of these values.

^{i}May use 0.7 times the wingspan plus 10 ft in lieu of these values.

^{j}May use 1.1 times the wingspan plus 10 ft in lieu of these values.

^{k}May use 0.6 times the wingspan plus 10 ft in lieu of these values.

SOURCE: Federal Aviation Administration [6].

The separation between a taxilane centerline and a parallel-taxilane centerline or a fixed or movable object in the terminal area is predicated on a wing-tip clearance of approximately one-half that required for an apron taxiway. This reduction in clearance is based on the consideration that taxiing speed is low in this area, taxiing is precise, and special guidance techniques and devices are provided. This requires a wing-tip clearance or wing-tip-to-object clearance of 0.1 times the wingspan of the most demanding aircraft plus 10 ft. Therefore, this establishes a minimum separation between the taxilane centerlines of 1.1 times the wingspan of the most demanding aircraft plus 10 ft, and between a taxilane centerline and a fixed or mov-

TABLE 9-18 Taxiway Gradient Standards

	Aircraft approach category				
	A	B	C	D	E
Gradient, %					
Taxiway, shoulder, and safety area					
Longitudinal					
Maximum	2.0	2.0	1.5	1.5	1.5
Maximum change	3.0	3.0	3.0	3.0	3.0
Taxiway transverse					
Minimum	1.0	1.0	1.0	1.0	1.0
Maximum	2.0	2.0	1.5	1.5	1.5
Shoulder transverse					
Minimum	3.0	3.0	1.5*	1.5*	1.5*
Maximum†	5.0	5.0	5.0	5.0	5.0
Safety area transverse					
Minimum	3.0	3.0	1.5	1.5	1.5
Maximum	5.0	5.0	3.0	3.0	3.0
Vertical curve, ft					
Minimum length‡	100	100	100	100	100
Minimum distance between points of intersection§	100	100	100	100	100

*A minimum of 3 percent for turf.

†A slope of 5 percent is recommended for a 10-ft width adjacent to the pavement areas to promote drainage.

‡For each 1 percent of grade change.

§Distance is multiplied by the sum of the absolute grade changes in percent.

SOURCE: Federal Aviation Administration [6].

TABLE 9-19 Taxiway Dimensional Standards, m

	Aerodrome code				
	A	B	C	D	E
Width					
Pavement	7.5	10.5	15 *	18 †	23
Pavement and shoulder			25	38	44
Edge safety margin	1.5	2.25	3 ‡	4.5	4.5
Strip	27	39	57	85	93
Graded portion of strip	22	25	25	38	44
Minimum separation					
Taxiway centerline to:					
Taxiway centerline	21	31.5	46.5	68.5	81.5
Object	13.5	19.5	28.5	42.5	49
Aircraft stand taxilane to:					
Object	12	16.5	24.5	36	42.5

*18 m if used by aircraft with a wheelbase equal to or greater than 18 m.

†23 m is used by aircraft with an outer main gear wheel span equal to or greater than 9 m.

‡4.5 m if intended to be used by an airplane with a wheelbase equal to or greater than 18 m.

SOURCE: International Civil Aviation Organization [2, 3, 4].

TABLE 9-20   ICAO Taxiway Gradient Standards

	Aerodrome code				
	A	B	C	D	E
Gradient, %					
Pavement longitudinal					
Maximum	3.0	3.0	1.5	1.5	1.5
Maximum change	4.0	4.0	3.33	3.33	3.33
Pavement transverse					
Maximum	2.0	2.0	1.5	1.5	1.5
Strip					
Maximum transverse					
Graded portion					
Upward	3.0	3.0	2.5	2.5	2.5
Downward	5.0	5.0	5.0	5. 0	5.0
Ungraded portion					
Upward	5.0	5.0	5.0	5.0	5.0
Vertical Curve, m					
Minimum length*	25	25	30	30	30

*For each 1 percent of grade change.

SOURCE: International Civil Aviation Organization [2, 3, 4].

able object of 0.6 times the wingspan of the most demanding aircraft plus 10 ft [6]. Therefore, when dual parallel taxilanes are provided in the terminal apron area, the taxilane object-free area becomes 2.3 times the wingspan of the most demanding aircraft plus 30 ft.

The separation criteria are presented in Table 9-17. The values indicated in this table are based upon the above specifications and using the largest wingspan in each airplane design group. As noted in this table, the required separations may be reduced to those that would result from using the actual wingspan of the design aircraft.

ICAO separation criteria.   The separation criteria adopted by ICAO are also predicated upon the wing tips of the aircraft for which the taxiway and taxilane system has been designed and provision of a minimum wing-tip clearance on these facilities. But consider as well a minimum clearance between the outer main gearwheel and the taxiway edge. The required separation between taxiways or between a taxiway and a taxilane necessitates a minimum wing-tip clearance C_1 of 3 m for aerodrome code A and code B runways, 4.5 m for aerodrome code C runways, and 7.5 m for aerodrome code D and code E runways. The minimum clearance between the edge of each taxiway and the outer main gearwheels, called the *taxiway edge safety margin* U_1, is given in Table 9-19. This clearance provides a minimum separation between a taxiway centerline and a parallel-taxiway centerline or taxilane centerline given by

$$S_{TT} = \text{WS} + 2U_1 + C_1 \qquad (9\text{-}2)$$

where S_{TT} = minimum taxiway-to-taxiway or taxiway-to-taxilane separation
WS = wingspan of most demanding aircraft
U_1 = taxiway edge safety margin
C_1 = minimum wing-tip clearance

Therefore, an ICAO aerodrome code E runway, e.g., which accommodates aircraft with wingspans up to 65 m, requires separation between a taxiway centerline and a taxiway centerline or a taxilane centerline from Eq. (9-2) of 65 + 2(4.5) + 7.5 = 81.5 m.

The required separation between a taxiway centerline or an apron taxiway centerline and a fixed or movable object is found from Eq. (9-3).

$$S_{TO} = 0.5(\text{WS}) + U_1 + C_2 \qquad (9\text{-}3)$$

where S_{TO} = minimum separation between taxiway or apron taxiway and a fixed or movable object
C_2 = required clearance between wing tip and object

The required clearance between a wing tip and an object C_2 is 4.5 m for aerodrome code A runways, 5.25 m for aerodrome code B runways, 7.5 m for aerodrome code C runways, and 12 m for aerodrome code D and code E runways.

The required separation between an aircraft stand taxilane centerline and a fixed or movable object is

$$S_{ATO} = 0.5(\text{WS}) + U_2 + C_1 \qquad (9\text{-}4)$$

where S_{ATO} = minimum separation between aircraft stand taxilane and fixed or movable object
U_2 = aircraft stand safety margin

Since aircraft moving on the aircraft stand taxilane are moving at low speed and are often under positive ground guidance, the aircraft stand safety margin is less than that on the taxiway system. The value for this safety margin, U_2, is 1.5 m for aerodrome code A and code B airports, 2 m for aerodrome code C airports, and 2.5 m for aerodrome code D or code E airports.

The taxiway and taxilane separation criteria adopted by ICAO are given in Table 9-19.

As noted earlier, the computer program airport design 3.2 can be used to assist airport planners in determining the dimensions of these critical taxiway and taxilane areas as well as other surfaces for airfields serving various types of aircraft under specific operating conditions. The results obtained from using this program are illustrated in Example Problem 9-2.

TABLE 9-21   Taxiway and Taxilane Design Criteria Based
upon Airport Reference Code from Airport Design
Computer Program for Example Problem 9-2

Design aircraft: Boeing 767-200		
Fuselage length	159.0 ft	
Wingspan	156.1 ft	
Undercarriage width	34.25 ft	
Tail height	52.90 ft	
Aircraft approach category	C	
Airplane design group	IV	
Taxiway		
Width	75	ft
Edge safety margin	15	ft
Shoulder width	25	ft
Safety area width	171	ft
Object-free area width	259	ft
Wing-tip clearance	44	ft
Taxilane		
Object-free area width	225	ft
Wing-tip clearance	27	ft
Runway centerline to:		
Parallel taxiway/taxilane centerline	400	ft
Edge of aircraft parking	500	ft
Taxiway centerline to:		
Parallel taxiway/taxilane centerline	215	ft
Fixed or movable object	129.5 ft	
Taxilane centerline to:		
Parallel taxilane centerline	198	ft
Fixed or movable object	112.5 ft	

SOURCE: Federal Aviation Administration [6].

Example Problem 9-2 It is necessary to determine the taxiway and taxilane design dimensional criteria for an airfield designed to accommodate a Boeing 767-200 aircraft as the critical design aircraft. The airfield is 1000 ft above mean sea level.

The airfield will consist of a single runway served by two parallel taxiways. The apron area consists of two piers with dual taxilanes.

The use of the specifications presented in Tables 9-14 and 9-17 or the computer program yields the results shown in Table 9-21. The design criteria presented in this table are developed based upon the specifications for an airfield designed to accommodate all airplanes in the aircraft approach category and airplane design group to which the Boeing 767-200 is assigned, i.e., FAA airport reference code C-IV.

Figures 9-11 through 9-13 show the dimensions of the various taxiway and taxilane components shown in Table 9-21 for this example problem.

Sight distance and longitudinal profile

As in the case of runways, the number of changes in longitudinal profile is limited by the sight distance and the minimum distance between vertical curves. These requirements are given in Tables 9-18 and 9-20.

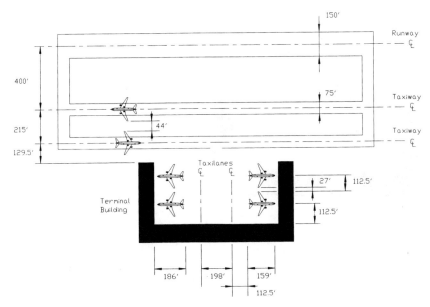

Figure 9-11   Taxiway and taxilane dimensions for Example Problem 9-2.

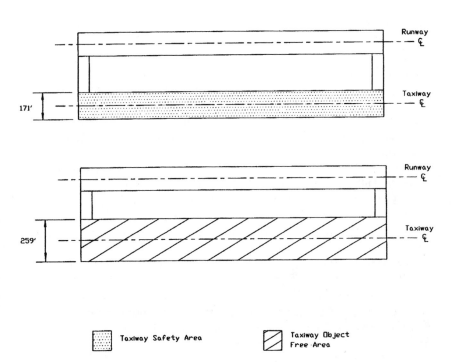

Taxiway Safety Area   Taxiway Object
                             Free Area

Figure 9-12   Taxiway safety area and taxiway object-free area for Example Problem 9-2.

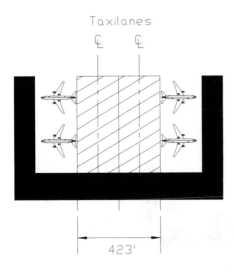

Taxilanes

423'

Figure 9-13 Taxilane object-free area for Example Problem 9-2.

Taxilane Object Free Area

The FAA does not specify line-of-sight requirements for taxiways other than those discussed earlier related to runway and taxiway intersections. However, the sight distance along a runway from an intersecting taxiway needs to be sufficient to allow a taxiing aircraft to enter or cross the runway safely. The FAA specifies that, from any point on the taxiway centerline, the difference in elevation between that point and the corresponding point on a parallel runway, taxiway, or apron edge must be no greater than 1.5 percent of the shortest distance between the points.

The ICAO requires that the surface of the taxiway be seen for a distance of 150 m from a point 1.5 m above the taxiway for aerodrome code A runways, for a distance of 200 m from a point 2 m above the taxiway for aerodrome code B runways, and for a distance of 300 m from a point 3 m above the taxiway for aerodrome code C, D, or E runways.

In regard to the longitudinal profile of taxiways, the ICAO does not specify the minimum distance between the points of intersection of vertical curves. The FAA specifies that the minimum distance for both utility and transport airports not be less than the product of 100 ft multiplied by the sum of the absolute percentage values of change in slope.

Geometry of Exit Taxiway

The function of exit taxiways, or runway turnoffs, as they are sometimes called, is to minimize runway occupancy by landing aircraft. Exit taxiways can be placed at right angles to the runway or at some other angle to the runway. When the angle is on the order of 30°, the term *high-speed exit* is often used to denote that it is designed for higher speeds than other exit taxiway configurations. In this chapter specific dimensions for high-speed exit, rapid-exit, and right-angle exit (regular) taxiways are given. The dimensions presented here are the results obtained from research conducted many years ago [13] and subsequent research conducted by the FAA.

The earlier tests [13] were conducted on wet and dry concrete and asphalt pavement with various types of civil and military aircraft to determine the proper relationship between exit speed and radii of curvature and the general configuration of the taxiway. A significant finding of the tests was that at high speeds a compound curve was necessary to minimize tire wear on the nose gear, and so the central or main curve radius R_2 should be preceded by a much larger-radius curve R_1.

Aircraft paths in the test approximated a spiral. A compound curve is relatively easy to establish in the field and begins to approach the shape of a spiral, thus the reason for suggesting a compound curve. The following pertinent conclusions were reached as a result of the tests [13]:

1. Transport and military aircraft can safely and comfortably turn off runways at speeds on the order of 60 to 65 mi/h on wet and dry pavements.

2. The most significant factor affecting the turning radius is speed, not the total angle of turn or passenger comfort.

3. Passenger comfort was not critical in any of the turning movements.

4. The computed lateral forces developed in the tests were substantially below the maximum lateral forces for which the landing gear was designed.

5. Insofar as the shape of the taxiway is concerned, a slightly widened entrance gradually tapering to the normal width of taxiway is preferred. The widened entrance gives the pilot more latitude in using the exit taxiway.

6. Total angles of turn of 30° to 45° can be negotiated satisfactorily. The smaller angle seems to be preferable because the length of the curved path is reduced, the sight distance is improved, and less concentration is required on the part of the pilots.

7. The relation of turning radius and speed, expressed by Eq. (9-5), will yield a smooth, comfortable turn on a wet or dry pavement when f is made equal to 0.13.

$$R_2 = \frac{V^2}{15f} \tag{9-5}$$

where V is the velocity in miles per hour, and f is the coefficient of friction.

8. The curve expressed by the equation for R_2 should be preceded by a larger-radius curve R_1 at exit speeds of 50 to 60 mi/h. The larger-radius curve is necessary to provide a gradual transition from a straight-tangent-direction section to a curved-path section. If the transition curve is not provided, tire wear on large jet transports can be excessive.

9. The length of the transition curve can be roughly approximated by the relation

$$L_1 = \frac{V^3}{CR_2} \tag{9-6}$$

where V is in feet per second, R_2 is in ft, and C was found experimentally to be on the order of 1.3.

10. Sufficient distance must be provided to comfortably decelerate an aircraft after it leaves the runway. It is suggested for the present that this distance be based on an average rate of deceleration of 3.3 ft/s². This applies only to transport aircraft. Until more experience is gained with this type of operation, the stopping distance should be measured from the edge of the runway.

A chart showing the relationship of exit speed to radii R_1 and R_2 as well as the length of transition curve L_1 is given in Fig. 9-14. The ICAO has indicated the relationship between the aircraft speed and the radius of curvature of taxiway curves shown in Table 9-22. For high-speed exit taxiways it recommends a minimum radius of curvature for the taxiway centerline of 275 m (900 ft) for aerodrome code 1 and code 2 runways and 550 m (1800 ft) for aerodrome code 3 and code 4 runways. This will allow exit speeds under wet conditions of 64 km/h (40 mi/h) for aerodrome code 1 and code 2 runways and 93 km/h (60 mi/h) for aerodrome code 3 and code 4 runways. It also recommends a straight tangent section after the turnoff curve to allow exiting aircraft to come to a full stop while clear of the intersecting taxiway when the intersection is 30°. This tangent distance should be 35 m (115 ft) for aerodrome code 1 and code 2 runways and 75 m (250 ft) for aerodrome code 3 and code 4 runways [2, 4].

A rapid turnoff exit with a turnoff angle of 45° is shown in Fig. 9-15. This type of exit is recommended at airports designed for air-

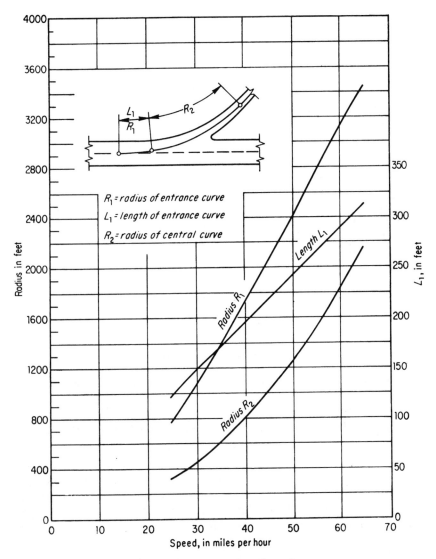

Figure 9-14 Radii of curvature and entrance curves for taxiways.

craft in approach categories A and B. A configuration for an exit speed of 60 mi/h and a turnoff angle of 30° is shown in Fig. 9-16. This type of high-speed exit is recommended for airports serving aircraft in approach categories C, D, and E. Note in Fig. 9-16 that the centerline of the taxiway exit is offset 3 ft from the centerline of the runway at a distance of 200 ft from the point where the exit curve begins the point of curvature (PC). The FAA also recommends that the taxiway center-

TABLE 9-22   Radii of Curvature for Transport Aircraft

Taxiing speed		Radius of exit curve	
mi/h	km/h	ft	m
10	16	50	15
20	32	200	60
30	48	450	135
40	64	800	240
50	80	1250	375
60	96	1800	540

SOURCE: International Civil Aviation Organization [4].

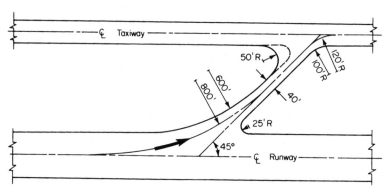

Figure 9-15   A 45° high-speed taxiway exit for aircraft in categories A and B (see Table 9-27 for dimensions) (*Federal Aviation Administration [6]*).

line circular curve be preceded by a 1400-ft spiral to smooth the transition from the runway centerline to the taxiway exit circular curve. The ICAO recommends the same geometry for both these high-speed exits. Right-angle, or 90°, exit taxiways, although not desirable from the standpoint of minimizing runway occupancy, are often constructed for other reasons. The configuration for a 90° exit is shown in Fig. 9-17. Common taxiway intersection configurations and details are shown in Fig. 9-18.

In the design of runway exits, it is strongly recommended that the computer program airport design 3.2, which is discussed later in this chapter, be used to verify the adequacy of the design for the critical aircraft to be used on the airfield.

Considerable research on the characteristics of high speed runway exits continues to be conducted, particularly to determine runway exit geometry which results in reductions in runway occupancy time [9]. This research presents findings for different high-speed exit turnoff angles and exit speeds relative to the required lateral separations between runway and taxiway centerlines.

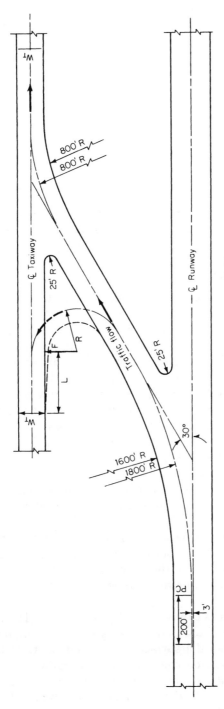

Figure 9-16 A 30° high-speed exit taxiway for aircraft in categories C, D, and E (see Table 9-27 for dimensions) *(Federal Aviation Administration [6]).*

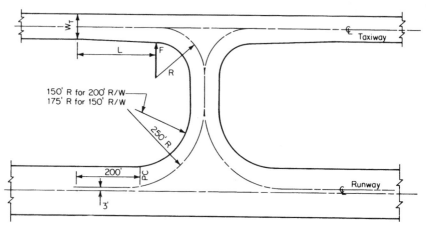

Figure 9-17 A 90° taxiway exit (see Table 9-27 for dimensions) (*Federal Aviation Administration [6]*).

Location of exit taxiways

The location of exit taxiways depends a great deal on the mix of aircraft, approach and touchdown speeds, point of touchdown, exit speed, rate of deceleration, which in turn depends on the condition of the pavement surface, i.e., dry or wet, and number of exits.

While the rules for flying transport aircraft are relatively precise, a certain amount of variability among pilots is bound to exist, especially in respect to braking force applied on the runway and the distance from runway threshold to touchdown. The rapidity and manner of processing arrivals through air traffic control are an extremely important factor in establishing the location of exit taxiways. The location of exit taxiways is also influenced by the location of the runways relative to the terminal area.

Mathematical analyses or models have been developed for optimizing exit locations. These are described in assorted references [4, 6, 7, 13]. While these analyses have been useful in providing and understanding the significant parameters affecting location, their utility to planners has been limited because of the complexity of the analyses and a lack of knowledge of the inputs required for the application of the models. As a result, greater use is made of much more simplified methods.

The landing process can be described as follows. The aircraft crosses the runway threshold and decelerates in the air until the main landing gear touches the surface of the pavement. At this point the nose gear has not made contact with the runway. It may take as much as 3 s to do so. No form of braking can be applied until the nose gear has made contact with the pavement. When it does, reverse thrust or wheel brakes or a combination of both is used to reduce the

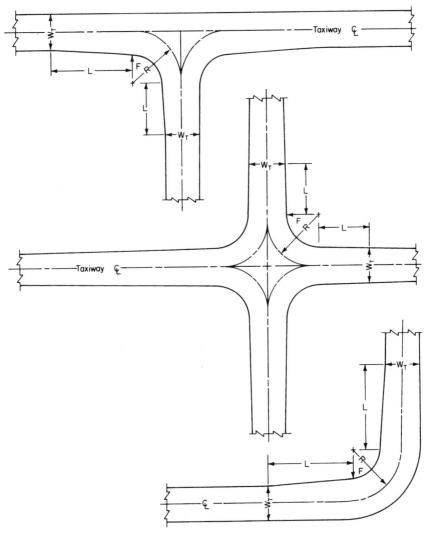

Figure 9-18 Common taxiway intersection details (see Table 9-27 for dimensions) (*Federal Aviation Administration [6]*).

forward speed of the aircraft to exit velocity. On the basis of observations [13], the average deceleration of an air carrier aircraft on the runway is about 5 ft/s².

In the simplified procedure, an aircraft is assumed to touchdown at 1.3 times the stall speed for a landing weight corresponding to 85 percent of the maximum structural landing weight. In lieu of computing the distance from threshold to touchdown, touchdown distances are assumed as fixed values for certain classes of aircraft. Typically these values range from 500 to 1500 ft from the runway threshold. To these distances are added the distances to decelerate to exit speed. These relationships may be approximated by Eqs. (9-7) and (9-8).

$$D = D_{td} + D_e \qquad (9\text{-}7)$$

where D = distance from runway threshold to exits
  D_{td} = distance from runway threshold to point where aircraft touches down
  D_e = distance from touchdown point to exit

$$D_e = \frac{V_{td}^2 - V_e^2}{2a} \qquad (9\text{-}8)$$

where V_{td}^2 = aircraft speed at touchdown
  V_e^2 = exit speed of aircraft
  a = deceleration of aircraft on runway

The simplified procedure is illustrated by the following example problem.

Example Problem 9-3 Suppose that the touchdown speed for a particular aircraft is 140 kn (161 mi/h) and that an exit is to be located at a distance from the runway threshold which will allow this aircraft to leave the runway at an exit speed of 52 kn (60 mi/h). Assume that the touchdown distance is 1500 ft from the threshold and that the average deceleration of the aircraft on the runway is 5 ft/s^2.

The distance from the threshold to reach the exit speed is given by Eq. (9-8)

$$D_e = \frac{V_{td}^2 - V_e^2}{2a}$$

$$= \frac{[161(5280/3600)]^2 - [60(5280/3600)]^2}{2(5)} = 4800 \text{ ft}$$

and from Eq. (9-7) we know that the distance from the threshold to the exit should be

$$D = D_{td} + D_e$$
$$= 1500 + 4800 = 6300 \text{ ft}$$

Although approach and touchdown speeds vary, they can be approximated for locating exit taxiways. At predominantly air carrier airports, air traffic control (ATC) authorities request general aviation aircraft to increase their speeds above normal to reduce the wide range in speed between air carriers and general aviation. At these airports, therefore, the normal approach speeds for general aviation are probably not applicable.

If it is assumed that (1) the distances to touchdown are 1500 ft for air carrier aircraft and 1000 ft for twin-engine general aviation aircraft, (2) a high-speed exit accommodates aircraft leaving the runway at an exit speed of 60 mi/h, and (3) a regular exit accommodates aircraft exiting the runway at 15 mi/h, then by using approximate touchdown speeds, the approximate exit locations for various types of aircraft may be found, as shown in Table 9-23.

These locations are derived by using standard sea-level conditions. Altitude and temperature can affect the location of exit taxiways. Altitude increases the distance on the order of 3 percent for each 1000 ft above sea level, and temperature increases the distance 1.5 percent for each 10°F above 59°F.

During runway capacity studies conducted for the FAA, data were collected on exit utilization at various large airports in the United States [18]. These data, which are tabulated in Table 9-24, indicate the cumulative percentage of each class of aircraft which have exited the runway at various distances from the arrival threshold. On the basis of these studies, runway exit ranges from the arrival threshold are used in runway capacity studies [5]. These exit ranges are given in Table 9-25. Comparisons between the approximate relationships given in Table 9-23 and the data given in Tables 9-24 and 9-25 indicate that fairly good correspondence results. Variations which occur are due to pilot technique and preference in choosing exits, the wide range of performance characteristics demonstrated by various aircraft in the aircraft approach categories, altitude and temperature considerations, and the amount of runway available for landing. The last factor is very important because if pilots recognize that the amount of runway available is near the minimum for a particular aircraft, they are most likely to touch down closer to the runway threshold and to apply larger than normal deceleration and braking to the aircraft.

TABLE 9-23   Approximate Taxiway Exit Location from Threshold, ft

Type of aircraft	Touchdown speed, kn	Exit speed, mi/h	
		60	15
Small propeller			
General aviation single-engine	60	2400	1800
General aviation twin-engine	95	2800	3500
Large turbojet			
Two-engine narrow-body	130	4800	5600
Three-engine narrow-body			
Heavy turbojet			
Four-engine narrow-body	140	6400	7100
Three-engine wide-body			
Four-engine wide-body			

TABLE 9-24 **Percentage of Aircraft Exiting at Various Distances from Runway Arrival Threshold**

Distance from threshold to exit, ft	Dry Runways							
	Regular exits				High-speed exits			
	Aircraft class*				Aircraft class*			
	A	B	C	D	A	B	C	D
0	0	0	0	0	0	0	0	0
1000	6	0	0	0	13	0	0	0
2000	84	1	0	0	90	1	0	0
3000	100	39	0	0	100	40	0	0
4000		98	8	0		98	26	3
5000		100	49	9		100	76	55
6000			92	71			98	95
7000			100	98			100	100
8000				100				

Distance from threshold to exit, ft	Wet Runways			
	Aircraft class*			
	A	B	C	D
0	0	0	0	0
1000	4	0	0	0
2000	60	0	0	0
3000	96	10	0	0
4000	100	80	1	0
5000		100	12	0
6000			48	10
7000			88	64
8000			100	93
9000				100

*The aircraft class is the classification of aircraft based upon the maximum certified takeoff weight [5].

SOURCE: Federal Aviation Administration [18].

TABLE 9-25 **Exit Range Appropriate to Runways Serving Aircraft of Different Arrival Mix Indices, ft**

Mix index*	Exit range from arrival threshold
0– 20	2000–4000
21– 50	3000–5500
51– 80	3500–6500
81–120	5000–7000
121–180	5500–7500

*Mix index is equal to the percentage of class C aircraft plus 3 times the percentage of class D aircraft, where a class C aircraft is an aircraft with a maximum certified takeoff weight greater than 12,500 lb and up to 300,000 lb and a class D aircraft is an aircraft with a maximum certified takeoff weight in excess of 300,000 lb.

SOURCE: Federal Aviation Administration [5].

It is recommended that the point of intersection of the centerlines of taxiway exits and runways, which are up to 7000 ft long and accommodate aircraft in approach categories C, D, and E, should be located about 3000 ft from the arrival threshold and 2000 ft from the stop end of the runway. To accommodate the average mix of aircraft on runways longer than 7000 ft, intermediate exits should be located at intervals of about 1500 ft. At airports where there are extensive operations with aircraft in approach categories A and B, an exit located between 1500 and 2000 ft from the landing threshold is recommended.

Planners often find that the runway configuration and the location of the terminal at the airport often preclude placing the exits at locations based on the previous analysis. This is nothing to be alarmed about since it is far better to achieve good utilization of the exits than to be too concerned about a few seconds lost in occupancy time. When one is locating exits, it is important to recognize local conditions such as frequency of wet pavement or gusty winds. It is far better to place the exits several hundred ft farther from the threshold than to have aircraft overshoot the exits a large amount of the time. The standard deviation in time required to reach exit speed is on the order of 2 or 3 s. Therefore, if the exits were placed down the runway as much as 2 standard deviations from the mean, the loss in occupancy time would be only 4 to 6 s. In planning exit locations at specific airports, one needs to consult with the airlines concerning specific performance characteristics of the aircraft intended for use at the airport.

The total occupancy time of an aircraft can be roughly estimated by using the following procedure. The runway is divided into four components: (1) flight time from threshold to touchdown of main gear, (2) time required for nose gear to make contact with the pavement after the main gear has made contact, (3) time required to reach exit velocity from the time the nose gear has made contact with the pavement and brakes have been applied, and (4) time required for the aircraft to turn onto the taxiway and clear the runway. For the first component, it can be assumed that the touchdown speed is 5 to 8 kn less than the speed over the threshold. The rate of deceleration in the air is about 2.5 ft/s^2. The second component is about 3 s, and the third component depends upon the exit speed. The time to turn off the runway will be on the order of 10 s. Thus the total occupancy time in seconds can be approximated by Eq. (9-9).

$$R_i = \frac{V_{ot} - V_{td}}{2a_1} + 3 + \frac{V_{td} - V_e}{2a_2} + t \qquad (9\text{-}9)$$

where R_i = runway occupancy time, s
  V_{ot} = over-the-threshold speed, ft/s
  V_{td} = touchdown speed, ft/s

TABLE 9-26   Runway Occupancy Time (s) of Aircraft Exiting at Various Distances from Runway Arrival Threshold

Dry Runways								
Distance from threshold to exit, ft	Regular exits Aircraft class*				High-speed exits Aircraft class*			
	A	B	C	D	A	B	C	D
0	24				19			
1,000	24	27			27	24		
2,000	34	27			35	24		
3,000	44	37	29		43	32	35	35
4,000	55	46	38	38		41	35	35
5,000	65	56	47	47		49	44	44
6,000	76	65	56	56			54	54
7,000	76	75	65	65			63	63
8,000	76	75	73	73				
9,000	76	75	82	82				
10,000	76	75	85	85				
11,000	76	75	90	90				

Wet Runways				
Distance from threshold to exit, ft	Aircraft class*			
	A	B	C	D
0	24			
1,000	24			
2,000	34	27		
3,000	44	37	30	
4,000	55	47	38	
5,000	65	56	47	47
6,000	76	65	56	56
7,000	99	99	65	65
8,000			73	73
9,000			82	82

*The aircraft class is the classification of aircraft based upon the maximum certified takeoff weight [5].

SOURCE: Federal Aviation Administration [18].

V_e = exit speed, ft/s
t = time to turnoff from runway after exit speed is reached, s
a_1 = average rate of deceleration in air, in ft/s^2
a_2 = average rate of deceleration on ground, ft/s^2

During the runway capacity studies cited earlier [18], data were also collected on the runway occupancy time. These data, which are tabulated in Table 9-26, indicate the total runway occupancy time of each class of aircraft which exited the runway at various distances from the arrival threshold. As may be observed in this table, typical runway occupancy times for 60 mi/h high-speed exits are 35 to 45 s.

TABLE 9-27 Taxiway Exit Geometry Recommended by FAA, ft

	Airplane design group					
	I	II	III*	IV	V	VI
Taxiway width	25	35	50 †	75	75	100
Edge safety margin	5	7.5	10 ‡	15	15	20
Centerline radius	75	75	100	150	150	170
Length of fillet lead-in	50	50	150	250	250	250
Fillet radius§						
Judgmental oversteering						
Symmetrical widening	62.5	57.5	68	105	105	110
One side widening	62.5	57.5	60	97	97	100
Centerline tracking	60	55	55	85	85	85

*If wheelbase is equal to or greater than 60 ft, use a fillet radius of 50 ft.

†If wheelbase is equal to or greater than 60 ft, the width is 60 ft.

‡For airplanes in airplane design group III with a wheelbase equal to or greater than 60 ft, the taxiway edge safety margin is 15 ft.

§These dimensions relate to the taxiway centerline radius specified. Figures 9-19 and 9-20 show the range of fillet designs which provide an acceptable edge safety margin for a range of wheelbase and undercarriage dimensions. For aircraft falling outside the acceptable range of safety margins, a custom-designed fillet is recommended.

SOURCE: Federal Aviation Administration [6].

The corresponding time for a 15 mi/h regular exit is 45 to 60 s for air carrier aircraft.

Design of taxiway curves and intersections

The basic designs of taxiway curves and intersections for three of the most common types of taxiway intersections have been developed by the FAA [6]. These designs have been taken from this reference and are shown in Fig. 9-18. The dimensions recommended by the FAA for the taxiway width, centerline radius, fillet radius (inner edge radius), and length of the fillet lead-in are given in Table 9-27. The dimensions given for the fillet radius in this table are related to the taxiway centerline radius. Figures 9-19 and 9-20 show the range of fillet designs which provide an acceptable edge safety margin for a range of aircraft wheelbase and undercarriage dimensions for each airplane design group for both judgmental oversteering and maintaining the cockpit over the centerline tracking methods. If the critical design aircraft falls outside the acceptable safety margin, found from Fig. 9-19 or 9-20, a custom-designed fillet is recommended. ICAO Annex 14 does not specify the minimum radii of fillets, which it terms *extra* taxiway width, but recommends that an analysis be conducted on taxiway turns to ensure that the minimum edge safety margin is attained for the design aircraft. Several methods of conducting this type of analysis are discussed below, and others are discussed in ICAO Annex 14 [4].

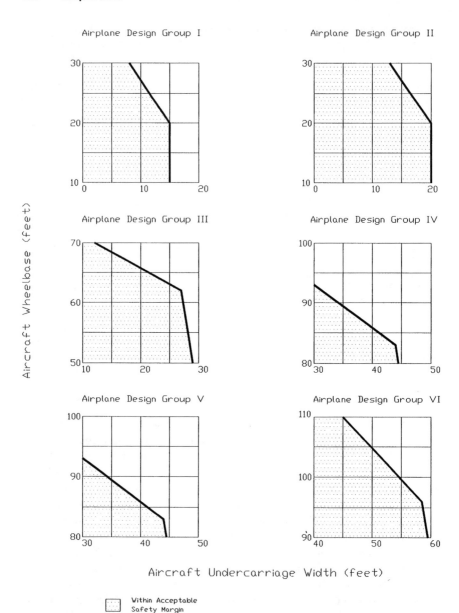

Figure 9-19 Acceptable safety margin for judgmental oversteering (*Federal Aviation Administration [6]*).

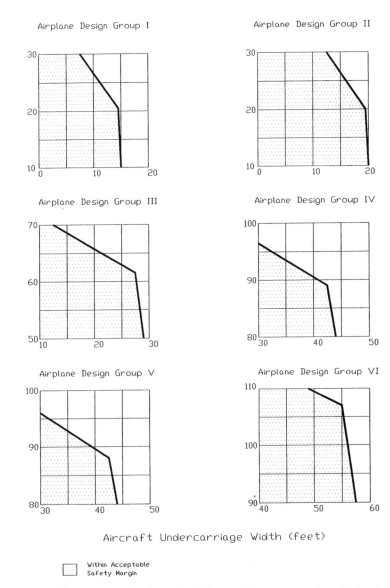

Figure 9-20 Acceptable safety margin for maintaining cockpit over centerline (*Federal Aviation Administration [6]*).

It is not necessary to use the specific dimensions shown in Figs. 9-15 through 9-18 or Table 9-27 for taxiway curves, since there are several methods for determining the path of the main undercarriage gear when the path of the nosewheel is known.

When an aircraft negotiates a turn with the nosewheel tracking a predetermined curved path, such as a taxiway centerline, the midpoint of the main undercarriage does not follow the same path as the nose gear because of the fairly large distance from the nose gear to the main undercarriage. The relationship between the centerline, which is being tracked by the nosewheel, and the position of the main undercarriage is shown in Fig. 9-21. At any point on the curve, the distance between the curved path followed by the nosewheel and the midpoint of the undercarriage of the main landing gear is referred to as the *track-in*. The track-in varies, increasing progressively during the turning maneuver. It decreases as the nose gear begins to follow the tangent to the curve. Given the path of the main gear, the radius of the fillet can be determined by adding an appropriate taxiway edge safety margin S between the outside edge of the tire on the main landing gear closest to the center of the path followed by the nosewheel and the edge of the pavement.

The nosewheel steering angle, or the *castor angle C,* is the angle formed by the longitudinal axis of the aircraft and the direction of movement of the nosewheel, or some other reference point or datum such as the location of the pilot in the cockpit. For preliminary design it is sufficiently accurate to assume that the datum is the nosewheel.

The size of the fillet depends not only on the wheelbase of the aircraft, the radius of the curve, the width of taxiway, and the total change in direction, but also on the path that the aircraft follows on the turn. There are three ways in which an aircraft can be maneuvered on a turn. One is to establish the centerline of the taxiway as the path of the nose gear. This is called *nosewheel-on-centerline* tracking. Another way is to establish the centerline of the taxiway as the path directly beneath the pilot and to assume that this path is fol-

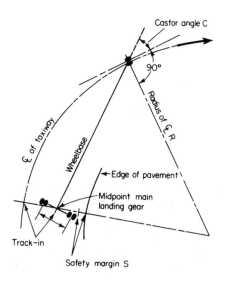

Figure 9-21   Path of main gear on curve.

lowed. This is called *maintaining cockpit over the centerline* tracking. The last way is to assume that the nose gear will follow a path offset outward of the centerline. This is called *judgmental oversteering* tracking. It will result in the least amount of taxiway widening but results in greater runway occupancy time and the possibility of pilot error in judging the turn to follow. While there is no agreement on which procedure is desirable, usually maintaining the cockpit over the centerline is preferred.

One method for designing fillets had its origin in the United Kingdom [4, 14], and although it is an approximation, it is sufficiently accurate for the purpose of preliminary design. The subject has also been given attention in Australia [11] and the United States [6, 7, 8, 12]. This method uses Figs. 9-22 through 9-25 to enable fillets for any aircraft to be designed in a relatively simple manner. The term *aircraft datum length* on these figures is the distance from the datum which is tracking the centerline to the midpoint of the main landing gear. For the preliminary design of fillets, it is usually sufficiently accurate to assume that the aircraft datum length is equal to the wheelbase of the aircraft. The maximum track-in, as a percentage of the datum length, is obtained from Fig. 9-22. It depends upon the datum length, radius of curvature of the centerline, and change in direction of the aircraft between tangent sections connected by the centerline curve. The maximum castor angle which must be attained by the nosewheel to negotiate the turn is obtained from Fig. 9-23. Care should be taken to ensure that the maximum castor angle does not exceed the normal operating limits of the aircraft. Also if the maximum castor angle necessary to complete the turn exceeds 50°, a larger-radius centerline curve should be used to eliminate excessive turning of the aircraft nose gear.

The castor angle of the nosewheel at the point where taxiway widening is no longer required is obtained from Fig. 9-24. The distance the nose gear has to travel along the tangent section on the centerline at the end of the curve to reduce the castor angle at the end of the turn to that at which taxiway widening is no longer required, is obtained from Fig. 9-25.

The use of this method is illustrated by the following example problem.

Example Problem 9-4 It is necessary to determine the size of the fillet for maneuvering a DC-10 with a wheelbase of 72.5 ft and a wheel tread of 35 ft on a taxiway curve. The taxiway changes direction by 90° and has a centerline radius of 250 ft and a width of 75 ft. The required taxiway edge safety margin is 15 ft.

The ratio of the taxiway centerline radius to the aircraft datum length R/D is $250/72.5 = 3.45$. Entering Fig. 9-22 with $R/D = 3.45$ and a 90° change in direction yields a maximum track-in of 14.5 percent of the aircraft datum length, or

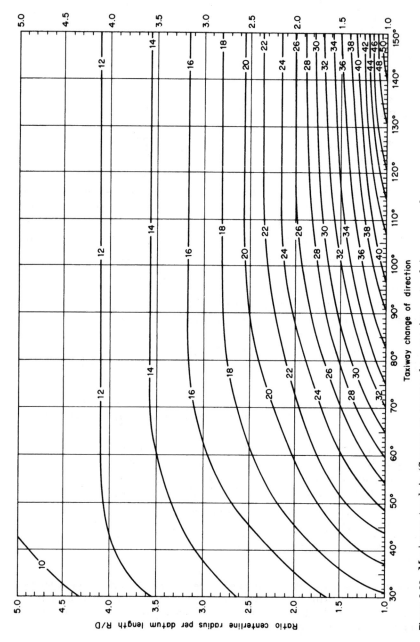

Figure 9-22 Maximum track-in (figures on curves show maximum track-in expressed as percentage of datum length) (*International Civil Aviation Organization and United Kingdom [4, 14]*).

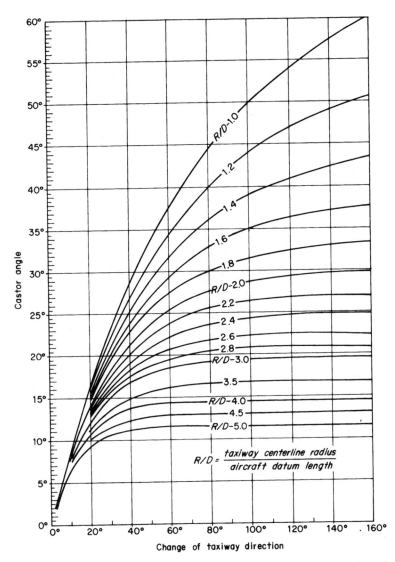

Figure 9-23 Increase of castor angle during a turn-in (*International Civil Aviation Organization and United Kingdom [4, 14]*).

$72.5 \times 0.145 = 10.5$ ft. If no widening is provided, the maximum allowable track-in on a taxiway is one-half the taxiway width minus the sum of one-half the wheel tread and the taxiway edge safety margin. Therefore, for this taxiway, the maximum allowable track-in T_{max} becomes

$$T_{max} = 0.5 \times 75 - (0.5 \times 35 + 15) = 5 \text{ ft}$$

From Fig. 9-24 this is equivalent to a maximum castor angle of 4°. Since the actual track-in of 10.5 ft exceeds the maximum allowable track-in T_{max}, the

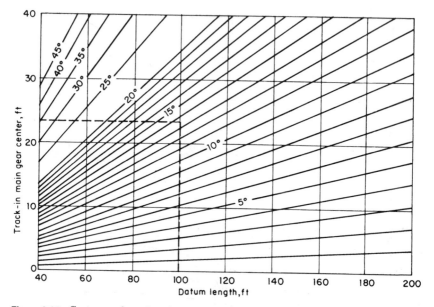

Figure 9-24 Castor angle and main gear center track-in (figures on curve show castor angles) (*International Civil Aviation Organization and United Kingdom [4, 14]*).

actual castor angle which must be attained by this aircraft in completing this turn is greater than 4°, and a fillet must be provided on the inside edge of the taxiway pavement. The required radius of the fillet F is equal to the centerline radius minus the sum of the actual track-in, one-half the wheel tread, and the required taxiway edge safety margin. Therefore, we have

$$F = 250 - [10.5 + 0.5(35) + 15] = 207 \text{ ft}$$

Figure 9-23 yields 17° as the castor angle attained at the end of the turn. The castor angles are converted to travel along a tangent section beyond the end of the curve by use of Fig. 9-25. A castor angle of 4° yields a distance of 205 ft, and a castor angle of 17° yields a distance of 100 ft. The difference between these distances, or 105 ft, is the required distance of travel of the nosewheel gear to reduce the castor angle from 17°, the castor angle that results in maximum track-in, to 4°, the castor angle that results in a track-in which is within the taxiway edge safety margin. The distance that the main wheels extend beyond the end of the curve is obtained by subtracting the aircraft datum length from this difference. That is, the distance L beyond the end of the curve at which no fillet is required is

$$L = 105 - 72.5 = 32.5 \text{ ft}$$

The results of these computations are shown in Fig. 9-26.

A simple graphical solution of sufficient accuracy is described in Refs. 4 and 7. The method consists of the following steps, which are shown in Fig. 9-27:

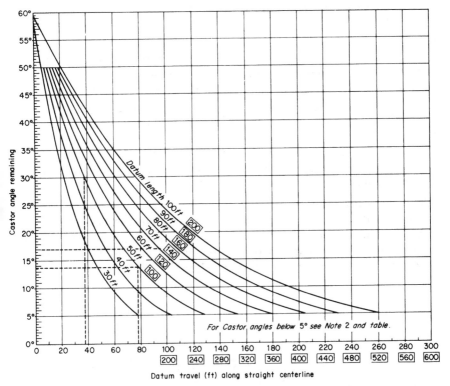

Figure 9-25 Decrease in castor angle during a turn (*International Civil Aviation Organization and United Kingdom [4, 14]*).

1. Draw the path followed by the nosewheel to some convenient scale. The scale should be large enough to obtain good accuracy.

2. Set a compass so that it represents the wheelbase of the aircraft at the scale chosen.

3. Mark the initial position of the wheelbase on the guiding line (such as M_1N_1 in the illustration).

4. Place the compass point at M_2, a short distance from M_1, and mark point N_2 on the guiding line. Distance M_1N_2 represents the first increment of the main gear movement, at the end of which the nosewheel is at N_2.

5. Place a straightedge in line with points M_2 and N_2. Now move the compass point a short distance along the straightedge to M_3 and mark N_3, the new position of the nosewheel on the path followed by the nosewheel.

6. Line up the straightedge with points M_3 and N_3, shift the compass point another short distance along the straightedge to N_4, and mark the corresponding position of the nosewheel on the path followed by the nosewheel.

7. Repeat this procedure until the path of the main gear midpoint, defined by M_1, M_2, M_3,..., is established for the desired distance. Note that in the initial stage of the turn the increments of main gear movement may be made reasonably large without appreciable loss of accuracy, but that shorter increments are desirable at later stages of the turn to limit the change in direction of the aircraft axis between successive positions of the nosewheel. After experience with the method, the size of the increments to yield reasonable accuracy becomes apparent.

8. Draw a curve through points M_1, M_2, M_3,..., to represent the path of the main gear midpoint. This in turn is superimposed on a taxiway design to establish the desired fillet dimensions.

The FAA has adopted a very precise method for the design of curves for runway exits and taxiway turns [6]. The method allows one to design for judgmental oversteering or maintaining the cockpit over the centerline. The method presented here treats only the case of maintaining the cockpit over the centerline.

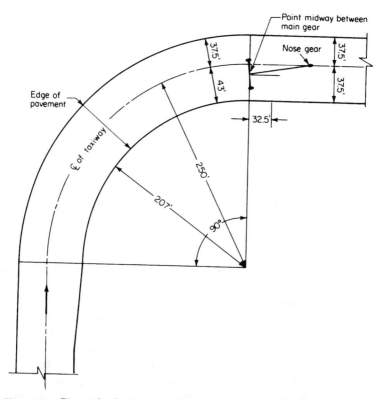

Figure 9-26 Pictorial solution of turn problem for Example Problem 9–4.

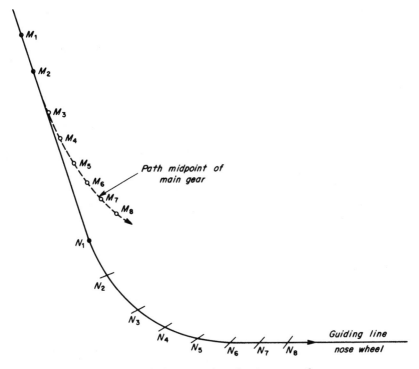

Figure 9-27 Graphical method of construction of main gear path.

The principal dimensions of the aircraft related to tracking a curve are given in Fig. 9-21. The geometry of the aircraft tracking the centerline curve from the point of curvature PC to the point of tangency PT and the various terms used in the equations below to define the movement of the aircraft are given in Fig. 9-28.

The maximum angle formed between the tangent to the centerline and the longitudinal axis of the aircraft will occur at the end of the curve when the nosewheel is at the point of tangency. This angle, called A_{max}, may be approximated by

$$A_{max} = \sin^{-1} \frac{d}{R} \tag{9-10}$$

where d is the distance from the nosewheel or the pilot cockpit position to the center of the main undercarriage (the wheelbase of the aircraft is often used to approximate this distance) and R is the radius that the nosewheel or the pilot is tracking on the curve.

The maximum nosewheel steering angle, the castor angle, the angle between the longitudinal axis on the nose gear and the longitudinal axis of the aircraft B_{max} is given by

$$B_{max} = \tan^{-1}\left(\frac{w}{d}\tan A_{max}\right) \qquad (9\text{-}11)$$

where w is the wheelbase of the aircraft.

The required fillet radius F is given by

$$F = (R^2 + d^2 - 2Rd \sin A_{max})^{0.5} - 0.5u - M \qquad (9\text{-}12)$$

where u is the undercarriage width, i.e., the distance between the outside tires on the main gear, and M is the minimum distance required between the edge of the outside tire and the edge of the pavement, i.e., the edge safety margin.

The length of the lead-in to the fillet is given by

$$L = d \ln\left(\frac{4d \tan 0.5A_{max}}{W - u - 2M}\right) - d \qquad (9\text{-}13)$$

where W is the taxiway width on the tangent, and ln represents the natural logarithm.

These equations may be solved for a given aircraft tracking a curve of radius R to find the necessary lead-in and fillet radius to maintain a minimum edge safety margin between the tire and the pavement edge. If the value of the maximum nosewheel steering angle B_{max} exceeds 50°, it is recommended that the radius of the centerline curve R which the nosewheel is tracking be increased.

The use of these equations in determining the critical taxiway curve design parameters is shown in Example Problem 9-5.

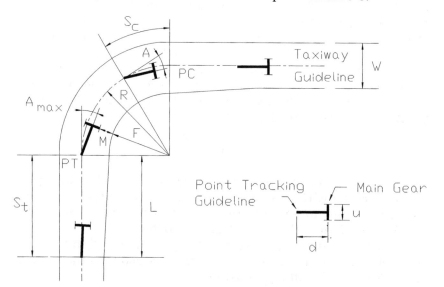

Figure 9-28 Taxiway fillet design geometry.

Example Problem 9-5 Determine the minimum lead-in and the radius of the fillet to maintain the cockpit over the centerline for an aircraft with a 156.1-ft wingspan, a wheelbase of 64.6 ft, an undercarriage width of 34.25 ft, and the distance between the main undercarriage and cockpit equal to 72.1 ft. The aircraft is moving between two parallel taxiways through a connecting taxiway which has a centerline perpendicular to the parallel taxiways.

With a wingspan of 156.1 ft, this aircraft is in airplane design group IV. From Table 9-27, the taxiway width is 75 ft, the minimum safety margin is 15 ft, the recommended centerline radius is 150 ft, the recommended length of the fillet lead-in is 250 ft and the recommended fillet radius is 85 ft.

To verify that these recommended values are acceptable, we can use Eqs. (9-10) through (9-13). From Eq. (9-10) we have

$$A_{max} = \sin^{-1} \frac{d}{R}$$

$$= \sin^{-1} \frac{72.1}{150} = 29°$$

From Eq. (9-11) we have

$$B_{max} = \tan^{-1} \left(\frac{w}{d} \tan A_{max} \right)$$

$$= \tan^{-1} \left(\frac{64.5}{72.1} \tan 29° \right) = 27°$$

Since this is less than 50°, the radius is adequate.

From Eq. (9-12) we have

$$F = (R^2 + d^2 - 2Rd \sin A_{max})^{0.5} - 0.5u - M$$
$$= [150^2 + 72.1^2 - 2(150)(72.1) \sin 29°]^{0.5} - 0.5(34.25) - 15$$
$$= 100.1 \text{ ft}$$

From Eq. (9-13) we have

$$L = d \ln \left(\frac{4d \tan 0.5A_{max}}{W - u - 2M} \right) - d$$

$$= 72.1 \ln \frac{4(72.1)(\tan 14.5°)}{75 - 34.25 - 2(15)} - 72.1$$

$$= 68 \text{ ft}$$

Therefore both the required fillet radius and the required length of the lead-in to the fillet for this specific aircraft are well within those recommended for the airplane design group to which this aircraft is assigned.

The computer program airport design 3.2 [6] allows the above equations to be solved for any type of runway exit or taxiway intersection. The program allows the designer to specify the characteristics of the critical design aircraft and the runway exit or taxiway intersection design. It computes and graphically displays the path followed by the outer edge of the wheel on the main undercarriage. The designer may specify the geometry of the runway exit in terms of tangent sections, single or compound circular curve sections, and spiral sections.

This computer program may also be used to design the taxiway curve computed in Example Problem 9-5. This is illustrated now.

Example Problem 9-6   Verify, using the airport design computer program, the minimum lead-in and the radius of the fillet required to maintain the cockpit over the centerline for an aircraft in Example Problem 9-5.

As before, from Table 9-27, the taxiway width is 75 ft, the minimum safety margin is 15 ft, the recommended centerline radius is 150 ft, the recommended length of the fillet lead-in is 250 ft and the recommended fillet radius is 85 ft.

The taxiway centerline curve will be designed with a tangent section beginning 250 ft from the point of curvature, followed by a circular curve of radius 150 ft turning through 90°, followed by another circular curve of radius 150 ft turning through an angle of 90°, and completed with another tangent section ending 250 ft from the point of tangency. The centerline curve geometry is shown in Fig. 9-29.

To provide a minimum safety margin of 15 ft from the edge of the main undercarriage gearwheel to the edge of the pavement, a minimum taxiway pavement width equal to one-half the undercarriage width plus 15 ft, or 32.12 ft, is required. This is, then, the minimum offset from the centerline of the taxiway.

The abridged results of using the computer program for the design of this taxiway curve are shown in Table 9-28. From this table we observe that the main undercarriage of the aircraft begins to track in from the centerline when the main undercarriage reaches a point 180 ft from the point of beginning of the section (POB). When the main undercarriage reaches the point of compound curvature (PCC), the minimum offset from the centerline of the taxiway to the pavement is 49.85 ft. Therefore, at this point the main undercarriage has tracked in a distance of 49.85 − 32.12 = 17.73 ft. It continues to track in until the main undercarriage reaches a point of 650.00 ft from the beginning of the section. At this point the minimum offset from the centerline of the taxiway to the edge of the pavement must be 50.49 ft, and the main undercarriage has tracked in a distance of 50.49 − 32.12 = 18.37 ft from the taxiway centerline at this point. From this point forward, the main undercarriage moves back toward the centerline of the taxiway. At the point of tangency PT, the maximum steering angle of 28.646° is reached. The main undercarriage continues to track back toward the centerline until it reaches a point 900 ft from the beginning of the section. From this point on, the main undercarriage will travel parallel to the centerline. From the point of tangency the steering angle is reduced until at the point of ending of the section (POE) the steering angle is less than 1°.

Based upon these results, the minimum fillet radius is equal to the radius of the taxiway centerline minus the maximum offset distance, or 150 − 50.49 = 100 ft. The minimum lead-in taper to the point of curvature is 250 − 180 = 70 ft. Both these values agree well with the results found in Example Problem 9-5. The minimum lead-out from the point of tangency to the point where the main undercarriage moves parallel to the taxiway centerline is 900 − 721.24 = 180 ft.

Therefore, the specifications for this curve requiring a minimum fillet radius of 85 ft and a minimum lead-in length of 250 ft more than satisfy the requirements for this aircraft.

If the taxiway were redesigned with a fillet radius of 100 ft and a lead-in distance of 180 ft, as shown in Fig. 9-30, this would meet the needs of this specific

TABLE 9-28   Minimum Offset Distances* from Taxiway Centerline to Left and Right Edges of Taxiway on a Taxiway Curve for Example Problem 9-6, ft

	Taxiway centerline		
Distance from beginning of section†	Left offset distance	Right offset distance	Steering angle, deg
POB 0.000	32.12	32.12	00.000
50.000	32.12	32.12	00.000
100.000	32.12	32.12	00.000
150.000	32.12	32.12	00.000
170.000	32.12	32.12	00.000
180.000	32.13	32.12	00.000
230.000	34.18	30.41	00.000
PC 250.000	37.46	27.90	00.000
300.000	43.51	21.07	13.799
350.000	46.74	17.60	20.827
400.000	48.50	15.78	24.495
450.000	49.45	15.00	26.445
PCC 485.619	49.85	15.00	27.253
500.000	49.97	15.00	27.492
550.000	50.25	15.00	28.057
600.000	50.41	15.00	28.364
650.000	50.49	15.00	28.530
700.000	48.87	15.77	28.621
PT 721.238	46.44	19.00	28.646
750.000	41.53	23.25	19.444
800.000	36.74	27.64	9.789
850.000	34.41	29.87	4.902
900.000	32.20	32.20	2.451
950.000	32.20	32.20	1.225
POE 971.238	32.20	32.20	0.913

*The offset distance is a perpendicular distance measured from the taxiway centerline. The hard surface needs to be widened at stations where the offset distance extends beyond the pavement surface, i.e., when the offset distance is more than half the taxiway width or 32.50 ft.

†POB = point of beginning; PC = point of curvature; PCC = point of compound curve; PT = point of tangency; POE = point of ending.

SOURCE: Federal Aviation Administration [6].

aircraft, require about 6100 ft² less of pavement area (shown by the shaded area in this figure), and result in construction cost savings.

Observe that the minimum distance between the taxiway centerlines for the physical accommodation of either of these exit designs is 300 ft, which is considerably more than the minimum distance of 215 ft specified in Table 9-17 for aircraft in design group IV.

The FAA recommends that an entrance spiral be used on runway exits to taxiways to decrease the amount of lead-in to fillets. For a 90° regular exit, it recommends at least a 300-ft entrance spiral from the point of curvature to the 30° deflection point. For a 30° high-speed exit, the FAA recommends a 1400-ft entrance spiral.

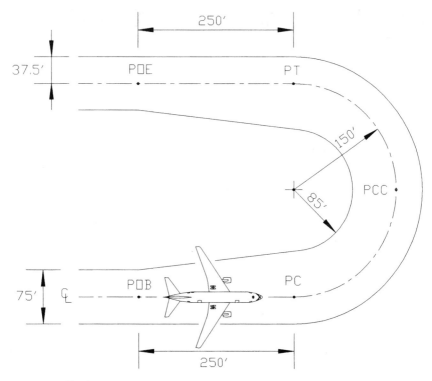

Figure 9-29 Taxiway geometry for Example Problem 9-6.

Aprons

Holding aprons

Holding aprons, holding pads, run-up pads, or holding bays, as they are sometimes called, are placed adjacent to the ends of runways. The areas are used as storage areas for aircraft prior to takeoff. They are designed so that one aircraft can bypass another whenever necessary. For piston-engine aircraft, the holding apron is an area where the aircraft instrument and engine operation can be checked prior to takeoff. The holding apron also allows a trailing aircraft to bypass a leading aircraft if the takeoff clearance of the latter must be delayed for some reason or if it experiences some malfunction. There are many configurations of holding aprons; two are shown in Fig. 9-31. The important design goals are to provide adequate space for aircraft to maneuver easily onto the runway irrespective of the position of adjacent aircraft on the holding apron and to provide sufficient room for an aircraft to bypass parked aircraft on the holding apron. The recommendations for the minimum separation between aircraft on holding aprons are

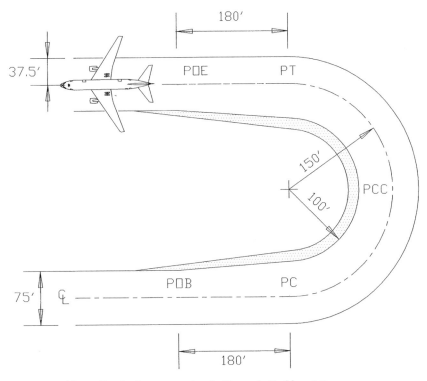

Figure 9-30 Alternative taxiway geometry for Example Problem 9-6.

the same as those specified for the taxiway object-free area. Space for four aircraft is usually provided.

Even when holding aprons are provided, often aircraft must queue on the taxiway leading to the end of the runway, since it may not be practical in most cases to construct holding aprons large enough to accommodate peak demands. Studies have been conducted for O'Hare International Airport to evaluate the potential for installing large flow-through holding pads at the departure end of several runways, to alleviate both airfield congestion and taxiway gridlock and to provide some additional ATC flexibility in maneuvering aircraft awaiting takeoff clearance [20]. The potential locations studied for these holding pads are shown in Fig. 9-32. The design of a typical flow-through holding pad studied is shown in Fig. 9-33. Holding pads must be designed for the largest aircraft which will use the pad. The holding pad should be located so that all aircraft using the pad will be located outside both the runway and the taxiway object-free area and in a position so as to not interfere with critical instrument landing system (ILS) signals.

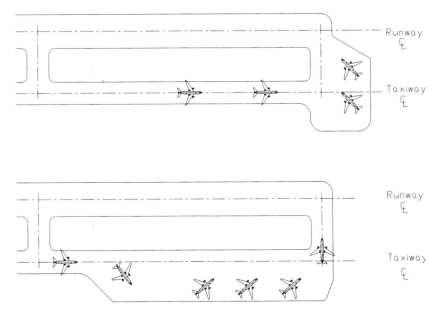

Figure 9-31 Typical holding pad configurations.

Terminal aprons and ramps

Aircraft parking positions, also called aircraft *gates* or aircraft *stands,* on the terminal apron or ramp are closely related to the terminal design concept used, as discussed subsequently in Chap. 10. These parking positions are accessed through taxilanes which run through the terminal apron area to the aircraft gates. The parking positions must be sized for the geometric properties of the aircraft, including wingspan, fuselage length, and turning radii, and for the requirements for aircraft access by the vehicles servicing the aircraft at the gates. Both the FAA and the ICAO recommend the minimum clearances between any part of an aircraft and other aircraft or structures in the apron area given in Table 9-29.

Example Problem 9-7 illustrates the determination of the terminal apron requirements for aircraft.

Example Problem 9-7   Design a terminal apron with two parallel concourses to accommodate gates for one wide-bodied aircraft and three narrow-bodied aircraft on the face of each of the concourses. The gate design aircraft for the wide-bodied gates is the Boeing 767-200, and the gate design aircraft for the narrow-bodied gates is the McDonnell-Douglas MD-87. Aircraft will park nose-in at each gate and will use the gates in a power-in, push-out mode of operation.

The fuselage length and wingspan of these aircraft can be found in Table 3-1. From this table, the Boeing 767-200 has a fuselage length of 159 ft 2 in and a wingspan of 156 ft 1 in, which places it in airplane design group IV, and the McDonnell-Douglas MD-87 has a fuselage length of 130 ft 5 in and a wingspan of 107 ft 10 in, which places it in airplane design group III.

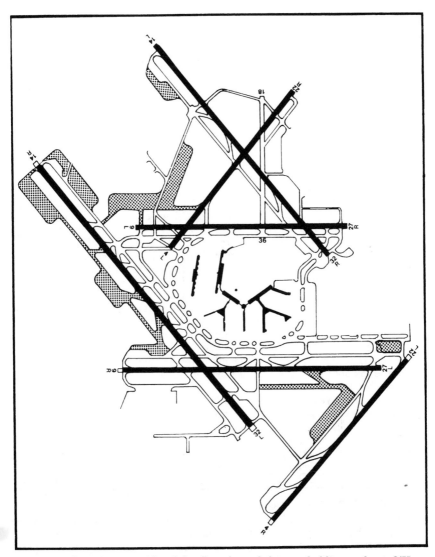

Figure 9-32 Locations considered for flow-through bypass holding pads at O'Hare International Airport (*Landrum and Brown, Inc.*).

If the aircraft are arrayed at the concourses as shown in Fig. 9-34, then the size of the terminal apron and the size of each gate position may be determined by referencing the specifications requiring specific separations between aircraft operating on the taxilanes and between aircraft parked at the concourse gates. In this problem the FAA specifications are used, and the design is based upon the actual dimensions of the aircraft rather than upon the dimensions of the largest aircraft in the airplane design groups to which these aircraft are assigned. Therefore, the relevant separations are contained in the footnotes in Table 9-17.

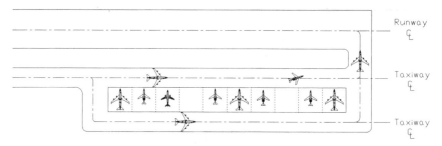

Figure 9-33 Flow-through bypass holding pad.

The Boeing 767-200 is the greater-wingspan aircraft and, therefore, the most demanding aircraft for taxilane dimensions. The separation between taxilane centerlines is equal to 1.1 times the wingspan of the most demanding aircraft plus 10 ft. So the recommended separation equals $1.1 \times 156.1 + 10 = 182$ ft. A ground access vehicle lane will be provided behind each aircraft for the use of aircraft service vehicles. This lane will be 12 ft wide. The distance from the centerline of each taxilane to a fixed or movable object is equal to 0.6 times the wingspan plus 10 ft. Therefore, this distance is $0.6 \times 156.1 + 10 = 104$ ft. Considering the ground access vehicle lane, the distance from the centerline of each taxilane to the tail of the aircraft is equal to $104 + 12 = 116$ ft.

Table 9-29 indicates that the recommended clearance between the face of each concourse and the nose of the 767-200 aircraft is 25 ft, since this aircraft is in airplane design group IV. The length of the 767-200 is 159 ft 2 in and therefore the distance from the face of each concourse to the tail of the aircraft is equal to $25 + 159.2 = 184$ ft.

By considering each of these recommended separations, the width of the terminal apron or ramp between the concourses is found to be 782 ft, as shown in Fig. 9-34.

The clearance between aircraft wing tips or between the aircraft wing and fixed or movable objects is recommended to be 0.1 times the wingspan plus 10 ft. Therefore, the Boeing 767-200 requires a clearance between its wing tips and fixed or movable objects of 0.1 times the wingspan plus 10 ft, or $0.1 \times 156.1 + 10 = 25.6$ ft. This results in a parking position for this aircraft that is $156.1 + 25.6 = 182$ ft wide by 182 ft long. The MD-87 requires a clearance of $0.1 \times 107.84 + 10 = 20.8$ ft. This results in a parking position for this aircraft that is $107.84 + 20.8 = 129$ ft wide. From Table 9-29 this aircraft may be parked as close as 15 ft to the concourse since it is in airplane design group III. So a parking position for the MD-87 is $130.42 + 15 = 146$ ft long. However, since the parking position for the 767-200 is longer, the actual length provided for the MD-87 is 182 ft.

For aircraft to be moved from their parking positions onto the apron taxilane, a separation from the inside edge of the last gate position to the terminal building equal to the fuselage length plus 0.1 times the wingspan plus 10 ft is required. This should allow the aircraft to turn onto the centerline of the taxilane from its gate position and allow another aircraft to gain access to the gate position before the last aircraft at that gate leaves the apron area. The parking arrangement shown in Fig. 9-34 shows the MD-87 as the closest aircraft to the terminal building. Therefore, the distance from the centerline of the last gate position to the terminal building is $130.42 + 54 + 0.1 \times 107.83 + 10 = 206$ ft.

TABLE 9-29 Minimum Clearance between Aircraft and Fixed or
Movable Objects at Terminal Apron Parking Positions

Airplane design group or aerodrome code		Minimum clearance*	
		ft	m
I	A	10	3.0
II	B	10	3.0
III	C	15	4.5
IV	D	25	7.5
V or VI	E	25	7.5

*The FAA recommends the wing-tip separation at parking positions to be
0.1 times the wingspan plus 10 ft.

SOURCES: Federal Aviation Administration [6, 19] and International Civil
Aviation Organization [4].

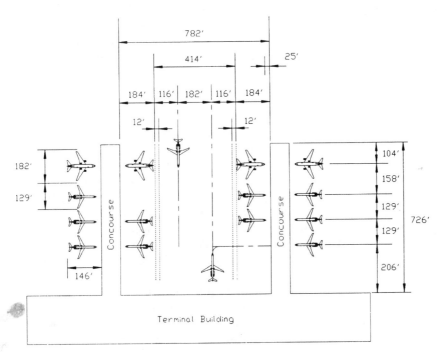

Figure 9-34 Terminal apron requirements for Example Problem 9-7.

The design must ensure that the clearances provided are adequate, based
upon the turning capability of the design aircraft. The analyst should verify, by
using the procedures discussed earlier in this chapter, that the tracking in of
the aircraft while making a turn from the taxilanes to the parking position will
not compromise safe clearances between aircraft. The analyst must also verify
for each aircraft that in turning to access gate positions no part of the aircraft
will compromise these safe clearances.

Using the above dimensions gives a ramp length of 726 ft. The dimensions for all aircraft parking positions or gates are shown in Fig. 9-34.

It is interesting that the terminal apron area for these eight aircraft equals $782 \times 726 = 567,800$ ft^2, which is equal to about 13 acres, or a total ramp area per aircraft of about 1.6 acres. A gate for a Boeing 767-200 is $184 \times 182 = 33,500$ ft^2, which is about 0.77 acre. The required gate for an MD-87 is $129 \times 146 = 18,800$ ft^2, which is slightly more than 0.4 acre. Useful rules of thumb which allow one to estimate the terminal apron gate requirements are a total ramp area of 1.5 to 2.0 acres per aircraft gate, an area of from 0.75 to 1.0 acre per gate for a wide-bodied aircraft gate position, and an area of about 0.5 acre per gate for a narrow-bodied aircraft gate position.

Terminal apron surface gradients

For fueling, ease of towing, and aircraft taxiing, apron slopes or grades should be kept to the minimum consistent with good drainage requirements. Slopes should never exceed 2 percent for utility airports and 1 percent for transport airports. At gates where aircraft are being fueled, every effort should be made to keep the apron slope within 0.5 percent.

Control Tower Visibility Requirements

At airports with a permanent ATC tower, the runways and taxiways must be located and oriented so that a clear line of sight is maintained to all traffic patterns, the final approaches to all runways, all runway structural pavements, all apron taxiways, and other operational surfaces controlled by the ATC tower. A clear line of sight to all taxilane centerlines is desirable. Operational surfaces not having a clear, unobstructed line of sight from the tower are designated as uncontrolled or nonmovement areas. At airports without a permanent ATC tower, the runways and taxiways should be located and oriented so that a future tower may be sited in accordance with the continuous-visibility requirements. This requirement may be satisfied where adequate control of aircraft exists by other means. [6]

A typical ATC tower site requires between 1 and 4 acres of land. The site must be large enough to accommodate current and future building needs including employee parking. Tower sites must afford maximum visibility to traffic patterns and clear, unobstructed, and direct lines of sight to the runway approaches, the landing area, and all runway and taxiway surfaces. Most towers penetrate the FAR part 77 surfaces and thus are obstructions to aviation and may be a hazard to air navigation unless an FAA study determines otherwise. The tower must not derogate the signal generated by any existing or planned electronic navigational aid or ATC facility.

References

1. *Advisory Circular Checklist,* Advisory Circular AC 00-2.6, Federal Aviation Administration, Washington, October 15, 1992, annual.
2. *Aerodromes, Annex 14 to the Convention on International Civil Aviation,* vol. 1: *Aerodrome Design and Operations,* 1st ed., International Civil Aviation Organization, Montreal, Canada, July 1990.
3. *Aerodrome Design Manual,* pt. 1: *Runways,* 2d ed., Doc. 9157-AN/901, International Civil Aviation Organization, Montreal, Canada, 1984.
4. *Aerodrome Design Manual,* pt. 2: *Taxiways, Aprons and Holding Bays,* 2d ed., International Civil Aviation Organization, Montreal, Canada, 1983.
5. *Airport Capacity and Delay,* Advisory Circular AC 150/5060-5, Federal Aviation Administration, Washington, 1983.
6. *Airport Design,* Advisory Circular AC 150/5300-13, Federal Aviation Administration, Washington, 1989.
7. *A Mathematical Model for Locating Exit Taxiways,* R. Horonjeff et al., Institute of Transportation and Traffic Engineering, University of California, Berkeley, 1959.
8. "Calculation of Aircraft Wheel Paths and Taxiway Fillets," J. W. L. van Aswegen, Graduate Student Report, Institute of Transportation and Traffic Engineering, University of California, Berkeley, July 1973.
9. "Characteristics of High Speed Runway Exit for Airport Design," A. A. Trani, A. G. Hobeika, B. J. Kim, H. Tomita, and D. Middleton, *International Air Transportation,* Proceedings of the 22d Conference on International Air Transportation, American Society of Civil Engineers, New York, 1992.
10. *Criteria for Approving Category I and Category II Landing Minima for FAR Part 121 Operators,* Advisory Circular AC 120-29 including Changes 1 through 3, Federal Aviation Administration, Washington, 1974.
11. "Determination of the Path Followed by the Undercarriage of a Taxiing Aircraft," Paper prepared by Department of Civil Aviation, Melbourne, Australia.
12. "Determination of Wheel Trajectories," E. Hauer, *Transportation Engineering Journal,* vol. 96, no. TE4, November 1970.
13. *Exit Taxiway Location and Design,* R. Horonjeff et al., Report prepared for the Airways Modernization Board by the Institute of Transportation and Traffic Engineering, University of California, Berkeley, 1958.
14. "Movement of Aircraft and Vehicles on the Ground—Taxiway Fillets," Working Paper no. 102 (prepared by the United Kingdom), 5th Air Navigation Conference of the International Civil Aviation Organization, Montreal, Canada, October 1967.
15. *Optimization of Runway Exit Locations,* E. S. Joline, R. Dixon Speas and Associates, Manhasset, N.Y.
16. *Report of the Department of Transportation Air Traffic Control Advisory Committee,* vols. 1 and 2, Washington, December 1969.
17. *Runway Length Requirements for Airport Design,* Advisory Circular AC 150/5325-4A with Change 1, Federal Aviation Administration, Washington, 1990.
18. *Supporting Documentation for Technical Report on Airport Capacity and Delay Studies,* Rep. No. FAA-RD-76-162, Federal Aviation Administration, Washington, 1976.
19. *The Apron and Terminal Building Planning Report,* Rep. FAA-RD-75-191, Federal Aviation Administration, Washington, 1975.
20. "The Design and Use of Flow-Through Hold Pads," Douglas F. Goldberg, *International Air Transportation,* Proceedings of the 22d Conference on International Air Transportation, American Society of Civil Engineers, New York, 1992.
21. *United States Standard for Terminal Instrument Operations (TERPS),* 3d ed., FAA Order 8260.3B including Changes 1 through 12, Federal Aviation Administration, Washington, July 1976.

Planning and Design of the Terminal Area

The *terminal area* is the major interface between the airfield and the rest of the airport. It includes the facilities for passenger and baggage processing, cargo handling, and airport maintenance, operations, and administration activities. The passenger processing system is discussed at length in this chapter. Baggage processing, cargo handling, and apron requirements are also discussed relative to the terminal system.

The Passenger Terminal System

The passenger terminal system is the major connection between the ground access system and the aircraft. The purpose of this system is to provide the interface between the airport access mode and aircraft; to process the passenger for origination, termination, or continuation of an air transportation trip; and to convey the passenger and baggage to and from the aircraft.

Components of the system

The passenger terminal system has three major components. These components and the activities that occur within them are as follows:

1. The *access interface* where the passenger transfers from the access mode of travel to the passenger processing component. Circulation, parking, and curbside loading and unloading of passengers are the activities that take place within this component.

2. *Processing,* where the passenger is processed in preparation for starting, ending, or continuation of an air transportation trip. The primary activities in this component are ticketing, baggage check-in, baggage claim, seat assignment, federal inspection services, and security.

3. The *flight interface* where the passenger transfers from the processing component to the aircraft. The activities that occur here include assembly, conveyance to and from the aircraft, and aircraft loading and unloading.

A number of facilities are provided to perform the functions of the passenger terminal system. These facilities are indicated for each of the components identified above.

The access interface.   This component consists of the terminal curbs, parking facilities, and connecting roadways that enable originating and terminating passengers, visitors, and baggage to enter and exit the terminal. It includes the following facilities:

1. The enplaning and deplaning curb frontage provides the public with loading and unloading positions for vehicular access to and from the terminal building.

2. The automobile parking facilities provide short-term and long-term parking spaces for passengers and visitors as well as facilities for rental cars, public transit, taxis, and limousine services.

3. The vehicular roadways provide access to the terminal curbs, parking spaces, and public street and highway system.

4. The designated pedestrian walkways for crossing roads include tunnels, bridges, and automated systems that provide access between the parking facilities and the terminal building.

5. The service roads and fire lanes provide access to various facilities in the terminal and to other airport facilities, such as air freight, fuel truck stands, the post office, etc.

The ground access system at an airport is a complex system of roadways, parking facilities, and terminal access curb fronts. This complexity is illustrated in Fig. 10-1, which shows the various ground access system facilities and directional flows at Greater Pittsburgh International Airport.

The processing system.   The terminal is used to process passengers and baggage for the interface with aircraft and ground transportation modes. The processing system includes the following facilities:

1. The airline ticket counters and offices are used for ticket transactions, baggage check-in, flight information, and administrative personnel and facilities.

2. The terminal services space consists of the public and nonpublic areas such as concessions, amenities for passengers and visitors, truck service docks, food preparation areas, and food and miscellaneous storage.

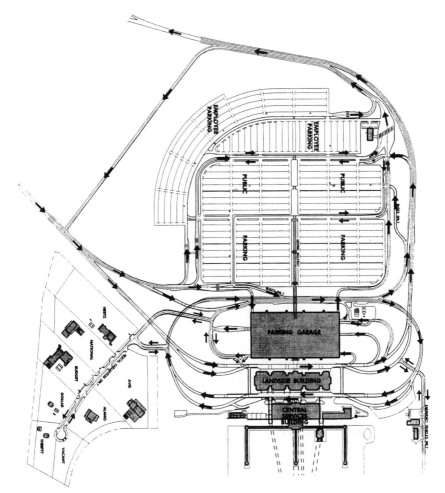

Figure 10-1 Ground access system configuration and directional flows for Greater Pittsburgh International Airport (*Tasso Katselas Associates and Michael Baker Jr., Inc. [32]*).

3. There is a lobby for circulation and passenger and visitor waiting.

4. Public circulation space for the general circulation of passengers and visitors consists of such areas as stairways, escalators, elevators, and corridors.

5. The outbound baggage space is a nonpublic area for sorting and processing baggage for departing flights.

6. The intraline and interline baggage space is used for processing baggage transferred from one flight to another on the same airline or on different airlines.

7. The inbound baggage space is used for receiving baggage from an arriving flight and for delivering baggage to be claimed by the arriving passenger.

8. Airport administration and service areas are used for airport management, operations, and maintenance facilities.

9. The federal inspection service facility is the area for processing passengers arriving on international flights, which is sometimes incorporated as part of the connector element.

The flight interface. The *connector* joins the terminal to parked aircraft and usually includes the following facilities:

1. The concourse provides for circulation to the departure lounges and other terminal areas.

2. The departure lounge or hold room is used for assembling passengers for a flight departure.

3. The passenger boarding device is used to transport enplaning and deplaning passengers between the aircraft door and the departure lounge or concourse.

4. Airline operations space is used for airline personnel, equipment, and activities related to the arrival and departure of aircraft.

5. Security facilities are used for the inspection of passengers and baggage and the control of public access to passenger boarding devices.

6. The terminal services area provides amenities to the public, and nonpublic areas are required for operations such as building maintenance and utilities.

The components of the passenger terminal system together with the corresponding specific physical facilities are shown in Fig. 10-2. The relative locations of the various physical facilities in the three-level landside building of the Midfield Terminal Complex at Greater Pittsburgh International Airport are shown in Figs. 10-3, 10-4, and 10-5. Figure 10-3 shows the enplaning roadway interface with the departure or ticketing level. This level also provides access to the commuter aircraft departure lounge. Figure 10-4 shows the transit level which provides airline baggage makeup space, passenger security processing, and access to the automated transit system, the interface between the landside building and the airside building. Figure 10-5 shows the deplaning roadway interface with the arrivals or baggage claim level, and it contains baggage claim facilities and rental-car facilities as well as a central service building with airport police offices and utility, maintenance, and storage space. The concourse

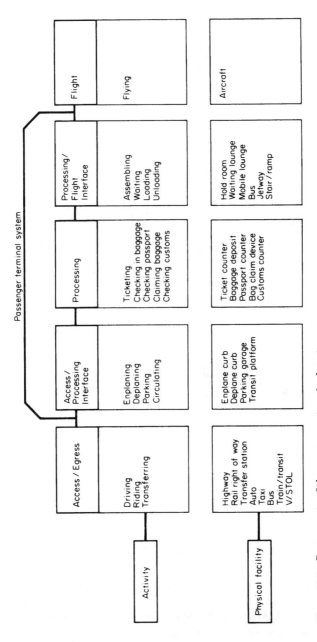

Figure 10-2 Components of the passenger terminal system.

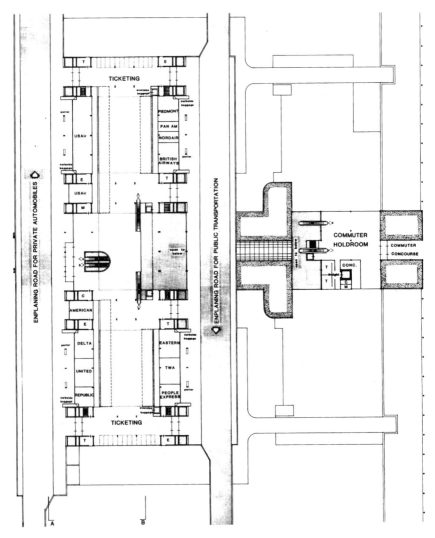

Figure 10-3 Landside building enplaning level at Greater Pittsburgh International Airport (*Tasso Katselas Associates and Michael Baker Jr., Inc. [32]*).

level of the airside building is shown in Fig. 10-6. On this level there is a common core with passenger amenities, and four piers providing the departure lounges and boarding devices at the gates providing the interface with aircraft. One of the four piers is for international arrivals and contains the sterile areas for customs and immigration functions required for international passenger processing. The apron level of the airside building is used for airline operations, and the lower level provides access to automated transit facility.

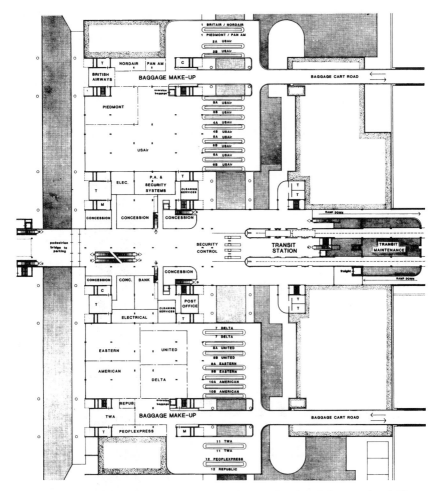

Figure 10-4 Landside building transit level at Greater Pittsburgh International Airport (*Tasso Katselas Associates and Michael Baker Jr., Inc. [32]*).

Design Considerations

In developing criteria for the design of the passenger terminal complex, it is important to realize that a number of different factors enter into a statement of overall design objectives. From these factors general and specific goals are established which set the framework on which design progresses. For example, in designing modifications to the apron and terminal complex at Geneva Intercontinental Airport, the general design objectives included the following [25]:

1. Development and sizing to accomplish the stated mission of the airport within the parameters defined in the master plan

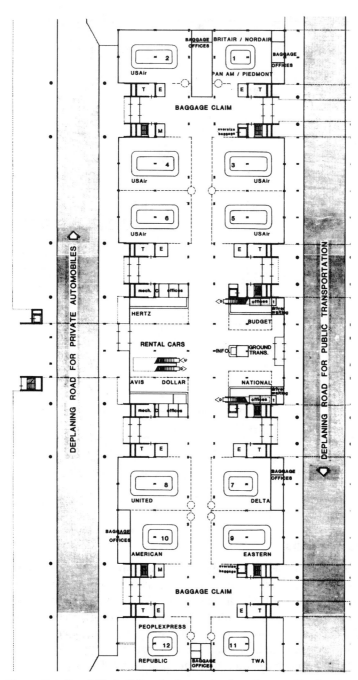

Figure 10-5 Landside building deplaning level at Greater Pittsburgh International Airport (*Tasso Katselas Associates and Michael Baker Jr., Inc. [32]*).

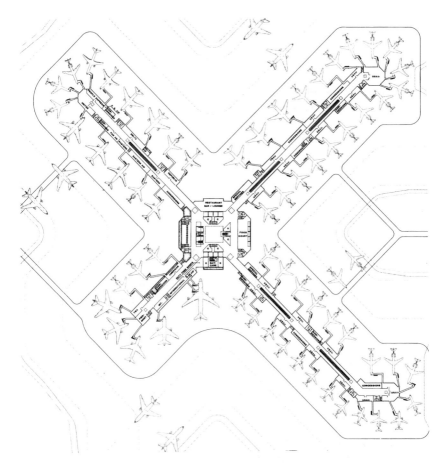

Figure 10-6 Airside building concourse level at Greater Pittsburgh International Airport (*Tasso Katselas Associates and Michael Baker Jr., Inc. [32]*).

2. Capability to meet the demands for the medium- and long-run time frames

3. Functional, practical, and financial feasibility

4. Maximization of use of existing facilities

5. Achievement of a balanced flow between access, terminal, and airfield facilities during the peak hour

6. Consideration of environmental sensitivity

7. Maintenance of the flexibility to meet future requirements beyond the current planning horizon

8. Capability to anticipate and implement significant improvements in aviation technology

Specific design objectives were derived from these general objectives which included the needs of the various categories of airport users. These included the following:

1. Passenger objectives
   a. Responsiveness to the needs of the people relative to convenience, comfort, and personal requirements
   b. Provision of effective passenger and access orientation through concise, comprehensive directional graphics
   c. Separation of enplaning and deplaning roadways and curb fronts to ensure maximum operational efficiency
   d. Provision of convenient access to public and employee parking facilities, rental-car areas, ancillary facilities, and on-site non-aviation facilities

2. Airline objectives
   a. Accommodation of existing and future aircraft fleets with maximum operational efficiency
   b. Provision of direct and efficient means of passenger and baggage flow for all passengers, including domestic and international originating, terminating, and transfer passengers
   c. Provision for economic, efficient, and effective security
   d. Provision of facilities which will embrace the latest energy conservation measures

3. Airport management objectives
   a. Maintenance of existing terminal operation, access system, runway system, and ancillary facilities during all stages of construction
   b. Provision of facilities which generate maximum revenues from concessionaires and other sources
   c. Provision of facilities which minimize maintenance and operating expenses

4. Community objectives
   a. Render a unique and appropriate expression and impression of the community
   b. Provision of harmony with existing architectural elements of the total terminal complex
   c. Coordination with existing and planned off-airport highway systems

The designer should consider the combination of these types of objectives in developing specific design criteria for the passenger terminal complex. These criteria should be used as performance measures for the evaluation of design alternatives. To generate performance measures, a detailed design should be provided. The analyst can then proceed to calculate the various performance measures, employing a number of analytical techniques. Some of these techniques are discussed later in this chapter.

Terminal demand parameters

The determination of space requirements at passenger terminals is strongly influenced by the quality of service desired by the various airport users and by the community. A review of passenger terminals in relation to passenger volumes at existing airports shows a wide range in the configuration and amount of area provided per passenger. However, some guidelines for the determination of space requirements can be outlined. The purpose of these guidelines is to give general orders of magnitude for values that are subject to change depending on the requirements of specific designs.

Follow these steps to determine terminal facility space requirements.

Identify access modes and modal splits. Vehicle volumes are normally derived from projections of passenger and aircraft forecasts. These volumes critically impact the design of highway access facilities; on-airport roadway and circulation systems; curb frontage requirements for private automobiles, buses, limousines, taxis, and rental cars; and parking. Surveys are normally conducted to determine the access modes of passengers and vehicle occupancy rates [26]. In the absence of such surveys, secondary sources may be investigated to ascertain the access characteristics of passengers in similar airport environments [24, 55].

The most important parameters to be obtained include the typical peak-hour volume of vehicles entering and leaving the airport on the design day, the access facilities used, and the duration of their use, including parking and curb front. Care should be taken to include employees and visitors as well as passengers in these access studies and to correlate the peaking characteristics and access modes of each group of airport traveler.

Identify passenger volumes and types. Passenger volumes can be obtained from forecasts normally done in conjunction with airport planning studies. Two measures of volume are used. The first is the annual passenger volume, which is used for preliminary sizing of the terminal building. The second measure is a more detailed hourly volume. It is customary to use typical peak-hour passengers as the hourly design volume for the passenger terminal design. This parameter is a design index and is usually in the range of 0.03 to 0.05 percent of the annual passenger volume, but it is significantly affected by the scheduling practices and fleet mix of the airlines.

The identification of passenger types is necessary because different types of passengers place different demands on various components. Passenger types are usually broadly classified into domestic and international passengers and then further grouped into originating, terminating, connecting or transfer, through, enplaning, and deplaning passengers. These various groups of passengers are made on the basis of the facilities within the terminal normally used by each type. Airports which are used as airline hubs and have a high proportion of connecting passengers require considerably fewer ground access and landside facilities than airports with a high proportion of originating and terminating passengers. Historical data and forecasts regarding the proportions of the total volumes that are made up by each of the different types of passengers are useful in obtaining estimates of the parameters needed for the design of the various facilities [4].

Identify access and passenger component demand. This is done by matching the passenger and vehicle types with facilities in the terminal area. The use of tabulations such as the one shown in Table 10-1 is quite helpful. This table shows which passengers are using each facility. By indicating the volume of each type of passenger in the rows corresponding to the facilities, it is possible to generate the total load on each facility. This is done by taking the row sums of the volumes entered.

Facility classification

The airport terminal facility may be classified by its principal characteristics relative to its functional role. In general, airports are classified as originating-terminating stations, transfer stations, or through stations. The facilities required vary considerably in magnitude and configuration for each.

An originating-terminating station processes a high level of passengers which are beginning or ending the air transportation trip at the airport. At such stations these passengers may be on the order of 70 to 90 percent of the total passengers. Such stations have a relatively long aircraft ground time, and the main flow of passengers is between the aircraft and the ground transportation system. Such stations have relatively high requirements for curb frontage, ticketing and baggage claim facilities, and parking. Typical data indicate that the hourly movement of aircraft per gate at such stations is on the order of 0.9 to 1.1.

In a transfer station or connecting airport, on the other hand, a high percentage of the total passengers are connecting between arriving and departing flights. Many airports in the United States today are connecting airports due to the creation of airline hubs. These sta-

TABLE 10-1 Determination of Demand for Various Types of Passenger Facilities

Facility j	Passenger type i, arriving			Passenger type i, departing			Total volume V
	Domestic, no bags, auto driver*	Domestic, with bags, auto passenger†	International, with bags, auto passenger	Domestic, with bags, auto passenger	Domestic, no bags, auto driver	International, with bags, auto driver	
Curb, arrivals	—	$V_{ij}^{\ddagger}$	V_{ij}	—	—	—	
Curb, departures	—	—	—	V_{ij}	V_{ij}	V_{ij}	
Domestic lobby	—	V_{ij}	—	V_{ij}	V_{ij}	—	
International lobby	—	—	—	—	—	V_{ij}	
Ticketing counter	—	—	—	V_{ij}	—	V_{ij}	
Assembly	—	—	—	V_{ij}	V_{ij}	V_{ij}	
Baggage check-in	—	—	—	V_{ij}	—	V_{ij}	
Security control	—	—	—	V_{ij}	V_{ij}	V_{ij}	
Customs, health	—	—	V_{ij}	—	—	—	
Immigration	—	—	V_{ij}	—	—	—	
Baggage claim	—	V_{ij}	V_{ij}	—	—	—	

*Auto driver = passenger driving a car to and from airport.

†Auto passenger = passenger driven to and from airport.

‡V_{ij} = design volume of passenger type i using facility type j.

tions need greater concourse facilities for the processing of connecting passengers and less ground access facility development. There are usually fewer airline ticketing positions and baggage claim facilities than with originating stations. However, usually there are more intraline and interline baggage facilities. Care must be exercised in planning such stations to locate the gate positions of airlines exchanging passengers close to each other to minimize central terminal flows and connecting times. Data indicate that such airports demonstrate aircraft activity at the rate of 1.3 to 1.5 aircraft per gate per hour in peak periods.

The through station combines a high percentage of originating passengers with a low percentage of originating flights. A high percentage of the passengers remain on the aircraft at such points. Aircraft ground times are minimal, averaging between 1.6 and 2.0 hourly movements per gate in peak periods. Departure lounge space is not as great at such stations, and there are fewer curb frontage, ticketing, and baggage facilities than at originating stations.

Overall space approximations

It is possible to estimate order-of-magnitude ranges for the overall size of a terminal facility prior to performing more detailed calculations for particular space needs. These estimates allow the planner to broadly define the scope of a project based upon information which summarizes the space provided of other existing facilities.

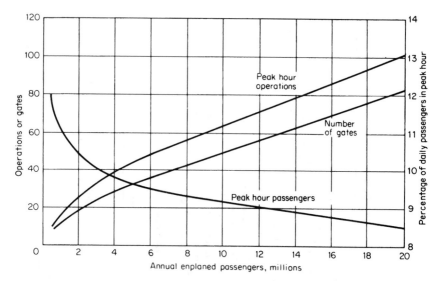

Figure 10-7 Estimated peak-hour passenger, operations, and gate requirements for intermediate-range planning (*Federal Aviation Administration [43]*).

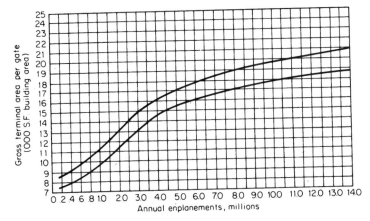

Figure 10-8 Gross terminal-area estimates for intermediate-range planning (*Federal Aviation Administration [49]*).

The FAA has indicated that gross terminal-area space requirements of between 0.08 and 0.12 ft^2 per annual enplaned passenger are reasonable. Another estimate is obtained by applying a ratio of 150 ft^2 per design-hour passenger [43]. Estimates of the level of peak-hour passengers, peak-hour aircraft operations, and gate positions are also obtained based upon the level of annual enplanements by using relationships such as those shown in Fig. 10-7. Others have provided estimating guidelines for total terminal space as shown, e.g., in Fig. 10-8 [43].

Approximations of the allocation of space for various purposes in a terminal building are also useful for preliminary planning. The FAA indicates that approximately 55 percent of terminal space is rentable and 45 percent is nonrentable [49]. An approximate breakdown of these space allocations typically allots 35 to 40 percent for airline operations, 15 to 25 percent for concessions and airport administration, 25 to 35 percent for public space, and 10 to 15 percent for utilities, shops, tunnels, and stairways. A final determination of the actual space allocations is made following detailed analyses of the performance of the elements of the system and consultation with the airlines and airport management as the design process proceeds from space programming through each of the subsequent phases.

Level-of-service criteria

There has been considerable research and discussion in the profession concerning the adoption of level-of-service standards and associated criteria to evaluate the level of service afforded in the design of land-side processing systems. Although it is relatively simple to develop relationships between aircraft delay on the airside and its economic consequences, such relationships are difficult to either define or develop on the airport landside. Much of the difficulty is related to the fact that the various constituent groups associated with airports view quality of service or level of service from different perspectives [37]. Airlines are concerned with such factors as on-time schedules, proper allocation of personnel, minimization of airport operating costs, and profitability. Passengers are concerned with the completion of an air transportation trip at a reasonable cost, with minimum delay and maximum convenience, without being subjected to excessive levels of congestion. The airport operator is interested in providing a modern airport facility which meets airline and passenger objectives in harmony with the expectations of the community where the airport is located. Given the number of possible measures of service quality and the differences in airports throughout the country, it is very difficult to adopt level-of-service criteria on a broad scale.

Many have examined level-of-service criteria for airports and some have attempted to define level-of-service standards [9, 11, 14, 16, 21, 29, 31, 36, 37, 41, 45, 56]. In general, the level-of-service measures commonly associated with the airport landside system include measures of congestion within the terminal building and the ground access system, passenger delays and waiting line lengths at various facilities within the terminal building, passenger walking distances, and total passenger processing time. Most of these parameters can be evaluated in a terminal design with the aid of mathematical modeling. However, the various measures of level of service from the perspective of airport users must be balanced in reaching some acceptable solution to the design problem.

As an illustration of the application of a level-of-service standard, let us say that an airline may desire to limit the percentage of its passengers who must spend more than some increment of time at an airport check-in facility. Through an extension of the mathematical formulation of this problem discussed later, one could develop a model which computes the percentage of passengers at the check-in facility for various durations of time when the number of check-in counters operated is varied for some peak-hour passenger demand [35]. The results could be shown graphically as in Fig. 10-9. From this illustration, if the criteria were to limit the percentage of passengers spending more than 5 min at this facility to 10 percent, then it would be necessary for the airline to operate 9 check-in counters at the airport during the peak hour.

Such formulations can be examined for various facilities within the airport landside to obtain quantitative measures of component and system performance. These are discussed further later in this chapter.

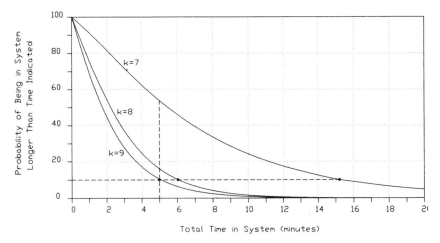

Figure 10-9 Impact of the number of check-in counters on passenger time at a check-in facility.

The Terminal Planning Process

The evolution and development of a terminal design are performed in a series of integrated steps. These may be identified as programming, concept development, schematic design, and design development. Terminal facilities are developed in conformity with the planned development of the airside facilities by considering the most effective use of the airport site, its potential for physical expansion and operational flexibility, its integration with the ground access system, and its compatibility with existing and planned land uses near the airport. The planning process explicitly examines physical and operational aspects of the system.

The programming phase defines the objectives and project scope including the rationale for the initiation of the study. It includes tentative implementation schedules; estimates of the anticipated level of capital investment as well as operating, maintenance, and administrative costs; and the space requirements program. In concept development, studies are undertaken to identify the overall arrangement of building components, functional relationships, and characteristics of the terminal building. Schematic design translates the concept and functional relationships to plan drawings which identify the overall size, shape, and location of spaces required for each function. Detailed budget estimates are prepared in schematic design so that comparisons may be made between the space requirements and costs. In design development, the size and character of the entire project are fixed, and detailed plans of the specific design and allocation of space within the complex are prepared. This phase forms the basis for the preparation of construction documents, bidding, construction, and final project implementation [50].

In the programming and concept development phases of a terminal design project, the following evaluation criteria are typically used to weigh alternatives:

1. Ability to handle expected demand
2. Compatibility with expected aircraft types
3. Flexibility for growth and response to technology changes
4. Compatibility with the total airport master plan
5. Compatibility with on-airport and adjacent land uses
6. Simplicity of passenger orientation and processing
7. Analyses of aircraft maneuvering routes and potential conflicts on the taxiway system and in the apron area
8. Potential for aircraft, passenger, and ground vehicle delay
9. Financial and economic feasibility

In the schematic and design development phases, more specific design criteria are examined, such as

1. The processing cost per passenger
2. Walking distances for various types of passengers
3. Passenger delays in processing
4. Occupancy levels and degree of congestion
5. Aircraft maneuvering delays and costs
6. Aircraft fuel consumption in maneuvering between runways and terminals
7. Construction costs
8. Administrative, operating, and maintenance costs
9. Potential revenue sources and the expected level of revenues from each source

Space programming

The space programming phase of terminal planning seeks to establish gross size requirements for the terminal facilities without establishing specific locations for the individual components. The nature of the processing components is such, however, that approximate locations are indicated for new and existing terminal facilities because of the sequential nature of the processing system. This section provides guidance concerning the spatial requirements to adequately accommodate the several functions carried out in various areas of the airport terminal.

The access interface system. The curb element is the interface between the terminal building and the ground transportation system. A survey of airport users will establish the number of passengers using each of the available ground transportation modes such as private automobile, taxi, limousine, courtesy car, public bus, rail, or rapid transit. Ratios may be established for both the passenger and the vehicle modal choice for airport access.

Terminal curb. The length of curb required for loading and unloading of passengers and baggage is determined by the type and volume of ground vehicle traffic anticipated in the peak period on the design day. Airports with relatively low passenger levels may be able to accommodate both enplaning and deplaning passengers from one curb front. More active airports may find it desirable to physically separate the enplaning from the deplaning passengers—horizontally if space permits or vertically if space is limited. There is a tendency at large airports to also separate commercial vehicle traffic from private vehicle traffic.

The determination of the curb space required is related to airport policies on the assignment of priorities to the use of curb front and the provision of staging areas for taxis, buses, and other public transport vehicles. The parameters required for a preliminary analysis of curb-front needs are the number and types of vehicles at the curb, vehicle length, and various occupancy times of different types of vehicles at the curb front for arriving and departing passengers.

Normally, a slot is considered to be about 25 ft for a private automobile, 20 ft for taxis, 30 ft for limousines, and 50 ft for transit buses. Reported dwell times for private automobiles range from 1 to 2 min at the enplaning curb and from 2 to 4 min at the deplaning curb. Taxi dwell times lie closer to the lower range of these values, whereas limousines and buses may be at the curb anywhere from 5 to 15 min. These dwell times are highly influenced by the degree of traffic regulation and enforcement in the vicinity of the curb, and they should be verified in specific studies. Normally a wide lane, on the order of 18 to 20 ft, is provided to accommodate direct curb access, maneuvering, and standing vehicles. This usually indicates a minimum of one and preferably two additional lanes in the vicinity of terminal entrances and exits to provide adequate capacity for through traffic. Rules of thumb which may be applied to determine curb-front needs indicate that the full length of the curb adjacent to the terminal plus about 30 percent of the maneuvering lane may be considered as the available curb front. Therefore, a 100-ft curb may be considered to provide 130 ft of curb front in 1 h or 7800 foot-minutes of vehicle occupancy. If 120 automobiles per hour demand curb space for an average dwell time of 2 min, then 6000 foot-minutes of curb front is required, or in the peak hour a curb length of 100 ft must be provided. Other methods for approximating curb frontage have been reported in the literature [12, 43, 52]. For example, at Geneva Intercontinental Airport, preliminary design guidelines indicated that curb frontage length should be made available at the rate of at least 0.5 ft per peak-hour enplaning passenger and 0.8 ft per peak-hour deplaning passenger [25].

Roadway elements. Determination of the vehicular demand for various on- and off-airport roadways is essential to ensure that adequate service levels are provided to airport users. The main components of the highway system providing access to airports from population and industrial centers are normally within the jurisdiction of federal, state, and local ground transportation agencies. However, coordination in areawide planning efforts is essential so that the traffic generation potential of airports may be included within the parameters necessary for the proper planning of regional transportation systems. Guidance on the level and peaking characteristics of airport-destined traffic may be found in the literature [55].

The provision of adequate feeder facilities from the regional transport network to the airport is largely within the jurisdiction of the airport operator or owner. Vehicle volumes and peaking characteristics are usually determined by correlating modal preference and occupancy rates with flight schedules. Normally roadway facilities are designed for the peak-hour traffic on the design day with adequate provision for the splitting and recirculation of traffic within the various areas of the airport property. The main roadway elements which must be considered are the feeder roads into the terminal area, enplaning and deplaning roadways, and recirculation roadways.

The *Highway Capacity Manual* [33] provides criteria for level-of-service design and quantitative methods for determining the volumes which can be accommodated by various types of roadway sections. Unfortunately little guidance is available for the level-of-service design of airport roadways. For preliminary planning, however, it is reasonable to assume that feeder roads on the airport property provide acceptable service when they are designed to accommodate from 1200 to 1600 vehicles per hour per lane. Roadways providing access to the enplaning and deplaning terminal systems provide adequate service when they are designed to accommodate from 900 to 1000 vehicles per hour per lane. Terminal frontage roads and recirculation roads, however, provide adequate service when they are designed to accommodate from 600 to 900 vehicles per hour per lane. It is recommended for preliminary planning purposes that the above ranges of values be used to establish bounds on the sizes of these facilities for a demand-capacity analysis. In schematic design, analysis of the flow characteristics of individual sections of the roadway elements will yield final design parameters.

Parking. Most large airports provide separate parking facilities for passengers and visitors, employees, and rental-car storage. In smaller airports, these facilities may be combined in one physical location. Passenger parking and visitor parking are often segregated into short-term, long-term, and remote parking facilities. Those parking facilities most convenient to the terminal are designated as short-term, and a premium rate is charged for their use. Long-term parking is usually near the main terminal complex, but not as convenient as short-term, and rates are usually discounted for long-term users. Remote parking is usually quite distant from the terminal complex, and provisions are normally made for courtesy vehicle transportation between these areas and the main terminal complex. The rates in these facilities are usually the most economical.

Short-term parkers are normally classified as those who park for 3 h or less, and these may account for about 80 percent of the parkers at an airport. However, these short-term parkers account for only 15 to 20 percent of the accumulation of vehicles in the parking facility [43]. Preliminary planning estimates of the number of parking spaces required at an airport may be obtained from Fig. 10-10. The range of public parking spaces provided at existing airports varies from 1000 to 3000 per 1 million originating passengers. It is recommended, however, that the range for preliminary planning be established between 1000 and 1400 parking spaces per 1 million originating passengers.

In planning for the Geneva airport, preliminary design criteria required 2 parking spaces per peak-hour passenger on the design day. More refined estimates of the total amount of parking required and of the breakdown of short- and long-term space are obtained from analyses performed in the schematic design phase of the terminal planning process.

Entrances to parking facilities through ticket-spitting devices are very common at airports. These devices can process anywhere from 400 to 650 vehicles per hour depending upon the degree of automation used as well as the continuity of the demand flow. It is recommended that the number of entrances be estimated on the basis of 500 vehicles per hour per device in preliminary planning. Parking revenue collection points at parking facility exits process from 150 to 200 vehicles per hour per position.

In parking garages, the capacity of ramps leading from one level to another is important during peak periods, when considerable search-

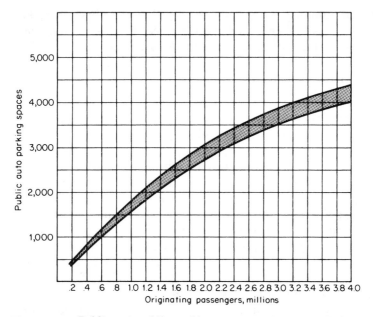

Figure 10-10 Public automobile parking space requirements (*Federal Aviation Administration [49]*).

ing for an available space may occur, or vehicles may be directed immediately to a particular level. One-way straight ramps can accommodate about 750 vehicles per hour. However, a reduction on the order of 20 percent should be considered when two-way ramps are utilized. Circular or helical ramps, often used for egress from parking facilities, accommodate about 600 vehicles per hour in one direction.

The precise volume of vehicles which may be accommodated by a particular design will depend to a large extent on the geometric characteristics of the design, continuity of flow, information systems installed, and characteristics of the vehicles and users of the particular facility [22, 40]. Most often, some type of analytical or simulation model is used in the schematic design phase of the project to test a preliminary design.

The passenger processing system. The passenger processing system consists of those facilities necessary and desirable for the handling of passengers and their baggage prior to and after a flight. It is the element which links the ground access system to the air transportation system. The terminal curbs provide the interface on the ground access side of the system, and the aircraft gates provide the interface on the airside of the system. In determining the particular needs of a specific component in this system, knowledge of the types of passengers and of the extent of visitor impact on each component is necessary.

Entryways and foyers. Entryways and foyers are located along the curb element and serve as weather buffers for passengers entering and leaving the terminal building. The size of an entryway or foyer depends upon its intended use. Since entrances and exits may be relatively small, sheltered public waiting areas should be provided and sized to meet local needs. Designs must accommodate the physically handicapped. These facilities are sized to process both passengers and visitors during the peak hour. Although the times of the enplaning and deplaning peaks may be different, it is likely that the deplaning peak will occur over a shorter time than the enplaning peak. It is often useful to subject a preliminary design proposal to both an average peak-hour demand and a peak 20- or 30-min demand, particularly for the deplaning elements of the system. Preliminary design processing rates for automated doors in the vicinity of the enplaning and deplaning curb front can be taken as 8 to 10 persons per min per unit. These values may be reduced by 50 percent if the doors are not automated.

Terminal lobby area. The functions of significance to an air traveler that are performed in a terminal lobby are passenger ticketing, passenger and visitor waiting, and baggage check-in and claim. Airports with less than 100,000 annual enplanements frequently carry out these functions in a single lobby. More active airports usually have separate lobbies for each function. The size of the lobby space depends on whether ticketing and baggage claim lobbies are separate, whether passenger and visitor waiting areas are provided, and the acceptable density of congestion. In general, the lobby area should provide for passenger queueing, circulation, and waiting. Waiting lobby areas are designed to seat from 15 to 25 percent of the design-hour enplaning passengers and visitors if departure lounges are provided for all gates, and from 60 to 70 percent if they are not provided [43, 50]. Usually about 20 ft^2 per person is provided for seating and for circulation.

Airline ticket counter and ticket office. The airline ticket counter and ticket office are where the airline and passenger make final ticket transactions and check in baggage for a flight. This includes the airline ticket counter, airline ticket agent service area, outbound baggage handling device, and support office area for the airline ticket agents. There are three types of ticketing and baggage check-in facilities: linear, pass-through, and island. Each of these facilities is shown in a typical arrangement in Fig. 10-11.

The ticket transaction takes place at the ticket counter, which is a stand-up desk. To the left and right of the ticket counter position, a low shelf is provided to deposit, check in, tag, and weigh baggage, if necessary, for the flight. Subsequently, the baggage is passed back by the agent to an outbound baggage conveyance device located near the

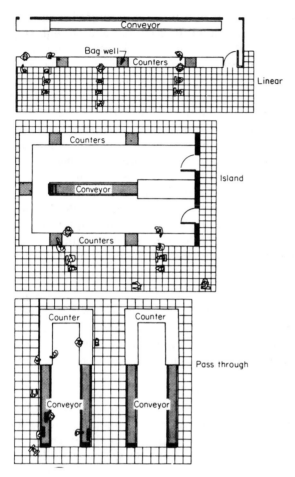

Figure 10-11   Typical ticket counter configurations (*Federal Aviation Administration [49]*).

counter. The total number of ticket counter positions required is a function of the level of peak-hour originating passengers, the types of facilities provided (i.e., multipurpose, express baggage check-in, and ticketing only), and the queues and delays acceptable to the airlines. In some markets, a considerable number of passengers may be pre-ticketed, and a higher percentage of express check-in positions may be warranted either within the terminal building or at the curb front. This is particularly true at airports serving tourist areas. A reasonable estimate of the number of positions required is gained by assuming that the peak loading on the facilities is about 10 percent of the peak-hour originating passengers and that a maximum queue length of 5 passengers per position is a desirable design goal. If this is the case, then 3000 peak-hour originating passengers translates to a 300 short-term peak-load, which requires 60 ticketing positions. The sizing of the counter length depends upon the mix of position types, but for preliminary estimating purposes, a counter length from 10 to 15 ft for two positions is reasonable. Therefore, in this case, the total linear footage of counter space will range from 300 to 450 ft. If a queueing depth of 3 ft per passenger is provided, then a minimum queueing depth of 15 ft is required. The counter itself requires a 10-ft depth, and a circulation aisle of from 20 to 35 ft is appropriate. Therefore, the area devoted to this function ranges from about 13,500 to 27,000 ft^2. If single-row lobby seating is provided in the area, this increases the area to between 16,500 to 30,000 ft^2. Approximations gained by other techniques yield similar results [12, 50].

In the preliminary design guidelines for Geneva, it was recommended that not less than 40 ft of queueing area be provided in front of the linear ticket counters. The reason for this criterion was to allow for increases in the demand at this airport without requiring major terminal modifications.

The above calculations are useful for linear-type counters where queues form in lines at each position. For corral-type queueing, similar results are obtained, the principal advantage of this type of queueing being less restriction to circulation in the ticket lobby area. The pass-through and island-type counters result in a similar number of positions and counter length, but in different geometric arrangements. The circulation and queueing areas may be modified as shown in Fig. 10-11. Normally, in pass-through ticketing and baggage check-in facilities, queueing is along the length of the counter, whereas in island types queueing is in lines at the various ticketing positions.

The final determination of the number and mix of ticketing positions is made through consultation with the various airlines to be served and through the use of analytical or simulation models [15].

The airline ticket office (ATO) support area may be composed of smaller areas for the operations and functions of accounting and safekeeping of tickets, receipts, and manifests; communications and information display equipment; and personnel areas for rest, personal grooming, and training. On the wall behind the ticket agents are posted information displays for the latest airline information on arriving and departing flights. Typical estimates of ATO space requirements may be obtained by taking the linear footage of the counter and multiplying by a depth of 20 to 25 ft for this area. This yields a value of about 6000 to 11,000 ft^2 for 3000 peak-hour originating passengers. As the level of peak-hour passengers increases, various economies of scale may be gained in the use of such facilities. For example, in moving from 3000 to 6000 peak-hour originations, the increase in ATO space may be only about 30 percent. Other estimating procedures yield similar results [43, 50]. Again, the final determination of the space requirements is made through consultation with the various airlines using the facility.

Security.   Security screening of all airline passengers is an extremely important function in an airport terminal. Security screening of passengers and of hand-carried articles is required prior to boarding airline aircraft. Depending upon the configuration of the terminal and the policy of the separate airlines, security screening may be conducted at various locations in a terminal within the area lying between the ticketing area and the aircraft boarding position. The area is considered a sterile area. Decisions to bar visitors from sterile areas must take into consideration both operational need and community reaction. In many cases, signing is used to discourage visitor entrance to sterile areas, but enforcement is purposely lax except in unusually heavy demand periods.

Security screening is typically performed in corridors leading to gates, or in some instances at the boarding gate. In most installations, the passenger and the visitor must walk through a magnetometer with hand-carried articles subject to manual search or x-ray screening.

A typical manual-search station with two tables to accept and two tables to return search articles plus the magnetometer requires approximately 144 ft^2 of space. A typical x-ray search unit requires approximately 120 ft^2 of space. These are shown in Fig. 10-12.

The processing rate of screening installations will range from 500 to 600 persons per hour. Care should be taken to locate the screening positions so that they do not impede the flow of passengers leaving the sterile area and that queues for processing passengers do not restrict other flows within the area. Normally, corridor widths in the vicinity of screening areas should be on the order of 30 to 40 ft to accommodate screening equipment as well as passengers leaving the screening area. Occasionally, screening equipment is placed in tandem along the length of the concourse for use in peak periods without further restricting flow rates or causing long queue lengths.

In many airports, automated facilities are beginning to be used for passengers exiting sterile areas. These facilities may consist of a series of revolving doors with pressure-sensitive floors to restrict unlawful entrance into the area. For preliminary planning, a revolving-door security exit facility will process about 500 to 750 persons per hour per unit.

Departure lounges.   The departure lounge serves as an assembly area for passengers waiting to board a particular flight and as the exit passageway for deplaning passengers. It is generally sized to accommodate the number of boarding passengers expected to be in the lounge 15 min prior to the scheduled departure time, assuming this is the time when aircraft boarding begins. A conservative estimate of the percentage of passengers in the lounge at this time is 90 percent of the boarding passengers. The space should accommodate seating for these passengers (although not all need be seated), space for airline processing plus passenger queues, and an exitway for deplaning passengers.

Processing queues should not extend into the corridor such that circulation is impaired. Lounge depths of 25 to 30 ft are considered reasonable for holding boarding passengers. Space for a departure lounge is proportioned on the basis of 10 to 15 ft^2 per boarding passenger. Therefore, if a departure lounge is to accommodate 100 boarding passengers for a flight, its area should range from 1000 to 1500 ft^2. To a greater and greater extent, common departure lounge areas are being utilized, and these are sized based upon the total peak-hour boarding passengers for the gates being served by the common lounge. Since it is likely that boarding for these aircraft will occur at different times in the peak hour, the total space required for separate lounges may be reduced by 20 to 30 percent for common lounges.

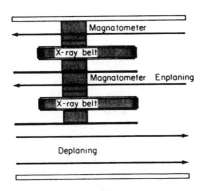

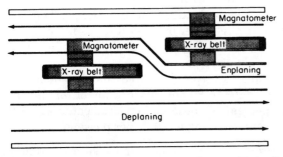

Figure 10-12 Typical security screening layouts (*Federal Aviation Administration [49]*).

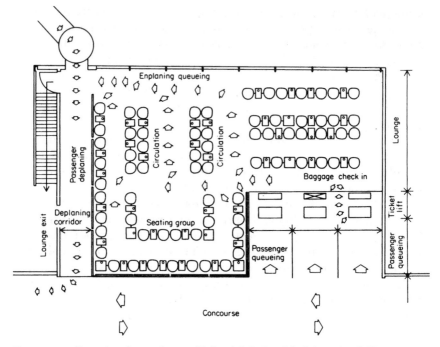

Figure 10-13   Departure lounge layout (*Federal Aviation Administration [50]*).

The corridor provided for deplaning passengers should be about 10 ft wide. The airline processing area should allow for at least two positions for narrow-bodied aircraft and up to four positions for wide-bodied aircraft, to minimize queue lengths extending into the corridor. Processing rates range from 1 to 2 passengers per minute at these positions, and peak arrival rates range from 10 to 15 percent of the boarding passengers. Therefore, queue depths of about 10 ft are reasonable values for preliminary design for individual departure lounges. For common lounges, the position of the processing area should be such as to not interfere with the circulation of passengers in the vicinity of the entrances and exits from the lounge. Typically these positions are located in the center of a satellite facility or at the end of the corridor in a pier facility. Figure 10-13 shows the layout for a departure lounge seating about 70 passengers.

Design criteria at the Geneva airport required that the departure lounge be sized to accommodate an 80 percent load factor for the gate design aircraft, with 80 percent of these passengers seated at 15 ft^2 per passenger and 20 percent standing at 10 ft^2 per passenger.

Corridors.   Corridors provide for passenger and visitor circulation between departure lounges and between departure lounges and the central terminal areas. Corridors should be designed to accommodate physically handicapped persons during peak periods of high-density flow. Studies have shown that a typical 20-ft-wide corridor will have a capacity ranging from 330 to over 600 persons per min. For planning purposes, corridor widths should be sized on the basis of about 16.5 passengers per 1-ft width per minute. The corridor width should be the width required at the most restrictive points, i.e., the minimum free-flow width in the vicinity of restaurant entrances, phone booth clusters, or departure lounge check-in points. This standard is based upon a width of 2.5 ft per person and a depth separation of 6.0 ft between people. The corridor width is adversely impacted by the peaking of deplaning passengers in platoons, but the decreased depth separations compensate for the decreased walking rates in these circumstances.

At the Geneva airport, public circulation corridors were required to be at least 20 ft wide, whereas those in airside connectors were required to be at least 30 ft wide [25]. Further guidance on corridor width design is given in the literature [12, 41, 43, 50].

Baggage claim facilities. The baggage claim lobby should be located so that checked baggage may be returned to terminating passengers in reasonable proximity to the terminal deplaning curb. At low-activity airports, checked baggage may be placed on a shelf for passenger claiming. More active airports have installed mechanical delivery and display equipment similar to that depicted in Fig. 10-14. The number of claim devices required is determined by the number and type of aircraft that will arrive during the peak hour, the time distribution of these arrivals, the number of terminating passengers, the amount of baggage checked on these flights, and the mechanism used to transport baggage from aircraft to the claim area. In the ideal situation, a baggage claim device should not be shared between flight arrivals at the same time, since this leads to considerable congestion in the vicinity of the device and to passenger confusion. Greater utilization of the devices is achieved when airlines time the sharing of claim devices for separate flights. Techniques similar to those used to construct ramp charts are useful in scheduling baggage claim devices and determining the level of congestion in the baggage claim area.

At present, except in the most unusual situations, passenger delays in the baggage claim area can be significant principally because passengers can travel from aircraft to claim areas much faster than baggage conveyance systems can transport the baggage from aircraft to claim areas. It is therefore essential that claim lobbies be designed to accommodate waiting passengers adequately and to permit rapid baggage claim once the baggage is transferred to the claim device. Estimating procedures have been provided by the FAA for sizing claiming devices based upon the equivalent peak 20-min aircraft

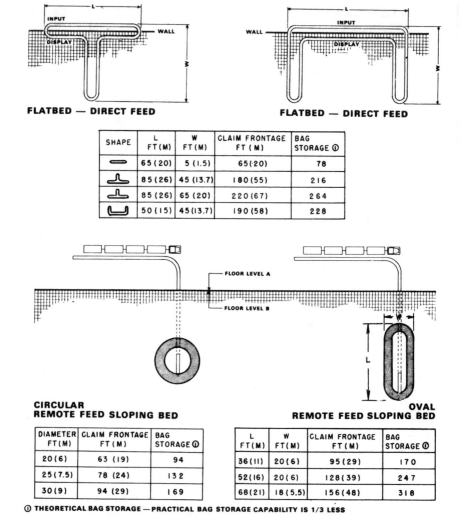

FLATBED — DIRECT FEED

FLATBED — DIRECT FEED

SHAPE	L FT(M)	W FT(M)	CLAIM FRONTAGE FT (M)	BAG STORAGE ①
⬭	65(20)	5 (1.5)	65(20)	78
⊥	85 (26)	45 (13.7)	180(55)	216
⊥	85 (26)	65 (20)	220(67)	264
⊔	50(15)	45(13.7)	190(58)	228

CIRCULAR
REMOTE FEED SLOPING BED

OVAL
REMOTE FEED SLOPING BED

DIAMETER FT(M)	CLAIM FRONTAGE FT (M)	BAG STORAGE ①
20(6)	63 (19)	94
25(7.5)	78 (24)	132
30(9)	94 (29)	169

L FT(M)	W FT(M)	CLAIM FRONTAGE FT (M)	BAG STORAGE ①
36(11)	20(6)	95(29)	170
52(16)	20(6)	128(39)	247
68(21)	18(5.5)	156(48)	318

① THEORETICAL BAG STORAGE — PRACTICAL BAG STORAGE CAPABILITY IS 1/3 LESS

Figure 10-14 Mechanized baggage claim devices commonly used at airports (*Federal Aviation Administration [43]*).

arrivals. Charts for this purpose are seen in Figs. 10-15 and 10-16. Observe that these charts are based upon checked baggage at the rate of 1.3 bags per person, and adjustments may be required as baggage exceeds this rate.

In preliminary planning studies, great care should be exercised in using the above guidelines. Particular attention should be given to the space provided around a claim device, and it is specifically recommended that a clear space of 13 to 15 ft be provided adjacent to the device for active and waiting claim, as well as claim-area circulation. In some

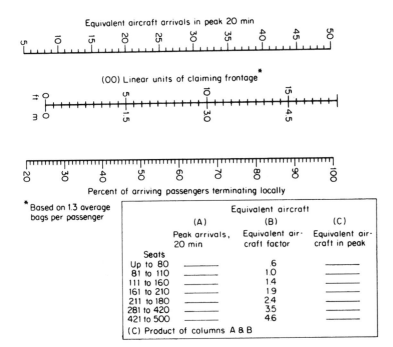

Figure 10-15 Estimating nomograph for baggage claim linear footage requirements (*Federal Aviation Administration [43]*).

instances, baggage lobbies have been designed with adequate waiting areas, and subsequently other facilities have been moved into these areas, considerably diminishing mobility. It is recommended that an additional 15 to 35 ft of circulation space be provided within the deplaning facility to allow for circulation between the claim devices, rental-car positions, and deplaning curbs. If ground transportation facilities are located in these areas, they should be physically positioned so as to not restrict passenger flow to and from the claim area.

In many airports, *positive claim* is being used to ensure the security of passenger baggage. This system requires that attendants examine baggage claim checks and compare them to baggage tags prior to allowing baggage to leave the facility. Positive claim is provided either by enclosing the claim area with a barrier and restricting exit points or by checking baggage at the exit doors from the baggage claim facility.

Intraairport transportation systems. The use of automated ground transportation systems within the terminal complex at airports is increasing as airports become larger and as both the distance and the time for passengers to travel through airports have become excessive. In most cases, automated ground transportation also provides a clear

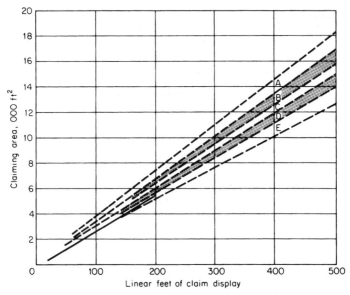

Areas for optimum configurations of :

A Round – sloping bed / remote feed
   Tee – flat bed / direct feed

B Tee and u-shape alternating @ 75'
   (flat bed / direct feed)

C Oval – flat bed / direct feed
   Oval – sloping bed / remote feed

D Tee and u-shape alternating @ 60'
   (flat bed / direct feed)

E ⊥⊥-shape flat bed / direct feed

Figure 10-16 Estimating chart for total area of baggage claim facilities (*Federal Aviation Administration [43]*).

delineation between the airside and landside functions of an airport and aids in the location of security processing facilities.

The new Denver International Airport and the Greater Pittsburgh International Airport are the latest U.S. airports to employ underground automated transit systems to convey passengers and visitors between the landside and the airside. At Denver trains move between the landside building to three parallel concourses, whereas at Pittsburgh trains move from the landside building to the core of the X-shaped airside building. Figure 10-17 shows a view of the Midfield Terminal Complex at Pittsburgh with an enclosed moving walkway connecting parking to the landside building and the commuter aircraft concourses attached to the landside building. Both airports employ fixed guideway trains with two or three 80-passenger cars which operate on headways of 90 s and transport passengers from the landside building to the remote airside buildings in under 5 min in Denver and under 2 min in Pittsburgh.

Figure 10-17 Mid-field terminal complex at Greater Pittsburgh International Airport (*Terry Clark Photography and Michael Baker Jr., Inc.*).

Various types of automated ground transportation systems are employed at airports ranging from moving walkways to small 16-passenger vehicles to multicar trains. These systems can be operated on a loop alignment or a shuttle alignment. Guidance on the planning of such facilities is given in recent literature [20, 46].

International facilities. Airports with international operations require space for the inspection of passengers, crew, baggage, aircraft, and cargo. The area required for customs, immigration, agriculture, and public health services may be in a separate facility or in the terminal building itself. These facilities should be designed so that (1) passenger flow between the aircraft and the initial processing station is unimpeded and as short as possible, (2) there is no possibility of contact with domestic passengers or any unauthorized personnel until processing is complete, (3) there is no possible way for an international arrival to bypass processing, and (4) there is a segregated area for in-transit international passengers.

The size of this facility is based on the projection of hourly passengers requiring processing. It is recommended that the appropriate officials and agencies be contacted during the preliminary design phase to determine specific design requirements. Some guidance on the processing rates and sizing of these facilities is found in the literature [43].

Other areas

Most terminals are developed to accommodate several other activities, and the space needs should be determined for each airport based upon local requirements. These activities are identified below.

Airline activities. The following airline activities may exist at some of or all terminal facilities and should be discussed with the airlines that plan on utilizing the facility.

1. Outbound baggage makeup and inbound baggage transfer and conveyance system

2. Cabin services and aircraft maintenance

3. Flight operations and crew ready rooms

4. Storage areas for valuable or outsized baggage

5. Air freight and mail pickup and delivery

6. Passenger reservations and VIP waiting areas

7. Administrative offices

8. Ramp vehicle and cart parking and maintenance

Passenger amenities.   The factors which influence the extent of passenger amenities include the passenger volume, community size, location and extent of off-airport services, interests and abilities of potential concessionaires, and rental rates. These generally include

1. Food and beverage services, newsstands, and tobacco stands
2. Drugstores; gift, clothing, and florist shops
3. Barber shops and shoe shine stands
4. Counters for car rental and flight insurance companies
5. Public lockers and public and courtesy telephones
6. Staffed or automated post offices
7. Amusement arcades and vending machines
8. Public rest rooms and nursery

Airport operations and services.   These facilities and services are usually found in most public buildings and include the following:

1. Offices for airport management and staff functions including police, medical and first aid, and building maintenance
2. Building mechanical systems such as heating, ventilation, and air conditioning
3. Communication facilities
4. Electrical equipment
5. Government offices for air traffic control, weather reporting, public health and immigration, and customs
6. Conference and press facilities

Overall space requirements

Guidelines have been presented above for the approximate space requirements of various components in passenger terminal facilities. Once the facilities have been estimated, one might compare the space requirements to the approximations given in Table 10-2. The values in this table represent overall space requirements which should provide a reasonable level of service and a tolerable occupancy level for the facilities indicated.

Concept development

In this phase of the process, the blocks of spaces determined in space programming are allocated in a general way to the terminal complex. There are a number of ways in which the facilities of the passenger

TABLE 10-2   Typical Terminal Building Space Requirements

Component	Space required, 1000 ft^2 or 100 m^2 per 100 typical peak hour passengers
Ticket lobby	1.0
Baggage claim	1.0
Departure lounge	2.0
Waiting rooms	1.5
Immigration	1.0
Customs	3.0
Amenities	2.0
Airline operations	5.0
Total gross area	
Domestic	25.0
International	30.0

terminal system are physically arranged and in which the various passenger processing activities are performed. Centralized passenger processing means all the facilities of the system are housed in one building and are used for processing all passengers utilizing the building. Centralized processing facilities offer economies of scale in that many of the common facilities may be used to service a large number of aircraft gate positions. Decentralized processing, on the other hand, means the passenger facilities are arranged in smaller modular units and are repeated in one or more buildings. Each unit is arranged around one or more aircraft gate positions and serves the passengers using those gate positions. There are four basic horizontal distribution concepts as well as many variations or hybrids which include combinations of these basic concepts. Each can be used with varying degrees of centralization. These concepts are discussed below.

Horizontal distribution concepts

The following terminal concepts should be considered in the development of the terminal area plan. Sketches of the various arrangements are shown in Fig. 10-18. Many airports have combined one or more terminal types.

Pier or finger.   In the pier concept, there is an interface with aircraft along piers extending from the main terminal area. Aircraft are usually arranged around the axis of the pier in a parallel or nose-in parking alignment. Each pier has a row of aircraft gate positions on both sides, with a passenger concourse along the axis which serves as the departure lounge and circulation space for both enplaning and deplaning passengers. This concept usually allows for the expansion of the pier to provide additional aircraft parking positions often

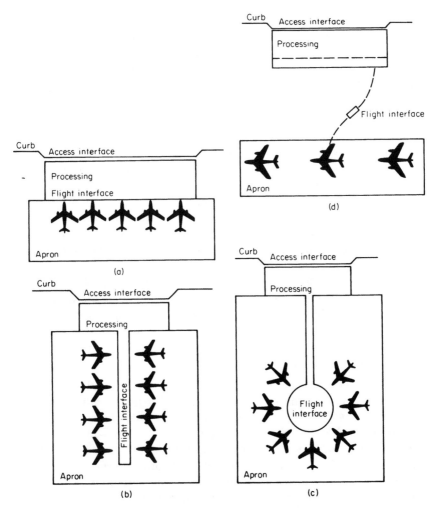

Figure 10-18 Horizontal distribution concepts for passenger terminals: (a) Linear; (b) pier; (c) satellite; (d) transporter.

without expansion of the central passenger and baggage processing facility. Access to the terminal area is at the base of the connector or the pier.

If two or more piers are employed, the spacing between the two piers must provide for maneuvering of aircraft on one or two apron taxilanes. When each pier serves a large number of gates, and when the probability exists that two or more aircraft may frequently be taxiing between two piers and will be in conflict with each other, then two taxilanes are advisable. Also access from this taxiway system by two or more aircraft may require two apron edge taxiways to avoid delays.

The chief advantage of this concept is its ability to be expanded in incremental steps as aircraft or passenger demand warrants. It is also relatively economical in terms of capital and operating cost. Its chief disadvantages are its relatively long walking distance from curb front to aircraft and the lack of a direct curb-front relationship to aircraft gate positions.

Satellite. The satellite consists of a building, surrounded by aircraft, which is separated from the terminal and is usually reached by a surface, underground, or aboveground connector. The aircraft are normally parked in radial or parallel positions around the satellite. It often affords the opportunity for simple maneuvering and taxiing patterns for aircraft, but it requires more apron area than other arrangements. It can have common or separate departure lounges. Since enplaning and deplaning from the aircraft are accomplished from a common and often remote area, mechanical systems may be employed to transport passengers and baggage between the terminal and the satellite.

The main advantages of this arrangement lie in its adaptability to common departure lounge and check-in functions and the ease of aircraft maneuverability around the satellite structure. However, the construction cost is relatively high because of the need to provide connecting concourses to the satellite. It lacks flexibility for expansion, and passenger walking distances are relatively long.

Linear, frontal, or gate arrivals. The simple linear terminal consists of a common waiting and ticketing area with exits leading to the aircraft parking apron. It is adaptable to airports with low airline activity which will usually have an apron providing close-in parking for three to six commercial passenger aircraft. The layout of the simple terminal should take into account the possibility of pier, satellite, or linear additions for terminal expansion. In the gate arrivals or frontal concept, aircraft are parked along the face of the terminal building. Concourses connect the various terminal functions with the aircraft gate positions. This arrangement offers ease of access and relatively short walking distances if passengers are delivered to a point near gate departure by vehicular circulation systems. Expansion may be accomplished by linear extension of an existing structure or by developing two or more terminal units with connectors.

Both concepts provide direct access from curb front to aircraft gate positions and afford a high degree of flexibility for expansion. They do not provide convenient opportunities for the use of common facilities, and as this concept is expanded into separate buildings, these concepts lead to high operating costs.

Transporter, open apron, or mobile conveyance. Aircraft and aircraft servicing functions in the transporter concept are remotely located from the terminal. The connection to the terminal is provided by vehicular transport for enplaning and deplaning passengers. The characteristics of the transporter setup include flexibility in providing additional aircraft parking positions to accommodate increases in schedules or aircraft size, capability to maneuver an aircraft in and out of a parking position under its own power, separation of aircraft servicing activities from the terminal, and reduced walking distances for the passenger.

This arrangement minimizes the level of capital costs since the building efficiently uses minimal departure lounge space and has gate positions for transporters rather than aircraft. It offers a high degree of flexibility in both operation and expansion. Aircraft maneuverability is very high, and a separation between the landside and airside is very obvious. The use of mobile lounges to convey passengers to and from aircraft can increase passenger processing time, and unless scheduling is carefully coordinated, unnecessary delays may result.

Concept combinations and variations. Combinations of concepts and variations are a result of changing conditions experienced from the initial conception of the airport throughout its lifespan. An airport may have many types of passenger activity, varying from originating and terminating passengers using the full range of terminal services to passengers using limited services on commuter or connecting flights. Each requires a concept that differs considerably from the others. In time, the proportion of traffic handled by flights may change, necessitating modification or expansion of the facilities. Growth of aircraft size or a new combination of aircraft types servicing the same airport will affect the type of concept. In the same way, physical limitations of the site may cause a pure conceptual form to be modified by additions or combinations of other concepts.

Combined concepts acquire certain of the advantages and disadvantages of each basic concept. A combination of concept types can be advantageous where more costly modifications would be necessary to maintain the original concept. For example, an airline might be suitably accommodated within an existing transporter terminal while an addition is needed for a commuter operation with rapid turnovers which would be best served by a linear extension. In this situation, combined concepts would be desirable. The appearance of concept variations and combinations in a total apron terminal plan may reflect an evolving situation in which altering needs or growth have dictated the use of different concepts. Illustrations of the generation of various horizontal distribution concepts in the concept development phase of Geneva Intercontinental Airport are shown in Fig. 10-19. Applications of several concepts to existing airports are shown in Fig. 10-20a through e.

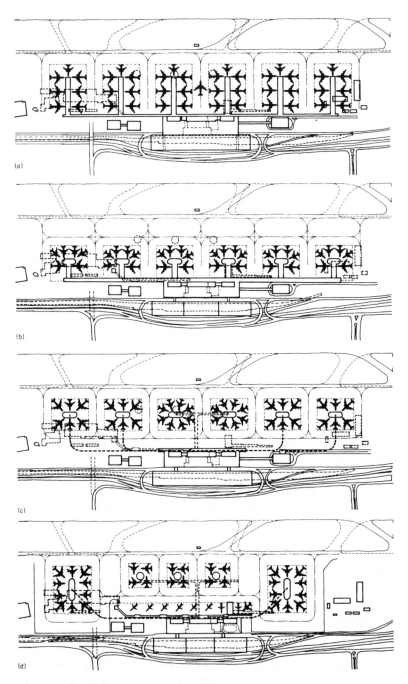

Figure 10-19 Conceptual designs for Geneva Intercontinental Airport (*Reynolds, Smith and Hills [25]*).

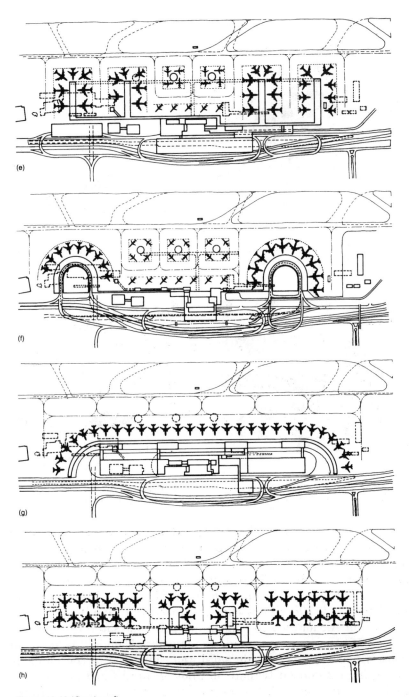

(e)

(f)

(g)

(h)

Figure 10-19 (*Continued*)

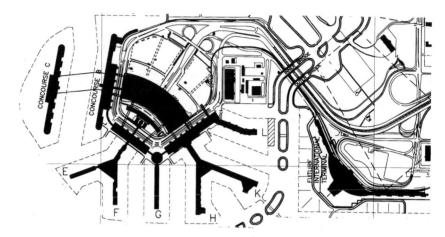

Figure 10-20a Linear, satellite, and pier concepts—O'Hare International Airport (*Landrum and Brown Aviation Consultants and the Department of Aviation, City of Chicago*).

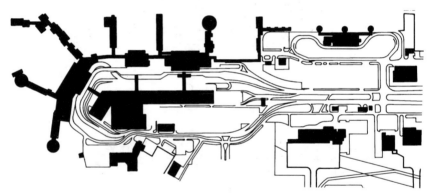

Figure 10-20b Pier and satellite concepts—Detroit Metropolitan Airport (*Landrum and Brown Aviation Consultants and Wayne County Department of Aviation*).

Figure 10-20*a* shows the terminal area layout at O'Hare International Airport which reflects the linear, satellite, and pier concepts. Concourse B is a linear concept; concourse C is a satellite concept connected to the main terminal building by an underground moving walkway; and concourses E, F, G, H, K, and L are pier concepts. Figure 10-20*b* shows the terminal area layout at Detroit Metropolitan Airport which consists of pier and satellite concepts. Figure 10-20*c* shows the terminal area layout at Tampa International Airport which has five airside satellites connected to the landside building by automated transit on aboveground fixed guideways. Figure 10-20*d* shows the terminal area at Greater Pittsburgh International Airport in which the commuter aircraft ramp area is a

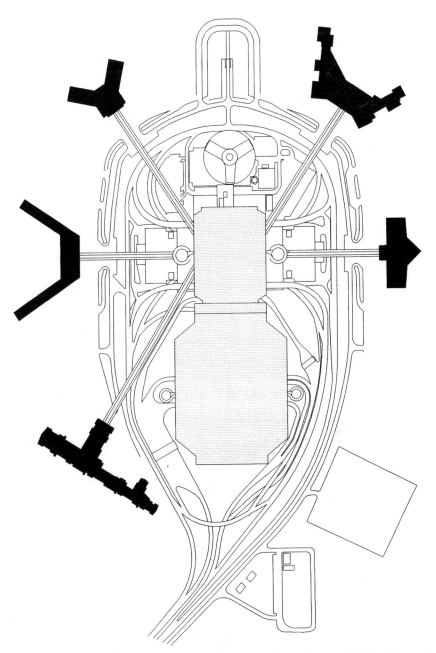

Figure 10-20c   Satellite concept—Tampa International Airport (*CH2M Hill and Hillsborough County Aviation Authority*).

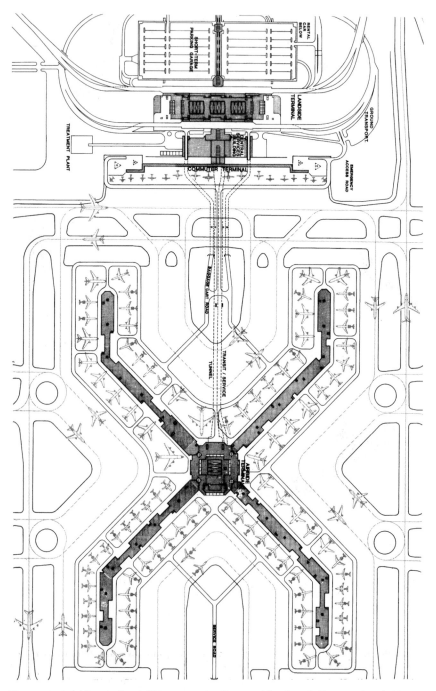

Figure 10-20d Pier and satellite concept—Greater Pittsburgh International Airport (*Tasso Katselas Associates, Michael Baker Jr., Inc., and County of Allegheny Department of Aviation*).

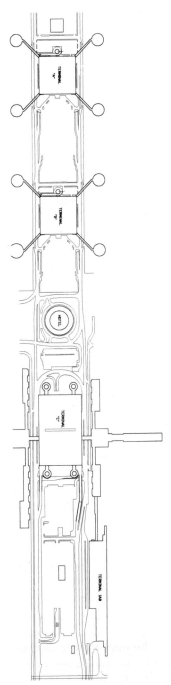

Figure 10-20e   Pier and satellite concept—Houston Intercontinental Airport (*Department of Aviation, City of Houston*).

pier concept and the air carrier ramp area is an airside satellite building connected to the landside building with an underground automated transit system on a fixed guideway. Piers extend outward from the central core of the airside satellite building. Figure 10-20*e* shows the terminal area at Houston Intercontinental Airport which consists of two satellite terminals, a pier terminal, and a linear terminal in the international arrivals building. All are connected by underground fixed-guideway transit.

Vertical distribution concepts

The motivation for distributing the primary processing activities in a passenger terminal among several levels is mainly to separate the flow of arriving and departing passengers. The decision concerning the number of levels that a terminal facility should have depends primarily on the volume of passengers and the availability of land for expansion in the immediate vicinity. The decision may also be influenced by the type of traffic, e.g., domestic, international, or commuter passengers being processed; by the terminal-area master plan; and by the horizontal processing concept chosen.

With a single-level system, all processing of passengers and baggage occurs at the level of the apron. Separation between arriving and departing passenger flows is achieved horizontally. Amenities and administrative functions may occur on a second level. With this system, stairs are normally used to load passengers onto aircraft. This system is quite economical and is suitable for relatively low passenger volumes. The single-level terminal is shown in Fig. 10-21*a*.

Two-level passenger terminal systems may be designed in a number of different ways. In one type, shown in Fig. 10-21*b*, the two levels are used to separate the passenger processing area and the baggage handling areas. Thus, processing activities including baggage claim occur on the upper level, while airline operations and baggage handling activities occur at the lower apron level. The advantage of raising the passenger handling level is that it becomes compatible with aircraft doorsill heights, allowing convenient interface with the aircraft. Vehicular access occurs on the upper level to facilitate the interface with the processing system.

Another articulation of the two-level system separates the arriving and departing passenger streams. In this case, departing passenger processing activities occur on the upper level, and arriving passenger processing including baggage claim occurs at the apron level. Airline operations and baggage handling also occur at the lower level. Vehicular access and parking occur at both levels, one for arrivals and the other for departures, and parking can be surface or structural. An example of this design is shown in Fig. 10-21*c*.

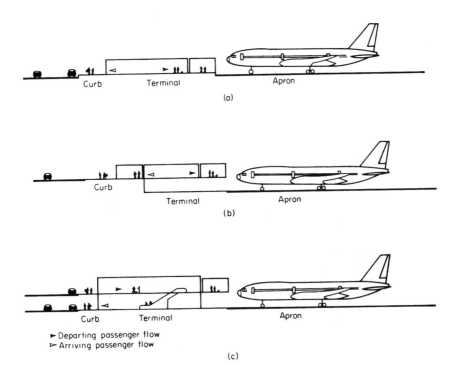

Figure 10-21 Vertical distribution concepts: (*a*) Single level; (*b*) second-level loading; (*c*) two-level system (Federal Aviation Administration [50]).

Variations in these basic designs may occur when traffic volumes or the type of traffic so requires. For example, for international airport terminals, a third level may be needed for international passengers. Also, at large airports where intraairport transportation systems operate, a special level may be needed to provide for these systems. Figure 10-22*a* shows a multilevel system with structural parking, intraairport transportation, and underground mass-transit access. Figure 10-22*b* shows a multilevel system with integrated structural parking. In this variation, more direct access to the processing component is attained by providing parking above the processing facility.

Based upon an examination of a large number of airports, it is possible to identify those concepts which are candidates for further consideration. By using the level of annual enplanements and the functional nature of the airport, as defined by the relative proportions of originating, terminating, through, or connecting passengers, Fig. 10-23 offers some guidance to the airport planner in the initial identification of appropriate horizontal and vertical distribution concepts. Note, however, that in many instances the constraints of existing terminal facilities, land availability, and the ground access system may restrict the options which are viable alternatives for terminal expansion.

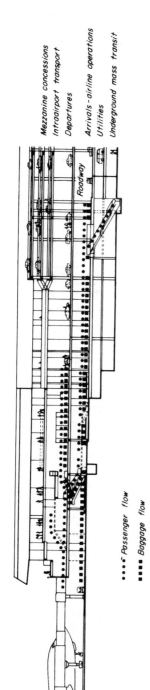

Mezzanine concessions
Intraairport transport
Departures
Arrivals - airline operations
Utilities
Underground mass transit

Roadway

••••◦ Passenger flow
■■■■ Baggage flow

Figure 10-22a Multilevel passenger processing system with structural parking adjacent to terminal (*Hamburg Airport Authority*).

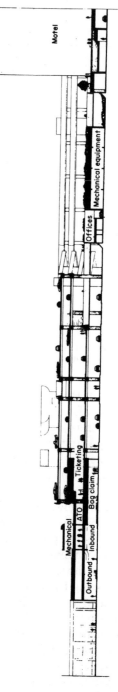

Motel

Mechanical equipment

Offices

Ticketing

ATO

Bag claim

Mechanical

Inbound

Outbound

Figure 10-22b Multilevel passenger processing system with structural parking above the processing area (*Reynolds, Smith and Hills*).

Airport size by enplaned pax/year	Concepts applicable				Physical aspects of concepts							
	Linear	Pier	Satellite	Transporter	Single level curb	Multi level curb	Single level terminal	Multi level terminal	Single level connector	Multi level connector	Apron level boarding	Aircraft level boarding
Feeder under 25,000	X				X		X				X	
Secondary 25,000 to 75,000	X				X		X				X	
75,000 to 200,000	X				X		X		X		X	
200,000 to 500,000	X	X			X		X		X		X	
Primary over 75% pax O/D 500,000 to 1,000,000	X	X	X		X		X		X	X	X	X
Over 25% pax transfer 500,000 to 1,000,000	X	X	X		X		X		X	X	X	X
Over 75% pax O/D 1,000,000 to 3,000,000		X	X	X	X	X	X		X	X	X	X
Over 25% pax transfer 1,000,000 to 3,000,000		X	X		X	X	X		X	X	X	X
Over 75% pax O/D over 3,000,000		X	X	X	X	X	X		X	X	X	X
Over 25% pax transfer over 3,000,000		X	X		X	X	X		X	X		X

Figure 10-23 Applicable concepts for airport design (*Federal Aviation Administration* [50]).

Prior to proceeding with the planning of the airport terminal system, the concepts which have evolved in the conceptual development phase of the project are evaluated, to identify those alternatives which should be brought into the schematic design and design development phase of the project. To do this, an overall rating of the various concepts relative to the design criteria is performed. The concept evaluation rating factors used for the Geneva airport are listed in Table 10-3.

Recent airport development projects

Denver International Airport. At the time this text was being prepared Denver International Airport was under construction to replace Denver Stapleton Airport. The airport is being constructed on a new site north of the existing airport and encompasses a total ground area

TABLE 10-3   Conceptual Development Rating Factors for Evaluation of Terminal Planning Concepts

Passenger convenience:
Walking distance from curb to aircraft
Walking distance for transfer passengers
Walking distance from parking to aircraft
Ease of passenger orientation
Ease of passenger processing
Operational effectiveness:
Efficient taxiing routes
Ground flow coordination of vehicles and aircraft
Apron area maneuverability
Apron adaptation to future aircraft
Vehicular access flows
Direct routes to ancillary facilities
Expansion adaptability:
Ancillary facilities, flexible land use
Staging adaptability
Visual character of increments
Gross terminal expandability
Expandability of terminal elements
Economic effectiveness:
Capital cost
Maintenance and operating costs
Ratio of revenue- to non-revenue-producing areas

SOURCE: Reynolds, Smith and Hills [25].

of 34,000 acres, or 53 mi^2. The total capital costs of the new airport have been estimated at $2.7 billion. The airfield has been planned to ultimately accommodate 12 runways with an annual capacity on the order of 1.2 million aircraft operations. These runways are oriented in quadrants such that there are three parallel runways in each quadrant with runway orientations in two directions: north-south and east-west. These runways are spaced so that both independent arrivals and independent departures can occur in two quadrants simultaneously, resulting in considerable hourly capacity. Initially, the airport will have six runways in operation, one of which will be 16,000 ft long to accommodate the Boeing 747-400 on long international stage lengths.

The landside terminal building and the airside parallel-gate concourses are located near the center of the airfield. The landside building, called the Great Hall, is similar to that existing at Orlando International Airport, and the airside concourses are similar in configuration to those at Atlanta Hartsfield International Airport. There will be three airside concourses, each of which can accommodate 60 aircraft gates. Connections between the landside building and the airside concourses are via an underground automated transit system operating on two parallel tracks. The landside building itself encompasses an area of about 15 acres. The ultimate passenger capacity of the new airport is estimated to be 110 million annual passengers, or almost twice the 1992 demand at O'Hare International Airport. Estimates are that 50 percent of the passengers using this airport will be connecting passengers. The ground access system in the vicinity of the landside building will be a three-level system with the middle level reserved for commercial vehicles. There will be 15,000 parking spaces in parking structures adjacent to the landside building.

Greater Pittsburgh International Airport. In October 1992, the new Midfield Terminal Complex at Greater Pittsburgh International Airport was opened. This complex had been in the planning stages since the early 1970s, and the airfield changes made since that time were made in recognition of the ultimate location of the new terminal complex. The terminal complex consists of about 1.2 million ft^2 and consists of a landside building, an airside building, and a commuter building. The airport was planned to accommodate commuter aircraft on a ramp adjacent to the landside building and large commercial aircraft in a remote airside building connected to the landside building by an underground automated transit system. The landside building consists of four levels: the mezzanine level for airport and airline management, the enplaning level, the deplaning level, and the middle level for access to the transit system. The airside building consists of a central core with concourses in the shape of the letter X emanating from the central core. The airside concourse initially will accommodate 62 air carrier gate positions with an ultimate expansion capability to 95 air carrier gate positions. Passengers can travel through the concourses on a moving walkway. The commuter building will accommodate 25 commuter aircraft gate positions.

The airport was designed to initially accommodate a demand of 24 million annual enplaned passengers and 6200 peak-hour enplaned passengers, which is expected to rise to about 32 million annual enplaned passengers and 9600 peak-hour enplaned passengers by the year 2003. About 61 percent of these passengers are expected to be connecting passengers. Airside demand is expected to rise from 471,000 annual operations at present to 544,000 annual operations in the year 2003. Total parking at the airport will ultimately consist of 27,000 parking spaces, including rental-car and employee parking, of which 5000 are in structural parking adjacent to the terminal complex. Travel from the parking areas to the landside building is by an enclosed moving walkway.

Schematic design

The schematic design process translates the concept development and overall space requirements to drawings which show the general size, location, and shape of various elements in the terminal plan. Functional relationships between the components are established and evaluated. The adequacy of the overall space program is evaluated by airport users relative to their specific needs. This phase of the process specifically examines passenger and baggage flow routes through the system and seeks to examine the adequacy of the facility from the point of view of flow levels and flow conflicts.

Modeling techniques are usually employed in this phase of the process to identify passenger processing, travel, and delay times and the generation of lines at processing facilities. The main purpose of analyzing passenger and baggage handling systems is to determine the extent and size of the facilities needed to provide a desired level of convenience to the passenger at reasonable cost. In this analysis, alternative layouts can be studied to determine which is the most desirable.

Analytical methods

A number of systems analysis techniques have proved useful for the analysis of facilities for passengers and baggage. These include network models, queueing models, and simulation models.

Network models. These models are particularly useful for representing and analyzing the interrelationships between various components of a passenger or baggage processing system. For example, passenger processing can be diagrammed as a network with the nodes representing service facilities and the links representing the travel paths and passenger splits. This type of representation allows estimation of delays to the passenger at various locations within the terminal.

An example of a network that has been applied to the evaluation of arriving passenger delay is the *critical-path model* (CPM) [47]. The CPM is used to coordinate the various activities in the system for handling both passengers and baggage. Nodes that represent critical activities, i.e., those that take the greatest time, are easily identified and can be analyzed in greater detail to determine their effect on overall system performance.

The service time and waiting time at each processor in a network model can be analyzed through either analytical queueing models or simulation.

Analytical queueing models

Queueing theory permits the estimation of delays and queue lengths for service facilities under specified levels of demand. The application of queueing theory yields useful estimates of processing and delay times from which the required sizes of facilities and operating costs may be derived.

Virtually all components of the passenger handling system can be modeled as service facilities using queueing models. The diagram in Fig. 8-20 and Example Problem 8-10 showed an example of the application of a deterministic queueing model to the operation of a runway system to determine aircraft delay. A similar type of analysis can be applied to passenger processing systems. It is possible to evaluate the impact of adding ticket agents on delays to passengers and on the size of queues. With this information it is possible to evaluate the feasibility of alternative operating strategies for the ticketing facility.

Diagrams similar to this one may be constructed for each processor for the passenger and baggage handling system and yield satisfactory results when the demand rate exceeds the service rate.

For the analysis of component delay and queue length when the average demand rate over some period is less than the average service rate, queueing theory is used to generate mathematical functions representative of the arrival and service performance of the system. To specify the mathematical formulation of this problem, it is necessary to define the arrival distribution, service distribution, number and use of servers, and service discipline. Many of the components which service passengers in the airport terminal exhibit a random or Poisson arrival process. The service characteristics are usually exponential, constant, or some general distribution defined by average service times and the variance of average service time. In most cases, there is more than one channel for the performance of passenger service, and the queueing mechanism is based on first come, first served. Extensive research has been carried out in recent years to determine mathematical formulations which adequately represent the processing system [2, 10, 14, 19, 21, 53, 56]. Because of the variability associated with passenger behavior at an airport, it is virtually impossible to obtain precise mathematical formulas for delay at processors. However, reasonable estimates of delay and corresponding queue lengths are possible by using simple formulations.

One such formulation [17, 35, 52, 56] is that of a multiple-station queueing system with a Poisson arrival distribution and a service time distribution characterized by the average service time and the variance of the average service time as shown in

$$W_s = \left(\frac{\sigma^2 + t^2}{2t^2} \right) \frac{\lambda^k t^{k+1}}{(k-1)!(k-\lambda t)^2 \displaystyle\sum_{n=0}^{n=k-1} [(\lambda t)^n/n!] + (\lambda t)^k/[(k-1)!(k-\lambda t)]} \tag{10-1}$$

where W_s = average delay per person

λ = average demand rate

t = average service time for a processor, the reciprocal of average service rate of processor μ

σ = standard deviation of average service time of a processor

k = number of processors

n = counter in equation

This equation is valid when the average demand rate on the system of processors λ is less than to total service rate of the processors $k\mu$; that is, the ratio of the average demand rate to the total service rate ρ is less than 1.

When the number of processors k is equal to 1, this equation reduces to

$$W_s = \left(\frac{\sigma^2 + t^2}{t^2}\right) \frac{\lambda}{2\mu(\mu - \lambda)} \tag{10-2}$$

or since ρ is equal to λ/μ,

$$W_s = \left(\frac{\sigma^2 + t^2}{t^2}\right) \frac{\rho}{2\mu(1 - \rho)} \tag{10-3}$$

For a single-server system which exhibits a Poisson arrival distribution and an exponential service time distribution, Eq. (10-1) reduces to

$$W_s = \frac{\lambda}{\mu(\mu - \lambda)} \tag{10-4}$$

or

$$W_s = \frac{\rho}{\mu(1 - \rho)} \tag{10-5}$$

For a single-server system which exhibits a Poisson arrival distribution and a constant service time distribution, Eq. (10-1) reduces to

$$W_s = \frac{\lambda}{2\mu(\mu - \lambda)} \tag{10-6}$$

or

$$W_s = \frac{\rho}{2\mu(1 - \rho)} \tag{10-7}$$

When demand is less than capacity, the following expression gives the average line length N over the period being analyzed and consists of those in service and those waiting for service at a processor.

$$N = \left(W_s + \frac{1}{\mu}\right)\lambda \tag{10-8}$$

When $\rho > 1$, there is statistical delay plus excess delay which is defined by a deterministic model. For design purposes it is sufficient to estimate delays in such a system in which the statistical delay W_s is computed from the appropriate equation above with $\rho = 0.90$ and the excess delay W_e is added to this from Eq. (10-9) to compute the total processor delay. The rationale for computing the statistical delay with $\rho = 0.9$ is that in reality as demand approaches capacity, the delay cannot become infinite because airline or airport operating practices will limit delay by utilizing additional servers.

$$W_e = \frac{T(\lambda - k\mu)}{2k\mu} \tag{10-9}$$

where W_e = delay when demand rate exceeds service rate
T = time period during which demand rate exceeds service rate
λ = total demand rate on system of processors
k = number of processors
μ = service rate of a processor

It is usually assumed that the time required to reduce the demand to capacity T is about one-half of the period being analyzed. Typically when demand exceeds capacity, the operator of the facility will increase the number of operating service facilities to alleviate the growth in both waiting time and queue length. However, the extent to which this is done is a function of airline operating policies and the availability of additional staff and physical facilities.

For a single processor system average line length N over the period analyzed, including those in service and those waiting for service, when demand exceeds capacity can be estimated by

$$N = \left(W_s + W_e + \frac{1}{\mu} \right)\mu \tag{10-10}$$

Estimates of the waiting times and queue lengths for multiple-service channels may be obtained by proportioning the demand equally among processors with the same service characteristics and utilizing single-server models. Better estimates may be obtained through the use of multiple-channel queueing models whose mathematical formulation was given above. A representation of the passenger delay at baggage claim facilities is given by the relationship [52]

$$W_t = E(t_2) + \frac{nT}{n+1} - E(t_1) \tag{10-11}$$

where W_t = average delay at baggage claim area
$E(t_2)$ = expected value of time when first piece of baggage arrives at claim area

$E(t_1)$ = expected value of time when passengers arrive at claim area

n = number of pieces of baggage to be claimed by each passenger

T = time from arrival of first bag until arrival of last bag at claim device

Others have formulated models which show the buildup and drop-off of passengers and baggage in the claim area [19].

The use of a generalized probability density function called the *Erlang distribution* is recommended as a mechanism for evaluating the service characteristics of various passenger component processors. By collecting a sample of data for a specific component, the constant in the Erlang distribution function can be calculated. This constant determines the particular functional relationships for the server. It is possible that this type of distribution may better describe the queueing characteristics of processors. This distribution has been used successfully in passenger terminal modeling [15]. Great care must be exercised in the application of mathematical models and the interpretation of the results. In most cases, the mathematical representation of the terminal system is best suited for comparison of alternatives and identification of those components in the system requiring detailed analyses.

The specification of the service time distribution for use in queueing equations is a function of the distribution of service times demonstrated by the service facility. In general, those processors which exhibit a requirement for small service times (e.g., flow-through type of facilities such as doors, security, and gates) are probably best represented by an exponential service time distribution. Those facilities which require a finite service time at a processor (such as ticketing, baggage check, seat selection, and rental-car checkout) are probably best represented by either a general service time distribution, which is characterized by the average service time and the standard deviation of the average service time, or a constant-service-time distribution.

In a network model, the expected value of the passenger processing time $E[T_p]$ is given by

$$E(T_p) = E(T_w) + E(T_s) + E(T_t) \qquad (10\text{-}12)$$

where $E(T_w)$ = expected value of the passenger delay or waiting time throughout the system of processors;

$E(T_s)$ = expected value of the passenger service time throughout the system of processors; and

$E(T_t)$ = expected value of the passenger travel time through the network of processors.

TABLE 10-4   Observed Service Times for Passenger Processing Facilities at Airports

Component type	Service rate per passenger, s	Standard deviation
Entrance and exit doors		
Automated with baggage	2.0–2.5	0.5
Automated without baggage	1.0–1.5	0.75
Manual with baggage	3.0–5.0	1.0
Manual without baggage	1.5–3.0	0.75
Stairways	3.0–4.0	1.0
Escalators	1.0–3.0	1.0
Moving sidewalks	1.0–3.0	1.0
Apron doors		
With stairs	4.0–8.0	2.0
Without stairs	3.0–7.0	1.5
Jetway	2.0–6.0	1.0
Ticketing and baggage		
Manual with baggage	180–240	60
Manual without baggage	100–200	30
Baggage only	30–50	10
Information	20–40	10
Automated with baggage	160–220	30
Automated without baggage	90–180	40
Security		
Hand-check baggage	30–60	15
Automated	30–40	10
Seat selection		
Single flights	25–60	20
Multiple flights	35–60	15
Rental car		
Check-in	120–240	60
Checkout	180–300	90
Automated check-in	60–90	20
Baggage claim		
Manual	10–15	8
Automated carousel	5–10	5
Automated racetrack	5–10	5
Automated tee	6–12	5

SOURCES: Various airport studies.

Estimates for the observed range of service time for many of the passenger processing components at an airport are given in Table 10-4.

Example Problem 10-1 illustrates the analysis of the enplaning system at an airport, by using a network model and the above queueing equations.

Example Problem 10-1   The terminal building layout for enplaning passengers at an airport is shown in Fig. 10-24.

Two airlines are servicing the airport, North American (NA) and Western Pacific (WP). The enplaning passenger processing facilities and their service rates are given in Table 10-5.

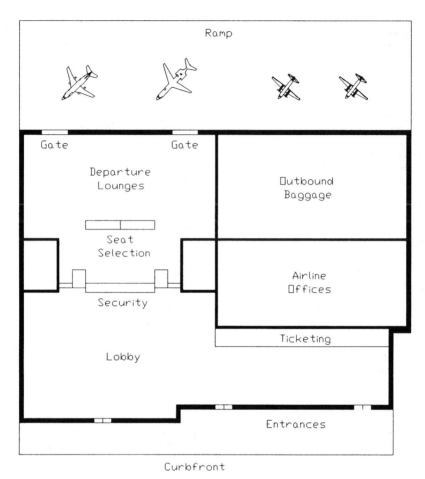

Figure 10-24 Layout of enplaning passenger processing system for Example Problem 10-1.

Demand data collected at the airport indicate that 10 percent of the enplaning passengers proceed directly from curb front to security, 20 percent of Western Pacific enplaning passengers use the express check-in, there is 0.5 visitor per passenger during the peak hour, and 50 percent of these visitors proceed beyond security.

The peak-hour enplaning passenger demand on the design day is expected to be 135 passengers, of which 53 percent use North American and 47 percent use Western Pacific. Typically the peak 30 min of the peak hour sees 57 percent of the peak-hour demand. Assume that average walking rates are 1.5 ft/s. Airline operating practices limit the periods when demand exceeds capacity to 15 min.

It is necessary to determine the average enplaning passenger delay and line length at each processor as well as the average passenger processing time during the peak 30 min of the peak hour on the design day at this airport.

The link-and-node diagram in Fig. 10-25 shows the relationship between the enplaning passenger processors at the airport, and it includes the passenger

TABLE 10-5   Enplaning Passenger Service Processors for Example Problem 10-1

Type	Airline code*	Number of processors	Average service time per passenger, s
Entrance doors	All	3	15
Regular ticketing	NA	3	210
	WP	3	180
Express check-in	WP	1	60
Security	All	2	30
Seat selection	NA	1	45
	WP	1	30
Ramp gates	NA	1	20
	WP	1	20

*NA = North American. WP = Western Pacific.

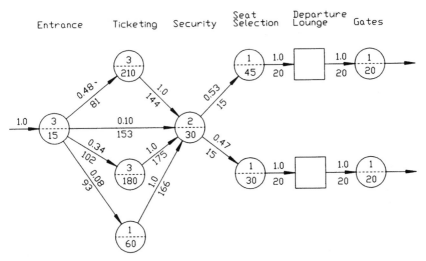

Figure 10-25   Link-and-node diagram representing enplaning passenger processing system for Example Problem 10-1.

split between processors, the number of processors, the processing time per passenger, and the distance between processors in ft.

Since the peak 30-min passenger arrival rate is 57 percent of the peak-hour arrival rate, the peak 30-min passenger demand into the enplaning passenger system is 0.57 × 135 = 77 passengers. Since there is 0.5 visitor per passenger, the peak 30-min visitor demand into the enplaning passenger system is 0.5 × 77 = 39 visitors. Combining these demand parameters with the passenger splits given in Fig. 10-25, recognizing that only 50 percent of the visitors proceed beyond security, we get the processor flow rates in the peak 30 min given in Fig. 10-26. In this figure, the numbers above the lines connecting processors represent the passengers flowing into the processor, and the numbers below

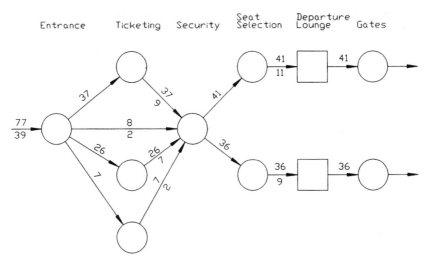

Figure 10-26 Link-and-node diagram representing passenger and visitor flows into passenger processing facilities for Example Problem 10-1.

TABLE 10-6　Service Processor Characteristics for Example Problem 10-1

Processor	Demand rate, persons/min	Service rate, persons/min	Processor utilization
Entrance doors	1.29	4.00	0.32
Regular ticketing			
North American	0.41	0.29	1.41
Western Pacific	0.29	0.33	0.88
Express check-in			
North American	0.23	1.00	0.23
Security	1.62	2.00	0.81
Seat selection			
North American	1.37	1.33	1.03
Western Pacific	1.20	2.00	0.60
Ramp gates			
North American	1.37	3.00	0.46
Western Pacific	1.20	3.00	0.40

these lines represent the visitors flowing into each processor when the processor processes visitors.

Assuming that all processors may be modeled by single-processor systems in which the demand is split evenly between processors, the demand rate in persons per min, the service rate in persons per min, and the ratio of the demand to capacity for each processor, or the *processor utilization*, are computed, as shown in Table 10-6.

The departure lounges are accumulating processors and are analyzed differently from the queueing processors. Also note that for the North American regular ticketing position and the North American seat selection components, demand exceeds capacity since the processor utilization is greater than 1.

Based upon an examination of airport processors, it was found that those processors which are essentially flow-through processors (such as entrance doors, security gates, and ramp gates) behave as Poisson arrivals and exponential service rate queueing mechanisms, whereas those processors which require a discrete service function (such as regular ticketing, express check-in, and seat selection) behave as Poisson arrival and constant-service-rate queueing mechanisms.

Therefore, for the entrance doors, which are Poisson arrivals and exponential services, the average passenger delay is, from Eq. (10-4),

$$W_s = \frac{1.29}{4.00(4.00 - 1.29)} = 0.12 \text{ min}$$

and from Eq. (10-8) the average line length is

$$N = \left(0.12 + \frac{1}{4.00}\right)1.29 = 0.48 \text{ person}$$

For the North American regular ticket counters, which exhibit Poisson arrivals and constant service times, since the demand rate is greater than the capacity, we have from Eq. (10-7) with $\rho = 0.9$

$$W_s = \frac{0.9}{2(0.29)(1 - 0.9)} = 15.52 \text{ min}$$

and from Eq. (10-9)

$$W_e = \frac{15 (0.41 - 0.29)}{2(0.29)} = 3.10 \text{ min}$$

Note that the value of k is 1 because the demand was allocated equally among the ticketing positions, and the excess delay period T is $0.5 \times 30 = 15$ min.

Therefore, $W = 15.52 + 3.10 = 18.62$ min of average delay time per passenger. Then the average line length is, from Eq. (10-10),

$$N = \left(18.62 + \frac{1}{0.29}\right)0.29 = 6.40 \text{ passengers}$$

Computing the delays and queue lengths for each processor yields the values in Table 10-7.

This table shows considerable delay and line formation at both North American and Western Pacific regular ticketing facilities, the security checkpoint, and the North American seat selection facility.

From the passenger splits shown in Fig. 10-25 and the distances between components, the expected values of the passenger delay time, passenger service time, and passenger travel time for the average airport passenger in the peak 30 min can be computed.

The expected values are then computed for the delay time (T_w), service time (T_s), and walking or travel time (T_t) for the average airport passenger, to find the average passenger processing time from Eq. (10-11).

TABLE 10-7   Average Passenger Delay Time, Service Time, and Processor Line Lengths in Peak 30 Min for Example Problem 10-1

Processor	Delay time, min		Service time, min	Line length, persons
	Statistical	Excess		
Entrance doors	0.12		0.25	0.48
Regular ticketing				
North American	15.52	3.10	3.45	6.40
Western Pacific	10.98		3.00	4.05
Express check-in				
Western Pacific	0.15		1.00	0.26
Security	2.13		0.50	4.26
Seat selection				
North American	3.38	0.23	0.75	5.80
Western Pacific	0.38		0.50	1.06
Ramp gates				
North American	0.28		0.33	0.84
Western Pacific	0.22		0.33	0.66

$$E(T_w) = 1.0(0.12) + 0.48(15.52 + 3.10) + 0.34(10.98) + 0.08(0.15)$$
$$+ 1.0(2.13) + 0.53(3.38 + 0.28) + 0.47(0.38 + 0.22)$$
$$= 17.2 \text{ min}$$

$$E(T_s) = 1.0(0.25) + 0.48(3.45) + 0.34(3.00) + 0.08(1.00) + 1.0(0.5)$$
$$+ 0.53(0.75 + 0.33) + 0.47(0.50 + 0.33)$$
$$= 4.5 \text{ min}$$

The average walking or travel distance (D_t) for a passenger is

$$E(D_t) = 0.48(81 + 144) + 0.34(102 + 175) + 0.08(93 + 166) + 0.10(153)$$
$$+ 0.53(55) + 0.47(55)$$

$$= 293 \text{ ft}$$

$$E(T_t) = \frac{293}{1.5(60)} = 3.3 \text{ min}$$

The average passenger processing time for the average airport passenger during the peak 30 min is then

$$E(T_p) = 17.2 + 4.5 + 3.3 = 25 \text{ min}$$

Since departure lounges are only waiting areas, there is no queueing system delay in these facilities. Generally speaking, these facilities are designed with an area requirement per passenger and visitor in the departure lounge. The area required is based upon the number of passengers and visitors who would be in the departure lounge at the moment the aircraft is allowed to be boarded. Typically the space requirement is from 15 to 25 ft^2 per person in the departure lounge at that point.

Given a typical arrival distribution at the airport, such as that shown in Table 10-8, it is assumed that the total number of passengers and visitors

TABLE 10-8 Typical Arrival Distribution of Originating Passengers at Airport for Domestic Flights for Example Problem 10-1

Time before scheduled departure, min	Percentage of passengers at airport	Time before scheduled departure, min	Percentage of passengers at airport
95	0	45	31
90	0	40	40
85	1	35	51
80	2	30	61
75	3	25	71
70	5	20	80
65	8	15	88
60	12	10	94
55	17	5	98
50	23	0	100

arriving at the airport at the moment the flight is called for boarding is also the number in the departure lounge at that time.

In this problem, let us assume that the flight is boarded 15 min before scheduled departure and that the airline departure lounge requirements are 20 ft² per passenger.

Therefore, at this time 88 percent of the passengers and their visitors would be in the departure lounge. For Western Pacific, this means that there are 0.88 × 0.47 × 135 passengers and 0.88 × 0.47 × 135 × 0.5 × 0.5 visitors in the departure lounge at this time. This totals 56 passengers and 14 visitors. Therefore, the departure-lounge size requirements are 20(56 + 14) = 1400 ft².

Simulation models

These models become particularly useful when the analysis of the passenger handling system is to be performed at a relatively detailed level or for extended periods. The models are useful for the analysis of both the whole system and parts of it [28, 34, 39]. When some important inputs to analysis are unobtainable, such as possible future flight schedules, it is possible with the use of computer simulation to analyze the operation of the system under randomly generated inputs.

Simulation is also particularly useful when analysis is to be repeated for varying operating conditions in order to perform sensitivity studies. Computers allow such repeated analysis which would otherwise be prohibitively expensive and very time-consuming. Most computer systems have standard simulation packages available which can be adapted to many physical planning problems including airports.

It is important to realize that computer simulation is not a substitute for analysis when information on the system is lacking. To construct a simulation model, nearly as much detailed information about system operation is necessary as for other analytic techniques. The main feature of simulation is the high speed with which computers can perform lengthy calculations. In the analysis of systems operations, computer simulation should be used with caution, and several runs should be made so that the statistical reliability of the results can be determined.

Simulation techniques have been studied by the FAA [10, 21, 24] and have been used in many studies to determine facility needs. An

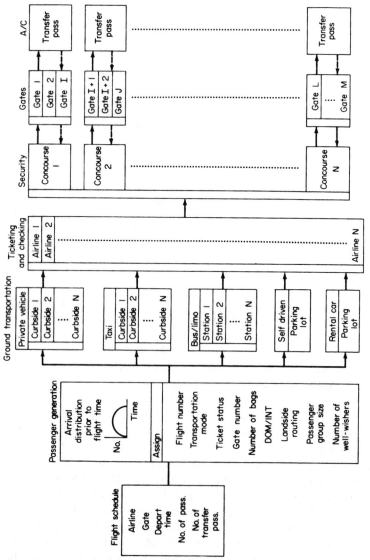

Figure 10-27 Flowchart for enplaning passenger simulation model (*Federal Aviation Administration [10]*).

outline of the flow of an enplaning passenger through the airport system which can be modeled by the FAA simulation model is given in Fig. 10-27. In the schematic design phase of the planning project at Fort Lauderdale–Hollywood International Airport, simulation was used to determine the parking requirements and the number of parking toll collection facilities needed at the exit [30]. Inputs into the simulation included the airline flight schedule, distribution of the total number of parkers relative to the flight schedule, and historical distribution of parking duration, obtained from the analysis of parking ticket stubs. Alternative flight schedules were used to generate the arrival distribution at the parking facility, and the parking duration distribution, shown in Fig. 10-28, was used to generate the random service times required by the arrivals. As a result of the simulation, it was possible to determine the peaking characteristics of arrivals and departures at the parking garage as well as the accumulation of vehicles within the parking facility.

The following steps are recommended in the design of a simulation model for application to airport terminal projects [50]:

1. Define the scope of the simulation in terms of the questions to be answered, components to be included, and level of detail required.

2. Specify the required output so that an interpretation of the results will solve the problems to be addressed.

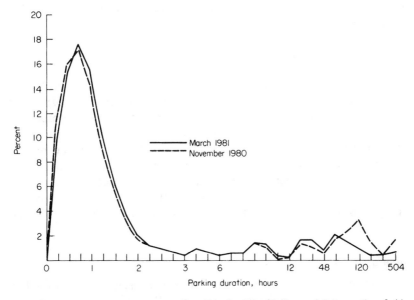

Figure 10-28 Parking duration at Fort Lauderdale–Hollywood International Airport (*Aviation Planning Associates, Inc. [30]*).

3. Structure the model so that the abstract representation of the components and the events and interactions between components are indicative of terminal performance.

4. Define the input data and their variability.

5. Once the model has been developed, verify through testing on actual systems.

6. Apply the model and modify the facility design in accordance with the model results.

7. Review the findings and design vis-à-vis the degree of variability in the output and through reasonableness checks.

Design development

The final planning phase in terminal projects is called *design development*. In this phase, the size and character of the project are fixed and checked against the findings and recommendations in prior phases of the project. Acceptance of the project by the airport owner, tenants, and airlines is the final product of this phase. Capital budgeting, operating, maintenance, and administrative costs over the lifetime of the project are determined, and a revenue plan is adopted. Agreements are drawn up concerning rate and charge structures for the airlines, concessionaires, and other tenants. The project moves toward implementation through the development of construction documents, bid letting and acceptance, and construction following this phase of planning.

The Apron Gate System

The apron provides the connection between the terminal buildings and the airfield. It includes aircraft parking areas, called *ramps,* and aircraft circulation and taxiing areas for access to these ramps. On the ramp, aircraft park in areas designated as gates. The discussion in this section is limited to the apron gate area for scheduled commercial aircraft operations. The size of the apron gate area depends on four factors: the number of aircraft gates, the size of the gates, the maneuvering area required for aircraft at gates, and the aircraft parking layout in the gate area. The layout of the apron area is discussed in Chap. 9.

Number of gates

As with other airport facilities, the number of gates is determined so as to accommodate a predetermined hourly flow of aircraft. Thus, the number of gates required depends on the number of aircraft to be handled during the design hour and on the amount of time each aircraft occupies a gate. The number of aircraft that need to be handled simultaneously is a function of the traffic volume at the airport. As mentioned earlier, it is customary to use the estimated peak hour volume as the input for estimating the number of gates required at the airport. However, to achieve a balanced airport design, this volume should not exceed the capacity of the runways.

The amount of time an aircraft occupies a gate is referred to as the *gate occupancy time*. It depends on the size of aircraft and on the type of operation, i.e., a through or turnaround flight. Aircraft parked at a gate are there for passenger and baggage processing and for aircraft servicing and preparation for flight. Large aircraft normally occupy gates a longer time than small aircraft. This is because large aircraft accommodate more passengers and require more time for aircraft servicing, preflight planning, and refueling. The type of operation also affects the gate occupancy time by affecting service requirements. Thus an aircraft on a through flight may require little or no servicing, and consequently the gate occupancy time can be as low as 20 to 30 min. On the other hand, an aircraft on a turnaround flight will require complete servicing, resulting in gate occupancy times ranging from 40 min to more than 1 h. Data on typical aircraft occupancy times at gates were given in Chap. 8. The table shown in Fig. 10-29 lists the activities that normally take place during a turnaround stop, together with a typical time schedule for these activities.

The methods used to estimate the peak-hour gate requirements are discussed in Chap. 8. However, the gate requirements are normally determined through the use of simulated design-day flight schedules or through direct requests from the airlines servicing the airport. If simulation is used, first a design-day schedule is forecast, and then based upon the design-day schedule and the practices of the airlines at the airport, a ramp chart based upon the design day is constructed to determine gate requirements.

The average daily gate utilization factor for all gates at an airport, discussed in Chap. 8, usually varies between 0.5 and 0.8. This factor accounts for the fact that it is unlikely that all the gates available at a terminal building will be used 100 percent of the time. This is caused by the fact that aircraft maneuvering into and out of a gate often block other aircraft attempting to move into or out of their gates and by the fact that aircraft schedules often lead to time gaps between the departure of one aircraft and the arrival of another using the same gate.

The *gate-use strategy* employed by the airlines at the airport also influences the average gate utilization factor. At airports where gates are used mutually by all airlines, a common gate-use strategy, the

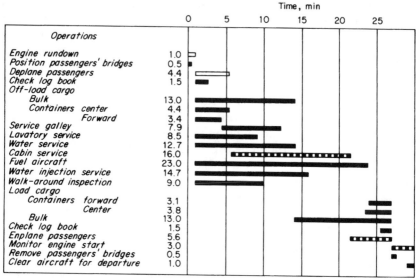

■■■ Critical time path

Figure 10-29 Typical time schedule of aircraft servicing activities at gate (*Ralph M. Parsons and Federal Aviation Administration [50]*).

gate utilization factor typically varies between 0.6 and 0.8. At airports where groups of gates are used exclusively by different airlines, an exclusive gate-use strategy, the utilization factor drops to about 0.5 or 0.6. The determination of the number of gates needed at an airport should be subjected to the analytical techniques given in Chap. 8 and to the gate-use strategies adopted by the tenant airlines.

The use of simulation to generate a design-day schedule from which a ramp chart can be constructed to determine design-day gate requirements is illustrated in Example Problem 10-2.

Example Problem 10-2 The current schedule of airline flights at an airport on a typical day in the peak month is given in Table 10-9. A forecast for this airport indicates that 10 additional flights will be added to the future-year design-day schedule at this airport. It is necessary to perform a simulation for the schedules of these additional flights.

We will assume in this problem that the flight schedule of the simulated flights will follow the current day flight schedule distribution.

First it necessary to obtain a probability distribution for both flight arrival times and gate occupancy times. These are obtained by determining the frequency distribution of each time range for flight arrivals and gate occupancy times and then integrating this frequency distribution to obtain the cumulative probability distribution function.

For simplicity in this problem, the flight arrival time distribution is found by grouping flight arrival times into 1-h increments, and the gate occupancy time distribution is found by grouping gate occupancy times into 15-min increments.

TABLE 10-9   Current Airline Schedule on Typical Day in the Peak Month for
Example Problem 10-2

Airline code*	Flight number	Arrival time	Departure time	Aircraft
AE	8/7	7:45 a.m.	9:30 a.m.	727
AE	353	10:30 a.m.	11:15 a.m.	727
AE	319/642	11:30 a.m.	1:00 p.m.	727
AE	421	12:00 p.m.	1:00 p.m.	727
AE	439	1:45 p.m.	2:30 p.m.	727
AE	889	1:45 p.m.	2:30 p.m.	727
AE	852	3:30 p.m.	4:00 p.m.	727
AE	422/660	3:45 p.m.	5:00 p.m.	727
AE	591/544	5:15 p.m.	6:15 p.m.	727
AE	310/390	6:00 p.m.	8:00 p.m.	727
AE	411/428	9:00 p.m.	10:15 p.m.	727
CL	64	7:15 a.m.	7:45 a.m.	737
CL	489	11:15 a.m.	11:45 a.m.	737
CL	41	11:30 a.m.	12:15 p.m.	737
CL	50	1:45 p.m.	2:15 p.m.	737
CL	936	1:45 p.m.	2:15 p.m.	737
CL	81	4:15 p.m.	5:00 p.m.	737
CL	493	8:30 p.m.	9:00 p.m.	737
RX	161	10:15 a.m.	10:45 a.m.	MD-8
RX	321/844	4:45 p.m.	5:45 p.m.	MD-8

*AE = Alpha Express. CL = Coastal Link. RX = Regional Express.

The flight schedule frequency distribution is then found by counting the number of flights in each 1-h increment beginning with 6:00 a.m. and ending with
11:59 p.m. The total numbers of flights arriving in the periods are shown in
Table 10-10.

The cumulative probability distribution function of flight arrival times is
computed by finding the probability that a flight will arrive in a given period or
later, beginning with the earliest period. These values are computed in Table
10-10 and plotted in Fig. 10-30.

By a similar technique, the gate occupancy times of the flights are grouped as
shown in Table 10-11, and the cumulative probability function of these gate
occupancy times is plotted in Fig. 10-31.

To simulate a flight arrival and a gate occupancy time for a flight, a table of
random digits must be used. If this is done, two sets of random numbers, one
representing the flight arrival time and the other representing the gate occupancy time, for each of the 10 flights to be simulated are found; these are shown
in Table 10-12.

The first simulated flight has an arrival time random number of 0.92 and a
gate occupancy time random number of 0.70. Using the flight arrival time random number of 0.92 with the cumulative probability function of flight arrival
times in Fig. 10-30, we find that the simulated arrival time of the first flight is
7:45 a.m. Using the gate occupancy time random number of 0.70 with the
cumulative probability function of gate occupancy times in Fig. 10-31, we find
that the simulated gate occupancy time of the first flight is 45 min. Both numbers are taken to the nearest 15-min increment for simplicity. Therefore, the
first simulated flight has an arrival time of 7:45 a.m. and a flight departure
time of 8:30 a.m.

TABLE 10-10   Aircraft Arrival Distribution for Example Problem 10-2

| Time period | | Number of flights | Arriving in period or later | |
From	To		Cumulative number	Cumulative percentage
6:00 a.m.	6:59 a.m.	0	20	1:00
7:00 a.m.	7:59 a.m.	2	20	1:00
8:00 a.m.	8:59 a.m.	0	18	0:90
9:00 a.m.	9:59 a.m.	0	18	0:90
10:00 a.m.	10:59 a.m.	2	18	0:90
11:00 a.m.	11:59 a.m.	3	16	0.80
12:00 p.m.	12:59 p.m.	1	13	0:65
1:00 p.m.	1:59 p.m.	4	12	0:60
2:00 p.m.	2:59 p.m.	0	8	0:40
3:00 p.m.	3:59 p.m.	2	8	0:40
4:00 p.m.	4:59 p.m.	2	6	0:30
5:00 p.m.	5:59 p.m.	1	4	0:20
6:00 p.m.	6:59 p.m.	1	3	0:15
7:00 p.m.	7:59 p.m.	0	2	0:10
8:00 p.m.	8:59 p.m.	1	2	0:10
9:00 p.m.	9:59 p.m.	1	1	0:05
10:00 p.m.	10:59 p.m.	0	0	0:00
11:00 p.m.	11:59 p.m.	0	0	0:00

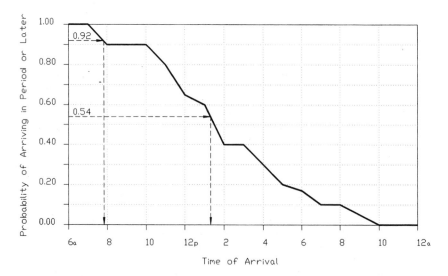

Figure 10-30 Flight arrival distribution for Example Problem 10-2.

This process is continued for each simulated flight. For example, the third simulated flight has an arrival time random number of 0.54 and a gate occupancy time random number of 0.19. Again, using the flight arrival time random number of 0.54 with the cumulative probability function of flight arrival times in Fig. 10-30, we find that the simulated arrival time of the third flight is 1:15 p.m. Using the gate occupancy time random number of 0.19 with the cumulative probability function of gate occupancy times in Fig. 10-31, we find that the simulated

TABLE 10-11   Aircraft Gate Occupancy Time Distribution for Example Problem 10-2

Gate occupancy time, min		Number of flights	Duration range or more	
From	To		Cumulative number	Cumulative percentage
0	14	0	20	1.00
15	29	0	20	1.00
30	44	7	20	1.00
45	59	5	13	0.65
60	74	3	8	0.40
75	89	2	5	0.25
90	104	1	3	0.15
105	119	1	2	0.10
120	134	1	1	0.05
135	149	0	0	0.00
150	164	0	0	0.00
165	179	0	0	0.00
180	194	0	0	0.00

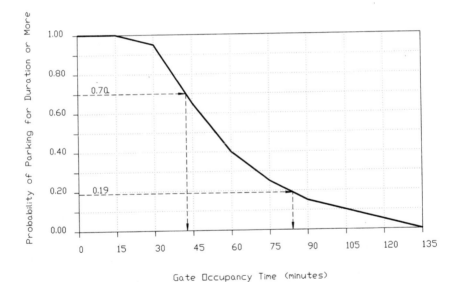

Figure 10-31   Flight gate occupancy time for Example Problem 10-2.

gate occupancy time of the third flight is 90 min. Therefore, the third simulated flight has an arrival time of 1:15 p.m. and a flight departure time of 2:45 p.m.

The above simulation process is shown by the dashed lines in Figs. 10-30 and 10-31.

Assuming a simulation was also performed for the airline and the aircraft for each flight, the results of the simulation are tabulated in Table 10-12. The new flight schedule in the design year, with both the current flights and the simulated flights, is shown in Table 10-13. In this table, the simulated flights are each given the flight number 9999 for reference purposes.

TABLE 10-12   Gate Simulation Results for Example Problem 10-2

	Aircraft arrivals			Gate occupancy time	
Simulated flight	Random number	Arrival time	Random number	Duration time, min	Departure time
1	0.92	7:45 a.m.	0.70	45	8:30 a.m.
2	0.88	10:15 a.m.	0.80	45	11:00 a.m.
3	0.54	1:15 p.m.	0.19	90	2:45 p.m.
4	0.18	5:30 p.m.	0.62	45	4:15 p.m.
5	0.82	10:45 a.m.	0.43	60	11:45 a.m.
6	0.87	10:15 a.m.	0.86	30	10:45 a.m.
7	0.75	11:15 a.m.	0.46	60	12:15 p.m.
8	0.28	4:15 p.m.	0.69	45	5:00 p.m.
9	0.59	1:00 p.m.	0.87	30	1:30 p.m.
10	0.60	1:00 p.m.	0.93	30	1:30 p.m.

TABLE 10-13   Simulated Airline Schedule on Typical Day in Peak Month in Design Year for Example Problem 10-2

Ref. no.	Airline code*	Flight number	Time		Aircraft
			Arrival	Departure	
1	AE	8/7	7:45 a.m.	9:30 a.m.	727
2	AE	9999	10:15 a.m.	10:45 a.m.	727
3	AE	9999	10:15 a.m.	11:00 a.m.	727
4	AE	353	10:30 a.m.	11:15 a.m.	727
5	AE	9999	10:45 a.m.	11:45 a.m.	727
6	AE	319/642	11:30 a.m.	1:00 p.m.	727
7	AE	421	12:00 p.m.	1:00 p.m.	727
8	AE	9999	1:00 p.m.	1:30 p.m.	727
9	AE	9999	1:00 p.m.	1:30 p.m.	727
10	AE	439	1:45 p.m.	2:30 p.m.	727
11	AE	889	1:45 p.m.	2:30 p.m.	727
12	AE	852	3:30 p.m.	4:00 p.m.	727
13	AE	422/660	3:45 p.m.	5:00 p.m.	727
14	AE	9999	4:15 p.m.	5:00 p.m.	727
15	AE	591/544	5:15 p.m.	6:15 p.m.	727
16	AE	9999	5:30 p.m.	6:15 p.m.	727
17	AE	310/390	6:00 p.m.	8:00 p.m.	727
18	AE	411/428	9:00 p.m.	10:15 p.m.	727
19	CL	64	7:15 a.m.	7:45 a.m.	737
20	CL	9999	7:45 a.m.	8:30 a.m.	737
21	CL	489	11:15 a.m.	11:45 a.m.	737
22	CL	9999	11:15 a.m.	12:15 p.m.	737
23	CL	41	11:30 a.m.	12:15 p.m.	737
24	CL	9999	1:15 p.m.	2:45 p.m.	737
25	CL	50	1:45 p.m.	2:15 p.m.	737
26	CL	936	1:45 p.m.	2:15 p.m.	737
27	CL	81	4:15 p.m.	5:00 p.m.	737
28	CL	493	8:30 p.m.	9:00 p.m.	737
29	RX	161	10:15 a.m.	10:45 a.m.	MD-8
30	RX	321/844	4:45 p.m.	5:45 p.m.	MD-8

*AE = Alpha Express. CL = Coastal Link. RX = Regional Express.

Ramp charts

A ramp chart is a graphical representation of the gate occupancy by aircraft throughout the day. Airlines and airports use ramp charts to display the actual gate assignment of aircraft for the flight schedule at the airport.

The ramp chart can also be used to determine the gate requirements at an airport. When used for this purpose, the ramp chart does not display the actual assignment of aircraft to specific gates for the design-day schedule; it indicates only the assignment of aircraft to gates for the determination of the number of gates required at the airport.

Several factors influence the gate requirements at an airport. Obviously the flight schedule and the gate occupancy time of aircraft are of paramount importance. However, the scheduling and ramp operating practices of the airlines and the gate-use strategy of the airlines are also important. As a result of the scheduling and ramp operating practices, a gate cannot be used 100 percent of the time, as discussed earlier. The gate-use strategy considers whether the gates will be exclusive-use, shared-use, or common gates. Exclusive-use gates are reserved for the exclusive use of one airline. A shared gate is shared by two or three airlines. A common-use gate is allocated by the airport based upon the demand for gates and may be used by any airline at the airport.

Gates are sized based upon the geometric properties of the aircraft that will occupy the gates. Therefore, gates may be called wide-bodied gates because they are sized to accommodate wide-bodied aircraft. These gates may also be used by narrow-bodied aircraft. Narrow-bodied gates can only be used by narrow-bodied aircraft. Many airports also have commuter gates which are sized to accommodate commuter aircraft.

The determination of the number of aircraft gates required at an airport is shown by constructing ramp charts in Example Problem 10-3.

Example Problem 10-3   Let us determine the number of gates required at an airport based upon the design-day schedule shown in Table 10-13. Let us determine the gate requirements under an exclusive gate-use strategy, a shared-gate-use strategy, and a common gate-use strategy. Let us assume that the scheduling and operating practices of the airlines require that a minimum time gap of 15 min must be allowed between the departure of a scheduled flight from a gate and the arrival of the next scheduled flight at that gate.

Under an exclusive gate-use strategy, each airline will have its own gates which cannot be used by any other airline. To find the number of gates required under this gate-use strategy, a graph is constructed with time on the X axis and the number of gates on the Y axis. To determine the minimum number of gates required, the flight schedule of each airline must be sorted by first arrival time and then departure time. That is, two flights which have the same arrival time are also sorted by the departure time.

In Table 10-13, the flight schedule of each airline has been sorted, and a reference number is placed next to each flight for each airline. This reference number is placed on the ramp chart for illustrative purposes.

To determine the gates required for Alpha Express (AE) Airlines, since the first scheduled flight is scheduled to arrive at 7:45 a.m. and scheduled to depart at 9:30 a.m., a gate is opened on the ramp chart, and the block of time from 7:45 a.m. to 9:30 a.m. is filled in on the ramp chart, to indicate that the gate is occupied during that period. The next scheduled flight by Alpha Express Airlines is scheduled to arrive at 10:15 a.m. and to depart at 10:45 a.m. Since the next scheduled arrival can be placed at the gate just opened up 15 min after the departure of the previous aircraft, this flight can be scheduled into that gate, and the block of time from 10:15 a.m. to 10:45 a.m. is filled in. The next flight of Alpha Express Airlines is scheduled to arrive at 10:15 a.m. and depart at 11:00 a.m. However, this flight cannot be scheduled for the first gate since that gate is occupied during part of that time. Therefore, a new gate is opened up on the ramp chart to accommodate this aircraft, and the block of time from 10:15 a.m. to 11:00 a.m. is filled in at that gate.

This process is continued until all the Alpha Express Airlines flights have been assigned to gates. As can be seen on the top portion of Fig. 10-32, to accommodate the flights of Alpha Express Airlines, 4 gates are required.

The same process is repeated for each airline, and as shown in the middle and bottom portions of Fig. 10-32, this results in 3 gates being required for Coastal Link (CL) Airlines and 1 gate being required for Regional Express (RX) Airlines.

Therefore, under an exclusive gate-use strategy, this flight schedule requires 8 gates to accommodate the airline schedule at the airport.

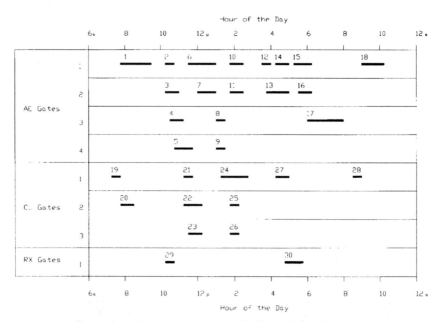

Figure 10-32 Ramp chart for exclusive gate use for Example Problem 10-3.

For the construction of a shared-gate-use strategy ramp chart, let us assume that Coastal Link and Regional Express Airlines will share gates at the airport and that Alpha Express Airlines will have exclusive-use gates. Therefore, the ramp chart for Alpha Express Airlines does not change, but the ramp chart for the shared-gate-use airlines must be constructed by first sorting all the flights of Coastal Link and Regional Express Airlines together by arrival time and departure time, as shown in Table 10-14. By using the procedure above, Fig. 10-33 shows the required ramp chart. Clearly, although Alpha Express Airlines requires 4 gates, Regional Express and Coastal Link airlines now require only a total of 3 gates. Therefore, the shared-gate-use strategy results in 1 less gate at the airport to accommodate the design-day flight schedule.

For the common gate-use strategy, all the flights of all the airlines are sorted together by arrival time and departure time, and the ramp chart is similarly constructed. If this is done, the ramp chart in Fig. 10-34 results which shows that under a common gate-use strategy only 5 gates are required at the airport to accommodate the design-day schedule.

The peak hour for gate use occurs around 1:00 p.m. and the peak-hour gate utilization is defined as the gate time demanded divided by the gate time supplied in the peak hour. The aircraft demand is 210 min of gate time in the peak hour, and the gate time supplied is 480, 420, or 300 min for the exclusive, shared, and common gate-use strategies, respectively. Therefore, the peak-hour gate utilization is 0.44, 0.50, or 0.70 by each of the gate-use strategies.

All the aircraft on the design day schedule in this example problem are narrow-bodied aircraft. If some were wide-bodied aircraft, the sorting would be done separately for the wide-bodied and narrow-bodied aircraft. To minimize the number of wide-bodied gates, first the wide-bodied aircraft would be assigned to the ramp chart, and then the narrow-bodied aircraft would be assigned, in recognition that a narrow-bodied aircraft may use a wide-bodied gate.

Gates at most airports vary within the range of 3 to 5 gates per 1 million annual passengers. The total number of gates may have to be

TABLE 10-14 Simulated Airline Schedule on Typical Day in Peak Month in Design Year for Coastal Link and Regional Express Airlines for Shared-Gate-Use Strategy for Example Problem 10-3

Ref. no	Airline code*	Flight number	Time		Aircraft
			Arrival time	Departure time	
1	CL	64	7:15 a.m.	7:45 a.m.	737
2	CL	9999	7:45 a.m.	8:30 a.m.	737
3	RX	161	10:15 a.m.	10:45 a.m.	MD-8
4	CL	489	11:15 a.m.	11:45 a.m.	737
5	CL	9999	11:15 a.m.	12:15 p.m.	737
6	CL	41	11:30 a.m.	12:15 p.m.	737
7	CL	9999	1:15 p.m.	2:45 p.m.	737
8	CL	50	1:45 p.m.	2:15 p.m.	737
9	CL	936	1:45 p.m.	2:15 p.m.	737
10	CL	81	4:15 p.m.	5:00 p.m.	737
11	RX	321/844	4:45 p.m.	5:45 p.m.	MD-8
12	CL	493	8:30 p.m.	9:00 p.m.	737

*CL = Coastal Link. RX = Regional Express.

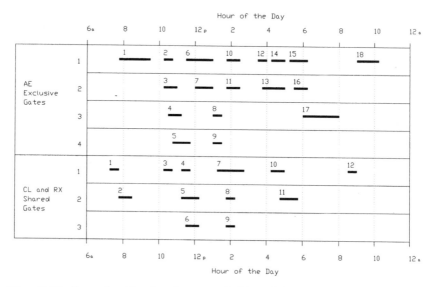

Figure 10-33 Ramp chart for shared gate use for Example Problem 10-3.

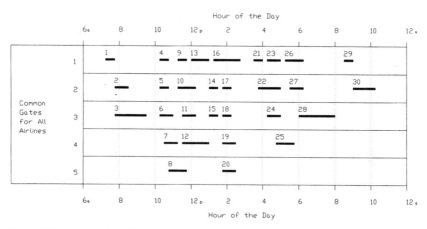

Figure 10-34 Ramp chart for common gate use for Example Problem 10-3.

modified if not all gates can handle all types of aircraft. This is particularly important at airports where the aircraft mix includes a considerable number of large and small aircraft. In such situations, and when data are available, it is preferable to compute gate requirements separately for the different types of aircraft, keeping in mind that large gates can be used to handle small aircraft, while the reverse is not true. It is also desirable to calculate the gate requirements separately for different types of traffic. For example, at a large international airport, separate calculations may be performed for domestic gates, for international gates, and for charter gates.

Gate size

The size of a gate depends on both the size of aircraft which it is to accommodate and the type of parking used, i.e., nose-in, parallel, or angled parking. The size of the aircraft determines the space required for parking as well as for maneuvering. Furthermore, the size of the aircraft determines the extent and size of the servicing equipment that needs to be provided to service the aircraft. The type of parking used at the gates affects the size since the area required to maneuver into and out of a gate varies depending on how the aircraft is parked.

In view of the large number of factors that affect the size and exact layout of gates, it is desirable to consult the airlines at an early stage in the design process to determine the manner in which they plan to maneuver aircraft and the types of servicing facilities they plan to use.

The design of the gates can be worked out with the aid of procedures and dimensions provided by the FAA [6, 7], the ICAO [3], and the International Air Transport Association [12]. Included in these references are diagrams which show various dimensions required for different types of aircraft and various parking and maneuvering conditions. Chapter 9 discusses the layout of ramp areas to accommodate aircraft.

While detailed design of aircraft gates requires the aid of charts such as those found in the airplane characteristics manuals published for aircraft, it is usually sufficient for preliminary planning to adopt uniform dimensions between centers of gates and to use these for sizing the apron gate area. The dimensions depend on the type of aircraft. The typical dimensions for the case where aircraft enter a gate under their own power and are pushed out by a tractor are given in Table 10-15.

Aircraft parking type

Aircraft parking type refers to the manner in which the aircraft is positioned with respect to the terminal building and to the manner in which aircraft maneuvers in and out of parking positions. It is an important factor affecting the size of the parking positions and consequently the apron gate area. Aircraft can be positioned at various angles with respect to the terminal building line and can maneuver into and out of parking positions either under their own power or with the aid of towing equipment. With aircraft towing it is possible to reduce the size of parking positions. It is advisable in choosing

TABLE 10-15 Comparison of Apron Parking Envelope Dimensions for Aircraft Push-out and Taxi-out Gate Use for Nose-in Configuration

Aircraft Group	Push-out*			Taxi-out*		
	L	W	Area, yd²	L	W	Area, yd²
A						
FH-227	103 ft 1 in	115 ft 2 in	1319	148 ft 10 in	140 ft 2 in	2318
YS-11B	106 ft 3 in	124 ft 11 in	1474	171 ft 0 in	149 ft 11 in	2850
BAC-111	123 ft 6 in	113 ft 6 in	1557	130 ft 0 in	138 ft 6 in	2001
DC-9-10	134 ft 5 in	109 ft 5 in	1634	149 ft 2 in	134 ft 5 in	2228
B						
DC-9-21, 30	149 ft 4 in	113 ft 4 in	1880	149 ft 0 in	138 ft 4 in	2290
727 (all)	173 ft 2 in	128 ft 0 in	2463	194 ft 0 in	153 ft 0 in	3298
737 (all)	120 ft 0 in	113 ft 0 in	1507	145 ft 4 in	138 ft 0 in	2228
C						
B-707 (all)	172 ft 11 in	165 ft 9 in	3188	258 ft 0 in	190 ft 9 in	5468
B-720	156 ft 9 in	150 ft 10 in	2627	228 ft 0 in	175 ft 10 in	4454
D						
DC-8-43, 51	170 ft 9 in	162 ft 5 in	3081	211 ft 10 in	187 ft 5 in	4411
DC-8-61, 63	207 ft 5 in	168 ft 5 in	3882	252 ft 4 in	193 ft 5 in	5423
E						
L-1011	188 ft 8 in	175 ft 4 in	3676	263 ft 6 in	200 ft 4 in	5865
DC-10	192 ft 3 in	185 ft 4 in	3959	291 ft 0 in	210 ft 4 in	6801
F						
B-747	241 ft 10 in	215 ft 8 in	5795	328 ft 0 in	240 ft 8 in	8771

*L = perpendicular to face of building; W = parallel to face of building.

SOURCE: Federal Aviation Administration [50].

among alternative parking types to consult with the airline in question, since different airlines have different preferences for the available systems. It is also advisable in adopting a parking type to take into consideration the objective of protecting passengers from the adverse elements of noise, jet blast, and weather, as well as the operating and maintenance costs of needed ground equipment.

The aircraft parking types which have been successfully used at a variety of airports and should be evaluated in any airport planning study include nose-in, angled nose-in, angled nose-out, and parallel parking. These parking types are shown in Fig. 10-35 and are discussed separately below.

Nose-in parking. In this configuration, the aircraft is parked perpendicular to the building line with the nose as close to the building as permissible. The aircraft maneuvers into the parking position under its own power. To leave the gate, the aircraft has to be towed out a sufficient distance to allow it to proceed under its own power. The advantages of this configuration are that it requires the smallest gate area for a given aircraft, creates less noise since there is no powered turning movement near the terminal building, sends no jet blast toward the building, and facilitates passenger loading since the nose is near the building. The disadvantages include the need for towing

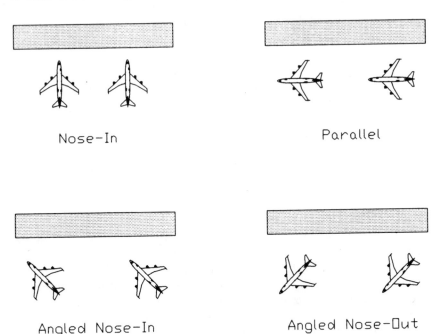

Figure 10-35 Aircraft parking types.

equipment and that the nose is too far from the building to effectively use the rear doors for passenger loading.

Angled nose-in parking. This configuration is similar to the nose-in configuration except that the aircraft is not parked perpendicular to the building. The configuration has the advantage of allowing the aircraft to maneuver into and out of the gate under its own power. However, it requires a larger gate area than the nose-in configuration and creates more noise.

Angled nose-out parking. In this configuration, the aircraft is parked with its nose pointing away from the terminal building. Like the angled nose-in configuration, it has the advantage of allowing aircraft to maneuver into and out of gate positions without towing. It does require a larger gate area than the nose-in position, but less than angled nose-in parking. A disadvantage of this configuration is that the breakaway jet blast and noise are pointed toward the building when the aircraft starts its taxiing maneuver.

Parallel parking. This configuration is the easiest to achieve from the aircraft maneuvering standpoint. In this case, the noise and jet blast are minimized, since no sharp turning maneuvers are required. It does require, however, a larger gate position area, particularly along the terminal building frontage. Another advantage of this configuration is that both forward and aft aircraft doors can be used for loading and unloading passengers, although relatively long loading bridges may be required.

It is evident that no one parking type can be considered ideal. For any planning situation, all the advantages and disadvantages of the different systems have to be evaluated by taking into consideration the preference of the airline that will be using the gates. The trend, however, is toward nose-in parking because of the saving in area and the reduction of noise and blast. Recent experimentation with power-out from the nose-in position is yielding savings in operating and maintenance costs for ground support equipment.

Apron layout

Another factor that affects apron size and installation requirements is the apron layout. This refers to the manner in which the apron is arranged around the terminal building. The apron layout depends directly on the way the aircraft gate positions are grouped around the buildings and on the circulation and taxiing patterns dictated by the relative locations of the terminal buildings and the airfield system.

Aircraft are grouped adjacent to the terminal building in a variety of ways depending on the horizontal terminal concept used. These groupings are referred to as *parking systems* and are classified as the frontal or linear system, the finger or pier system, the satellite system, and the open-apron or transporter system. These were discussed and illustrated earlier.

Naturally the choice of aircraft parking system is strongly influenced by the horizontal passenger processing concept adopted. For each, positive and negative attributes must be weighed. While the open-apron system has the advantage of separating the aircraft and the terminal building from each other, it does require buses or mobile lounges to convey passengers between them. These vehicles use the apron, and their circulation patterns need careful planning to avoid interference with the flow of aircraft and other service vehicles. While the finger system allows for the efficient expansion of gate positions and the efficient use of terminal building space, it may lead to long passenger walking distances if it is allowed to become excessively long. The frontal system is suitable for the gate arrival processing concept. Other features of these systems were discussed earlier in this chapter.

Apron circulation

In designing the apron layout, it is important to take into account aircraft circulation, particularly the movement of aircraft within the apron gate area and from this area to the taxiways. When the traffic volume is high, it is desirable to provide a taxilane on the periphery of the apron. It is also important to allow sufficient space to permit easy access of aircraft to gates. This is particularly important when pier fingers are used for aircraft parking and the fingers are parallel to each other. Sufficient space must be provided between the fingers to allow aircraft ready access to the gates. The separation between fingers depends on their length and on the size of aircraft to be accommodated. The longer the finger, the more aircraft gates can be accommodated. However, the increase in the number of gates may necessitate the provision of two taxilanes instead of one between the fingers to provide circulation without excessive delay. One taxilane will probably suffice when there are no more than 5 or 6 gates on each side of a finger. A large number of gates may require two taxilanes.

Passenger conveyance to aircraft

Depending on the passenger processing system used, the type of aircraft parking, and the parking system layout, any of three methods of conveyance can be used between the building and the aircraft: walking on the apron, walking through aircraft building connectors such as boarding bridges, and mobile conveyance using a variety of apron vehicles.

The first method can be employed with all processing and parking systems. However, as the number of parking positions and the apron size increase, it becomes impractical to expect passengers to walk. The economic appeal of this method is overcome by the need to protect passengers from the elements and from the hazards of walking on the apron.

The second method can be employed for all systems other than where open-apron parking is used. A variety of fixed and movable loading systems have been developed for passenger conveyance. Most common among these are the nose bridges, which are short connectors suitable for use when the aircraft door comes close to the building such as with nose-in parking. Another common system is the telescoping loading bridge. This has the flexibility of extending from the building to reach the aircraft door and of swinging to accommodate different types of aircraft. A typical boarding bridge is shown in Fig. 10-36.

The third system is suitable when open-apron parking is used. Here passengers can be conveyed by either buses or mobile lounges. With buses, passengers have to walk upstairs to aircraft doors. With the mobile lounges, this need is eliminated, since the lounge is capable of direct interface with aircraft doors through vertical movement. The mobile lounge thus provides for complete passenger protection. There is, of course, a considerable cost difference between the mobile lounge and the conventional bus. A typical mobile lounge is shown in Fig. 10-37.

Apron utility requirements

Aircraft need to be serviced at their respective gates. Thus certain fixed installations may be required on the apron. Apron congestion is always a problem; hence, there is a definite trend at larger airports toward replacing mobile servicing equipment with fixed facilities.

Aircraft fueling. Aircraft are fueled at the apron by fuel trucks, fueling pits, and hydrant systems. At the smaller and even the larger airports, the use of fuel trucks is prevalent, but the pattern is changing

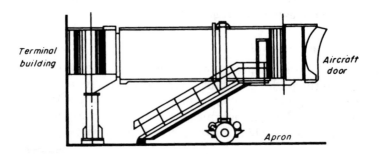

Figure 10-36 Typical aircraft loading bridge.

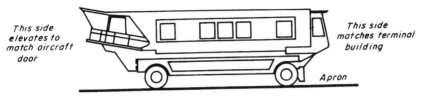

Figure 10-37 Typical mobile lounge (*Ralph M. Parsons and Federal Aviation Administration [50]*).

in favor of the hydrant system at airports requiring large amounts of fuel.

The principal advantage of fuel trucks is their flexibility. Aircraft can be fueled anywhere on the apron, the units can be added or taken away according to need, and the system is relatively economical insofar as airport management and airline operations are concerned. There are, however, disadvantages associated with the use of fuel trucks. Large jet transports require a lot of fuel, from nearly 8000 gal (U.S.) for the Boeing 727-100 to almost 50,000 gal for the Boeing 747-100. Two refueler units are normally required, one under each wing. For the large jets, standby units are sometimes required if the fuel requirements are in excess of two units. This means that there are a large number of vehicles on the apron during peak periods, creating a potential hazard of collision with personnel, other vehicles, and aircraft. Since each truck carries a lot of fuel, it also constitutes a potential fire hazard while moving around on an apron where a number of other activities are taking place. Trucks are large and awkward and take up valuable space in the operations area. When a truck is empty, it must return to the storage area for refueling before it can be used again. Thus extra trucks must be provided for use during the time when other trucks are being reloaded. When refueling trucks are not in use, parking space must be provided for them. Modern refuelers are approximately 40 ft long and weigh as much as 83,000 lb. The capacity of the larger trucks is approximately 8000 gal. For the larger refuelers, axle loads are in excess of the legal limits on highways; consequently, the airport designer must provide adequate pavement strengths to support these vehicles.

The hydrant system is used at most large airports. In this system, a large fuel storage area, often called a *fuel farm,* is located on the airport property. Fuel is transferred from the fuel storage area to aircraft gate positions through a system of pipes located below the pavement surface. A special valve is mounted in a box in the pavement flush with the pavement surface at each gate position. A special vehicle, a hydrant dispenser, with a hose, meter, filter, and air eliminator is used to connect the fuel supply to the aircraft. At one end of the hose is a specially designed valve which is coupled to the valve installed in the pavement. This hose feeds into the meter, filter, and air eliminator, from which another hose, usually on a reel, is led to the fuel intakes on the aircraft.

The principal advantages of the hydrant system are that a continuous supply of fuel is available at the gates, fuel is safely carried underground, and fuel trucks are eliminated from the apron. The principal disadvantage is that vehicles are not entirely removed from the apron. However, because of their small size, hydrant dispensers minimize possible collision damage.

The amounts of fuel required at many airports are so large that, regardless of the type of fueling system used, a central fuel storage area in the vicinity of the landing area is required. If the hydrant system is used, provision must be made for installing pipes from the storage area to the apron.

The location of the hydrant valves at an individual gate will depend upon the location of the fueling connections in the wings of the aircraft occupying the gates. It is desirable that the hose line from the hydrant dispenser to the intakes in the wings not exceed 20 to 30 ft. If a wide variety of aircraft are to be serviced at a gate position, the precise spacing of the hydrant valves should be established in consultation with the airlines. The number of hydrants required per gate position depends not only on the type of aircraft but also on the number of grades of fuel required. Each grade of fuel requires a separate layout.

At a number of airports, hydrant systems are installed by oil companies which contract for fuel with a particular airline or airlines. It is not uncommon to have the hydrant system and fuel trucks operate simultaneously at the same airport. The trend at large airports is definitely toward the hydrant system.

Electric power. Electric power is required on the apron for the servicing of aircraft prior to engine starting. External electric power is often required for starting the engines. Power requirements vary widely for different aircraft. Consequently it is necessary to consult with the airlines concerning this matter. Power can be supplied by mobile units or by fixed installations in the pavement. The latter is preferable since it removes the need for a vehicle and to some extent reduces the noise that emanates from a motor generator set. For a fixed installation, the most satisfactory technique is to bury conduits under the apron, terminating them at supply points some distance from the hydrant valves but convenient to the aircraft.

Recently, there has been a trend toward fixed ground power and air conditioning systems using terminal power sources. The need for these facilities has grown because of the costs of providing power and conditioned air to aircraft during servicing times at the apron gate by using the power generated from the auxiliary power unit (APU) on the aircraft. Considerable operating cost economies have been reported in the use of such systems [27].

Aircraft grounding facilities. Grounding facilities will be required on the apron to provide protection of parked aircraft and fuel trucks from static discharge, particularly during fueling operations. The location of the grounding facility will be governed by the location of the hydrant valves. With high fueling rates it is essential that grounding facilities be provided.

Apron lighting and marking. Adequate lighting and marking is essential on an apron. Wherever possible, each gate should be floodlighted. Floodlighting removes the need for mobile equipment to use headlights, which, experience has shown, cause confusion and glare. A system of elevated lights appears to offer the best method of providing apron illumination. Where pier fingers are utilized, the lights can be attached to the fingers. Lighting should be located so as to provide uniform illumination of the apron area yet not cause glare to the pilot.

When personnel are servicing an aircraft, there is a need for lighting its underside and far side, if the floodlights do not provide the necessary illumination. This can be accomplished by installing flush lights in the pavement. When lights of this type are installed, they should be arranged so as not to confuse the pilot insofar as guidance to the gate position is concerned.

Painted guidelines have proved very desirable as aids to maneuvering aircraft accurately on the apron. The best guide appears to be a single line—usually the color is yellow—which is followed by the nosegear of the aircraft. A typical layout is shown in Fig. 10-38. It is recognized that a single line will not provide precise guidance for a variety of different aircraft. Usually the guideline is painted for the most critical aircraft using a particular gate position. Smaller aircraft can use the same lines and maneuver without difficulty, especially if personnel on the ground are available to direct the pilot. Because of possible fuel spillage, it is desirable to paint guidelines with special resistant paint in areas where spillage might occur.

References

1. *Access to Commercial Service Airports: The Planning and Design of On-Airport Ground Access System Components,* F. X. McKelvey, Final Report, Federal Aviation Administration, Washington, 1984.
2. "A Decision Tool for Airport Terminal Building Capacity Analysis," B. F. McCullough and F. L. Roberts, Council for Advanced Transportation Studies, University of Texas at Austin, Presented at 58th annual meeting of the Transportation Research Board, Washington, January 1979.

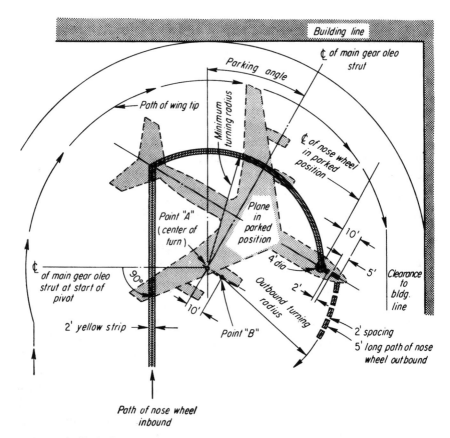

Figure 10-38 Typical painted guidelines at gate positions.

3. *Aerodrome Design Manual,* pt. 2: *Taxiways, Aprons and Holding Bays,* 2d ed., International Civil Aviation Organization, Montreal, Canada, 1983.
4. *Aircraft Movement and Passenger Data for 100 U.S. Airports,* Air Transport Association of America, Washington, periodic.
5. *Airline Aircraft Gates and Passenger Terminal Space Approximations,* AD/SC Rep. 4, Air Transport Association of America, Washington, July 1977.
6. *Airport Capacity and Delay,* Advisory Circular AC 150/5060-5, Federal Aviation Administration, Washington, 1983.
7. *Airport Design,* Advisory Circular AC 150/5300-13, Federal Aviation Administration, Washington, 1989.
8. "Airport Gate Position Estimation under Uncertainty," S. Bandara and S. C. Wirasinghe, *Transportation Research Record,* no. 1199, Transportation Research Board, Washington, 1988.
9. *Airport Ground Access Planning Guide,* Rep. No. FAA-EM-80-9, Department of Transportation, Transportation Systems Center, and Federal Aviation Administration, Washington, 1980.
10. *Airport Landside: The Airport Landside Simulation Model (ALSIM),* Reps. FAA-EM-80-8 and DOT-TSC-FAA-84-4, Department of Transportation, Transportation Systems Center, and Federal Aviation Administration, Washington, 1982.

11. "Airport Landside Level of Service Estimation: Utility Theoretic Approach," K. F. Omer and A. M. Khan, *Transportation Research Record,* no. 1199, Transportation Research Board, Washington, 1988.
12. *Airports Terminal Reference Manual,* 7th ed., International Air Transportation Association, Montreal, Canada, September 1989.
13. "Airport Terminal Designs with Automated People Movers," L. David Shen, *Transportation Research Record,* no. 1273, Transportation Research Board, Washington, 1990.
14. *Airport Terminal Flow Simulation Model,* Transport Canada, Ottawa, Canada, 1988.
15. *An Airport Passenger Processing Simulation Model,* S. Hannig-Smith, Aviation Planning Associates, Inc., Cincinnati, Ohio, January 1981.
16. "Analysis of Factors Influencing Quality of Service in Passenger Terminal Buildings," N. Martel and P. N. Seneviratne, *Transportation Research Record,* no. 1273, Transportation Research Board, Washington, 1990.
17. "Analysis of Passenger and Baggage Flows in Airport Terminal Buildings," R. Horonjeff, *Journal of Aircraft,* vol. 5, no. 5, 1969.
18. "Analysis of Passenger Delays at Airport Terminals," S. Yager, *Transportation Engineering Journal,* vol. 99, no. TE4, November 1973.
19. "Analytical Models for the Design of Aircraft Terminal Buildings," J. D. Pararas, Master's thesis, Massachusetts Institute of Technology, Cambridge, January 1977.
20. "Applications for Intra-Airport Transportation Systems," F. X. McKelvey and W. J. Sproule, *Transportation Research Record,* no. 1199, Transportation Research Board, Washington, 1988.
21. *A Review of Airport Terminal System Simulation Models,* F. X. McKelvey, Final Report, Department of Transportation, Transportation Systems Center, Cambridge, Mass., November 1989.
22. *Automobile Parking Systems Handbook,* Airports Association Council International-North America, Washington, 1991.
23. *Chicago Midway Airport Master Plan Study, Determination of Facility Requirements,* Working Paper no. 1, Draft, Landrum and Brown Aviation Consultants, Chicago, 1990.
24. *Collection of Calibration and Validation Data for an Airport Landside Dynamic Simulation Model,* Wilbur Smith and Associates, Federal Aviation Administration, Washington, January 1980.
25. *Conceptual Studies for Geneva Intercontinental Airport,* Reynolds, Smith and Hills, Jacksonville, Fla., March 1981.
26. *Demand-Capacity Analysis Ground Access Study,* Draft Report, Terminal Support Working Group, Chicago O'Hare International Airport, Landrum and Brown Aviation Consultants, Chicago, 1990.
27. *Design Guidebook—400 Hz Fixed Power Systems,* Air Transport Association of America, Washington, January 1980.
28. "Designing an Improved International Passenger Processing Facility: A Computer Simulation Analysis Approach," V. Gulewicz and J. Browne, *Transportation Research Record,* no. 1273, Transportation Research Board, Washington, 1990.
29. "Evaluating Performance and Service Measures for the Airport Landside," S. A. Mumayiz, *Transportation Research Record,* no. 1296, Transportation Research Board, Washington, 1991.
30. *Fort Lauderdale–Hollywood International Airport Parking Analysis,* Working Paper, Aviation Planning Associates, Inc., Cincinnati, Ohio, September 1981.
31. *Ground Transportation Facilities Planning Manual,* Rep. no. AK-69-13-000, Airport Facilities Branch, Transport Canada, Ottawa, Canada, 1982.
32. *Greater Pittsburgh International Airport Expansion Program, New Terminal Complex Schematic Refinement Phase,* Final Report, Tasso Katselas Associates, Inc., and Michael Baker, Jr., Inc., Pittsburgh, Pa., 1986.
33. *Highway Capacity Manual,* Special Report 209, Transportation Research Board, Washington, 1985.
34. "Interactive Airport Landside Simulation: An Object-Oriented Approach," S. A. Mumayiz and R. K. Jain, *Transportation Research Record,* no. 1296, Transportation Research Board, Washington, 1991.

35. *Introduction to Transportation Engineering and Planning,"* E. K. Morlok, McGraw-Hill, New York, NY, 1978.
36. "Level of Service Design Concept for Airport Passenger Terminals: A European View," N. Ashford, *Transportation Research Record*, no. 1199, Transportation Research Board, Washington, 1988.
37. *Measuring Airport Landside Capacity,* Special Report 215, Transportation Research Board, Washington, 1987.
38. "Opportunities for Fixed Rail Service to Airports," W. J. Sproule, International Air Transportation, Proceedings of the 22d Conference on International Air Transportation, American Society of Civil Engineers, New York, 1992.
39. "Overview of Airport Terminal Simulation Models," S. A. Mumayiz, *Transportation Research Record,* no. 1273, Transportation Research Board, Washington, 1990.
40. *Parking Structures: Planning, Design, Construction, Maintenance, and Repair,* A. P. Chrest, M. S. Smith, and S. Bhuyan, Van Nostrand and Reinhold, New York, 1989.
41. *Pedestrian Planning and Design,* J. Fruin, Metropolitan Association of Urban Designers and Environmental Planners, New York, 1971.
42. *"Pier Finger Simulation Model,"* E. E. Smith and J. T. Murphy, Graduate Report, University of California, Berkeley, 1972.
43. *Planning and Design Guidelines for Airport Terminal Facilities,* Advisory Circular AC 150/5360-13, Federal Aviation Administration, Washington, 1988.
44. *Planning and Design of Airport Terminal Facilities at Non-Hub Locations,* Advisory Circular AC 150/5360-9, Federal Aviation Administration, Washington, 1980.
45. *Planning Guide for Airport Ground Transportation Facilities,* Rep. AK-69-13, Airport Facilities Branch, Transport Canada, Ottawa, Canada, 1982.
46. "Planning of Intra-Airport Transportation Systems," W. J. Sproule, Ph.D. dissertation, Michigan State University, East Lansing, 1985.
47. "Simulating the Turnaround Operation of Passenger Aircraft Using the Critical Path Method," J. B. Braaksma, Doctoral thesis, Waterloo University, Waterloo, Canada, 1970.
48. *Survey of Airport Ground Access,* U.S. Aviation Industry Working Group, Washington, June 1981.
49. *The Apron and Terminal Building Planning Report,* Rep. FAA-RD-75-191, Federal Aviation Administration, Washington, July 1975.
50. *The Apron Terminal Complex,* Ralph M. Parsons Company, Federal Aviation Administration, Washington, September 1973.
51. "The Design of the Airside Concourses (The New Denver International Airport)," J. M. Suehiro, E. K. McCagg and J. M. Seracuse, International Air Transportation, Proceedings of the 22d Conference on International Air Transportation, American Society of Civil Engineers, New York, 1992.
52. *The FAA's Airport Landside Model,* Analytical Approach to Delay Analysis, Rep. FAA-AVP-78-2, Federal Aviation Administration, Washington, January 1978.
53. "The Movement of Air Cargo between Cargo Terminals and Passenger Aircraft Gates—Airport Planning Considerations," R. J. Roche, Graduate Report, Institute of Transportation and Traffic Engineering, University of California, Berkeley, 1972.
54. "The Planning of Passenger Handling Systems," A. Kanafani and H. Kivett, Course Notes, University of California, Berkeley, 1972.
55. *Trip Generation,* 5th ed., Institute of Transportation Engineers, Washington, 1991.
56. "Use of an Analytical Queueing Model for Airport Terminal Design," F. X. McKelvey, *Transportation Research Record,* no. 1199, Transportation Research Board, Washington, 1988.

Heliports, STOL Ports,
and Vertiports

For a number of years, there has been much research and development on different types of air transportation vehicles which could hover, or, in effect, "stand still while flying." This type of vehicle is called a *rotorcraft,* and the most widely known of this family of vehicles is the helicopter. A rotorcraft is a rotary-wing aircraft that can lift vertically and can sustain forward flight by power-driven rotor blades turning on a vertical axis. The helicopter is a vehicle which essentially can take off from and land in a nearly vertical direction. This is known as *vertical takeoff and landing* (VTOL). Since the helicopter is by far the most advanced and utilized of the vertical-takeoff aircraft, the emphasis on this chapter is on ground facilities for helicopters and other rotary-wing aircraft, referred to as *heliports.*

In recent years, considerable research and development effort has also been devoted to vehicles capable of *short takeoff and landing* (STOL) for intercity transportation. The most prevalent application of STOL aircraft is for commuter aircraft feeding major airline hub airports from outlying airports in smaller communities. These aircraft are unable to lift vertically, but are designed to take off and land on very short runways and to climb and descend at steeper angles than conventional fixed-wing aircraft. A transport-category aircraft requiring less than 3000 ft of runway is usually considered a STOL aircraft. Another way to define a STOL aircraft is that it is a vehicle requiring "powered lift" for its operation rather than relying entirely on "mechanical lift" as conventional fixed-wing aircraft do. Powered lift means that propulsion from the engines is used to pass air over parts of the aircraft body, improving the lift, particularly at low speeds. Mechanical lift relies solely on lift generated by trailing- and leading-edge flaps and other devices on the wings. The argument for the use of STOL aircraft is that they are less noisy and less costly to operate than VTOL aircraft and that they require less land for runways than conventional aircraft. Thus the runway lengths at general aviation airports may be adequate for STOL aircraft, whereas conventional aircraft could not be accommodated at these airports. Similarly, dedicated short runways at major airports could be used exclusively for STOL aircraft. Facilities for the operation of STOL aircraft are called *STOL ports*. No passenger STOL aircraft with a capacity of 50 or more passengers were in service at the time this text was prepared, but many concepts have been studied. The characteristics of these concepts and the corresponding ground requirements are briefly discussed in this chapter.

Considerable research and development efforts have also been directed to the study of the tilt-rotor aircraft, tilt-wing aircraft, and fan-in-wing aircraft. A tilt-rotor aircraft has fixed wings but can take off and land vertically and can fly in cruise as a conventional fixed-wing aircraft does. The axes of the propeller blades on the engines of these aircraft are capable of being pivoted to vertical for vertical take-off and landing and to horizontal to derive lift from the wings while cruising. In a tilt-wing aircraft, the wing chord and the axes of the power-driven propeller blades are capable of pivoting from vertical, for vertical takeoff and landing, to horizontal, to derive lift from the wing in cruise. Fan-in-wing aircraft are fixed-wing aircraft with rotor fans in the wing to permit vertical operations. *Vertiports* are facilities for the landing or takeoff of rotorcraft, tilt-rotor, tilt-wing, or fan-in-wing aircraft.

Heliports

A *heliport* is an identifiable area on land or water or structures, including buildings or facilities thereon, used or intended to be used for the landing and takeoff of helicopters or other rotary-wing aircraft [10]. A *helideck* is a heliport located on a floating or an offshore structure. A *helistop* is an area developed and used for helicopter landings and takeoffs to drop off or pick up passengers or cargo. A *helipad* is a paved or other surface used for parking helicopters at a heliport.

The nature of helicopter transportation

Transportation by helicopter can be classified into two general categories, namely, private operations and commercial operations. Private operations are of the same nature as general aviation, and commercial operations are similar to scheduled air carrier activity.

In the United States there were approximately 7000 registered helicopters operating about 2.7 million flight hours and conducting 10 million helicopter operations in 1990. Forecasts by the FAA indicate that by the year 2010 the number of registered helicopters would increase to nearly 20,000 that would operate 8.4 million flight hours in 20 million helicopter operations [9, 16]. Most of the helicopters used in private operations have a capacity of 1 to 5 persons and have maximum gross weights between 3000 and 6000 lb. Helicopters in commercial operations have greater capacity, typically between 10 and 50 passengers, and have maximum gross weights between 10,000 and 50,000 lb. Primarily because of the difference in size, heliport facilities for private operations are normally much smaller than those for commercial operations.

Private operations include construction, forest and police patrol, crop dusting, advertising, emergency medical service and rescue, and transport to offshore oil well locations. Commercial operations may be classified into two types: (1) transportation in large metropolitan areas between several airports in the region and between airports and the business center or centers in the region and (2) intercity transportation between cities not necessarily in the same metropolitan area.

Experience to date shows that the helicopter has achieved its greatest success in the first type of service, where the operating service area is within a radius of about 50 mi. The continued development of the second type of service will depend on the ability of helicopters to compete favorably with fixed-wing aircraft with respect to speed and economics over stage lengths up to about 300 mi.

Despite the growth of transport by helicopter, it accounts for only a small percentage of the total number of persons traveling by air. Helicopter operating costs have gradually been reduced, but they are still considerably higher than those for fixed-wing aircraft.

Characteristics of helicopters

A helicopter is a powered aircraft which gains its lift from the rotary motion of airfoil surfaces. The distinctive characteristic of a helicopter is its ability to hover through application of power to the rotating airfoils. The practical consequences of this characteristic are a much greater range of flight speeds and flight attitudes than is the case with conventional aircraft and the ability to land on and take off from comparatively small areas. When on the ground, helicopters have the ability of taxiing under their own power. Helicopters in private operation typically have cruise speeds between 90 and 130 kn, ranges between 300 and 400 nmi, and passenger capacities between 2 and 10. Helicopters in the transport category typically have cruise speeds from 100 to 150 kn, ranges between 300 and 700 nmi, and passenger capacities between 10 and 50.

While helicopters can ascend vertically from the ground, prolonged vertical ascents severely restrict load-carrying capacities. The usual procedure is to employ vertical ascent only to initiate the takeoff. As with other aircraft, takeoffs are usually made into the wind. The initial vertical rise for takeoff is aided by a ground cushion built up by the pressure of the air directed against the ground by the revolving rotors. After a few feet of vertical ascent, horizontal acceleration is begun until the climb-out speed is reached. Prior to reaching climb-out speed, the helicopter can be flown in a horizontal path or in a slightly ascending path. Climb-out and descent speeds vary from 30 to 60 kn. Just before touchdown, the helicopter hovers momentarily 5 to 10 ft above the landing pad.

From a safety standpoint, the operation of single-engine helicopters requires that emergency landing areas be available along the entire flight path. In case of engine failure, safe landings by using autorotation can be made if space is available. *Autorotation* is the continuation of rotor rotation in flight after cessation of power. Sufficient height must be reached by the helicopter in order to utilize the principle of autorotation. Twin-engine helicopters, however, are designed to permit continuation of flight and even a moderate rate of climb if one engine fails. For these types of helicopters, it is not necessary, from a safety standpoint, to have space available for emergency landing along the entire route. Virtually all small helicopters in private operation are single-engine. Most helicopters, with the exception of a few models, are also single-engine. Twin-engine helicopters are not, however, designed to hover with only one engine in operation. On takeoff, therefore, if an engine fails before the helicopter has reached a one-engine-out flight speed, then a landing must be made. To take care of this eventuality, sufficient space must be provided ahead of the landing area for an emergency landing. The single-engine helicopter also needs this area, but in addition requires space for emergency landings all along the flight path. If helicopters are designed to hover with one engine out, space ahead of the landing area is not required.

Because it is not economically practical for a helicopter to ascend and descend vertically, unobstructed approach-departure paths leading to the heliport are required. To protect the approach-departure path, obstructions are not permitted to extend above a prescribed inclined plane, an approach surface, beginning at the heliport and extending a specified distance from the heliport. The obstruction clearance requirements specified by the FAA are discussed later, and the ICAO has adopted similar recommendations.

Helicopters can be either single-rotor or tandem-rotor and can be powered by one or two engines. The landing gear can consist of pontoons for landing on water, skids, or wheels equipped with rubber tires. When wheels are used, the landing gear normally consists of

Figure 11-1 Port Authority of New York and New Jersey heliport (*courtesy of Bell Helicopter Textron, Inc.*).

two main wheels and a single nosewheel or tail wheel, or four wheels. Figure 11-1 shows several small utility helicopters at the Port Authority of New York and New Jersey heliport on the East River in New York City. The principal dimensions of representative helicopters used for private and commercial operations are shown in Fig. 11-2 and tabulated Table 11-1.

Factors related to heliport site selection

The selection of a heliport site in an urban area requires consideration of many factors, the most important of which are as follows:

1. The best locations to serve potential traffic

2. The provision of minimum obstructions in the approach and departure areas

3. The provision of minimum disturbance from noise and desirable location with respect to adjacent land use

4. The provision of adequate access to surface transportation and parking

5. The minimization of cost to acquire and develop

6. The provision of two approach paths separated by at least 90° and oriented with respect to prevailing winds

7. The avoidance of traffic conflicts between helicopters and other air traffic

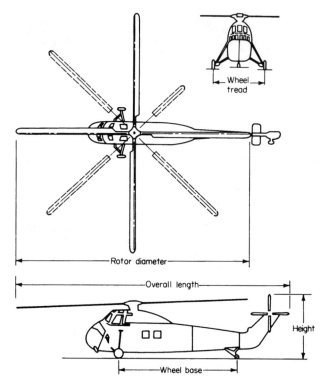

Figure 11-2 Dimensional definitions for helicopters.

8. The consideration of turbulence and visibility restrictions presented by nearby buildings

9. The provision of emergency landing areas along the entire route for single-engine helicopters

Final selection of a heliport site will usually require a compromise among these factors. The most severe problems can be expected in large, highly developed metropolitan areas. In large urban areas, heliports should be planned on a regional basis. The first step is to prepare an estimate of the demand for helicopter services and the origins and destinations of this demand. The second step is to select a heliport site or sites which can reasonably satisfy the demand and yet meet the requirements cited above.

The principal market for commercial helicopter transportation has been in large urban areas between the central business district and one or more airports. Therefore, it is essential that the downtown heliport be centrally located near the hotel area and the business district. Likewise adequate provision for helicopters should be made at

TABLE 11-1 Dimensions of Typical Commercial Helicopters

Aircraft	Rotor Diameter, ft	Overall length, ft	Height, ft	Wheelbase, ft	Wheel tread, ft	Gross weight, lb	Maximum passengers
Aerospatiale 330J	49.5	59.8	16.9	13.2	7.9	16,315	19
Aerospatiale 332L	51.2	61.4	16.2	17.2	9.8	18,410	24
Bell-212	48.0	47.3	13.0	7.6	8.3	11,200	14
Bell-214ST	52.0	62.2	13.2	8.1	8.3	17,500	18
Boeing-Vertol 107II	50.0	83.3	16.9	24.9	12.9	20,000	25
Boeing-Vertol 234	60.0	99.0	18.7	25.8	10.5	48,500	44
Boeing-Vertol 360	83.7	49.7	19.4	32.7	11.4	36,160	30
Sikorsky S-61N	62.0	73.0	18.9	23.5	14.0	20,500	28
Sikorsky S-64	72.0	88.5	25.4	24.4	19.8	42,000	45
Sikorsky S-76B	44.0	52.5	14.5	16.4	8.0	11,400	12
Westland 30300	42.5	52.1	16.3	17.8	9.3	16,000	19

SOURCES: International Civil Aviation Organization [3] and Federal Aviation Administration [10].

airports. In extremely large urban complexes, e.g., Los Angeles, where, in addition to a central core, there are outlying smaller centers, secondary heliports are needed so that the benefits of air transportation can extend to these centers. A heliport must have good access to streets, highways, and public transit facilities so that passengers using buses, personal vehicles, or mass transit can easily reach the facility.

Noise.   The noise caused by helicopter operations within or adjacent to built-up urban areas is, and will continue to be, an extremely important factor in planning for helicopter transport, as it has been with fixed-wing aircraft. Manufacturers are aware of this problem and continue to study ways in which noise can be minimized.

A heliport should be located so that the noise generated by helicopters will not cause excessive disturbance to surrounding developments. The noise factor is most critical underneath the flight path on takeoff and landing. The amount of sound that can be tolerated by the average person is dependent upon a number of factors, including the overall noise level, its frequency, and its duration; the type of development (residential, industrial, etc.) surrounding the source of the noise; and the ambient sound level in the area. A greater amount of noise can be tolerated in industrial areas than in residential areas. Docks and other waterfront sites offer some of the best possibilities for heliport location in large, congested urban centers. Approach and noise problems can usually be overcome by using water areas for heliport location. The downtown heliport in New York City is an example.

Noise generated by small two- and three-seat helicopters can be tolerated in business and industrial areas, but the noise generated by large multiengine helicopters powered by turbine engines can exceed tolerable levels, even in business and industrial areas. It is well to check with the manufacturers concerning the latest information on the levels of noise generated by the several transport-type helicopters.

To minimize noise, it is desirable, whenever possible, to orient the landing pad so that landings and takeoffs are made over areas where noise would be least objectionable. Considerably more latitude can be exercised in this respect for helicopters than for fixed-wing aircraft.

Protection of approach and departure paths.   Zoning is necessary both to control the location of heliport sites for maximum benefit to the community and to provide safety in helicopter operations by protection of the surrounding airspace. The dimensions of the approach-departure paths for various types of heliport operations are discussed below.

Turbulence and visibility.   Another factor in heliport site selection is the effect of turbulence over roof surfaces and downdrafts near buildings. This factor is of particular importance for rooftop heliports. If there is doubt in the planner's mind, the site should be flight-checked with a helicopter.

Poor visibility can be an important factor to consider for sites on tall buildings, i.e., those of 100 ft or more high. The cloud deck seldom reaches the ground, but at higher levels the heliport might find itself enveloped in fog when the ground is clear.

Physical characteristics of a heliport

A heliport is a facility which is intended to be used for the landing and takeoff of helicopters, and it may include space for helicopter parking, buildings, servicing facilities, and vehicular parking. The *final approach and takeoff* (FATO) *area* is a defined area over which the final phase of the approach maneuver to a hover or a landing is completed and from which the takeoff maneuver is commenced. The *touchdown and liftoff* (TLOF) *area* is a hard-surfaced load-bearing area typically located within the final approach and takeoff area on which a helicopter may touch down or lift off. Functionally, the terminal-area requirements for parking, servicing, and fueling of helicopters and processing of passengers and ground vehicles are no different from the requirements for fixed-wing aircraft.

Heliports are usually classified according to use:

Military heliport.   Facilities are operated by one of the branches of the armed services. The design criteria are specified in the branch of the service and usually prohibit nonmilitary uses.

Federal heliport.   Facilities are operated by a nonmilitary agency or department of the federal government. They are used to carry out the functions appropriate to the agency.

Private-use heliport.   Facilities are restricted in use by the owner. These may be publicly owned, but their use is restricted, as in police or fire department use.

Public-use heliport.   Facilities are open to the general public and do not require the prior permission of the owner to land. The extent of the facilities available may limit operations to helicopters of specified sizes or weights.

Commercial service heliport.   These public-use and public-owned facilities are designed for the use of helicopters in commercial passenger or cargo service which enplane 2500 passengers annually and receive scheduled passenger service with helicopters.

Personal-use heliport.   Facilities are used exclusively by the owner.

The principal components of a heliport are the final approach and touchdown area, the touchdown and liftoff area, and, for large heliports, taxiways, helicopter parking areas, and the terminal building area. The relationships between these components are shown in Fig. 11-3.

Final approach and takeoff area.   The final approach and takeoff (FATO) area is any surface from which the helicopter can land or take off. The FAA allows the FATO area to be any shape as long as it is enclosed by a square of the dimensions indicated in Table 11-2. The ICAO specifies that the FATO area is a circle. Its size depends primarily on the overall length of the largest helicopter to be accommodated by the heliport. Because of the dust that can be created by the rotor of a helicopter, it is necessary to prepare the surface of the land-

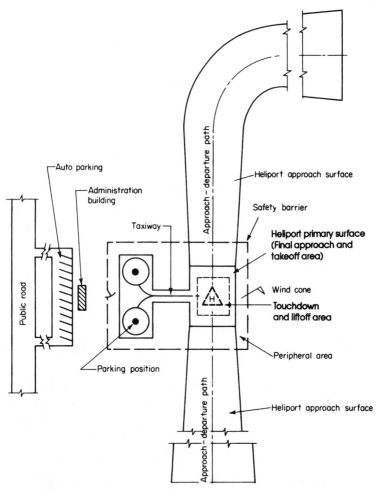

Figure 11-3 Typical heliport layout (*Federal Aviation Administration*).

TABLE 11-2   Geometric Design Standards for Heliports[a]

	Private	FAA public utility	Commercial transport	ICAO Land	ICAO Water
Final approach and takeoff area					
Length	$1.5L^b$	$1.5L^b$	200 ft[b]	$1.5D^h$	$1.5D + 10\%^i$
Width	$1.5L^b$	$1.5L^b$	$2R^d$	$1.5D^h$	$1.5D + 10\%^i$
Clearance	$\frac{1}{3}R^c$	$\frac{1}{3}R^c$	30 ft	$0.25D^i$	$0.25D^j$
Touchdown and liftoff area					
Length and width	$1.5U$	$1.0L$	$1.0L^e$	$1.5U$	
Parking area					
Clearance[f]	$\frac{1}{3}R^g$	30 ft	30 ft	l	
Minimum width		$1.5U$	$1.5U$	l	
One-way taxiway route width					
Hover operations		$R + 60$ ft	$R + 60$ ft	$2R$	
Ground operations		$R + 40$ ft	$R + 40$ ft	7.5–20 m[j]	
Parallel-taxiway route width					
Hover operations		$R + 90$ ft	$R + 90$ ft		
Ground operations		$R + 70$ ft	$R + 70$ ft		
Taxiway pavement width		$2T$	$2T$		
Air transit route				$7R^k$	

[a]L is overall length of design helicopter; R is rotor diameter of design helicopter; U is maximum of undercarriage length or width of design helicopter; D is the overall length or width of the design helicopter, whichever is greater; T is wheel tread of design helicopter.

[b]May need to be adjusted for elevation; see *Heliport Design* [10].

[c]Minimum of 10 ft for private, 20 ft for public.

[d]100 ft for public-owned.

[e]Position on major axis of FATO area with its center at least 50 ft from end or edge of FATO area.

[f]Cannot lie under approach or climb path.

[g]Minimum of 20 ft.

[h]The overall length or width of the design helicopter, whichever is greater; for class 2 or class 3 helicopters on water heliports, this is $2D$; the FATO area described is circular with this diameter.

[i]For VFR; for IFR, the clearance should be at least 45 m on each side of the centerline and 60 m beyond the ends.

[j]Depending upon the main gear span; separation between parallel taxiways should be 60 m on the side and 90 m on the ends of the final approach and takeoff area.

[k]For daytime operations; $10D$ for nighttime operations.

[l]The same as for an aircraft parking area; see Chap. 9.

SOURCES: Federal Aviation Administration [10] and International Civil Aviation Organization [3].

ing and takeoff area so that it will be free of dust (e.g., turf or pavement). Paving is recommended. Stabilizing the soil of the takeoff and landing area is also recommended, to improve the load-carrying ability of the surface, minimize the erosive effects of rotor downwash, and facilitate surface runoff due to rain or snow.

The recommended dimensions of the final approach and takeoff area are given in Table 11-2. For precision-instrument operations the final approach and takeoff area is 300 ft wide by 1225 ft long and incorporates a *final approach reference area* (FARA) which is an obstruction-free area 150 ft by 150 ft located at the far end of the final approach and takeoff area.

Touchdown and liftoff area. Within the final approach and takeoff area, an area is designated for the normal everyday landing of helicopters. The touchdown and liftoff (TLOF) area is usually defined by a solid border painted on the pavement surface. The recommended dimensions of the touchdown and liftoff area are given in Table 11-2.

Peripheral area. A peripheral or clearance area surrounding the final approach and takeoff area is recommended as an obstruction-free safety zone. The area should be kept free of objects hazardous to the operation of helicopters. The clearance from the edges of the final approach and takeoff area required for this area is also given in Table 11-2.

Effect of wind. Although helicopters can maneuver in much higher crosswinds than fixed-wing aircraft, the takeoff and landing area should preferably be oriented as nearly as possible to permit operation into the wind. At present it appears that the crosswind characteristics of the helicopter will be such that for a majority of cases a rectangular takeoff and landing area should be oriented in one direction only.

Terminal area. At heliports where the volume of traffic is relatively small, the loading and unloading of passengers can be accomplished within the final approach and takeoff area. As traffic increases, it becomes necessary to provide additional space for the parking of helicopters and for passenger processing. This is usually accomplished on a helipad, which is an area adjacent to the terminal building for processing passengers. This area provides one or more parking spaces for helicopters and is similar in nature and function to the gate or ramp area provided on the apron adjacent to airport terminal facilities. Clearances between adjacent helicopter parking positions ensure a separation between the rotor planes of helicopters, as shown in Table 11-2.

The helipad is connected to the final approach and takeoff area by taxiways and taxilanes. Helicopters may traverse taxiways in a hover or ground mode on single- or parallel-taxiway routes. The widths of these taxiway routes and the taxilanes adjacent to the helipad are given in Table 11-2. The taxilane widths are the same as the taxiway routes in a ground mode of operation.

Approach and takeoff climb path. Heliports must have at least one approach and takeoff climb path that is free of obstructions which should be established on the basis of the direction of prevailing winds and the access route that has the fewest obstacles in the flight path. As conditions permit, additional approach and takeoff climb paths should be established to facilitate operations when winds come from other directions. At private-use heliports, it is recommended that these paths flare out in the horizontal plane from the final approach and takeoff area at the rate of 1 to 20 and slope upward at the rate of 8 to 1. These paths terminate when the design helicopter attains a safe en route altitude. These paths may curve to avoid objects or noise-sensitive areas when necessary. The FAA is in the process of developing design criteria for curved visual approaches and recommends that the FAR part 77 specifications be applied in the meantime. For a commercial service airport, it is recommended that at least one instrument approach and takeoff climb path be established. For public-use heliports, the dimensions of the final approach and climb paths correspond to those specified under FAR part 77, as discussed below.

Obstruction clearance requirements. Imaginary obstruction clearance surfaces are established for each class of heliport in FAR part 77 [10]. For heliports, the principal surfaces are the approach and departure surfaces, the transitional surfaces, and the heliport protection zone. The *heliport protection zone* is the area on the ground below the approach surface from the edge of the final approach and takeoff area to the point where the approach surface is 35 ft above the elevation of the final approach and takeoff area. The horizontal surface required for airports is not necessary for heliports. The approach surface requirements for visual, non-precision-instrument, and precision-instrument operations specified by the FAA are given in Table 11-3. Similar requirements are specified by the ICAO [3]. Precision-instrument operations by commercial helicopters are very limited, and thus visual or non-precision-instrument operation criteria will suffice for most heliports. The various FAR part 77 surfaces are shown in Fig. 11-4. Note that curved-path approaches and departures are currently not permitted under IFR conditions, but with the implementation of microwave landing systems (MLS) this is subject to change. The FAA has developed specifications for the MLS critical areas and siting requirements which should be consulted if the installation of an MLS is contemplated. The specifications for IFR operations are a function of the nature of the navigational aids, and references [10] should be consulted prior to establishing landing and takeoff paths.

The various dimensions specified by the FAA for a commercial service heliport are illustrated in Example Problem 11-1.

TABLE 11-3   FAR Part 77 Approach Surface Dimensions for Heliports, ft

	Type of approach		
	Visual	Non-precision-instrument	Precision-instrument
Length	4,000	10,000	25,000*
Inner width	†	500	1,000
Outer width	500	5,000	6,000
Slope	8:1	20:1	34:1‡
Transitional			
Inner width	†	†	600
Outer width	250	600	1,500
Slope	2:1	4:1	7:1

*Begins 1225 ft from the far end of the final approach and takeoff area.

†Width of final approach and takeoff area.

‡For a 3° glide slope; 22.7 : 1 for a 4.5° glide slope; 17 : 1 for a 6° glide slope. Glide slope can be increased in 0.1° increments with corresponding corrections to approach slope.

SOURCE: Federal Aviation Administration [10].

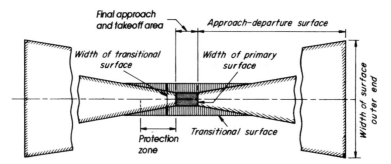

Figure 11-4   Obstruction clearance requirements for heliports (*Federal Aviation Administration [10]*).

Example Problem 11-1   Let us design the layout of a commercial service heliport for operations with a Boeing-Vertol 234 design helicopter. Let us assume that the helipad or parking apron adjacent to the passenger terminal building will require space for four helicopters. We assume that the heliport elevation is 800 ft above mean sea level. A ground taxiway route is to be provided from the touchdown and liftoff area to the helipad.

From Table 11-1 the design helicopter has a rotor diameter D of 60 ft, an overall length L of 99 ft, a height of 18.7 ft, a wheelbase of 25.8 ft, a wheel tread of 10.5 ft, a maximum gross weight of 48,500 lb, and a maximum capacity 44 passengers.

For a commercial service heliport, from Table 11-2 the final approach and takeoff area is required to be a minimum of 200 ft long and a minimum of 2 times the rotor diameter or 2 × 60 = 120 ft wide. An object-free area (OFA) width of at least 30 ft from the edges of the final approach and takeoff area is also required.

The length and width of the liftoff and touchdown area are equal to the overall length of the design helicopter or 99 ft. We use 100 ft for these dimensions. To provide for the length of the helicopter and the minimum OFA distance from the final approach and takeoff area, the minimum length of the taxiway leading to and from the touchdown and liftoff area is $100 + 30 = 130$ ft.

Parking positions must have a minimum width of 1.5 times the undercarriage length or undercarriage width, whichever is greater. The greater of these two dimensions is the wheelbase, and therefore the required minimum width of a parking position is $1.5 \times 25.8 = 39$ ft. Let us provide 40 ft. However, since the rotor diameter of the design helicopter is 60 ft, the minimum width of the parking position must be 60 ft, of which 40 ft will be paved. There must be also be a 30-ft clearance between the edges of adjacent parking positions. This makes the minimum distance between the centerlines of adjacent parking positions 90 ft.

A paved ground taxiway will be provided, and this must have a minimum width of 2 times the wheel tread or $2 \times 10.5 = 21$ ft. The ground taxiway route must also have a safety area width of the rotor diameter plus 40 ft or $60 + 40 = 100$ ft. Taxiways or taxilanes in the vicinity of the terminal building must also have this safety area, to provide clearances between the building and parked helicopters. Therefore, the minimum distance from the centerline of the taxiway or taxilanes is 50 ft.

The layout of the heliport with the corresponding dimensions is shown in Fig. 11-5.

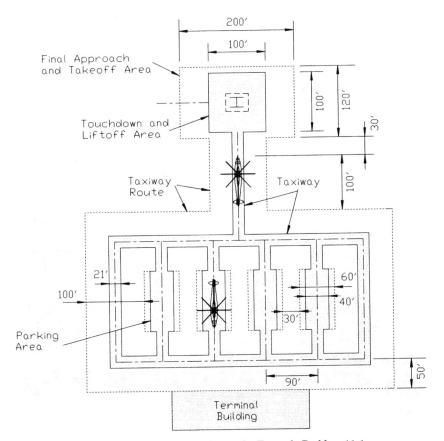

Figure 11-5 Commercial service heliport layout for Example Problem 11-1.

Marking of heliports. The primary purpose for marking heliports is to identify the area clearly as a facility for the use of helicopters. The requirements for marking heliports are specified by the FAA and the ICAO. Essentially these requirements consist of painting an equilateral square with an H in the center of the touchdown and liftoff area. For hospital heliports, a white cross is also inscribed within the square along with the letter H, as shown in Fig. 11-6 for the heliport located at the Alexian Brothers Medical Center in Dade County, Florida. The marking delineating the edges of the final approach and takeoff area should be broken white lines, whereas the edges of the touchdown and liftoff area should be delineated by continuous white lines. Taxiway centerlines are solid yellow lines, and taxi route centerline and apron edge markings should be solid yellow lines. Taxiway edges should be marked by a double solid yellow line. A painted yellow line is also recommended to define the centerline of parking positions, and when these positions vary in the clearance provided, a number enclosed by a circle should be painted on the entrance to the parking position to indicate the largest helicopter that can be accommodated.

Lighting of heliports. For operation during hours of darkness, various types of lights are suggested [10]. The amount of lighting depends on the character and volume of operations. Thus more lighting is required for scheduled air carrier operations than for private heliports seeing occasional use.

Figure 11-6 Heliport located at the Alexian Brothers Medical Center in Dade County, Florida (*Courtesy of Howard Needles Tammen & Bergendoff and Alexian Brothers Medical Center*).

The minimum recommendations for private-use and public-use heliports consist of lighting of the perimeter of either the final approach and takeoff area or the touchdown and liftoff area with lights with yellow lenses uniformly spaced at 25-ft intervals. At public-use heliports, green lights are used to define the taxiway and taxilane centerlines. For commercial service heliports, the touchdown and liftoff area is delineated by yellow lights located 10 ft from the outside edge. At such heliports, in-pavement green lights are recommended for taxiway and taxilane centerlines. Blue retroreflective markers are also used at these airports to identify taxiway entrance and exit points and to define taxiway edges. Perimeter lighting defining the touchdown and liftoff area of a commercial service heliport is shown in Fig. 11-7. All objects that penetrate the obstruction clearance surfaces should be lighted with red lights.

Other useful lighting aids are the landing direction lights, visual glide path indicators, and heliport identification beacons. The landing direction lights are miniature approach lights since they extend only 75 ft. The color is yellow. Visual glide path indicators are also recommended for visual operations at commercial service heliports. The lowest on-course signal should provide a 1° clearance over any object in the approach path within 10° horizontally on either side of the approach path centerline. The optimal location is on the extended runway centerline of the approach path such that it will bring the helicopter to a 3- to 8-ft hover distance over the touchdown and liftoff area. Heliport identification beacons—alternately flashing white-green-yellow lights—should be located within $\frac{1}{4}$ mi of a commercial service heliport.

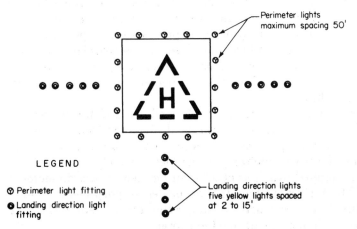

Figure 11-7 Heliport lighting configuration (*Federal Aviation Administration [10]*).

Elevated heliports. When ground-level sites are not available or are unsuitable, an elevated site may be practical. Elevated heliports may be located on piers or other structures over water as well as on buildings. The dimensions of the touchdown and liftoff area are the same as for heliports on the ground, but the final approach and takeoff area can be smaller and there is no need for peripheral areas. When a rooftop heliport is planned, a thorough study should be made of the air currents caused by the presence of adjacent buildings. Roof areas make it possible to locate the heliport closer to the center of business activities in a city, provided the facility is environmentally acceptable. Another advantage is that the land cost is partially absorbed by the tenancy of the lower floors of the building. However, it should be realized that if the operations are of any size, space on the floors below the takeoff and landing area may have to be devoted to uses such as lobby, freight, and baggage handling. The possible disadvantage of height with respect to visibility was mentioned earlier. A heliport 100 ft or more above the ground would require a higher cloud base than a ground heliport, to provide the same operating safety. A downtown commercial heliport would require, in addition to lobby space, car parking facilities relatively close by.

When heliports are built on elevated structures, the strength of the floor should be greater than the strength of the landing gear of the helicopter. The loads imposed by helicopters and recommendations concerning the structural design of elevated structures are discussed in the next section.

Structural design of heliports. Helicopters using facilities on land are usually supported on tubular skids or wheels equipped with rubber tires. Helicopters equipped with conventional landing gearwheels are normally supported by two main wheels and one tail wheel or nosewheel. For larger helicopters, each main landing gear consists of two wheels. Each main gear typically supports 40 to 45 percent of the weight of the helicopter, and the tail wheel or nosewheel supports the remainder of the weight, approximately 10 to 20 percent. If the helicopter is supported by tubular skids, 50 percent of the weight is supported by each skid.

The strength requirements for the touchdown and liftoff area are determined by considering the static load, dynamic load, and downwash load of the helicopter. Both the static load and the dynamic load are applied through the landing gear contact area, whereas the downwash load is applied over a contact area defined by the diameter of the rotors. The FAA recommends that for design purposes the touchdown and liftoff area be capable of supporting 150 percent of the maximum takeoff weight of the design helicopter.

Heliports at airports. A large number of helicopters will operate in airports to serve traffic from the downtown area and surrounding communities. Accordingly, provisions should be made at an airport for the landing and takeoff of helicopters. The takeoff and landing area should be located so as to

1. Provide maximum separation from fixed-wing aircraft traffic patterns, to avoid creating a conflict in takeoff and landing operations

2. Be as close as possible to passenger check-in areas for fixed-wing aircraft, to avoid long walking distances for passengers

3. Avoid as much as possible the mixing of taxiing fixed-wing aircraft and helicopters, since helicopters taxi at relatively low speeds

If simultaneous same-direction diverging helicopter and fixed-wing aircraft operations are to be conducted under visual flight rule conditions, a runway centerline to a final approach and takeoff area and a touchdown and liftoff area centerline separation of 700 ft are recommended. A 2500-ft minimum separation is required for radar departures under instrument flight rule conditions [10]. Helicopter parking apron areas should meet the same runway clearance standards as fixed-wing aircraft parking [5].

Alternative locations for heliports at an airport are the roof of the terminal building, the apron adjacent to the terminal building used by fixed-wing aircraft, and the area adjacent to the terminal building separate from the fixed-wing aircraft apron. There are advantages and disadvantages to all three locations. Normally, a ground-level site is preferred. The most convenient and least expensive means of accomplishing this is to reserve a part of the fixed-wing aircraft apron for the takeoff and landing of helicopters. If this is not convenient, a special pad for helicopter operations on the aircraft side of the terminal building should be provided.

Figure 11-8 shows the heliport at Miami International Airport.

STOL Ports

A STOL port is a facility designed for use by aircraft capable of short takeoff and landing (STOL). It may consist simply of a runway designed for such aircraft at an airport, or it may encompass an entire ground facility similar to an airport.

Status of STOL transportation

As discussed throughout this book, the continued growth of the domestic air travel industry, in particular, has placed a strain on the capacity of air carrier airports serving larger cities. The short-haul

Figure 11-8 Heliport at Miami International Airport (*Courtesy of Howard Needles Tammen & Bergendoff and Miami International Airport*).

traveler is particularly sensitive to airport and airspace congestion and delay. Although the development of VTOL aircraft offers a potential solution, their economics and noise generation have made them less attractive than STOL aircraft. The use of STOL aircraft has, therefore, been proposed for short-haul intercity air travel to provide better service to the passenger and to relieve airspace and ground congestion at busy airports. STOL aircraft, because of their high-lift capability, can operate on much shorter runways than conventional aircraft. In large metropolitan areas, this feature offers the potential for using existing general aviation airports in the region. The use of these airports can relieve strain on major airports and at the same time provide facilities closer to the origin and destination of travelers. Another argument for STOL aircraft is their capability for reducing noise on the ground because of higher approach and departure angles (7° as opposed to 3° for conventional aircraft approaches). Finally, STOL aircraft, because of their substantially lower speeds in the airport terminal airspace, are more maneuverable and can therefore make more efficient use of congested airspace.

Research and development efforts are continuing in the United States, Canada, and elsewhere for the evaluation of STOL aircraft as an alternative means of air transportation. These efforts encompass not only aircraft technology but also the market potential, environmental acceptance, ground facilities, aids to navigation, and economic incentives for the manufacturers and operators of STOL aircraft.

From these analyses, it is clear that the operating costs of STOL aircraft depend a great deal on the length of runway available. There is consensus that downtown STOL ports either on the ground or on the roof of a structure will be extremely difficult to develop because of noise and other environmental considerations. Therefore the need for very short runways on the order of 2000 ft has been questioned. Consequently, runway lengths of 3000 and even 4000 ft are considered viable for the development and operation of STOL aircraft. First-generation STOL aircraft primarily serve the commuter airline industry and provide short-haul intercity connections between smaller outlying areas and large airports for trips typically not exceeding 200 to 300 mi. These aircraft typically operate from air carrier runways, but they are capable of landing and taking off on runways as short as 2500 ft. At some airports, separate short runways similar to those provided for general aviation aircraft are utilized for STOL aircraft operations.

Characteristics of STOL aircraft

A number of STOL aircraft are presently used principally in commuter airline service applications. These aircraft are relatively small, typically accommodate about 25 to 45 passengers, are powered by turboprop engines, and operate on runways up to 4000 ft long. Among the first aircraft which can be considered STOL aircraft were the DeHavilland Twin Otter DHC-6, DHC-7, and DHC-8; the Dornier 228 and 328; and the Shorts 330. The principal characteristics of these aircraft were identified in Table 3-2.

These aircraft have been successfully used by commuter airlines for several years to connect outlying airports with large airports. The DeHavilland aircraft were designed for service from downtown STOL ports with very short runways. Passenger service on STOL aircraft has been provided in Canada on an experimental basis between Ottawa and Montreal to determine the market for STOL aircraft and to evaluate operational factors associated with such aircraft [6].

Under the auspices of the National Aeronautics and Space Administration (NASA), several aircraft manufacturers have made extensive studies of large STOL aircraft transportation with respect to market potential, aircraft technology, operating economics, ground facility requirements, and noise impact [15, 22]. In these analyses, the three principal variables were the runway length, number of passengers, and method of providing lift. Runway lengths varied from 2000 to 4000 ft. Aircraft capacities ranged from 150 to 300 passengers. Noise levels varied from 95 to 98 EPNdB at a distance 500 ft from the aircraft. The lift concepts studied included upper surface blowing, external blowing of flaps, augmenter wings, and mechanical

TABLE 11-4 Characteristics of Alternative Concepts for Future STOL Aircraft

Item	2000-ft runway, 150 passengers			3000-ft runway, 150 passengers		4000-ft runway, 150 passengers, MF
	EBF*	AW	USB	EBF	MF	
Wingspan†	149'8"	133'10"	164'8"	108'1"	141'8"	124'0"
Overall length‡	153'6"	153'7"	159'10"	139'9"	164'6"	131'8"
Height†	54'0"	54'2"	58'6"	41'4"	45'8"	39'0"
Number and type of engines	4TF†	4TF	4TF	4TF	2TF	2TF
Maximum takeoff weight, lb	196,000	213,000	227,000	149,000	191,000	158,000
Wheel tread,§ ft	20	20	25	25	25	25
Cruise speed, Mach no.	0.74	0.78	0.76	0.69	0.70	0.72

*EBF = externally blown flap; AW = augmentor wing; USB = upper surface blowing; MF = mechanical flap.

†TF = turbofan.

‡Dimensions in feet and inches.

§Approximate only.

SOURCES: National Aeronautics and Space Administration [15] and McDonnell-Douglas Corporation [22].

flaps. The upper limit of range was 500 mi. The principal characteristics of a potential 150-passenger-capacity STOL aircraft are shown in Table 11-4. The data in this table were taken from a report prepared by one of the contractors for NASA. Table 11-4 is presented to show the influence of runway length and lift techniques on the physical dimensions and weight of potential large STOL aircraft.

This comparison indicates that the airfield dimensions for future STOL aircraft carrying up to 150 passengers will be similar to current short- to medium-range conventional aircraft. Pertinent conclusions reached from one study [15] indicate the following:

1. Runway lengths of 2000 ft are not economically viable while runway lengths of 3000 ft are economically viable.

2. A payload of 150 passengers is the preferred size.

3. Mechanical flap aircraft are competitive with propulsion-lift aircraft for runway lengths of 3000 to 4000 ft.

4. The noise criteria have a large effect on the economics of the aircraft.

5. A goal of 95 EPNdB at 500 ft is difficult to achieve, and some relaxation of this goal may be necessary.

Although these conclusions are the result of mathematical analyses, they indicate the direction in which STOL aircraft development is likely to proceed in the future.

Physical characteristics of a STOL port

Several years ago, both the FAA and the ICAO published design criteria for STOL ports [12, 18]. The FAA publication is no longer current, but it is useful to indicate past criteria for aircraft of the DHC-6 and DHC-7 type so that the reader may have a point of comparison with the requirements for conventional aircraft. The physical characteristics of a metropolitan-area STOL port were as indicated in Table 11-5 [12]. The dimensions indicated by the FAA for the obstruction surfaces were as follows [12]:

1. The primary surface length should be the runway length plus 100 ft at either end of the runway.

2. The primary surface width should be 300 ft.

3. The approach and departure surface length should be 10,000 ft.

4. The approach and departure surface inner width should be 300 ft.

5. The width of the approach and departure surface at the outer end should be 3400 ft.

TABLE 11-5   Dimensional Design Criteria for Metropolitan STOL Ports

Item	Recommended criteria, ft
Runway length	1500–1800*
Runway width	100
Runway safety area length	1700–2000
Runway safety area width	200
Taxiway width	60
Runway centerline to:	
Taxiway centerline	200
Edge of parked aircraft	250
Building line	300
Holding line	150
Separation between parallel runways	700
Taxiway centerline to fixed obstacle	100

*Correction for elevation and temperature required based upon aircraft performance characteristics.

SOURCE: Federal Aviation Administration [12].

6. The approach and departure surface slope should be 34 to 1.

7. The transitional surface slope should be 4 to 1.

8. The width of the transitional surface should be 1100 ft.

9. The length of the runway clear zone (runway protection zones) should be 750 ft.

10. The inner width of the runway clear zone should be 300 ft.

11. The outer width of the runway clear zone should be 532 ft.

Since STOL ports have the same characteristics as airports except that the runways are shorter, the planning criteria are the same as for conventional airports. For large STOL aircraft, therefore, runway widths would range between 100 and 150 ft, and taxiway widths would vary between 60 and 75 ft, both depending upon the aircraft approach category and the airplane design group to which the STOL aircraft was assigned. Runway and taxiway separations would be a function of the types of approaches made to the runways and the airplane design groups to which the aircraft were assigned. For small STOL aircraft, like the DHC-6 and DHC-7, the pavement widths and separations would be similar to those for utility airports.

STOL runways were recommended to be marked as such, as shown in Fig. 11-9. White markings would be used on the runways and yellow markings would be used on the taxiways. Lighting would consist of threshold lighting, runway edge lighting, and approach lights. The lighting, marking and signing criteria contained in Chap. 13 would apply to STOL ports.

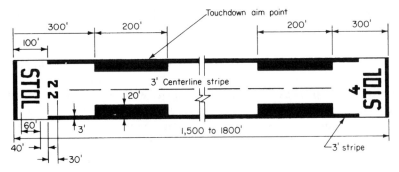

Figure 11-9 STOL runway marking (*Federal Aviation Administration [12]*).

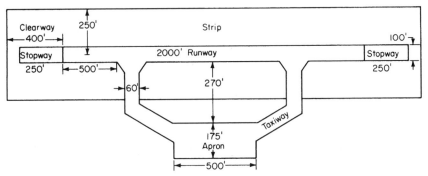

Figure 11-10 Typical STOL port layout for aircraft similar to DHC-6 twin otter (*International Civil Aviation Organization [18] and Canadian Air Transportation Administration [6, 14]*).

Figure 11-10 shows a typical STOL port layout to accommodate an aircraft similar to the DHC-6.

Vertiports

A *vertiport* is an identifiable ground or elevated area, including any buildings or facilities thereon, used for the takeoff and landing of tilt-rotor aircraft and rotorcraft. A *vertistop* is a vertiport designed solely for takeoff and landing of tilt-rotor aircraft and rotorcraft to drop off or pick up passengers and cargo.

Status of aircraft designed for vertiport operation

Tilt-rotor technology is thought to offer a viable alternative for air transportation between city centers. The potential use of such aircraft technology in short-haul air transportation can result in appreciable benefits in the operation of the air transportation system. By replacing many of the fixed-wing commuter aircraft presently operating at larger airports, using limited runway and airspace capacity, and accommodating passengers in tilt-rotor aircraft operating from vertiports and vertistops, it is thought that significant gains can be realized in air transportation system capacity. Vertiports can be located at airports where such aircraft could operate on simultaneous and nonconflicting approach and departure paths with fixed-wing aircraft traffic or in urban areas near city centers, alleviating to some extent the need for commuter aircraft with destinations in the urban area. Studies conducted in the New York metropolitan area have indicated that a replacement of only one-half of the commuter flights of 300 mi or less at the three major airports in this area could result in a capacity gain of over 20 percent with relatively little construction expenditure and significant reductions in airport delay [25].

The characteristics of existing and planned aircraft designed for use at vertiports are given in Table 11-6. Figure 11-11 shows a tilt-rotor aircraft, and Fig. 11-12 shows a tilt-wing aircraft.

Physical characteristics of vertiports

Final approach and takeoff area. The final approach and takeoff area for vertiports and vertistops can be any shape, but its size must be

TABLE 11-6   **Approximate Vertiport Aircraft Characteristics (Current and Future)**

Aircraft	Takeoff weight, lb	Overall length, ft	Rotor span, ft	Rotor diameter, ft	Height, ft	Landing gear, ft Tread	Landing gear, ft Base	Passenger capacity
CTR-800	15,750	41	58	26	15	8	15	8
CTR-1900	22,800	47	65	28	17	13	20	19
CTR-22A/B	45,120	57	85	38	18	15	22	31
CTR-22C	46,230	69	85	38	21	11	30	39
CTR-22D	49,260	72	86	38	23	11	30	52
CTR-7500	79,820	84	109	46	28	17	28	75
EUROTOR	18,000	52	76	33	20	13	28	19
EUROTOR	36,000	65	86	36	21	13	28	30
TW-68	16,500	39	41	17	13	9	15	11–16
V-2000	4,500	32	29	10	10	7	14	6

SOURCE: Federal Aviation Administration [25].

Figure 11-11   Civil tilt-rotor aircraft (*courtesy of Bell Helicopter Textron, Inc.*).

Figure 11-12   TW-68 tilt-wing aircraft (*courtesy of Ishida Group*).

such that it will circumscribe a square with at least 250-ft sides. To compensate for elevation at locations with elevations greater than 1000 ft above mean sea level (AMSL), the final approach and takeoff area must be increased in size by 50 ft for each 1000 ft of elevation AMSL. The final approach and takeoff area must be free of objects which would adversely affect the takeoff and landing operation. The major axis should be oriented to provide maximum wind coverage.

Touchdown and liftoff area.   The touchdown and liftoff area is a hard or paved surface capable of supporting the heaviest tilt-rotor aircraft or rotorcraft using the facility. The touchdown and liftoff area for vertiports and vertistops is typically located within the final approach and takeoff area, and it must have dimensions such that it will circumscribe a square with 100-ft sides for VFR operations and 150-ft sides for IFR operations. An elongation of the touchdown and liftoff area up to 400 ft offers significant economic and operational benefits and is recommended whenever site conditions permit. Normally, the touchdown and liftoff area is centered on the final approach course, and there must be at least a 75-ft clearance between the edges and ends of the touchdown and liftoff area and the edge of the final approach and takeoff area. For IFR operations, the touchdown and liftoff area must be located in the final approach and takeoff area, with its primary axis centered on the final approach course.

Taxiways.   Taxiways or hover taxiways connect the touchdown and liftoff area with the ramp or apron area used for passenger service and any other apron areas used for aircraft maintenance or storage, refueling, or aircraft tie-downs. A ground taxiway requires a safety area of 150 ft, and a hover taxiway must have a safety area of 250 ft—both centered on the centerline of the taxiway. Paved taxiways should be 75 ft wide. For economy, the center portion of a paved taxiway may be constructed with the required load-bearing capacity to a width equal to the landing gear width plus 20 ft of the most demanding aircraft using the facility.

On a taxiway, a clearance of at least one-half the rotor span but not less than 25 ft should be provided between the limits of the rotor span of an aircraft and a fixed or movable object when the undercarriage of the aircraft is located 15 ft from the edge of the taxiway. The *rotor span* is the distance between the extreme edges of the planes generated by the spinning rotors or prop rotors of the design aircraft. For a hover taxiway, there should be 75 ft of clearance between the limits of the rotor span of an aircraft and a fixed or movable object. When parallel taxiways are provided, the minimum distance between the centerline of the touchdown and liftoff area and the centerline of a parallel taxiway or between the centerlines of parallel taxiways should be 200 ft.

Aprons and ramps.   Aircraft parking positions should provide a minimum clearance of one-half the rotor span but not less than 25 ft between the limits of the rotor-span planes of adjacent aircraft or between the limits of the rotor-span planes of an aircraft and a fixed or movable object. To compensate for the dynamic forces created by tilt-rotor aircraft and rotorcraft, pavement design should be based upon a load equal to 150 percent of the maximum takeoff weight of the design aircraft with 75 percent of that load distributed equally to the main landing gear.

Airspace requirements.   A primary surface is established in the horizontal plane which has the size and shape of the final approach and takeoff area. Its elevation is the same as the elevation of the highest point in the touchdown and liftoff area.

For VFR operations, an approach surface begins at the edge of the primary surface and extends outward a distance of 4000 ft at a slope of 8 to 1. The inner width of the approach surface is the same as that of the primary surface, and the outer width is 650 ft. Transitional surfaces extending outward from the edges of the primary surface and the approach surface for a distance of 325 ft at slope of 2 to 1 are required on each side of the primary surface and the approach surface. Curved approach surfaces are permitted, but the final 1200-ft segment must be straight and aligned with the centerline of the touchdown and liftoff area.

The primary surface for non-precision-instrument approaches must be at least 300 ft long and 300 ft wide. The approach surfaces begin at the edge of the primary surface and extend outward a distance of 5000 ft at a slope of 20 to 1. The inner width of the approach surface is 500 ft, and the outer width is 2000 ft. Transitional surfaces extending outward from the edges of the primary surface and the approach surface for a distance of 350 ft at slope of 4 to 1 are required on each side of the primary surface and the approach surface. The transitional surface provided to the approach surface gradually narrows to the extent of the approach surface at the far end of the approach surface.

For IFR the final approach and takeoff area is 300 ft wide and 1225 ft long. Precision-instrument operations can be conducted at vertiports with 4°, 6°, or 9° glide slopes. For a 4° or 6° glide slope, the approach surface is a trapezoid. It begins 1225 ft from the far end of the final approach and takeoff area and extends outward a distance of 25,000 ft at a slope of 17 to 1 for a 6° glide slope and 25.5 to 1 for a 4° glide slope. The inner width of the approach surface is 1000 ft, and the outer width is 6000 ft. Transitional surfaces extending outward from the edges of the approach surface for a distance of 1500 ft at a slope of 7 to 1 are required on each side of the approach surface. The transitional surface provided to the approach surface gradually narrows to 600 ft at a distance of 1225 ft from the beginning of the approach surface. Transitional surfaces also extend outward from the edges of the final approach and takeoff area a distance of 350 ft at a slope of 7 to 1.

For a 9° glide slope, the approach surface is a trapezoid. It begins at the edge and end of the primary surface with a width of 250 ft and extends outward a distance of 10,000 ft at a slope of 10 to 1. The inner width of the approach surface is 250 ft, and the outer width is 3000 ft. Transitional surfaces extending outward from the edges of the approach surface for a distance of 960 ft at slope of 7 to 1 are required on each side of the approach surface. The transitional surface provided to the approach surface gradually narrows to 600 ft at the beginning of the approach surface. Transitional surfaces also extend outward from the edges of the primary surface area a distance of 600 ft at a slope of 7 to 1.

The dimensions of the approach surfaces for vertiports are tabulated in Table 11-7.

Marking of vertiports.　Painted lines, numerals and letters, or raised boundary markers are used to identify vertiport and vertistop surfaces. The extent of the final approach and takeoff area should be delineated with either in-ground markers or raised markers placed at the corners and at 125-ft intervals along the edges. The edges of a paved final approach and takeoff area are delineated by white painted lines. The edges of the touchdown and liftoff area are also delineat-

TABLE 11-7   FAR Part 77 Approach Surface Dimensions for Vertiports, ft

			Type of approach	
			Precision-instrument	
			Glide slope angle	
Dimension	Visual	Non-precision-instrument	4° or 6°	9°
Length	4,000	5,000	25,000*	10,000
Inner width	†	500	1,000	250
Outer width	600	2,000	6,000	3,000
Slope	8:1	20:1	17:1 ‡	10:1
Transitional				
Inner width	†	350	600	600
Outer width	325	0	1,500	960
Slope	2:1	4:1	7:1	7:1

*Begins 1225 ft from the far end of the final approach and takeoff area.
†Width of final approach and takeoff area.
‡For a 6° glide slope; 25.5:1 for a 4° glide slope.
SOURCE: Federal Aviation Administration [25].

ed by white painted lines. When the touchdown and liftoff area has a length more than 3 times its width, the centerline should be marked with 50-ft-long lines separated by 25-ft intervals. The centerline begins 25 ft from the numerals indicating the magnetic direction of the approach. A distinctive marking recommended to identify the touchdown and liftoff area of a vertiport is shown in Fig. 11-13. If there are restrictions on the maximum weight or rotor span, this restriction is indicated in the near right corner of the touchdown and liftoff area by a bar, with the number above the bar representing the maximum takeoff weight accommodated in thousands of pounds and the number below the bar indicating the maximum rotor span accommodated in tens of feet.

Taxiway centerlines are delineated by solid yellow lines, and taxi route centerline and apron edge markings should be solid yellow lines. Taxiway edges should be marked by a double solid yellow line. The portion of the taxiway pavement which is not full-strength pavement should be marked with solid yellow lines perpendicular to the taxiway centerline. Raised yellow markers are used to delineate the safety area for hover taxiway routes. A painted yellow line is also recommended to define the centerline of parking positions.

Lighting of vertiports

For operation during hours of darkness, various types of lights are suggested [25]. The limits of the ground final approach and takeoff

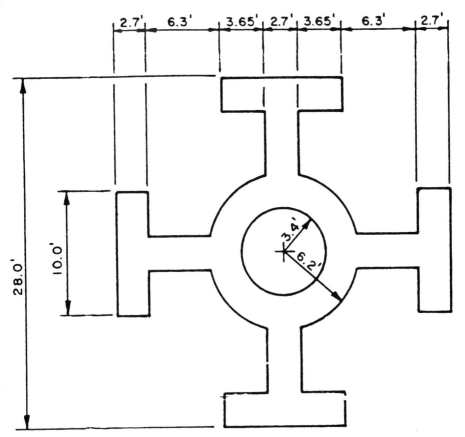

Figure 11-13 Recommended vertiport marking (*Federal Aviation Administration [25]*).

area need not be lighted if the touchdown and liftoff area is lighted. On elevated vertiports or vertistops, an odd number of red double-lens obstruction lighting fixtures at 50-ft intervals are used to mark the edges of the building. The touchdown and liftoff area is delineated by at least five omnidirectional yellow lights located at least 10 ft from the outside edge. This lighting configuration is also used to delineate the extent of the final approach and takeoff area. At verti-ports, blue taxiway lights define taxiway edges. Alternatively, inset green bidirectional lights may be used to delineate taxiway center-lines including taxiways on aprons or ramps. The edges of hover taxi-ways are also delineated by blue taxiway lights. Yellow bidirectional inset lights may also be used to delineate guidance routes for apron or ramp parking positions.

For precision-instrument operations, the basic lighting system for the final approach and takeoff area and the touchdown and liftoff area must be enhanced with high-intensity lights and a high intensity approach lighting system which has a length of 1000 ft. These systems are described by the FAA [25].

References

1. *A Canadian STOL Air Transport System—A Major Program,* Rep. 11, Canadian Science Council, Ottawa, Canada, 1970.
2. *Aerodrome Design Manual,* pt. 2: *Taxiways, Aprons, and Holding Bays,* 2d ed., International Civil Aviation Organization, Montreal, Canada, 1983.
3. *Aerodromes, Annex 14 to the Convention on International Civil Aviation,* vol. 2: *Heliports,* International Civil Aviation Organization, Montreal, Canada, 1990.
4. *A Guide to STOL Transportation System Planning,* The DeHavilland Aircraft of Canada, Limited, Ottawa, Canada, January 1970.
5. *Airport Design,* Advisory Circular AC 150/5300-13, Federal Aviation Administration, Washington, 1989.
6. *Canadian STOL Demonstration Service Montreal STOL Port Master Plan,* ST-71-8, Canadian Air Transportation Administration, Ottawa, Canada, March 1972.
7. *Certification and Operations of Scheduled Air Carriers with Helicopters,* pt. 127, Federal Aviation Regulations, Federal Aviation Administration, Washington, 1974.
8. *Guide for the Planning of Small Airports,* Roads and Transportation Association of Canada, Ottawa, Canada, 1980.
9. *Helicopter Annual,* Helicopter Association International, Alexandria, Va., 1992.
10. *Heliport Design,* Advisory Circular AC 150/5390-2A, Draft, Federal Aviation Administration, Washington, May 1992.
11. *Objects Affecting Navigable Airspace,* pt. 77, Federal Aviation Regulations, Federal Aviation Administration, Washington, 1989.
12. *Planning and Design Criteria for Metropolitan STOL Ports,* Advisory Circular AC 150/5300-8, Federal Aviation Administration, Washington, April 1975.
13. "Planning STOL Facilities," L. Schaefer, Paper no. 690421, *Proceedings,* Society of Automotive Engineers, New York, 1969.
14. *Provisional Criteria for STOL Port Zoning,* Canadian Air Transportation Administration, Ottawa, Canada, August 1973.
15. *Quiet Turbofan STOL Aircraft for Short-Haul Transportation,* Contractor Reps. NASA CR 114612 and CR 114613, National Aeronautics and Space Administration, Washington, June 1973.
16. *Rotorcraft Master Plan,* Federal Aviation Administration, Washington, November 1990.
17. *STOL Aircraft Future Trends,* Transport Aircraft Council, Aerospace Industries Association of America, Inc., Washington, May 1971.
18. *STOL Port Manual,* 1st ed., Doc. 9150-AN/899 with Amendment 1, International Civil Aviation Organization, Montreal, Canada, 1988.
19. *STOL-VTOL Air Transportation Systems,* C. Hintz, Jr., Civil Aeronautics Board, Washington, 1970.
20. *Studies in Short Haul Air Transportation—Effects of Design Runway Length, Community Acceptance, Impact on Return on Investment and Fuel Cost Increases,* R. S. Shevell and D. W. Jones, Jr., Stanford University, National Aeronautics and Space Administration, Ames Research Center, Moffett Field, Calif., July 1973.
21. *Study of Aircraft in Intraurban Transportation Systems, San Francisco Bay Area,* Boeing Company, Contractor Rep. NASA CR-114347, National Aeronautics and Space Administration, Ames Research Center, Moffett Field, Calif., 1970.
22. *Study of Quiet Turbofan STOL Aircraft for Short-Haul Transportation,* McDonnell-Douglas Corporation Rep. MDC-J4371, for National Aeronautics and Space Administration, Moffett Field, Calif., June 1973.

23. *Study of Short-Haul Aircraft Operating Economies,* Contractor Rep. CR-137685, Ames Research Center, National Aeronautics and Space Administration, Moffett Field, Calif., September 1975.
24. *United States Standard for Terminal Instrument Procedures (TERPS),* FAA Order 8260.3B with Changes 1 through 12, Federal Aviation Administration, Washington, DC, 1992.
25. *Vertiports,* Advisory Circular AC 150/5390-3, Federal Aviation Administration, Washington, 1991.
26. "V/STOL Aircraft: The Future Role in Urban Transportation as a Pickup and Distribution System," R. H. Miller, Proceedings of Symposium on Transportation and the Prospects for Improved Efficiency, National Academy of Engineering, Washington, October 12, 1972.

12

Structural Design
of Airport Pavements

In this chapter various methods for designing pavements are briefly described. The term *structural design of pavements* as used in this text refers to the determination of the thickness of the pavement and its components, not to the design of materials in the pavement (e.g., asphalt concrete or portland cement concrete mixes).

Pavement or *pavement structure* is a structure consisting of one or more layers of process materials. A pavement consisting of a mixture of bituminous material and aggregate placed on high-quality granular materials is referred to as *flexible*. When the pavement consists of a slab of portland cement concrete, it is referred to as *rigid.**

The pavement is intended to provide a smooth and safe all-weather riding surface, and the thickness of each layer must be adequate to ensure that the applied loads will not lead to distress† in it or the underlying layers.

Figure 12-1 illustrates the type of structural pavement section suitable for heavy-duty pavements for airports.

A flexible pavement may consist of one or more layers of material classified as *surface course, base course,* and *subbase course,* resting on a *prepared subgrade* layer. This latter material, as seen in Fig. 12-1, is the material on which the pavement structure rests and may be either an embankment or an excavation. The surface course con-

---

*The terms *flexible* and *rigid* have been retained in the text since this terminology is still used in practice. To consider a thick asphalt concrete layer on subgrade as flexible or a portland cement concrete slab as rigid does not appear "reasonable" at this time. Accordingly the authors urge that pavements with a substantial thickness of asphalt concrete in the structural section be referred to as asphalt concrete pavements, while those containing portland cement concrete slabs be referred to as a portland cement concrete pavement.

†The term *distress* is preferred to *failure.*

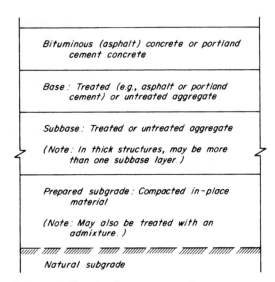

Figure 12-1 Structural pavement section.

sists of a mixture of bituminous material (generally asphalt) and aggregate ranging in thickness from a minimum of 3 or 4 in (for heavy-duty airfields) to 12 in or more. Its principal functions are to provide for smooth and safe traffic operations, to withstand the effects of applied loads and environmental influences for some prescribed period of operation,* and to distribute the applied load to the underlying layers. The base course may consist of treated (e.g., portland cement or asphalt) or untreated granular material. Like the surface course, it must be adequate to withstand the effects of load and environment and to distribute the applied loads to the underlying layers. This point is illustrated in Fig. 12-2, which shows the influence of base thickness as well as stiffness (modulus) of the various pavement components on the stress at the surface of the subgrade. The subbase course may be composed of treated or untreated material; quite often unprocessed pit-run material or material selected from a suitable excavation on the site is utilized. Its function is the same as that of the base. Note that not every flexible pavement requires a subbase course. On the other hand, very thick pavements may be composed of several subbase courses.

A rigid pavement consists of a slab of portland cement concrete, 8 to 24 in thick, placed on a prepared layer of imported materials. For heavily trafficked pavements in particular, it is desirable that the

---

*This may be expressed in terms of the number of load repetitions or some period of time.

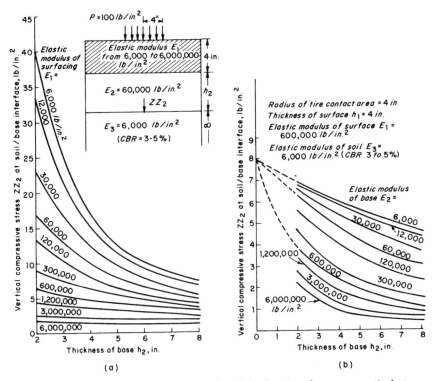

Figure 12-2 Influence of base thickness and moduli of various layers on vertical stress at surface of subgrade (*Whiffin and Lister [65]*).

upper 4 to 6 in of this material be treated with portland cement or asphalt to minimize pumping.* This layer directly under the slab is sometimes referred to as a subbase rather than a base because its quality would not necessarily have to be as high as that of a material directly under a comparatively thin bituminous layer.

Several methods of design for airport pavements are available. Although there is no one method for flexible-pavement design accepted by all agencies, a method in widespread use is that developed by the Corps of Engineers and based on the *California bearing ratio* (CBR) test. Similarly, for rigid pavements, thickness selection based on solutions developed originally by Westergaard is used extensively. These methods are described in this text. Also included are procedures used by the FAA for both flexible and rigid pavements. A method adopted by the FAA and the ICAO, the aircraft classification number–pavement classification number (ACN-PCN), is discussed.

---

*Pumping is described in this chapter in the section "Design of Rigid Pavements."

Procedures for the design of asphalt pavements, which treat the structure as a layered elastic solid, developed by Shell Oil Company and by the Asphalt Institute are also briefly summarized.

The design of overlay pavements is treated in some detail since it is considered to be an important aspect of airfield operations in the future. In this regard, existing procedures as well as some new developments having the potential to improve overlay design are included.

Details of the aircraft for which the designs have been developed are included in Chap. 3.

CBR Method of Design for Flexible Airport Pavements

Historical background

The CBR method of design was developed by the California Division of Highways in 1928.* The method subsequently was adopted for military airport use by the Corps of Engineers, U.S. Army, shortly after the outbreak of World War II. The outbreak of the war required that a decision be made with little delay concerning a design method. At the time, there were no methods available specifically developed for airport pavements. It was apparent that the time required to develop a completely new method of design would preclude its use in a war emergency program. Consequently, it was decided to review all available methods for the design of highway pavements and to select one which readily could be adopted for airfield use. The criteria for selecting a method were many. Among the more important were (1) simplicity in procedures for testing the subgrade and the pavement components, (2) a record of satisfactory experience, and (3) adaptation to the airport problem in a reasonable time. After several months of investigating of suggested methods, the CBR method was tentatively adopted. Application of the CBR method enables the designer to determine the required thickness of the subbase, base, and surface course by entering a set of design curves with the result of a relatively simple soil test.

The CBR test

The CBR test expresses an index of the shearing strength of soil. Essentially the test consists of compacting about 10 lb of soil into a 6-in-diameter mold, placing a surcharge on the surface of the sample, immersing the sample in water for 4 days, and penetrating the soaked sample with a steel piston approximately 2 in in diameter at a speci-

---

*The person most closely associated with its development was O. J. Porter.

fied rate of loading. The resistance of the soil to penetration, expressed as a percentage of the resistance for a standard crushed stone, is the CBR of the soil. Thus, a CBR of 50 means that the stress necessary for the piston to penetrate the soil sample a specified distance is one-half that required for the piston to penetrate the same distance in the standard crushed stone. The relationship is usually based on a penetration of the piston of 0.1 in, with 1000 lb/in^2 used as the stress required to penetrate the crushed stone at 0.1-in penetration.

The weight placed over the sample prior to immersion in water is commonly referred to as the *surcharge*. The unit weight of the surcharge corresponds to an estimate of the unit weight of the pavement structure.

The 4-day soaking period was chosen because a majority of soils approach complete saturation in this period within the depth affected by the piston. Thus, a soaked sample represents the worst condition of the soil regarding its ability to support a pavement structure. Soaking of the sample is not required where it can be conclusively shown that with the passage of time additional moisture will not be absorbed into the pavement structure.

If the soil in its natural state, for one reason or another, cannot be improved by compaction, then the CBR test is performed on an undisturbed sample. But if stability can be improved by subsequent compaction, the soil is placed in the CBR mold in five equal layers, and each layer is compacted with a 10-lb weight falling from a height of 18 in. The standard number of blows per layer is 55 (corresponding to the modified AASHO* compactive effort), although fewer blows are also used to establish the CBR for design.

As will be seen subsequently, empirical relationships for various wheel loads have been established between the CBR test value and the thickness of the pavement structure.

Adaptation of the CBR procedure to airfield pavements

Investigations made by the California Highway Department from 1928 to 1942 on both adequate and inadequate pavements furnished data from which the empirical relationship of CBR versus thickness, shown in Fig. 12-3, was developed. Curve *B* indicates the minimum thickness of the pavement structure for light traffic, and curve *A* the thickness for average high traffic conditions. Further analysis of the data from which the two curves were derived indicated that curve *A*

---

*American Association of State Highway Officials (now the American Association of State Highway and Transportation Officials).

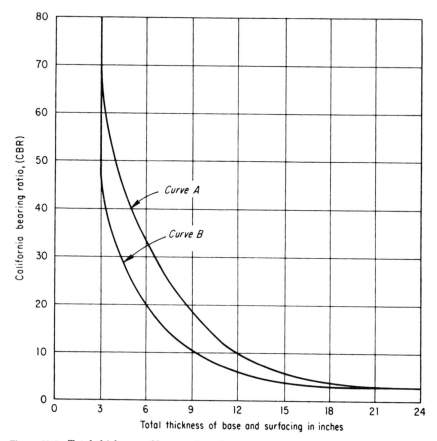

Figure 12-3 Total thickness of base and surfacing in relation to CBR values (*Corps of Engineers*).

was the more reliable of the two and that it was reasonable to assume that it represented a 9000-lb truck wheel load. Because aircraft tires are operated at much larger deformations than truck tires, and because highway traffic is much more channelized, it was reasoned that the 9000-lb truck wheel load was equivalent to a 12,000-lb aircraft wheel load. Thus, curve A (Fig. 12-3) was assumed to represent a 12,000-lb aircraft wheel load.

At the time the CBR procedure was first adopted by the Corps of Engineers, aircraft tire pressures were on the order of 60 lb/in^2, and single-wheel loads ranged from 25,000 to 70,000 lb. Because of the war emergency program, an attempt was made to utilize soil mechanics theory to extrapolate from the 12,000-lb wheel load to larger loads. The following procedure was used.

With a contact pressure of 60 lb/in^2, contact areas were computed for 12,000-, 25,000-, 40,000-, and 70,000-lb wheel loads. It was

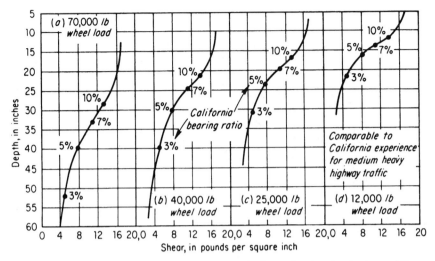

Figure 12-4 Extrapolation of highway pavement thickness by the elastic theory (*Corps of Engineers*).

assumed that the contact areas were circular. Shear stresses were then computed as shown in Fig. 12-4. The thicknesses of the pavement structure corresponding to CBRs of 3, 5, 7, and 10 on curve *A* (Fig. 12-3) were plotted on the shear stress curve for the 12,000-lb wheel load, and the corresponding stresses were noted (for example, 5 lb/in² for a CBR of 3). These stress values were located on the curves for the 25,000-, 40,000-, and 70,000-lb wheel loads, and the corresponding depths noted. The depths, which represent thicknesses, were then plotted on a graph of thickness versus CBR, and a chart similar to the one shown in Fig. 12-5 was developed.

From a strictly theoretical point of view, the conditions assumed in the calculations have several limitations. One is the assumption of a homogeneous mass for the pavement structure. Nevertheless the analysis was a good beginning, and it proved to be in substantial agreement with the thicknesses developed later from full-scale test tracks.

Concurrent with the theoretical approach, a comprehensive investigation program involving the construction of a number of full-scale test tracks was initiated. The results of these investigations indicated that the curves established from theoretical considerations appeared to be conservative for the higher CBR values, and for the heavier wheel loads they were not sufficient for the lower CBR values. Accordingly, the basic curves shown in Fig. 12-5 were adjusted to reflect the results of the investigations.

During the greater part of World War II, heavy bombardment aircraft were supported by two main landing gears, each gear consisting

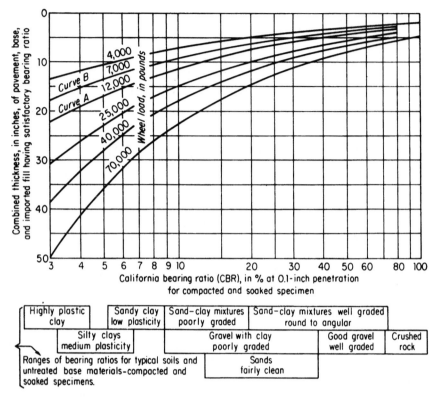

Figure 12-5 Tentative design of foundation for flexible pavements (*Corps of Engineers*).

of a single wheel. Toward the end of the war, the B-29 entered into service. Its landing gear consisted of dual wheels. An analysis of the effect of this type of gear on pavement thickness and the development of appropriate thickness charts to reflect the new type of landing gear were required.

Subsequent to the introduction of the B-29, as aircraft grew in size, it was necessary to spread the load more and more to keep the thickness of the pavement structure within tolerable limits, with the result that the number of wheels in each landing gear was increased from two to four (dual-in-tandem arrangement). The analysis developed for the B-29 assembly was extended to develop thickness design relationships for new aircraft with this type of gear. This methodology was used until the middle 1950s, at which time the Corps of Engineers reanalyzed their data and ascertained that the thicknesses were slightly unconservative.

The discussion to follow details the results of a part of this reanalysis, since it is used in the present methodology to develop design thicknesses for new aircraft as they are developed.

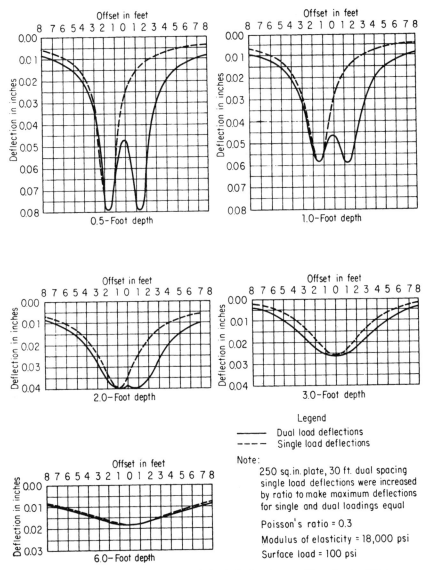

Figure 12-6 Comparison of single- and dual-deflection profiles theory (*Waterways Experiment Station, Corps of Engineers*).

One of the principal causes of distress in a pavement is an undesirable amount of movement of material. This movement is manifested as strain or deflection. It was therefore reasoned that an acceptable criterion of failure would be strain or deflection. Since few or no data on strain were available, the curve of the slope or rate of change of deflection versus offset (Fig. 12-6) was considered a reasonable index of critical strain.

Deflection-versus-offset curves were computed for single and dual wheels by use of Boussinesq's theory. Some test data on deflection profiles were also available. The test data confirmed the validity of the theoretical computations. It was found, from both the theoretical work and the test data, that without exception the slopes of the deflection-versus-offset curves for the single loads were equal to or steeper than those for dual wheels at equal depths, as shown in Fig. 12-6. From such an analysis it was demonstrated that a *single-wheel load, which yields the same maximum deflection as a multiple-wheel load, would produce equal or more severe strains in the foundation in comparison with the multiple-wheel load.* For purposes of design, the single-wheel load could be considered equivalent to the multiple-wheel load; thus the concept of the *equivalent single-wheel load* (ESWL), was introduced. The contact area of this ESWL is equal to the contact area of one of the wheels of the multiple-wheel assembly.

To develop the thickness requirements for a multiple-wheel assembly of known dimensions and total load, ESWLs were computed at various depths by use of the theory of elasticity (Boussinesq's theory as expanded by A. E. H. Love). For each depth there is a different equivalent single-wheel load. The ESWLs are determined for several depths; a curve relating the ESWL to depth is drawn.

To illustrate the procedure, the following example is provided. Assume a dual-in-tandem assembly, as shown in Fig. 12-7. Total load on the assembly is 130,000 lb, and the contact pressure is 140 lb/in^2.

The total contact area is equal to 928 in^2, and the contact area of each wheel is 232 in^2. Suppose it is desired to find the maximum ESWL at a depth equal to 3 times the radius of the contact area of each wheel.

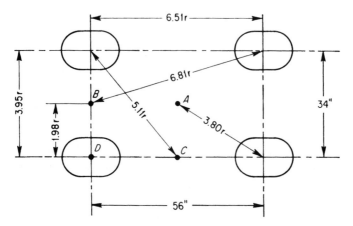

Figure 12-7 Dual-in-tandem example—reanalysis.

The calculations are as follows:

$$r = \sqrt{\frac{A}{\pi}} = \sqrt{\frac{232}{\pi}} = 8.6 \text{ in}$$

$$3r = 25.8 \text{ in}$$

where r = radius of contact area, in
A = contact area of wheel, in^2

The problem becomes one of finding the location where the maximum deflection at a depth of 25.8 in will occur. As a trial, four locations were investigated, as shown in Fig. 12-7.* The center of the assembly is designated as location A; midway on the 34-in dimension as location B; midway on the 56-in dimension as location C; and under one wheel of the assembly as location D.

In an elastic medium, the deflection W is expressed by

$$W = \frac{prF}{E_m} \tag{12-1}$$

where p = load intensity
E_m = modulus of elasticity
F = deflection factor obtained from Fig. 12-8

By using subscripts s and d to denote single and dual wheels, we may write

$$W_s = \frac{r_s}{E_m} p_s F_s \quad \text{and} \quad W_d = \frac{r_d}{E_m} p_d F_d$$

Since $W_s = W_d$ and $r_s = r_d$,

$$\frac{p_s}{p_d} = \frac{F_d}{F_s}$$

The contact area of the single wheel is equal to the contact area of one wheel of the assembly. Thus Eq. (12-2) results:

$$\frac{P_s}{P_d} = \frac{F_d}{F_s} \tag{12-2}$$

or the ratio of the ESWL (P_s) to one wheel of the dual assembly (P_d) is the inverse of the ratio of the maximum deflection factors.

---

*Tire imprints in Fig. 12-7 are assumed to be circles.

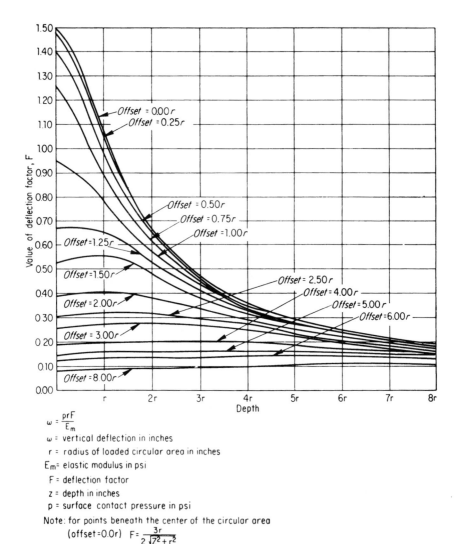

$$\omega = \frac{prF}{E_m}$$

ω = vertical deflection in inches

r = radius of loaded circular area in inches

E_m = elastic modulus in psi

F = deflection factor

z = depth in inches

p = surface contact pressure in psi

Note: for points beneath the center of the circular area

$$(\text{offset} = 0.0r) \quad F = \frac{3r}{2\sqrt{z^2 + r^2}}$$

offsets measured from origin along x-axis.

Figure 12-8 Deflection factor F for uniform load of radius r at points beneath the x axis; Poisson ratio = 0.5 (*Waterways Experiment Station, Corps of Engineers*).

The deflection factors are obtained from Fig. 12-8 and are as follows. For the single wheel, the factor at a depth of $3r$ is 0.47 (directly under the axis of load). For the dual-in-tandem assembly, the factors must be computed as shown in Table 12-1.

To determine the deflection factors from Fig. 12-8, the distances from each location (A, B, C, and D) to each wheel of the assembly must be expressed in terms of radius r.

TABLE 12-1

Depth	Deflection factors for dual-in-tandem assembly			
	At A	At B	At C	At D
$3r$	0.21	0.34	0.25	0.47
$3r$	0.21	0.34	0.25	0.20
$3r$	0.21	0.10	0.16	0.12
$3r$	0.21	0.10	0.16	0.10
Total	0.84	0.88	0.82	0.89

TABLE 12-2

Depth in	Single-wheel deflection factor	Dual-wheel deflection factor	Load ratio	
			Single to one wheel of assembly	Entire assembly
25.8	0.47	0.89	1.90	0.475

In the example, the maximum deflection occurs under the axis of one of the wheels of the assembly (at D). The corresponding deflection factor is 0.89. The various deflection factors are summarized in Table 12-2. The ESWL at a depth of 25.8 in is then

$$0.475(130,000) = 61,750 \text{ lb}$$

or it may be computed by multiplying the deflection factor of 1.90 by the load on a single wheel of the assembly (32,500 lb).

Analyses of this type, together with the single-wheel load curves developed over the years (Fig. 12-9 for loads with 100 lb/in^2 tire pressure and Fig. 12-10 for loads with tire pressures of 200 lb/in^2), permit the development of design curves for aircraft with multiple wheels. The method is shown later.

In 1958 an analysis of all available service-behavior data for test sections and prototype airfields indicated that the CBR design criteria for single-wheel loads could be expressed in these parameters: thickness per square root of contact area ($t/\sqrt{A}$) and CBR per tire pressure, which separated failures and nonfailures for near-capacity operations (approximately 5000 coverages). The mathematical expression for the relationship was

$$t = \sqrt{\frac{P}{8.1(\text{CBR})} - \frac{A}{\pi}} \tag{12-3}$$

where t = design thickness, in
P = single-wheel load, lb
A = tire contact area, in^2

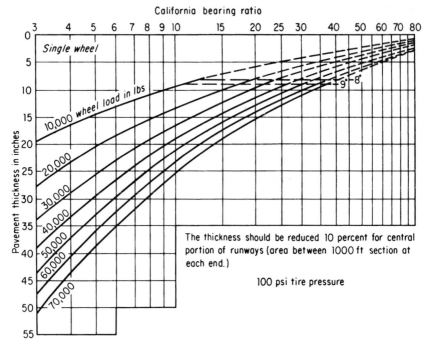

Figure 12-9 Flexible-pavement design curves for taxiways (*Corps of Engineers*).

Data from the curves of Figs. 12-9 and 12-10, plotted in this form, are shown in Fig. 12-11. Note that the expression appears valid for CBR values less than about 12. Thicknesses at CBR values greater than 12 are larger because of durability and other requirements.

To account for load repetitions and multiple-wheel gear configurations, the above equation was modified to the following:

$$t = f \sqrt{\frac{ESWL}{8.1(CBR)} - \frac{A}{\pi}} \qquad (12\text{-}4)$$

where f = percent of design thickness $(0.23 \log C + 0.15)$
ESWL = equivalent single-wheel load (determined as described previously)
C = coverage (sufficient wheel passes to cover every point of traffic lane once)

It is by this procedure that many of the design charts for commercial jet aircraft have been developed. Thickness versus CBR curves for a number of representative jet aircraft are shown in Fig. 12-12. The thicknesses are for taxiways and runway ends (first 1000 ft).

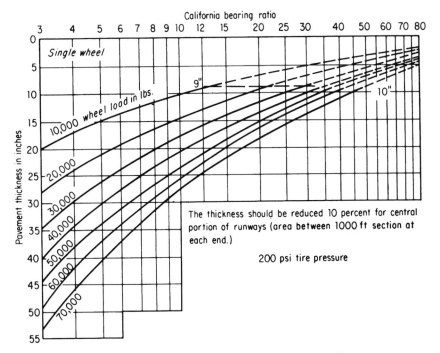

Figure 12-10 Flexible-pavement design curves for taxiways (*Corps of Engineers*).

Thicknesses for runway interiors are less than those shown by the design relationships [33].*

Recent studies of test pavements subjected to loads representative of multiple-wheel heavy-gear-load aircraft (for example, B-747) have indicated that thicknesses developed at larger repetitions are somewhat conservative if the equation of the preceding paragraph is utilized. Accordingly the present method uses the formula [6]

$$t = \alpha_i \sqrt{\frac{\text{ESWL}}{8.1(\text{CBR})} - \frac{A}{\pi}} \qquad (12\text{-}5)$$

where α_i = load repetition factor, which is dependent on the number of wheels for each main landing gear used to compute ESWL; for example, for a B-747, eight wheels should be used, as shown in Fig. 12-13.

The load repetition factor α_i is based on aircraft passes, whereas the previous relationships were based on coverages.† Relationships

---

*The Corps of Engineers uses a reduced load for a particular aircraft to design the runway interior.

†A *pass* is one operation of an aircraft.

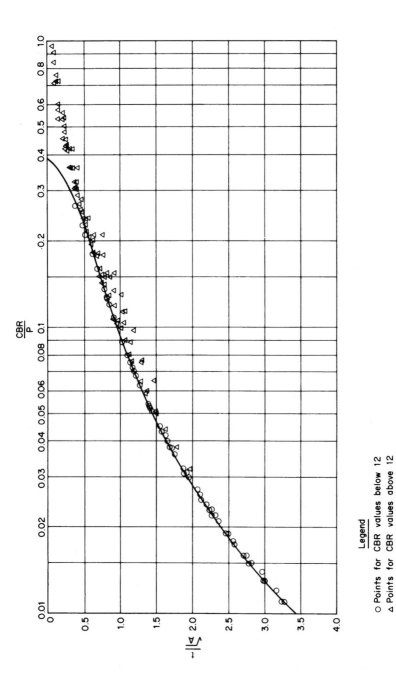

Figure 12-11 Consolidated CBR design criteria (*Waterways Experiment Station, Corps of Engineers*).

Legend

○ Points for CBR values below 12
△ Points for CBR values above 12

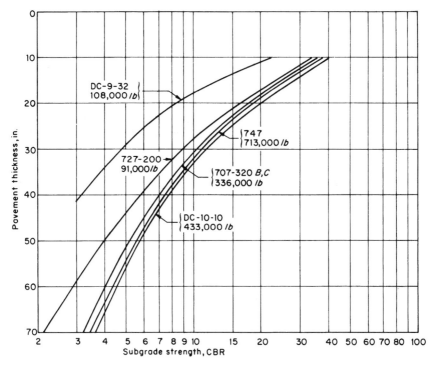

Figure 12-12 Flexible-pavement thickness—Corps of Engineers design method. Weights represent maximum ramp weights with the main landing gears loaded to their maximum.

between coverages and passes can be found [9, 45]. This equation provides reasonable thicknesses up to CBR values of 15. For CBR values in excess of 15, durability considerations as well as other factors govern pavement thickness.

Thus to develop a CBR-versus-thickness relationship for any aircraft, the steps may be summarized as follows:

1. For a particular depth t, determine the ESWL according to the procedure described previously.

2. For the desired number of aircraft operations (design period), such as 100,000, select α_i from Fig. 12-13.

3. Solve Eq. (12-5) for CBR.

4. Repeat for other depths so that curves of the type shown in Fig. 12-12 can be developed.

Similarly, curves of ESWL versus pavement thickness may be developed for ranges of CBR. The procedure for a particular aircraft is as follows:

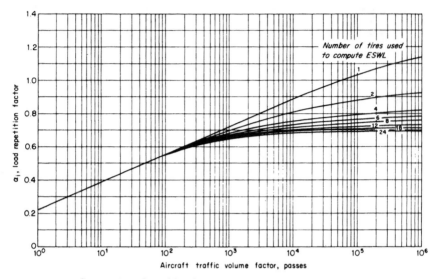

Figure 12-13 Composite plot of load repetition factors versus passes (*Waterways Experiment Station, Corps of Engineers*).

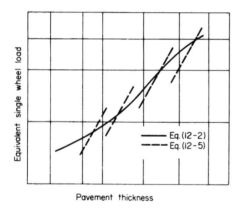

Figure 12-14 Illustration of the determination of the flexible-pavement design curve for an aircraft by ESWL method.

1. Determine the ESWL for various depths.

2. Plot the ESWL versus the depth.

3. Determine the pavement thickness t required for each ESWL and CBR.

4. Plot the ESWL-versus-thickness curve on the same graph as in step 2 for each CBR.

5. The point of intersection between the curves determines the required thickness.

This procedure is shown in Fig. 12-14.

Design example

To illustrate the thickness selection procedure, assume that a pavement is to be designed for a set of conditions where frost and special subgrade treatment are not considerations.

Assume that a taxiway is to be designed for a DC-10-10 by using the relationship shown in Fig. 12-12. The design CBR values for the prepared (compacted) subgrade and structural section materials are as follows.

Material	Design CBR
Compacted subgrade soil	5
Select material subbase	25
Crushed-rock base	80

The total thickness of pavement is governed by the CBR of the compacted soil. From Fig. 12-12 the required thickness of the pavement structure is 55 in; it is composed of subbase, base, and an asphalt concrete course. Above the subbase a thickness of 17 in is required; thus the subbase thickness is 38 in. If an 8-in asphalt concrete layer is used, the base layer thickness will be 9 in, the difference between 17 and 8 in. For a pavement section of this thickness, it is possible that more than one subbase layer might be utilized.

Materials

Bituminous surface courses for airfields should be of sufficient thickness to ensure that (1) excessive stresses will not be transmitted to the underlying layer and (2) the stresses in the layer itself will not lead to premature cracking. According to recommendations by the Corps of Engineers, the thickness should be not less than 3 in.*

The base course immediately under the bituminous surface course should be able to withstand the stresses produced in the zone directly under the wheels of an aircraft. The required stability depends on the type and thickness of the surface course and the magnitude of loading. For the surface course noted in the example, the CBR of a crushed stone base course should be in the range of 80 to 100, depending on the load. The minimum thickness of a crushed stone base suggested by the Corps of Engineers is 6 in. A greater thickness is normally preferred.

Recent tests by the Corps of Engineers indicate the advantage of stabilized bases (asphalt or portland cement) in improving the perfor-

---

*The authors' experience would suggest larger thicknesses for the bituminous course, particularly for many repetitions of heavier aircraft.

mance of pavements subjected to frequent repetitions of heavy loads. Accordingly it would appear desirable to ensure that the upper 9 to 15 in of the pavement structure is composed of a bituminous surface layer plus a stabilized course. Care should be exercised in the analysis of the benefits of structurally improved layers to consider the economics associated with applications to particular airport projects [18, 21, 27].

Strength characteristics of subbases vary widely, from 20 to 50 CBR. Above 50 CBR, materials are generally regarded as base courses.

Compaction requirements

Field studies by the Corps of Engineers have shown that compaction produced in a given layer by traffic is a function of the total load, arrangement of tires, tire pressure, number of repetitions, and depth of a given layer. These parameters can be combined by the use of the design CBR. The term *compaction index* C_i was introduced; this value is the design CBR value for the corresponding depth and load.

Curves of the degree of compaction (based on the modified AASHO compaction test) versus the compaction index for cohesive and cohesionless soils are shown in Fig. 12-15; these curves represent relationships established from field data.

To illustrate the use of the compaction index curves to establish compaction requirements for a specific aircraft, consider the DC-10-10 curve shown in Fig. 12-12. Since the design CBR equals the compaction index, by combining the data of Figs. 12-12 and 12-15 it is possible to develop the density requirements for a specific aircraft, as shown in Fig. 12-16. From this curve note that any cohesionless material in the upper 3 ft of the pavement section should be compacted to at least 100 percent of the density obtained in the modified AASHO compaction test.

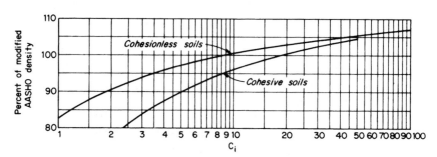

Figure 12-15 Compaction requirements for flexible pavements. Compaction index is the design CBR value for the corresponding depth and load (*Waterways Experiment Station, Corps of Engineers*).

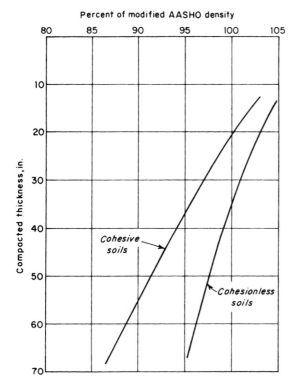

Figure 12-16 Density requirements for pavement subjected to DC-10-10 loading.

Design of Rigid Pavements

The design of portland cement concrete pavements for airports is based upon a vast amount of data which have been compiled during many years of research and observation of the behavior of pavements. The Corps of Engineers, the Portland Cement Association (PCA), and the FAA have contributed much to our knowledge of concrete pavements. The material presented here, much of which has been obtained from publications of these three agencies, briefly summarizes the design procedures that have been developed from research and practice.

Stress in a concrete pavement is induced in several ways, as follows:

1. Wheel loads

2. Change of shape of slab due to differential in temperature between top and bottom of the slab (commonly referred to as *curling*)

3. Change of shape of slab due to differential in moisture between top and bottom of the slab

4. Friction developed between slab and foundation when the slab changes its volume

The first consideration in the design of any pavement is the load which it is to carry. When the load is known, the next step is to determine the critical stresses developed in the slab. Several methods of computing stresses have been proposed, the best known of which is the analysis developed by Westergaard [46]. A number of organizations have conducted comprehensive studies to determine the validity of the analysis.

Approximate solutions for stresses due to curling and friction have also been developed and are summarized in the succeeding paragraphs.

Westergaard's analysis

Westergaard assumed the pavement slab to be a thin plate resting on a special subgrade which is considered elastic in the vertical direction only. That is, the reaction is proportional to the deflection of the subgrade $p = kz$, where $z =$ deflection and $k =$ a soil constant, referred to as the *modulus of subgrade reaction*. Other assumptions are that the concrete slab is a homogeneous, isotropic elastic solid and that the wheel load of an aircraft is distributed over an elliptical area. Although these assumptions do not satisfy theory in a strict sense, the results have compared reasonably with observations. The Westergaard analysis can be used to evaluate stress in a pavement slab and the deflection of the slab. The analysis, however, is not applicable for the determination of stresses and deflections in the foundation material.

For airports Westergaard developed formulas for stresses and deflections (1) in the interior of a slab and (2) at an edge of a slab or at a joint that has no capacity for load transfer. These formulas for stress are as follows [46].

Case 1.   The load is applied in the interior of the area of a panel at a considerable distance from any edge or joint. The total load P is distributed uniformly over an elliptical area defined by

$$\frac{x^2}{a^2} + \frac{y^2}{b^2} = 1$$

The maximum principal tensile stress σ_i at the bottom of the slab under the center of the load can be computed by

$$\sigma_i = \frac{P}{d^2} \left\{ 0.275(1 + \mu) \log \frac{Ed^3}{k[(a+b)/2]^4} + 0.2391(1 - \mu) \frac{a - b}{a + b} \right\} \quad (12\text{-}6)$$

Case 2. The load is applied adjacent to an edge or joint that has no capacity for load transfer. The total load P is distributed uniformly over an elliptical area, the prolate edge of which is tangent to the edge or joint and is defined by the equation

$$\frac{x^2}{a^2} + \frac{(y - b)^2}{b^2} = 1$$

The tensile stress σ_e at the bottom of the slab, along the edge or joint, directly under the point of tangency of the ellipse, can be computed from

$$\sigma_e = \frac{2.2(1 + \mu)P}{(3 + \mu)d^2}\log \frac{Ed^3}{100k[(a + b)/2]^4} + \frac{3(1 + \mu)P}{\pi(3 + \mu)d^2}\left[1.84 - \frac{4}{3}\mu \right.$$

$$\left. + (1 + \mu)\frac{a - b}{a + b} + 2(1 - \mu)\frac{ab}{(a + b)^2} + 1.18(1 + 2\mu)\frac{b}{l} \right] \quad (12\text{-}7)$$

where P = load transmitted through a tire to a panel of pavement, lb

   a, b = semiaxes of an ellipse which represents footprint of tire. If the load is next to an edge or joint, a is the semiaxis parallel to the edge or joint. Either a or b may be the greater semiaxis, depending on whether the joint is longitudinal or transverse.

   x, y = horizontal rectangular coordinates. If the load is next to an edge or joint, the axis of x is along the line of the edge or joint; if the footprint of the tire is represented by an ellipse, the x axis is in the direction of semiaxis a.

   d = thickness of slab, in

   E = modulus of elasticity of concrete, lb/in^2

   μ = Poisson's ratio of concrete

   k = modulus of subgrade reaction, lb/in^3

   l = radius of relative stiffness, in

To obtain the allowable stress, it is customary to divide the modulus of rupture of concrete by a factor of safety, for which the PCA suggests the values tabulated below [25].

Installation	Safety factor
Aprons, taxiways, runway ends	1.7–2.0
Runways (central portion)	1.4–1.7

These safety factors are related to the fatigue of concrete. It has been found that if concrete is stressed repeatedly at a value approximately one-half of the ultimate flexural strength, concrete can stand an unlimited number of load repetitions. As the stress under repeated load approaches the ultimate flexural strength, the number of repetitions of load which can be applied without causing failure becomes progressively fewer. On aprons, taxiways, and runway ends, the traffic is more channelized than in the central portion of runways. Hence, the safety factor is greater for the former group of pavements.

The modulus of rupture used for design is normally the flexural strength at 90 days, although the 28-day value increased by a factor of 1.10 to 1.14 can also be used.

The charts for determining stress which were prepared by the PCA [25] are based on the assumption that the wheel load is in the interior of a panel (case 1). This assumption stems from the argument that all interior joints have adequate load transfer devices so that conditions at any point within the pavement panel are nearly the same, as though the point under consideration were near the middle of a large slab.

The charts prepared by the Corps of Engineers [59] are derived from Westergaard's edge formula, on the basis that such a loading condition does occur at the edges of joints, and, moreover, it is more critical. However, to recognize the action of load transfer devices, it is assumed that when the load is placed at an edge of a slab, one-quarter of the load is transferred to the adjacent slab. The remaining load ($0.75P$) is increased by 30 percent to take care of curling stresses, load repetitions, and other contingencies. The net load [$0.75(1.3) = 0.98P$] is used in the Westergaard edge load formula.

Figures 12-17 through 12-20 contain design charts prepared by the PCA for a number of different types of jet aircraft. These design relationships were developed by using a computer program based on the Westergaard analysis [16]. This program can be used for a range in gear configurations and loads, making it possible for the designer to develop designs for aircraft to which a specific pavement might be subjected.

In the design charts, gear load rather than gross aircraft weight is used. This information is available from the aircraft manufacturers' data.

An alternative approach to the use of the PCA computer program is to use the influence charts developed by Pickett and Ray [38]. Their application is discussed in Ref. 25. A sample chart is shown in Fig. 12-21 with an aircraft gear superimposed. Note that the scale of the chart is expressed in terms of l, the radius of relative stiffness. The scale to which the tire imprint of the multiple-wheel assembly is drawn is based on the ratio of the graphical l on the chart to the actual l. That is, the graphical dimensions are equal to the actual dimen-

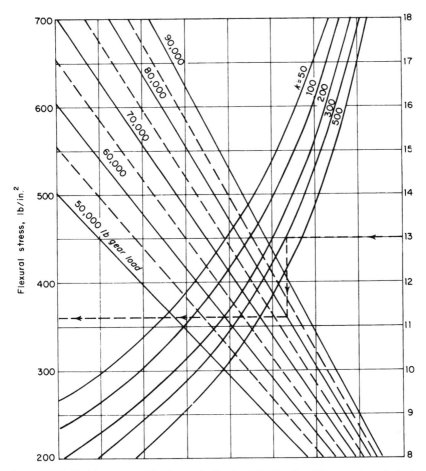

Figure 12-17   Rigid-pavement thickness for Boeing 727 (*Portland Cement Association*).

sions multiplied by the ratio of the graphical l to the actual l of the pavement. The objective is to place the wheels on the chart so that the maximum number of blocks is incorporated within the tire imprint area. The bending moment M is expressed as

$$M = \frac{ql^2N}{10,000} \qquad (12-8)$$

where q = tire contact pressure, lb/in^2
   l = radius of relative stiffness of slab
   N = number of blocks enclosed

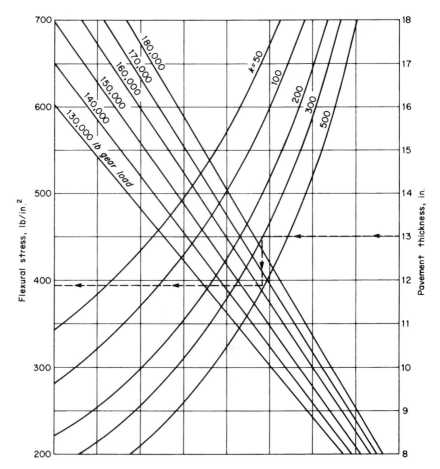

Figure 12-18 Rigid-pavement thickness for DC-8-62, 63 (*Portland Cement Association*).

Application of fatigue concept to traffic analysis

The fatigue procedure developed by the PCA can be applied to the design and evaluation of pavements at airports serving large volumes of heavy multigeared aircraft of various types as an alternative to the procedure described earlier utilizing the safety factor concept. According to the PCA, its applications include

1. Design for specific volumes of mixed traffic

2. Evaluation of future traffic capacity of existing pavements or of an existing pavement's capacity to carry a limited number of overloads

3. Evaluation of the fatigue effects of future aircraft with complex gear arrangements

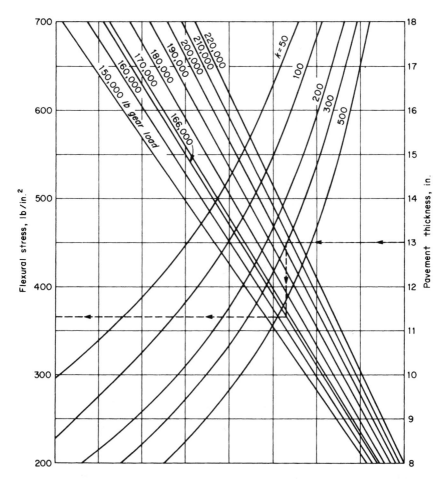

Figure 12-19 Rigid-pavement thickness for Boeing 747 (*Portland Cement Association*).

4. More precise definition of the comparative thicknesses of runways, taxiways, and other pavement areas depending on operational characteristics

Essentially the procedure involves the analysis of a selected slab thickness* for the expected traffic. For a particular aircraft, the stress corresponding to a specific gear load can be ascertained by either the computer program [16] or a chart like Fig. 12-18. To determine the damage produced by the anticipated repetitions of an aircraft, the *stress ratio,* defined as the ratio of the applied stress to the

---

*This may be either an existing slab thickness or one whose suitability is being ascertained.

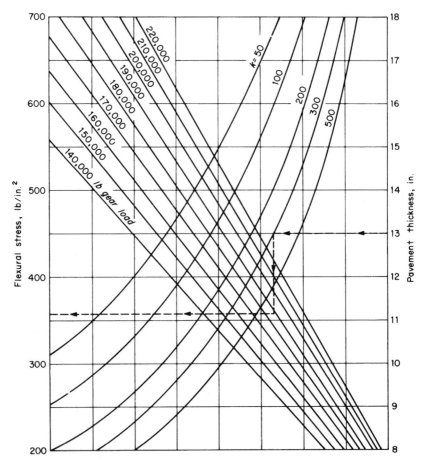

Figure 12-20 Rigid-pavement thickness for DC-10 (*Portland Cement Association*).

modulus of rupture, is determined. Fatigue tests on portland cement concrete have indicated that test data can be conservatively normalized in terms of this same stress ratio; Table 12-3 contains the basic concrete fatigue data used by the PCA. The structural capacity used up by the anticipated repetitions n_i of an aircraft class i is the ratio of n_i to the allowable number of repetitions to cause distress at the stress ratio N_i, produced by an aircraft of class i. The total structural capacity used is the sum of these cycle ratios for the mix of aircraft structural capacity used, namely,

$$\text{Structural capacity used} = \sum_{i=1}^{n} \frac{n_i}{N_i} \leq 1 \qquad (12\text{-}9)$$

where the term n_i is the actual repetitions of load of a specified aircraft of class i multiplied by a *load repetition factor* (LRF) which

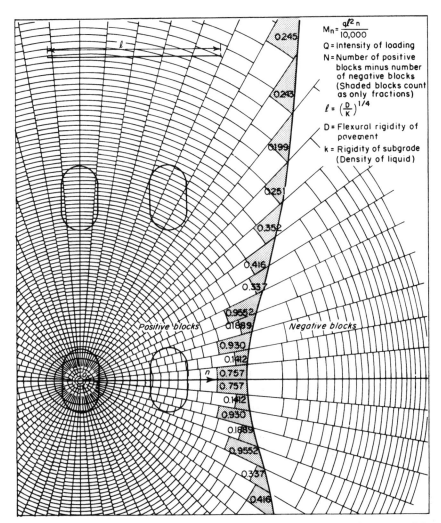

Figure 12-21 Influence chart for moment M_n due to load at interior of concrete slab. Subgrade assumed to be a dense liquid. Poisson ratio for pavement assumed to be 0.15 (*Portland Cement Association*).

accounts for the lateral wander of the wheels. Typical LRFs are 0.57 for the DC-10, 0.58 for the 747, 0.83 for the 707 and DC-8, and 0.41 for the 727 on taxiways, with a standard deviation of 24 in for the average path tracked. Other values for use on runways are contained in Ref. 25. Values for other aircraft are also listed. Practically, this summation should not exceed unity.

Provision is included in the PCA procedure to recognize some lateral distribution of aircraft on both taxiways and runways, variation in concrete strength, and the influence of weak foundation support ($k < 200 \text{ lb/in}^3$).

TABLE 12-3   Stress Ratios and Allowable Load Repetitions

Stress* ratio	Allowable repetitions	Stress ratio	Allowable repetitions
0.51†	400,000	0.63	14,000
0.52	300,000	0.64	11,000
0.53	240,000	0.65	8,000
0.54	180,000	0.66	6,000
0.55	130,000	0.67	4,500
0.56	100,000	0.68	3,500
0.57	75,000	0.69	2,500
0.58	57,000	0.70	2,000
0.59	42,000	0.71	1,500
0.60	32,000	0.72	1,100
0.61	24,000	0.73	850
0.62	18,000	0.74	650

*Load stress divided by modulus of rupture.

†Unlimited repetitions for stress ratios of 0.50 or less.

SOURCE: Portland Cement Association [67].

Determination of the modulus of subgrade reaction

The modulus of soil reaction k of the subgrade is determined by a field plate bearing test. The test should be made on representative areas of the foundation materials which are to support the pavement. Normally, the procedure consists in applying loads by means of a hydraulic jack through a jacking frame onto a steel plate 30 in in diameter. By loading the plate, a load-versus-deformation curve is obtained. By definition,

$$k = \frac{\text{pressure in lb/in}^2 \text{ to cause a deformation of 0.05 in}}{0.05 \text{ in}}$$

If the tests are made during the dry season, there might be no indication of how much reduction in k would result if the foundation material became saturated. The Corps of Engineers, recognizing that most soils exhibit a marked reduction in the modulus of soil reaction with an increase in moisture content, have developed a more refined procedure for determining k. With a 30-in diameter plate, first the plate is seated, and then a load of 10 lb/in² is applied (7070 lb), and the corresponding deflection is observed. The uncorrected modulus of subgrade reaction is

$$k_\mu = \frac{10 \text{ lb/in}^2}{\text{observed deflection corresponding to 10 lb/in}^2}$$

The value k_μ is further corrected for bending of the plate and for the assumption of full saturation of the foundation material. The correction for saturation is made by use of the consolidation test. The procedure is described in detail in Ref. 59.

The choice of a 30-in-diameter plate stems from observations which showed that, for plates less than 26 in in diameter, the size of the plate has a significant influence on k, while for larger plates k is relatively independent of the plate diameter.

Actually, k is not a significant variable in the Westergaard formulas. Fairly large variations in k do not produce large variations in stress. Therefore, for preliminary evaluation, k need not be measured but can be estimated by reference to available correlations with other soil strength indices, such as the CBR. The general range of k values is as follows [25]:

Description of foundation material	k
Very poor	Less than 150
Fair to good	200–250
Very good	300 and higher

Foundation for rigid pavements

An imported foundation material to support a rigid pavement is often referred to as a *subbase,* because its quality, in most cases, need not be as high as a base course for flexible pavement. Imported granular and treated material is placed over the native soils to serve one or more of the following functions:

1. To prevent or reduce pumping

2. To prevent damage from frost action

3. To prevent detrimental action due to swell and shrinkage in high-volume soils

4. To improve the supporting capacity of the native soil (very rarely justified)

5. To produce a smooth stable platform during construction

When certain types of fine-grain subgrade soils are saturated, repeated deflection of the pavement causes a mixture of soil and water to be ejected at or near joints and cracks. This is referred to as *pumping.* Three conditions must be present to produce pumping: (1) water, (2) soil that will go into suspension, and (3) traffic. Soils most conducive to pumping are those in which the silt and clay fractions predominate. The best way to minimize pumping is to employ a treated subbase using portland cement or asphalt, particularly for heavily trafficked airport pavements.

Stabilized subbases offer a number of benefits in addition to preventing pumping:

1. Provide an impermeable, uniform, and strong support for the pavement

2. Eliminate traffic compaction influences in the material directly under the slab

3. Improve load transfer at joints

4. Expedite construction since stabilized layers will minimize shutdowns due to adverse weather

5. Provide firm support for slip-form paver or side forms, contributing thereby to the construction of smoother pavements

Guidelines for the design of cement-stabilized subbases are contained in Refs. 21 and 25.

When granular or stabilized layers are used on top of the subgrade, there will be an increase in the k value. If it is not feasible to test such a construction, an estimate of the k value can be made from Figs. 12-22 and 12-23.

Stresses in rigid pavements due to temperature and moisture differentials

Whenever the top and bottom surfaces of a concrete pavement simultaneously have different temperatures, the slab tends to assume a bent or warped shape which induces flexural or so-called *curling* or *warping* stresses. If the slab were weightless and could change shape freely, there would be no warping stress. The stress is caused by the resistance of the slab (due to its weight) to change shape. Because of the insulating effects of slab thickness, the bottom surface will react to the change more slowly than the top surface.

The subject of warping stresses is complicated and has been treated theoretically by Westergaard [10] and Kelley [11]. Stresses due to temperature warping can, under certain conditions, be equal to and additive to stresses due to load. In the interior, the maximum stress due to load causes tension at the bottom of the slab; the same is true of temperature warping.

Warping stresses can also be caused by a differential in moisture content between the top and bottom of the slab. The reason is that the slab tends to dry out faster at the top than on the bottom. The greater the moisture content, the more the slab tends to elongate. The follow-

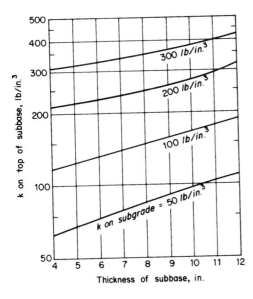

Figure 12-22 Effect of granular subbase thickness on k value (*Portland Cement Association*).

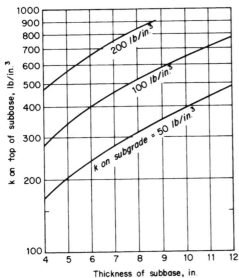

Figure 12-23 Effect of cement-treated subbase on k value (*Portland Cement Association*).

ing pertinent statement is quoted from a report prepared by the Bureau of Public Roads (now the Federal Highway Administration):

> The annual cyclic variation in moisture conditions within the concrete produces a warping of the slab surface similar to that caused by temperature. The edges of the slab reach their maximum position of upward warping from this cause during the summer and the maximum portion of downward warping during the winter, the extent of the upward movement apparently exceeding that of the downward movement appreciably.

While sufficient information is not available to permit an estimate to be made of the magnitude of the stresses arising from restrained moisture warping, it appears that at the time of year when the stresses from restrained temperature warping are a maximum (summer months), any stresses caused by restrained moisture warping will be of opposite sense and will thus tend to reduce rather than to increase the state of stress caused by restrained temperature warping.

Rarely do the conditions in practice approach those assumed by theory. In a majority of cases, warping stresses are not computed. Warping stresses can be reduced substantially by avoiding long slabs, e.g., introducing joints.

Neither the PCA nor the Corps of Engineers suggests that warping stresses be computed, for it is recognized that only a few times will these stresses be additive. There is usually sufficient margin between the actual stress developed by the wheel load and the ultimate flexural strength of the concrete to take care of warping stresses.

Frictional stress

Changes in temperature tend to change the length of a slab. Each half of the slab tends to move with reference to the subgrade. If the slab expands, the movement is from the center of the slab toward the free ends. When the slab contracts, the reverse is true. Friction between the foundation and the slab partially resists this movement, and in this way, stress is induced in the slab. The stress is compressive when expansion occurs and tensile when the slab contracts. In Fig. 12-24, horizontal forces are created at the bottom of the slab, inducing stress in the slab (in this case, because of contraction of the slab). If w is the unit weight of the slab in pounds per cubic foot and c is the coefficient of subgrade friction, the force developed per linear foot and width of slab is wc; the total force is $wcL/2$. If the slab is 1 ft wide, the amount of steel required to resist this force per 1-ft width of slab is

$$A_s f_s = wc \, \frac{L}{2}$$

or

$$A_s = \frac{wcL}{2f_s} \qquad (12\text{-}10)$$

where L = length of slab between joints
f_s = allowable tensile stress in reinforcement
c = coefficient of subgrade (or subbase) resistance to slab movement
A_s = cross-sectional area of steel

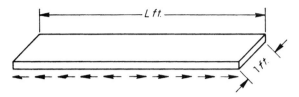

Figure 12-24 Stress induced by subgrade friction (*Courtesy of R. D. Bradbury, Wire Reinforcement Institute [56]*).

The PCA states [25]:

> The resistance coefficient c is sometimes referred to as the coefficient of friction between the slab and subbase (or subgrade). The situation is more complex than pure sliding friction since shearing forces in the subbase or subgrade and warped slabs may be involved in the resistance. For subgrades and granular subbases, coefficients of resistance range from 1 to 2 depending on type of material and moisture conditions. Coefficients for stabilized subbases are likely to be slightly greater. Research indicates that the coefficient also varies with respect to slab length and thickness. While these variations may be taken into account..., use of a coefficient other than 1.5 does not seem justified by pavement performance at this time.

Amounts and size of steel used in practice

The advantages claimed for placing steel in portland cement concrete pavements are as follows:

1. Reduces number of joints and, hence, cost of joint maintenance
2. Reduces pumping by keeping cracks tightly closed
3. Prolongs service life when pavement is overloaded
4. Maintains aggregate interlock for load transfer
5. Reduces pavement deflection

The majority of airport pavements constructed in this country are plain slabs without reinforcement. Joints have been spaced no more than 25 ft, obviating the necessity for steel insofar as crack control is concerned. Steel represents an extra expense to the designers, which they must justify. The justification for the use of steel has not been readily apparent for one reason of another. As with all engineering problems, however, the conditions at each airport must be evaluated separately. At one location steel may be justified, while at another it may not. If steel is used, the spacing between contraction joints suggested for plain pavements can be increased. For example, the PCA suggests the distance may be increased up to 50 ft for thick slabs, while the FAA recommends distances between contraction joints in the range 45 to 75 ft for reinforced-concrete slabs [9].

Generally, the pavement thickness for reinforced pavements should be the same as for unreinforced pavements, since distributed steel does not significantly increase flexural strength when used in quantities that are practical from an economic standpoint. This is particularly true for civil airports. For military airfields where traffic delays for routine maintenance may not be as critical, the Corps of Engineers permits a reduction in concrete thickness if steel is used. The procedure is outlined in detail in [59]. Steel placed in concrete pavements is in the form of bar mats or sheets or welded wire. The Corps of Engineers recommends that the steel be placed at a depth of $\frac{1}{4}h + 1$ in from the top of the slab, where h is the thickness of the slab in inches.

One should always keep in mind that the purpose of distributed steel is not to prevent cracking, but to hold the cracks closed after they have formed.

Joints and joint spacing

Joints are placed in concrete pavements to permit expansion and contraction of the pavement, thereby relieving flexural stresses due to curling and friction and to facilitate construction. There are three general types of joints: expansion, construction, and contraction. These joints are shown in Fig. 12-25. Their functions are as follows.

Expansion joint. The primary function of an expansion joint is to provide space for the expansion of the pavement, thereby preventing the development of very high compressive stresses, which can cause the pavement to buckle.

Construction joint. Construction joints are required to facilitate construction. The spacing between longitudinal joints is dictated by the width of the paving machine and by the pavement thickness.

Contraction joint. Contraction or weakened-plane (dummy) joints are provided to relieve the tensile stresses due to temperature, moisture, and friction, thereby controlling cracking. If contraction joints were not installed, random cracking would occur on the surface of the pavement.

The spacing between contraction joints is dependent on the thickness of the slab, the character of the aggregate, and whether the slab is plain or reinforced. On the basis of experience, it has been found that for plain slabs 8 to 10 in thick, the spacing should be in the range of 15 to 20 ft. For thicker slabs the spacing can be increased to 25 ft. The Corps of Engineers omits placing of contraction joints in a longitudinal direction, provided the thickness of the slab is 8 in or more. If

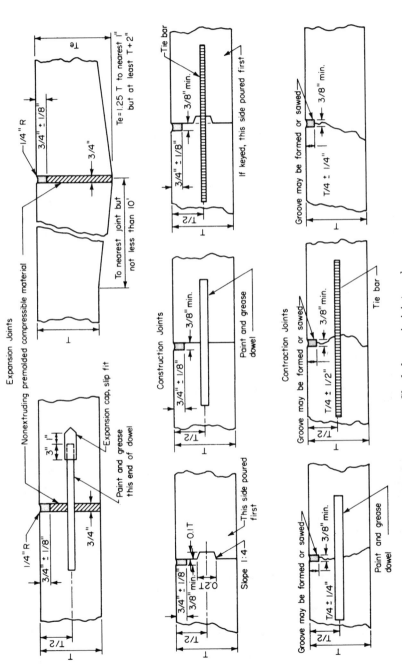

Figure 12-25 Details of joints in rigid pavements. Shaded area is joint sealer. (*Federal Aviation Administration [9]*).

the slab is less than 8 in thick and the spacing between longitudinal construction joints is 20 or 25 ft, a contraction joint is recommended midway between the construction joints. The FAA recommends a longitudinal contraction joint in accordance with the specifications listed in Table 12-4. As a rule of thumb, the joint spacing in feet should not exceed twice the slab thickness in inches to any significant degree.

The spacing between construction joints is primarily dictated by the width of the paving machine. Recent developments in paving equipment permit joint widths up to 50 ft. This allows a selection of joint spacings to satisfy specific design thicknesses.

Generally expansion joints are not used in airfield pavements except under special conditions. They are necessary where the pavement intersects other structures or where two pavements intersect each other at an angle. Experience has shown that expansion joints have been a source of trouble; consequently, every effort is made to minimize their number. For thick pavements (10 in or more in thickness), the Corps of Engineers suggests that expansion joints can be omitted entirely if the concrete is placed during warm weather. Expansion joints may be necessary, particularly if the thickness of the pavement is less than 10 in and if it is placed during cold weather. Thus, the need for and spacing of expansion joints are primarily a function of factors such as slab thickness, thermal properties of the concrete, seasonal temperature, and temperatures during construction.

Contraction joints are either grooved while the concrete is fresh or sawed after the concrete has hardened. Construction joints are formed by the slab form. In Fig. 12-25, note that expansion joints can be either thickened at the edge without load transfer devices or of uniform thickness with a load transfer device. Although dowels have been shown in the contraction joints, the practice varies. The Corps of Engineers omits dowels from all transverse contraction joints except for the last three joints at the runway and taxiway ends. On the other hand, the FAA recommends doweling of all transverse contraction joints for the entire length of a taxiway and at the ends of the runways. Doweling of all contraction joints in reinforced-concrete pavements is also recommended by the FAA.

TABLE 12-4   FAA-Recommended Maximum Joint Spacing, ft

Slab thickness, in	Joint orientation	
	Transverse	Longitudinal
Less than 9	15	12.5
9–12	20	20
Greater than 12	25	25

SOURCE: Federal Aviation Administration [9].

Construction joints are doweled, keyed, or both.

All joints in concrete pavements should be sealed with a sealing compound to prevent infiltration of water or foreign material into the joint spaces.

This joint sealant must be capable of withstanding repeated extension and compression as the pavement changes volume. Moreover, if the sealant is poured, the shape of the reservoir is critical. Guidelines for joint reservoirs are provided in Ref. 25.

Dowels are load transfer devices which permit joints to open but which prevent differential vertical displacement. Usually dowels are solid, round steel bars, although pipe may be used. Tie bars are deformed steel bars which are used to hold certain joints in contact with each other. Tie bars are not designed to transfer load from one slab to another.

Tie bars are designed in a manner similar to distributed steel, i.e.,

$$A_s = \frac{wcb}{f_s} \tag{12-11}$$

where b = distance between joint and the nearest untied joint or free edge, in feet, and the other terms have the same meaning as before.

The length of the tie bar may be determined from

$$L_t = \frac{1}{2} \frac{f_s d_b}{u} + 3 \tag{12-12}$$

where L_t = length of tie bar, in
   f_s = allowable tensile stress in tie bar, lb/in^2
   d_b = diameter of tie bar, in
   u = allowable bond stress (350 lb/in^2)

The FAA recommends that tie bars be $\frac{5}{8}$ in in diameter, and 30 in long and be placed 30 in on centers.

Several different analyses have been proposed for the design of dowels. The spacing of dowels depends on the thickness of the pavement, modulus of subgrade reaction, and size of the dowel. Table 12-5 contains recommendations for dowel sizes and spacings by the Corps of Engineers and the FAA.

Continuously reinforced concrete pavements

A *continuously reinforced concrete pavement* (CRCP) is one in which transverse joints have been eliminated* and the longitudinal reinforc-

---

*Except where the pavement intersects or abuts existing pavements or structures.

TABLE 12-5   Dimensions and Spacing of Steel Dowels

Thickness of slab, in	Diameter, in	Length, in	Spacing, in
	Corps of Engineers		
Less than 8	$\frac{3}{4}$	16	12
8–11	1	16	12
12–15	$1\frac{1}{4}$	20	15
16–20	$1\frac{1}{2}$	20	18
21–25	2	24	18
Over 25	3	30	18
	FAA		
6–7	$\frac{3}{4}$	18	12
8–12	1	19	12
13–16	$1\frac{1}{4}$	20	15
17–20	$1\frac{1}{2}$	20	18
21–24	2	24	18

SOURCES: Federal Aviation Administration [9] and Corps of Engineers [59].

ing steel is continuous throughout the length of the pavement. Due to the comparatively large amount of steel reinforcement in the longitudinal direction, the pavement develops transverse cracks at relatively close intervals, averaging between 3 and 7 ft.

Design for this type of pavement consists of providing sufficient reinforcing steel to keep the cracks tightly closed and adequate pavement thickness for the applied loads. Detailed information on design and construction of this type of pavement is provided in Refs. 1 and 19.

Thickness selection for CRCP is done in the same way as for plain concrete pavement slabs, i.e., with use of charts, such as Figs. 12-17 through 12-20. No reduction in thickness because of the presence of the steel is recommended.

The amount of reinforcing steel required to control volume changes is dependent primarily on the slab thickness, concrete tensile strength, and yield strength of the steel. While several procedures have been proposed for estimating the required amount of steel, experience indicates that it should be about 0.6 percent of the gross cross-sectional area and that the yield strength should be at least 60,000 lb/in^2. The minimum amount may be determined from

$$p_s = \frac{f_t'}{f_s - nf_t'} (1.3 - 0.2c)100 \qquad (12\text{-}13)$$

where p_s = percentage of steel, area of steel/area of concrete
f_t' = tensile strength of concrete (assumed equal to 0.4 × modulus of rupture), lb/in^s
f_s = allowable tensile stress in steel, lb/in^2
n = E_s/E_c, ratio of elastic modulus of steel to that of concrete
c = coefficient of subgrade friction

The FAA recommends that the cross-sectional area of the reinforcing steel A_s be obtained from

$$A_s = \frac{3.7L \sqrt{Lt}}{f_s}$$ (12-14)

where A_s = area of steel, in^2
L = width of slab, ft
t = slab thickness, in

Recommendations for the size and spacing of this longitudinal steel are summarized in Ref. 25. Steel placement in the slab ranges from the middepth (maximum) to one-third depth (minimum) with a minimum cover of $2\frac{1}{2}$ in. Transverse steel is designed in the same way as tie bars.

Terminal provisions are an important part of the design features for CRCP due to large seasonal end movements. End treatments are discussed in Refs. 19 and 23.

Pavement Design Using Elastic Layer Theory

While many of the materials comprising pavement structures exhibit nonlinear and rate-dependent stress-strain characteristics, the response of pavements to moving wheel loads appears "elastic" in nature since the majority of deformation resulting from a single moving wheel load application is recoverable.

Over the years, engineers have used elastic theory to assist in the analysis and design of pavement structures. For example, the Corps of Engineers used such theory, as already noted, to extend the design curves established for highway loading conditions to those representative of airfields. Moreover, elastic theory is used in the current CBR procedure to compute the ESWL for any gear arrangement by representing the pavement structure as a semi-infinite elastic solid.

The reader has also observed that design procedures for portland cement concrete pavements utilize elastic theory by considering the concrete slab as an elastic plate resting on a dense liquid subgrade.

Recently design procedures have been developed for asphalt-type pavements in which the structural section is represented as a multilayer elastic solid. Two of these procedures now in use for airfield pavements were developed by Shell [62] and the Asphalt Institute [66].

As will be seen subsequently, multilayer theory also has been used in the overlay design for airfield pavements [51].

Figure 12-26 contains a schematic representation of a pavement structure as a layered elastic solid with the lowest layer semi-infinite in extent. Computer programs are available to permit the calculation of stresses, strains, and displacements at any point in a particular layer for one load [47] or a number of loads [30, 35, 39]. Loads are applied to circular areas with uniform contact pressures. The layers are assumed to be infinite in lateral extent, and usually (although not necessarily) full continuity is assumed at each of the interfaces.

In using this approach, engineering judgment is required in the selection of representative material properties: critical stresses, strains, and deformations; limiting values for these quantities, termed *distress criteria*; and relationships between the distress criteria and performance of the pavement. Both the Shell and the Asphalt Institute procedures are summarized within such a context.

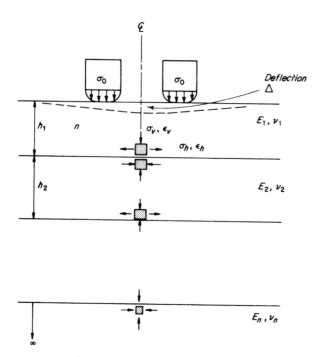

Figure 12-26 Representation of pavement structure as a multilayer elastic system.

Shell method

This method of design is applicable for pavement structures consisting of asphalt concrete, untreated granular base, and prepared subgrade or asphalt concrete resting directly on subgrade.

The structure is represented by a three-layer elastic system (full friction at layer interfaces). Critical conditions for design are as follows:

1. *Horizontal tensile strain on the underside of the asphalt-bound layer* (ϵ_h, layer 1, Fig. 12-26). If excessive, cracking may occur in the asphalt layer. The value of the tensile strain is dependent on the fatigue characteristics of the asphalt mixture with an allowable strain value of 2.30×10^{-4} in/in associated with 10^6 strain repetitions.[*]

2. *Vertical compressive strain in the surface of the subgrade* (ϵ_V, layer n, Fig. 12-26). If this value exceeds a specific limit, depending on traffic, permanent deformation may occur at the top of the subgrade, leading in turn to permanent deformation (rutting) at the surface of the pavement. The limiting value of vertical compressive strain, which is also dependent on the number of load applications, has been established as 10.3×10^{-4} in/in at 10^6 repetitions.

Materials in each of the layers are assumed to exhibit linear elastic behavior.

For asphalt concrete, time of loading and temperature dependence are recognized. Tensile strains in the asphalt concrete are determined at a stiffness E_1 of 900,000 lb/in^2 (which corresponds to a temperature of 50°F and a time of loading of 0.02 s). For determining subgrade strain, the air temperature is assumed to be 95°F, and an effective stiffness modulus in the range of 100,000 to 200,000 lb/in^2 (depending on the thickness of asphalt concrete) is selected.

The modulus of the untreated granular base is a function of the subgrade modulus E_3 and is dependent on the thickness of this layer h_2; it ranges from 2 to 3 times E_3.

The modulus of the subgrade soil has been related to the CBR based on dynamic vibratory tests in situ by the relation

$$E_3 = 1500(\text{CBR}) \qquad \text{lb/in}^2$$

---

[*]The fatigue characteristics of asphalt concrete mixtures can be represented by an equation of the form

$$N_f = k \left(\frac{1}{\epsilon_t} \right)^n$$

where N_f = applications to failure
   ϵ_t = tensile strain
   k, n = experimentally determined coefficients dependent on mixture characteristics

The Poisson ratio ν for all layers has been assumed to be 0.35.

Figure 12-27 illustrates the distribution of tensile strains on the underside of an asphalt-bound layer subjected to B-707 aircraft, computed by using the BISTRO program [39]. From analyses such as these it is possible to establish thicknesses of the asphalt-bound and granular layers which will ensure that the limiting values for strain are not exceeded. One such design chart is illustrated in Fig. 12-28 for a B-747.

When the subgrade modulus has been established (e.g., by the CBR test), the thickness of granular layer and asphalt concrete, or of asphalt concrete only, can be selected from a chart like that in Fig. 12-28.

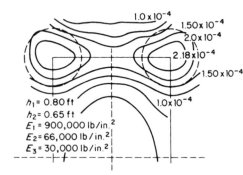

$h_1 = 0.80$ ft
$h_2 = 0.65$ ft
$E_1 = 900,000$ lb/in.^{2}
$E_2 = 66,000$ lb/in.^{2}
$E_3 = 30,000$ lb/in.^{2}

Figure 12-27 Asphalt strain pattern under B-707. h_1 = asphalt surface course; h_2 = base course (*Shell Oil Co.*).

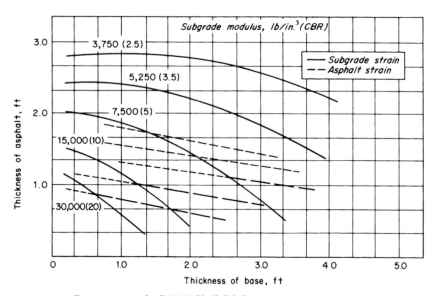

Figure 12-28 Design curves for B-747 (*Shell Oil Co.*).

Asphalt Institute method

This procedure is limited to asphalt concrete resting directly on a pre-pared subgrade (two-layer system). As in the Shell method, thickness design is based on limiting (1) the horizontal tensile strain on the underside of the asphalt layer to minimize fatigue cracking in the asphalt concrete and (2) the vertical compressive strain on the sub-grade surface to minimize the potential for surface rutting. A flow diagram detailing the procedure is shown in Fig. 12-29, and the steps used to select the design thickness are shown schematically in Fig. 12-30. Determining the design thickness requires knowing the sub-grade modulus, mean annual air temperature, and projected forecast of aircraft traffic mix.

Essentially the design procedure then consists of determining the following:

1. The allowable traffic volume N_a, which is the number of equiva-lent DC-8-63F strain repetitions that an asphalt concrete layer of specified thickness can withstand for the specific subgrade modulus and environmental condition at the site.

2. The predicted traffic value N_p, which is the number of equivalent DC-8-63F strain repetitions (based on aircraft traffic forecasts) that will actually occur during the design life of the pavement.[*]

3. The thickness of asphalt concrete T_A required to satisfy the strain criteria for the design input values of subgrade modulus, mean air temperature, and traffic mix. This thickness is determined by a simul-taneous graphical solution of N_a and N_p, as shown in Fig. 12-30. In the illustration, the design thickness is controlled by the asphalt concrete tensile strain since this results in the larger of the two thicknesses.

This method contains a number of innovative concepts for design. For example, like the fatigue procedure it permits an airfield pave-ment to be designed for the mixed traffic conditions representative of most large civil airports. In addition, environmental influences are con-sidered in a more realistic fashion for asphalt-bound layers since the influence of temperature on asphalt concrete stiffness is recognized.

Effect of Frost on Pavement Thickness

Frost action, if severe, results in nonuniform heave of pavements dur-ing the winter because of ice segregation and in loss of supporting capacity of the subgrade during periods of thaw.

---

[*]N_p varies with depth.

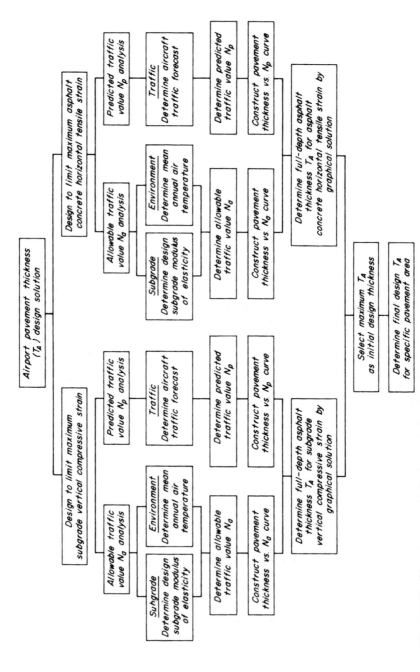

Figure 12-29 Airport pavement thickness design flowchart (Asphalt Institute).

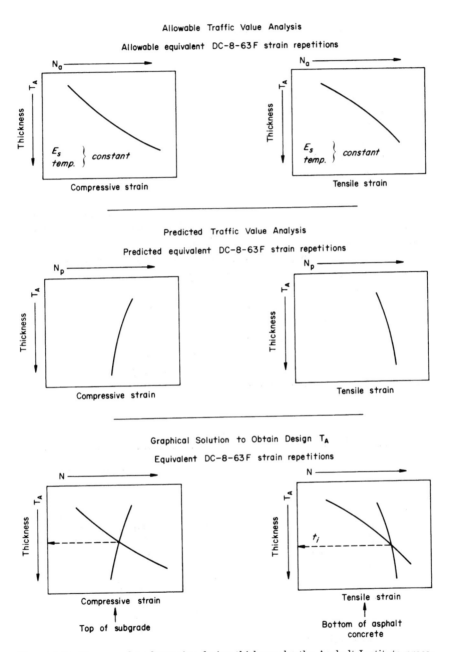

Figure 12-30 Steps used to determine design thickness by the Asphalt Institute procedure (*Asphalt Institute*).

According to the Corps of Engineers, the process of ice segregation can be described as follows [50]:

> The water in void spaces below a critical size can be supercooled and remains liquid below the normal freezing temperatures of water. A strong attraction exists between this water and ice crystals which form in the larger voids, and the water flowing to the crystals solidifies on contact. Crystal growth leads to formation of an ice lens. A lens continues to grow in thickness in the direction of heat transfer and at the same time laterally, until ice formation at a lower elevation cuts off the source of water, or until the temperature of the soil just below the surface of ice formation rises above the normal freezing point.

During the spring and early summer, the ice lenses begin to melt, and the water which is released cannot drain through the still-frozen soil at greater depths. Thus, lack of drainage results in loss of strength in the subgrades. It is also possible that a reduction in stiffness will occur in subgrade soils during the thaw period, even though ice lenses may not have formed [12].

The Corps of Engineers has developed a comprehensive design procedure which is briefly summarized here [50].

Some soils are very susceptible to frost, others are only mildly affected, and still others can be considered as non-frost-susceptible. With respect to susceptibility to frost action, the Corps of Engineers has classified soils as shown in Table 12-6.

TABLE 12-6   Frost Design Soil Classification

Frost group	Kind of soil	Percentage finer than 0.02 mm by weight	Typical soil types under unified soil classification system
F1	Gravelly soils	3–10	GW, GP, GW-GM, GP-GM
F2	Gravelly soils	10–20	GM, GW-GM, GP-GM
	Sands	3–15	SW, SP, SM, SW-SM, SP-SM
F3	Gravelly soils	Over 20	GM, GC
	Sands, except very fine silty sands	Over 15	SM, SC
	Clays, PI > 12	—	CL, CH
F4	All silts	—	ML, MH
	Very fine silty sands	Over 15	SM
	Clays, PI < 12	—	CL, CL-ML
	Varved clays and other fine-grained, banded sediments	—	CL and ML; CL, ML, and SM; CL, CH, and ML; CL, CH, ML, and SM

SOURCE: Corps of Engineers.

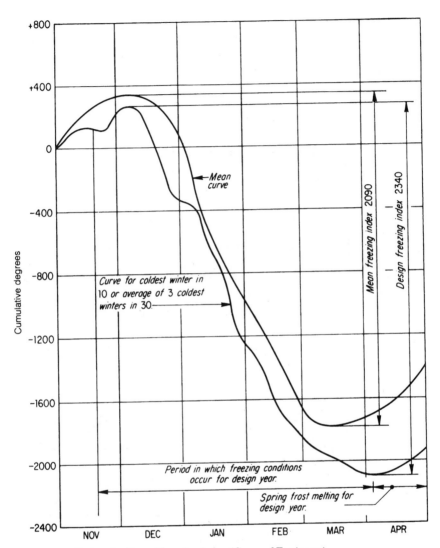

Figure 12-31 Determination of freezing index (*Corps of Engineers*).

Depths of frost penetration have been measured in many localities. These depths have been correlated with a number called the *freezing index,** which can be determined from a chart similar to the one shown in Fig. 12-31. The *degree-day* is the difference between the average daily temperature and 32°F. The degree-day is minus when

---

*The freezing index is the number of degree-days between the highest and the lowest points on a curve of cumulative degree-days versus time for one freezing season.

the average daily temperature is below 32°F; it is plus when the average daily temperature is above 32°F. The freezing index used in design is the coldest winter in a 10-year period of record, or the average of the three coldest winters in 30 years of record.

Two methods have been developed for designing pavements on frost-susceptible subgrades.* One method is to provide a sufficient thickness of pavement to insulate the subgrade, not completely, but substantially so. The other method is to allow freezing of the subgrade and to design the pavement on the basis of reduced supporting capacity during the period of thaw. The choice depends on the type of subgrade material, the economics of construction at a particular site, and the amount of nonuniform heave permitted.

In the first method, a correlation between frost penetration and freezing index has been established, as shown in Fig.12-32. If complete protection of the subgrade is to be provided, the thickness of pavement corresponds to the depth of frost penetration. A normal amount of frost penetration into the subgrade is acceptable; thus, the problem becomes one of determining how much. From its extensive work in the field of frost action, the Corps of Engineers has determined the accept-

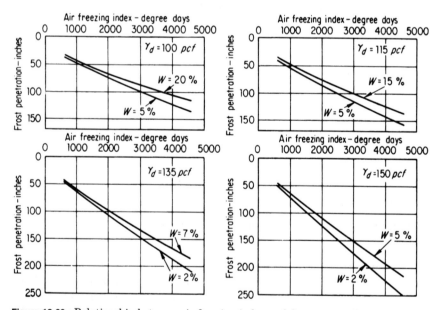

Figure 12-32 Relationship between air freezing index and frost penetration into granular non-frost-susceptible soil beneath pavements kept free of snow and ice (*Corps of Engineers*).

---

*Complete protection, which is used only in exceptional situations, can be considered as a third method.

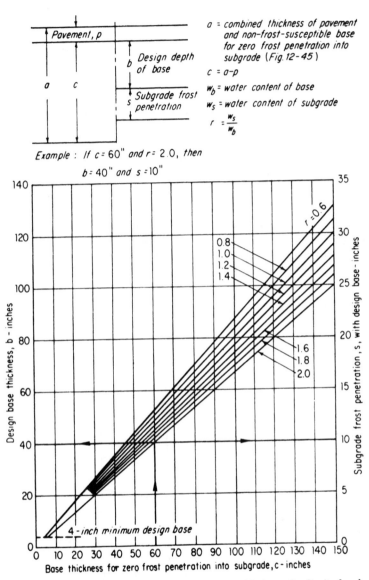

Figure 12-33 Design depth of non-frost-susceptible base for limited subgrade frost penetration (*Corps of Engineers*).

able amount of penetration of frost into the subgrade, depending on the thickness of the base course. This relationship is shown in Fig. 12-33. The abscissa of the chart represents the thickness of the base for a flexible pavement, assuming no frost penetration into the subgrade, and is obtained by subtracting the thickness of the surface course from the depth of frost penetration, obtained from Fig. 12-32. The ordinate yields the thickness of the base to be used in design and the acceptable amount of frost penetration into the subgrade. If r is greater than 2, use the curve marked 2.0. The pavement thickness obtained in this manner must be compared with the thickness that would be required if frost had not been a consideration. The larger of the two thicknesses should be used for design.

In the second method, charts have been developed for flexible pavements that relate wheel load to thickness of pavement for various classes of subgrade soil. A typical chart for a dual-in-tandem assembly is shown in Fig. 12-34. This procedure is usually not applicable for the F4 type of subgrade soils or for any other soils which will exhibit nonuniform heave.* For such soils, only the first method is applicable.

For rigid pavements, the chart shown in Fig. 12-35 has been prepared. The procedure is to compute the thickness of a concrete slab with the appropriate k value, assuming no frost. The next step is to assume that for frost action a thickness of non-frost-susceptible mate-

---

*Permissible for F4 subgrades in nonimportant areas where aircraft speeds are very slow and nonuniform heave can be tolerated. In Figs. 12-34 and 12-35, use F3 curve for F4 subgrades.

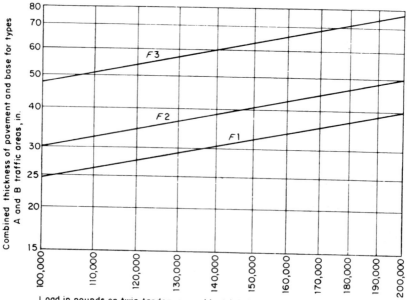

Figure 12-34 Frost condition; reduced subgrade strength design curves for flexible pavements (*Corps of Engineers*).

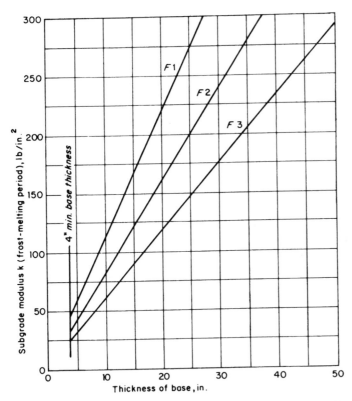

Figure 12-35 Frost condition; reduced subgrade strength design subgrade modulus curves for rigid pavements (*Corps of Engineers*).

rial equal to the thickness of the slab is required (labeled *base* on the chart). By reference to the chart at this thickness of base, the reduced modulus k is obtained, and the thickness of the slab is recomputed on the basis of the reduced k. As for flexible pavements, this procedure usually applies only to F1, F2, and F3 soils.

If the thickness of the pavement is less than the depth of frost penetration, it is desirable to have the bottom 4 in of the base consist of non-frost-susceptible gravel, sand, or screenings to act as a soil filter between the subgrade soil and the main body of the base.

The FAA Method of Design
for Flexible and Rigid Airport Pavements

The FAA has recently changed its methods of pavement design to incorporate the unified soil classification system instead of the soil classification system used in the past [9, 58]. To provide continuity in pavement design, the former system may be used to evaluate the characteristics of pavements designed on the basis of the former system. The former system was based upon a gradation analysis of the soil as well as the plastic and liquid limit. It was found that this system was not as reliable or as well known as the unified system, and therefore, the change was made.

Soil investigations and evaluation

Accurate identification and evaluation of pavement foundations are essential to the proper design of the pavement structure. The subgrade supports the pavement and the loads placed upon the pavement surface. The function of the pavement is to distribute the loads to the subgrade, and the greater the capability of the subgrade to support the loads, the less the required thickness of the pavement.

A soil investigation will consist of a soil survey to determine the arrangement of the different layers of soil in relation to the subgrade elevation, a sampling and testing of the various layers of soil to determine the physical properties of the soil with respect to in-place density and subgrade support, and a survey to determine the availability and suitability of local materials for use in the construction of the subgrade and pavement. Surveys and sampling are usually accomplished through soil borings to determine the soil or rock profile and its lateral extent. General criteria for the frequency and depth of soil borings for the different elements of a pavement at an airport are contained in Ref. 9. Note that local site conditions determine the extent of the required boring operations. The sampled materials are then tested to determine soil types, gradation of particle sizes, liquid and plastic limits, plasticity index, moisture-density relationships, shrinkage factors, permeability, organic content, and strength properties, including the CBR and the modulus of subgrade reaction.

In the unified system, soils are initially divided by a separation between coarse- and fine-grained soils and highly organic soils. The distinction between coarse- and fine-grained soils is based upon the amount of material retained on the no. 200 sieve. The coarse-grained materials are further subdivided into gravels and sands on the basis of the amounts retained on the no. 4 sieve. Then the gravels and sands are classified according to whether fine material is present. Fine-grained soils are subdivided into two groups based upon the liquid limit. A separate division is made for highly organic soils not suitable for construction purposes. A flowchart illustrating the procedure for the classification of soils by the unified system is given in Fig. 12-36. The uses of the various soil materials for pavement foundations are described in Table 12-7.

The strength properties of the soil for use in the FAA design method consist of the CBR value for flexible-pavement design and the

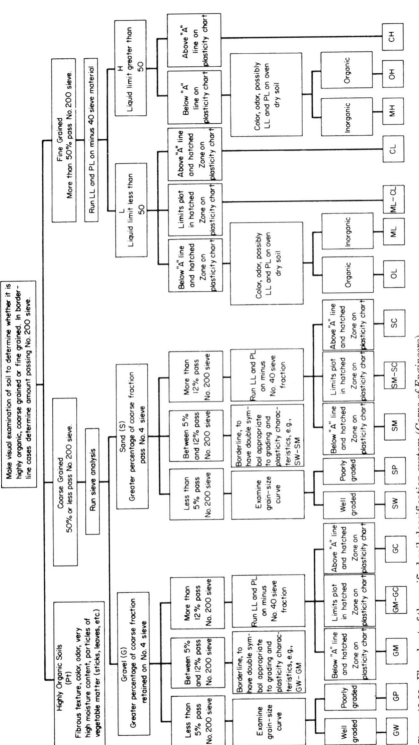

Figure 12-36 Flowchart of the unified soil classification system (*Corps of Engineers*).

TABLE 12-7 Characteristics of Soil Related to Airport Pavements

Major divisions (1)	(2)	Letter (3)	Name (4)	Value as foundation when not subject to frost action (5)	Value as base directly under wearing surface (6)	Potential frost action (7)	Compressibility and expansion (8)	Drainage characteristics (9)	Compaction equipment (10)	Unit dry weight (pcf)* (11)	Field CBR (12)	Subgrade modulus (pci)† (13)
Coarse-grained soils	Gravel and gravelly soils	GW	Gravel or sandy gravel, well graded	Excellent	Good	None to very slight	Almost none	Excellent	Crawler-type tractor, rubber-tired equipment, steel-wheeled roller	125–140	60–80	300 or more
		GP	Gravel or sandy gravel, poorly graded	Good to excellent	Poor to fair	None to very slight	Almost none	Excellent	Crawler-type tractor, rubber-tired equipment, steel-wheeled roller	120–130	35–60	300 or more
		GU	Gravel or sandy gravel, uniformly graded	Good	Poor	None to very slight	Almost none	Excellent	Crawler-type tractor, rubber-tired equipment	115–125	25–50	300 or more
		GM	Silty gravel or silty sandy gravel	Good to excellent	Fair to good	Slight to medium	Very slight	Fair to poor	Rubber-tired equipment, sheepsfoot roller, close control of moisture	130–145	40–80	300 or more
		GC	Clayey gravel or clayey sandy gravel	Good	Poor	Slight to medium	Slight	Poor to practically impervious	Rubber-tired equipment, sheepsfoot roller	120–140	20–40	200–300
	Sand and sandy soils	SW	Sand or gravelly sand, well graded	Good	Poor	None to very slight	Almost none	Excellent	Crawler-type tractor, rubber-tired equipment	110–130	20–40	200–300
		SP	Sand or gravelly sand, poorly graded	Fair to good	Poor to not suitable	None to very slight	Almost none	Excellent	Crawler-type tractor, rubber-tired equipment	105–120	15–25	200–300
		SU	Sand or gravelly sand, uniformly graded	Fair to good	Not suitable	None to very slight	Almost none	Excellent	Crawler-type tractor, rubber-tired equipment	100–115	10–20	200–300

	Symbol	Name						Compacting equipment	Unit weight*	†	
	SM	Silty sand or silty gravelly sand	Good	Poor	Slight to high	Very slight	Fair to poor	Rubber-tired equipment, sheepsfoot roller; close control of moisture	120–135	20–40	200–300
	SC	Clayey sand or clayey gravelly sand	Fair to good	Not suitable	Slight to high	Slight to medium	Poor to practically impervious	Rubber-tired equipment, sheepsfoot roller	105–130	10–20	200–300
Fine-grained soils	ML	Silts, sandy silts, gravelly silts, or diatomaceous soils	Fair to poor	Not suitable	Medium to very high	Slight to medium	Fair to poor	Rubber-tired equipment, sheepsfoot roller; close control of moisture	100–125	5–15	100–200
Low compressibility, LL <50	CL	Lean clays, sandy clays, or gravelly clays	Fair to poor	Not suitable	Medium to high	Medium	Practically impervious	Rubber-tired equipment, sheepsfoot roller	100–125	5–15	100–200
	OL	Organic silts or lean organic clays	Poor	Not suitable	Medium to high	Medium to high	Poor roller	Rubber-tired equipment, sheepsfoot	90–105	4–8	100–200
High compressibility, LL >50	MH	Micaceous clays or diatomaceous soils	Poor	Not suitable	Medium to very high	High	Fair to poor	Rubber-tired equipment, sheepsfoot roller	80–100	4–8	100–200
	CH	Fat clays	Poor to very poor	Not suitable	Medium	High	Practically impervious	Rubber-tired equipment, sheepsfoot roller	90–110	3–5	50–100
	OH	Fat organic clays	Poor to very poor	Not suitable	Medium	High	Practically impervious	Rubber-tired equipment, sheepsfoot roller	80–100	3–5	50–100
Peat and other fibrous organic soils	Pt	Peat, humus, and other	Not suitable	Not suitable	Slight	Very high	Fair to poor	Compaction not practical			

SOURCE: Corps of Engineers.

*pcf = pounds per cubic foot.

†pci = pounds per cubic inch.

modulus of subgrade reaction k for rigid-pavement design. These two measures of soil strength were described earlier.

Pavement design considerations

The design of pavements is a complex engineering problem which involves the consideration of a great number of variables. Flexible-pavement design is based upon the CBR method, an essentially empirical method. Rigid-pavement design is based upon the Westergaard analysis of edge loading, which has been modified to simulate a jointed edge condition. These methods represent a departure from previous FAA pavement design policy and yield slightly different thicknesses. The required parameters for the design of pavements include the gross takeoff weight of the aircraft using the airfield, the landing gear configuration and dimensions, tire contact areas and pressures, and the traffic volume. Separate design curves are presented for single-gear, dual-gear, dual-tandem-gear, and wide-body aircraft.

The first step of the procedure is to determine the forecast of the annual departures by each type of aircraft and to group them into narrow-body by gear configuration and wide-body by aircraft type. The maximum takeoff weight of each aircraft is used, and 95 percent of this weight is assigned to the main landing gear. The design aircraft is then determined as that aircraft which requires the greatest pavement thickness. It is not necessarily the heaviest aircraft which will operate from the airport. Since the aircraft operating from the airport will have different landing gear configurations, it is necessary to determine the equivalent annual departures of each aircraft in terms of the gear configuration of the design aircraft. These equivalencies are given in Table 12-8.

TABLE 12-8 **Factors for Converting Annual Departures by Aircraft to Equivalent Annual Departures by Design Aircraft**

Actual aircraft landing gear	Design aircraft landing gear	Multiplier for actual departures to obtain equivalent departures
Single-wheel	Dual-wheel	0.8
	Dual-tandem	0.5
Dual-wheel	Single-wheel	1.3
	Dual-tandem	0.6
Dual-tandem	Single-wheel	2.0
	Dual-wheel	1.7
Double-dual-tandem	Dual-wheel	1.7
	Dual-tandem	1.0

SOURCE: Federal Aviation Administration [9].

The equivalent annual departures of the design aircraft are determined by summing the equivalent annual departures of each aircraft in the group:

$$\log R_1 = (\log R_2)\left(\frac{W_2}{W_1}\right)^{0.5} \tag{12-15}$$

where R_1 = equivalent annual departures by design aircraft
 R_2 = annual number of departures by an aircraft in terms of design aircraft landing gear configuration
 W_1 = wheel load of design aircraft
 W_2 = wheel load of aircraft being converted

Since wide-body aircraft have radically different landing gear configurations from other aircraft, the relative effects on the pavement are considered by assigning a gross takeoff weight of 300,000 lb to any wide-body aircraft and considering it to have a dual-tandem landing gear for the analysis of equivalent departures. The actual curve for the design aircraft is used, however, for the determination of pavement requirements.

Research has shown that aircraft traffic is normally distributed over a lateral range of the pavement surface during operations [32]. Likewise, during a portion of an operation on the runway system, the aerodynamic properties of the aircraft will decrease the actual pavement loading. Therefore, the FAA permits pavement thickness modifications on different surfaces as follows [9]:

1. Full-depth thickness T is required where departing aircraft will use the pavement, such as aprons, holding areas, and the center portions of the taxiway and runway widths.

2. Pavement thickness of $0.9T$ is required where arriving aircraft will use the surface, such as high-speed runway turnoffs.

3. Pavement thickness of $0.7T$ is required where aircraft traffic is unlikely, such as the outer edges of taxiway and runway pavements.

The FAA also specifies the degree of frost protection required for airport pavements based upon the uniformity of soil conditions, the nature of the airport (large, medium, or small hub), and the particular pavement element (runway, taxiway, or apron). The methods of design to protect from frost effects include the following [9]:

1. The complete frost protection method involves the removal of all frost-susceptible material to the depth of frost penetration and replacement with non-frost-susceptible material.

2. The limited subgrade frost penetration method allows the frost to penetrate a limited depth into the frost-susceptible material, holding deformations to minimal values.

3. The reduced subgrade strength method provides adequate capacity during the frost melt period and is applicable to situations where aircraft speeds are low and the effects of frost heave are less objectionable.

4. The reduced subgrade frost penetration method is statistically based and may be used where aircraft speeds are low and some frost heave can be tolerated.

The FAA method also allows for the stabilization and treatment of subgrade, subbase, and base materials to improve pavement performance and economics. Equivalency factors are used to determine the reduction in pavement materials possible under stabilization conditions [9]. Stabilized base and subbase courses are required for new flexible pavements designed to accommodate jet aircraft weighing 100,000 lb or more.

Flexible-pavement design

Flexible pavements consist of a bituminous wearing surface placed upon a base course and, where required by subgrade conditions, a subbase. The entire flexible-pavement structure is ultimately supported by the subgrade. The wearing course prevents the penetration of surface water to the base course, provides a smooth, well-bonded surface free of loose particles, resists the shearing stresses caused by aircraft loading, and furnishes a texture of nonskid qualities not causing undue tire wear. The course must also be resistant to fuel spillage and other solvents in areas where maintenance may occur. The base course is the major structural element of the pavement; it has the function of distributing the wheel loads to the subbase and subgrade. It must be designed to prevent failure in the subgrade, withstand the stresses produced in the base course, resist vertical pressures tending to produce consolidation and deformation of the wearing course, and resist volume changes caused by fluctuations in its moisture content. The function of the subbase, when required, is similar to that of the base course; but since the subbase is further removed from the area of load application, it is subjected to lower stress intensities. The subgrade soils are subjected to the lowest loading intensities, and the controlling stresses are usually at the top of the subgrade since stress decreases with depth. However, unusual subgrade conditions, such as layered subgrade materials, can alter the location of controlling stresses. The FAA presents techniques for considering the support strength of layered subgrade materials [9].

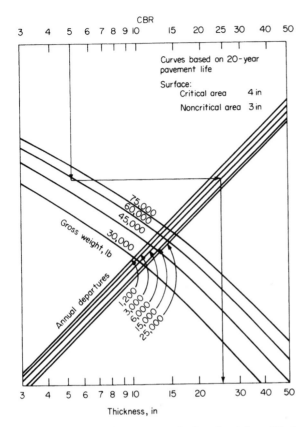

Figure 12-37 Flexible-pavement design chart for critical areas, single-wheel gear (*Federal Aviation Administration [9]*).

Examples of the design charts for flexible-pavement design are shown in Figs. 12-37 to 12-40. To determine the thickness of the total pavement thickness (wearing course, base, and subbase), the CBR of the subgrade is used together with the gross takeoff weight and the equivalent annual departures of the design aircraft. To find the required thickness of the wearing course and base, these charts are used with the CBR of the subbase and the gross takeoff weight and the equivalent annual departures of the design aircraft. The minimum base course required, however, is subject to the limitations found from Fig. 12-41. The base must have a minimum thickness of 6 in in critical areas. The thickness of the wearing course is indicated on the design charts for critical, full-depth pavement and for noncritical areas. If the thickness of the wearing and base courses exceeds that determined from the design chart, the subbase may be proportionately decreased.

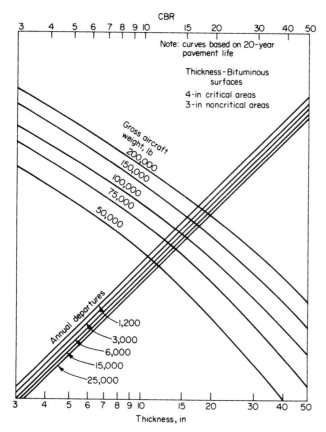

Figure 12-38 Flexible-pavement design chart for critical areas, dual-wheel gear (*Federal Aviation Administration [9]*).

The design charts are used by entering the chart at the CBR axis, proceeding vertically to the gross takeoff weight curve, then horizontally to the equivalent annual departure curve, and finally vertically to the pavement thickness axis.

Stabilized base and subbase courses may be substituted for granular base and subbase courses in accordance with the equivalency factors given in Table 12-9. Restrictions in the use of these equivalency factors due to construction and other limitations are contained in [9].

Rigid-pavement design

Rigid pavements for airports consist of a portland cement concrete slab placed upon a granular or treated subbase course resting on a compacted subgrade. A subbase is not required under certain conditions related to the adequacy of drainage and the susceptibility to

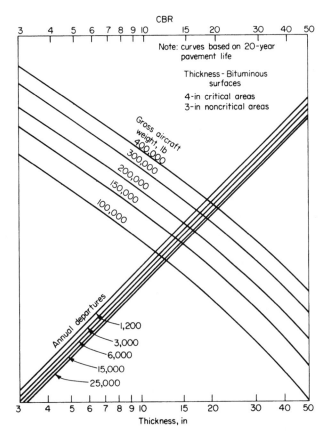

Figure 12-39 Flexible-pavement design chart for critical areas, dual-in-tandem gear (*Federal Aviation Administration*).

frost [9]. The concrete slab must provide an acceptable nonskid surface, prevent the infiltration of surface water, and provide the structural support required for aircraft. The subbase provides a uniform stable support for the pavement slab and allows for drainage. It is recommended that the minimum subbase under a concrete slab be at least 4 in, except when not required. The subbase may be increased above this minimum value to increase the modulus of subgrade reaction and thereby decrease the thickness of the slab. Stabilized subbases, required in new pavements for aircraft weighing more than 100,000 lb, reflect structural benefits in an increased modulus of subgrade reaction. The effect of a stabilized base on the subgrade modulus is shown in Fig. 12-42. Generally speaking, the most efficient combination of rigid-pavement thickness and stabilized subbase thickness for structural capacity is a ratio of 1 to 1 [9, 27].

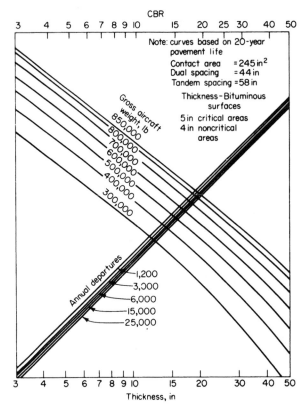

Figure 12-40 Flexible-pavement design chart for critical areas, for Boeing 747-100, SR, 200, B, C, F wide-body aircraft only (*Federal Aviation Administration [9]*).

Examples of the design curves for rigid pavements are given in Figs. 12-43 to 12-46. These curves are based upon a jointed edge loading assumption in which the load is tangent to the concrete joint. The use of the design curves is similar to that for flexible pavements in that the gross takeoff weight and the equivalent annual departures of the design aircraft are required in addition to the flexural strength of concrete and the modulus of subgrade reaction for the subgrade. The curves shown give only the thickness of the slab required, and independent methods must be used to obtain the thickness of the subbase. The 90-day flexural strength of concrete is required; however, the 28-day strength may be used and increased by a factor of 1.10 to approximate 90-day strength. Reduction in thickness in noncritical sections to 0.90 of the critical-section thickness is allowed.

The design procedure also allows for modification of the pavement thickness determined from the design charts for high traffic volumes. The increases in thickness are given in Table 12-10.

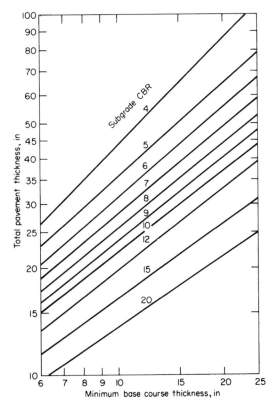

Figure 12-41   Minimum base course requirements for
flexible pavement (*Federal Aviation Administration
[9]*).

Pavements for light aircraft

Light-aircraft pavements are defined as landing areas intended for
personal or other small aircraft engaged in nonscheduled flying activities. Such pavements are not required to accommodate aircraft which
have gross takeoff weights not in excess of 30,000 lb. In many cases
these aircraft will not exceed 12,500 lb. Landing facilities may be
turfed, aggregate-turfed, flexible- or rigid-pavement surfaces. The
FAA has established design procedures for flexible and rigid pavements, similar to those for heavier air carrier type of aircraft. No
reduction in pavement thickness is allowed for noncritical pavement
surfaces.

The design of flexible pavements for light aircraft is performed by
using Fig. 12-47. The determination of the total flexible structure
thickness is found by using the CBR of the subgrade and the gross
takeoff weight of the heaviest aircraft using the facility. The thickness of the wearing and base courses is found by using the same chart

TABLE 12-9   Recommended Equivalency Factors for Stabilized Base and Subbase Materials in Pavements

| Material* | Equivalency factor range,† | |
	Subbase	Base
Bituminous surface course	1.7–2.3	1.2–1.6
Bituminous base course	1.7–2.3	1.2–1.6
Cold-laid bituminous base course	1.5–1.7	1.0–1.2
Mixed-in-place base course	1.5–1.7	1.0–1.2
Cement-treated base	1.6–2.3	1.2–1.6
Soil cement base course	1.5–2.0	NA
Crushed-aggregate base course	1.4–2.0	1.0‡
Subbase course	1.0§	NA

*See FAA AC 150/5370-10A for specifications.

†Divide granular material thickness by equivalency factor to obtain stabilized material thickness.

‡A CBR of 20 was used to establish relative factors.

§A CBR of 80 was used to establish relative factors.

NA = Not applicable.

SOURCE: Federal Aviation Administration [9].

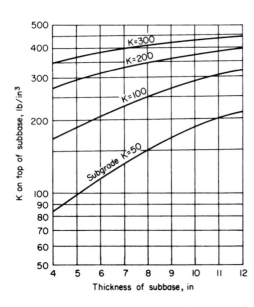

Figure 12-42 Effect of stabilized subbase on subgrade modulus (*Federal Aviation Administration [9]*).

with a CBR of 20 for the subbase. A minimum thickness of 2 in for the wearing surface and 3 in for the base course is recommended for these facilities [9]. The Asphalt Institute has also prepared design curves and recommendations for light aircraft [34].

Rigid-pavement design for light aircraft requires a minimum concrete slab thickness of 5 in for aircraft weighing 12,500 lb or less and 6 in for aircraft weighing between 12,500 and 30,000 lb. Design curves

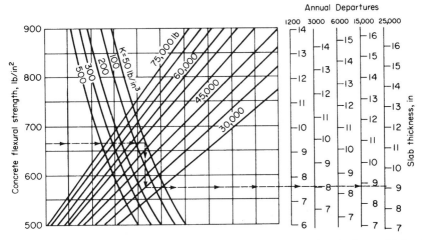

Figure 12-43 Rigid-pavement design chart for critical areas, single-wheel gear (*Federal Aviation Administration [9]*).

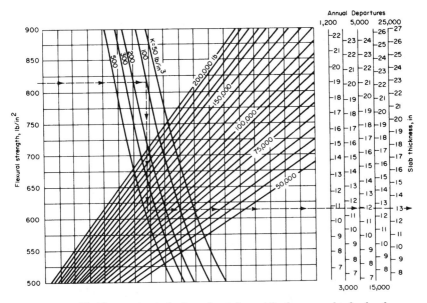

Figure 12-44 Rigid-pavement design chart for critical areas, dual-wheel gear (*Federal Aviation Administration [9]*).

are not presented since these are the only thickness requirements for rigid pavements designed to accommodate light aircraft. For aircraft weighing between 12,500 and 30,000 lb, a subbase of 4 in is required under the slab. A subbase of 5 in is required under the slab for aircraft weighing less than 12,500 lb in poor soil conditions [9].

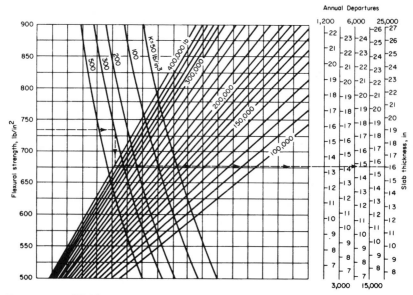

Figure 12-45 Rigid-pavement design chart for critical areas, dual-in-tandem gear (*Federal Aviation Administration [9]*).

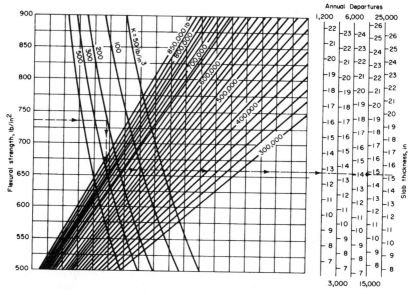

Figure 12-46 Rigid-pavement design chart for critical areas, for Boeing 747-100, SR, 200, B, C, F wide-body aircraft only (*Federal Aviation Administration [9]*).

TABLE 12-10 Rigid-Pavement Thickness for High Level of Departures as Percentage of 25,000 Departure Thickness

Annual departure level	Percentage of 25,000 departure thickness
50,000	104
100,000	108
150,000	110
200,000	112

SOURCE: Federal Aviation Administration [9].

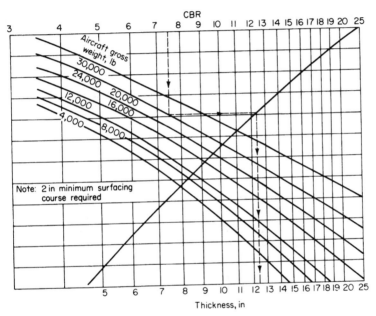

Figure 12-47 Flexible-pavement design chart for light aircraft (*Federal Aviation Administration [9]*).

Design of Overlay Pavements

Overlay pavements are required when existing pavements are no longer serviceable due to either a deterioration in structural capabilities or a loss in riding quality. They are also required when pavements must be strengthened to carry greater loads or increased repetitions of existing aircraft beyond those anticipated in the original design. Overlays also provide a solution for increased safety (i.e., improved skid resistance and reduced danger of hydroplaning). In this section the discussion is limited to overlays required for structural purposes.

Types of overlay pavements

There are several types of overlay pavements. For example, a concrete pavement can be overlaid with additional concrete, a bituminous surfacing, or a combination of aggregate base course and a bituminous surfacing. Likewise, a flexible type of pavement can be overlaid with concrete, a bituminous surfacing, or the combination cited above. The various types of overlay pavements are defined as follows:

1. *Overlay pavement:* the thickness of a rigid or flexible type of pavement placed on an existing pavement

2. *Portland cement concrete overlay:* an overlay pavement constructed of portland cement concrete

3. *Bituminous overlay:* an overlay consisting entirely of a bituminous surfacing

4. *Flexible overlay:* an overlay consisting of a base course and a bituminous surfacing

Overlay design procedures described in this section include those developed by the Corps of Engineers and the FAA. The procedures are based on the results of full-scale test sections together with observations of behavior of overlays in service. Also included are recently developed procedures which make use of layered elastic theory.

Construction details such as treatment of the existing pavement are important, and the reader is referred to references for guidelines for construction of overlays [9, 59].

Bituminous or flexible overlays on flexible pavements

The same approach is used by the Corps of Engineers and the FAA for determining the thickness of bituminous or flexible overlays. The thickness of pavement for the new wheel load is computed with the assumption that the existing pavement does not exist. The thickness of bituminous or flexible overlay is equal to the difference between the computed thickness and the thickness of the existing pavement. The Corps of Engineers recommends a minimum overlay of 4 in, and the FAA recommends that overlays for increasing strength be a minimum of 3 in. It is recommended that if the base course is less than these minima, the entire overlay should be of the bituminous type.

The FAA permits adjustments to the required thickness of overlay, depending on the condition of the existing pavement and the character of the materials in the overlay pavement [9]. No adjustments of this kind are permitted by the Corps of Engineers.

The Asphalt Institute procedure [34] is similar in principle to the Corps of Engineers and the FAA methods, except that the difference between the existing and the required pavement thicknesses (overlay) is limited to asphalt concrete, and the components of the existing pavement are expressed in equivalent thicknesses of asphalt concrete.

Portland cement concrete overlays on flexible pavements

In the design of portland cement concrete overlays, both the Corps of Engineers and the FAA treat the problem as new construction on a foundation consisting of the existing flexible pavement. The Westergaard analysis for determining the thickness of the overlay slab is used. Plate bearing tests are made on the existing flexible pavement to establish the value of the subgrade modulus k. The FAA considers the existing flexible pavement as a subbase for the slab. The subgrade on which the existing pavement rests is given a rating, and the thickness of the overlay is determined from the design charts. The minimum thickness of portland cement concrete overlay should not be less than 5 in.

Bituminous or flexible overlays on portland cement concrete pavements

From an analysis of the results of full-scale test sections constructed at several military airfields, the Corps of Engineers developed an empirical relationship which gives the thickness of a bituminous or flexible overlay on an existing rigid pavement. The relationship is

$$t = 2.5(Fh_d - C_b h) \tag{12-16}$$

where t = thickness of bituminous or flexible overlay, in*
h_d = thickness of portland cement concrete pavement, assuming that existing pavement did not exist, in
h = thickness of existing pavement, in
F = a factor which, when multiplied by h_d, is thickness of a concrete slab that would crack in time, but not so much as to cause undesirable distress in overlay pavement. The existing slab was assumed to break up into pieces of 5 to 7 ft² by the end of the overlay pavement life.
C_b = condition factor for existing base pavement. This factor ranges from 0.75 to 1.0.

Values of F are given in Fig. 12-48. In determining the thickness h_d, the flexural strength of the existing slab is used.

---

*The FAA considers the formula to yield thickness of flexible overlay only.

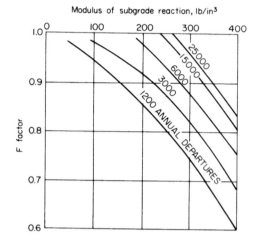

Figure 12-48 Relationship between overlay factor F and modulus of subgrade reaction for different traffic levels (*Federal Aviation Administration [9]*).

The Corps of Engineers recommends that the minimum thickness of overlay be 4 in and that for overlays less than 8 in thick a bituminous overlay be used. This implies that the thickness of the base course should not be less than 4 in. The FAA requires that the thickness of bituminous surfacing not be less than 3 in.

The Asphalt Institute uses the same procedure as for asphalt overlays on flexible pavements, assigning an equivalent thickness of asphalt concrete to the portland cement concrete depending on the condition of the concrete slab.

Portland cement concrete overlays on portland cement concrete pavements

The thickness of portland cement concrete overlay slabs is determined by both the Corps of Engineers and the FAA by the following formulas:

$$h_c = \sqrt[1.4]{h^{1.4} - Ch_e^{1.4}} \qquad (12\text{-}17)$$

or

$$h_c = \sqrt{h^2 - Ch_e^2} \qquad (12\text{-}18)$$

where h_c = thickness of overlay slab, in
h_e = thickness of existing slab, in
h = thickness of equivalent single slab placed directly on subgrade with a working stress equal to that of overlay slab, in
C = coefficient depending on condition of existing pavement; where $C = 1$, existing pavement in good condition; $C =$

0.75, existing pavement with initial corner cracks due to loading, but no progressive cracks; $C = 0.35$, existing pavement badly cracked or crushed (an overlay is not recommended)

These formulas assume that the flexural strength of concrete used for the overlay is approximately equal to that of the existing pavement. Intermediate values of C are not used in design. The change in the overlay thickness requirement normally is approximately 1 in for a change of C from 1.0 to 0.75 or from 0.75 to 0.35.

Although the flexural strength of concrete of the overlay does not appear in these formulas, it enters into the computations in the determination of h, the thickness of the equivalent single slab. The flexural strength of the existing slab does not enter into the computations, since it was found that a substantial difference in flexural strength in the two pavements would result in a very small change in thickness.

Overlay design using layered elastic theory

A method in which layered elastic theory has been adapted to the evaluation and design of overlays for both rigid and flexible pavements has been described by Vallerga et al. [44, 51]. Figure 12-49 outlines the various steps in this procedure. Briefly they are summarized in the following paragraphs:

Step 1. In the condition survey, the condition of the existing pavements, including the nature and extent of distress, is carefully determined and plotted on a large-scale plan of the airfield facilities. This information assists in the sampling and field-test phase as well as in establishing distress related to performance criteria.

Step 2. Deflection measurements* provide a measure of pavement response under known loading conditions and allow differentiation of areas of significant different structural responses.

Step 3. In the drilling and sampling phase, pavement cores, layer samples, and undisturbed subgrade samples to depths of 10 ft or more are obtained together with information on layer thicknesses, etc.

Step 4. Laboratory testing is primarily concerned with determining representative moduli by a form of dynamic rather than static testing and with establishing failure criteria where appropriate (e.g., fracture strength of cement-stabilized base).

Step 5. To determine the traffic, a detailed examination of the past and anticipated future aircraft traffic projections for each specific location (e.g., taxiway) is required. The concept of equating various

---

*Any one of a number of devices, such as the Benkelman beam, can be used.

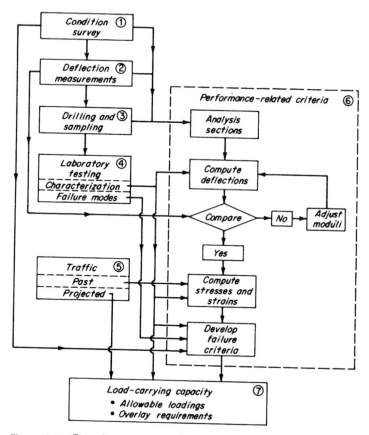

Figure 12-49 Steps in pavement evaluation method (*courtesy of B. A. Vallerga*).

aircraft to equivalent damage of a standard aircraft can be used effectively in this step [31].

Step 6.   From the information established in the previous steps, performance-related criteria are established. By utilizing laboratory properties measured in step 4 as initial input, deflections under the known loadings can be estimated. By comparing these values to those determined in step 2, adjustments in laboratory moduli values are made until predicted (theoretical) deflections agree with the measured deflections. Once this is accomplished, the critical performance parameters for the aircraft in question are determined from layered theory and related to "acceptable" and "not acceptable" performance areas found in step 1, as well as to the laboratory-developed failure criteria of step 4.

Step 7.   From the results of step 6, either the load-carrying capacity of the existing pavements or projected overlay requirements for future traffic of a given design life can be readily determined.

Aircraft and Airport Pavement Classification System

The FAA and ICAO have adopted a pavement classification system for reporting airfield strength [2, 61, 64] called the aircraft/pavement classification number. This system reports a unique *pavement classification number* (PCN) which indicates that an aircraft with an *aircraft classification number* (ACN) equal to or less than the PCN can operate on the pavement subject to any limitation on the tire pressure. The ACN is a number expressing the relative structural effect of an aircraft on either flexible or rigid pavements for a specified subgrade strength in terms of a standard single-wheel load. The PCN is a number which expresses the relative load-carrying capacity of a pavement in terms of a standard single-wheel load.

The bearing strength of the pavement is reported by indicating the PCN, pavement type used for the PCN determination, subgrade category, maximum allowable tire pressure, and basis for the evaluation. Different PCNs may be reported if the strength of the pavement is subject to seasonal variations due to frost or other conditions. A given aircraft will have a different ACN dependent on the pavement on which it operates, flexible (F) or rigid (R), and on the relative strength of the subgrade. The ACN is in subgrade strength units (meganewtons per cubic meter) and can be found from the pavement design charts or analytical equations. One of the primary advantages of this system is that it is independent of the technique used to evaluate pavement strength and relates only to the impact of the aircraft on the supporting subgrade. The PCN is reported for light aircraft, those with a maximum gross takeoff weight less than 5700 kg, in terms of the aircraft weight and tire pressure. For larger aircraft, the system reports the PCN as an ACN, the type of pavement, the subgrade category, the tire pressure category, and the evaluation method used to obtain the PCN.

The concept of the single-wheel load was adopted to determine the ACN so that the interaction between the landing gear and the pavement could be evaluated without reference to the pavement thickness. A ratio between the pavement thickness required for the aircraft and that required for a standard single-wheel load of 500 kg (1100 lb) at a standard tire pressure of 1.25 MPa (181 lb/in^2) defines the ACN. Methods for determining the ACN are presented in some of the references [3, 25, 61, 64].

The ACN for any aircraft may be obtained by any established pavement design technique. For example, the ACN for aircraft on rigid pavements can be obtained by modifying the PCA computer program for rigid-pavement design [16, 25]. The subgrade strength is classified as high (A), medium (B), low (C), or ultralow (D) based upon whether the CBR is equal to or greater than 13, 8, or 4, or is less than 4, respectively, for flexible pavements or the modulus of subgrade reaction k is equal to or greater than 400, 200, or 100 or is less than 100 lb/in³, respectively, for rigid pavements. The maximum allowable tire pressures for aircraft operations are classified as high (W) for no limit, medium (X) for a 217 lb/in² limit, low (Y) for a 145 lb/in² limit, and very low (Z) for a 73 lb/in² limit. If the evaluation is conducted by a specific technical study of the pavement characteristics and behavior, the evaluation method is classified as technical (T). If the evaluation is based upon experience on the type of runway with particular aircraft, it is termed as being performed by *aircraft experience* (U). The aircraft classification numbers for some typical aircraft are shown in Table 12-11. A flowchart illustrating the procedure for reporting the pavement-bearing strength is shown in Fig. 12-50. Figure 12-51 shows a chart used to determine the PCN of an aircraft with a dual-in-tandem landing gear on a flexible pavement. The illustration in this figure shows the computation of a PCN value of 37 for an aircraft with a dual-in-tandem landing gear weight of 225,000 lb acting on a flexible pavement with a subgrade strength of at least a CBR of 8 (code B) [61].

TABLE 12-11   Aircraft Classification Numbers for Selected Aircraft for High Subgrade Strength

Aircraft	Maximum takeoff weight, lb	Tire pressure, lb/in²	ACN Rigid pavement	ACN Flexible pavement
Airbus A-300-B2	313,000	179	37	40
Airbus A-310-300	346,000	217	45	57
Airbus A-320-100	150,000	163	18	18
BAC 1-11-475	87,500	135	25	22
Boeing 727-200	173,000	167	46	41
Boeing 737-200	111,000	148	27	25
Boeing 737-400	143,000	210	41	35
Boeing 747-100	738,000	225	44	46
Boeing 747-200	823,000	210	49	52
Boeing 747-400	872,000	206	53	57
Boeing 757-200	241,000	171	27	29
Boeing 767-200	317,000	191	33	37
Concorde	408,000	183	61	65
DC-8-62	353,000	187	47	49
DC-9-51	122,000	170	35	31
DC-10-40	558,000	170	44	53
L-1011-500	498,000	184	50	60
MD-81	141,000	171	41	36
MD-87	150,000	185	45	39
MD-11	605,000	206	56	64

Sources: Federal Aviation Administration [61] and International Civil Aviation Organization [2].

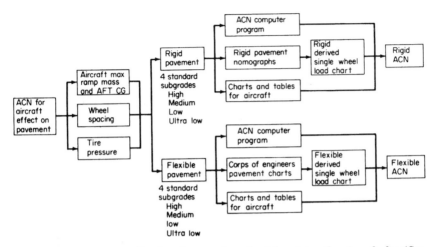

Figure 12-50 Flowchart indicating the procedure for determining the aircraft classification number for pavements (*Douglas Aircraft Company [64]*).

To illustrate the PCN reporting system, a PCN of 46/R/A/X/T indicates that the pavement may be used for unlimited operations by all aircraft with an ACN equal to or less than 46—a Boeing 727-200 with a maximum takeoff weight of 173,000 lb in this case—and that the pavement is rigid (R) with a modulus of subgrade reaction (A) greater than or equal to 400 lb/in^3 such that the maximum allowable tire pressure (X) is 217 lb/in^2. This pavement classification number was determined by a technical evaluation (T) of the pavement properties.

The ICAO permits some overloading of the pavement by aircraft with ACNs slightly larger than the reported PCN [2]. This allows airport managers to assess the optimal operating criteria for airport pavements by considering such factors as traffic demand and pavement life. It provides aircraft operators with information essential to the development of effective operating plans related to the aircraft fleet, route network, and strength properties of airports served. Finally, it provides information to aircraft manufacturers relative to the development of aircraft which will operate with the constraints of existing airports.

Pavement Management Systems

A *pavement management system* (PMS) is a mechanism for providing consistent, objective, and systematic procedures for evaluating pavement condition and for determining the priorities and schedules for pavement maintenance and rehabilitation within available resource

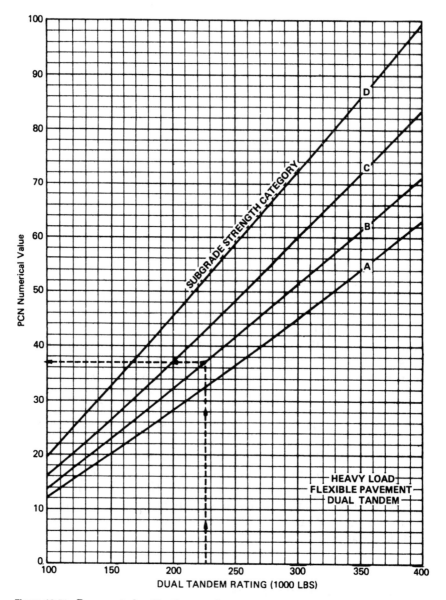

Figure 12-51 Pavement classification number for dual-in-tandem landing gear aircraft on a flexible pavement (*Federal Aviation Administration [61]*).

and budgeting constraints. The PMS can also be used to maintain records of pavement condition and to provide specific recommendations for actions which may be required to maintain a pavement network at an acceptable condition while minimizing the cost associated with pavement maintenance and rehabilitation. The PMS is a tool that is used for assisting engineers and airport managers in making cost-effective decisions concerning pavement maintenance and rehabilitation.

A pavement management system evaluates present pavement condition and predicts future condition through the use of a pavement condition indicator. By projecting the rate of deterioration in the pavement condition indicator and adopting some minimum acceptable level for this indicator, a life-cycle cost analysis can be performed for various maintenance and rehabilitation alternatives, and a determination can be made of the optimal time for the application of the most appropriate alternative. As shown in Figure 12-52, the rate of deterioration of a pavement accelerates with time. By implementing a maintenance or rehabilitation strategy to upgrade the pavement condition at the proper time, the overall cost of maintenance and rehabilitation can be minimized. As noted by the FAA, the total annual cost to maintain or rehabilitate a pavement in relatively poor condition can be 4 to 5 times that of maintaining or rehabilitating a pavement in relatively good condition.

Components of a pavement management system

An effective PMS for use at airports should include the following components [61]:

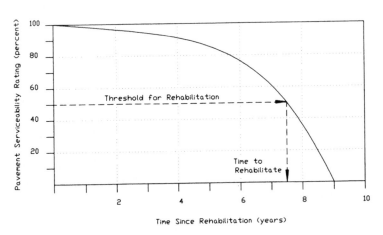

Figure 12-52 Pavement serviceability rating versus time.

1. A systematic mechanism for regularly collecting, storing, and retrieving the necessary data associated with pavement use and condition

2. An objective system for evaluating pavement condition at regular intervals

3. Procedures for identifying alternative maintenance and rehabilitation strategies

4. Mechanisms for predicting and evaluating the impact of pavement maintenance and rehabilitation strategies and alternatives on pavement condition, serviceability, and useful service life

5. Procedures for estimating and comparing the costs of various strategies and alternatives

6. Techniques for identifying the optimal strategy or alternative based upon relevant decision criteria

Essential to an effective PMS is the maintenance of a pavement database, which should include

1. Information about the pavement structure, including when it was originally constructed, the structural components, the soil conditions, a history of subsequent maintenance and rehabilitation, and the cost of these actions.

2. A record of the airport pavement traffic including the number of aircraft operations by various types of aircraft using the pavement over its life.

3. The ability to regularly track pavement condition, including measures of pavement distress and the causes of distress. A pavement rating system should be developed based upon the quantity, severity, and type of distress affecting the pavement surface condition. This rating system measures pavement surface performance and has implications for structural performance. The periodic collection of condition-rating data is essential to tracking pavement performance.

The FAA has developed a computerized PMS called Micro-PAVER [42, 43]. The program allows for the storage of the pavement condition history, of nondestructive testing data, and of the construction and maintenance history of the pavement including cost data. The database provides the capability to evaluate current pavement, to predict future pavement condition, and to identify maintenance and rehabilitation needs. It allows for pavement inspection, scheduling, economic analysis, and budget planning. To evaluate the pavement condition, the *pavement condition index* (PCI) is used [36]. The PCI is a number reflecting the structural integrity and surface condition of the pavement. It is based on an objective measurement of the type, severity, and quantity of pavement distress indicators. By projecting the rate of deterioration based upon the pavement condition history, a life-cycle cost analysis can be performed for various maintenance and rehabilitation alternatives, in which the best alternative and the optimal time for application of this alternative are determined. The program can be used at the network level, in which decisions are made for an entire pavement network under the jurisdiction of management, or at the project level, in which decisions are made regarding the most appropriate maintenance and rehabilitation alternative for a pavement identified at the network level.

References

1. "A Design Procedure for Continuously Reinforced Concrete Pavements for Highways." ACI Subcommittee VII, Title 69-32, *Journal of the American Concrete Institute,* vol. 69, pp. 309–319, 1972.
2. *Aerodrome Design Manual,* pt. 3: *Pavements,* 2d ed., Doc. 9157-AN/901, International Civil Aviation Organization, Montreal, Canada, 1983.
3. *Aerodromes, Annex 14 to the Convention on International Civil Aviation,* vol. 1: *Aerodrome Design and Operations,* 1st ed., International Civil Aviation Organization, Montreal, Canada, 1990.
4. *Aerodromes, Annex 14 to the Convention on International Civil Aviation,* vol. 2: *Heliports,* 1st ed., International Civil Aviation Organization, Montreal, Canada, 1990.
5. *Aircraft Loading on Airport Pavements, ACN-PCN, Aircraft Classification Numbers for Commercial Turbojet Aircraft,* Aerospace Industries Association of America, Inc., Washington, 1983.
6. "Aircraft Pavement Loading: Static and Dynamic," R. C. O'Massey, *Research in Airport Pavements,* Special Report 175, Transportation Research Board, Washington, 1978.
7. *Airfield Pavement Requirements for Multi-Wheel Heavy Gear Loads,* Rep. FAA-RD-70-77, Federal Aviation Administration, Washington, 1971.
8. *Airport Design,* Advisory Circular AC 150/5300-13, Federal Aviation Administration, Washington, 1989.
9. *Airport Pavement Design and Evaluation,* Advisory Circular AC 150/5320-6C with Changes 1 and 2, Federal Aviation Administration, Washington, 1988.
10. "Analysis of Stresses in Concrete Pavements due to Variations of Temperature," H. M. Westergaard, *Proceedings,* 6th Annual Meeting, Highway Research Board, Washington.
11. "Applications of the Results of Research to the Structural Design of Pavements," E. F. Kelley, *Journal of the American Concrete Institute,* 1939.
12. "Characterization of Subgrade Soils in Cold Regions for Pavement Design Purposes," A. T. Bergan and C. L. Monismith, *Highway Research Record,* no. 431, Highway Research Board, Washington, 1973.
13. *Computer Aided Design for Flexible Airfield Pavements,* Computer Program FAD (FAA version F806FAA), U.S. Army Corps of Engineers, Waterways Experiment Station, Vicksburg, Miss., 1992.
14. *Computer Aided Design for Rigid Airfield Pavements,* Computer Program RAD (FAA version R805FAA), U.S. Army Corps of Engineers, Waterways Experiment Station, Vicksburg, Miss., 1992.
15. *Computer Aided Evaluation for Airfield Pavements,* Computer Program PCN, U.S. Army Corps of Engineers, Waterways Experiment Station, Vicksburg, Miss., 1992.
16. *Computer Program for Airport Pavement Design,* R. G. Packard, Portland Cement Association, Skokie, Ill., 1967.

17. *Computer Program Supplement to Thickness Design—Asphalt Pavements for Air Carrier Airports,* Manual Series, no. MS-11A, The Asphalt Institute, Lexington, Ky., 1987.

18. *Design and Construction and Behavior under Traffic of Pavement Test Sections,* Rep. FAA-RD-73-198-I, Federal Aviation Administration, Washington, 1974.

19. *Design and Construction, Continuously Reinforced Joint and Crack Sealing Materials and Practices,* NCHRP Report, no. 38, National Cooperative Highway Research Program, Highway Research Board, Washington, 1967.

20. *Design and Construction of Airport Pavements on Expansive Soils,* Rep. FAA-RD-76-66, Federal Aviation Administration, Washington, 1976.

21. *Design and Construction of MESL,* Rep. FAA-RD-73-198-III, Federal Aviation Administration, Washington, 1974.

22. "Design Considerations for Multi-Wheel Aircraft," W. R. Barker and C. R. Gonzalez, *International Air Transportation,* Proceedings of the 22nd Conference on International Air Transportation, American Society of Civil Engineers, New York, 1992.

23. *Design Manual for Continuously Reinforced Concrete Pavements,* Rep. FAA-RD-74-33-III, Federal Aviation Administration, Washington, 1974.

24. *Design of Civil Airfield Pavement for Seasonal Frost and Permafrost Conditions,* Rep. FAA-RD-74-30, Federal Aviation Administration, Washington, 1974.

25. *Design of Concrete Airport Pavement,* R. G. Packard, Engineering Bulletin, Portland Cement Association, Skokie, Ill., 1973.

26. *Design of Flexible Airfield Pavements for Multiple-Wheel Landing Gear Assemblies,* Rep. 2: *Analysis of Existing Data,* Tech. Memo 3-349, U.S. Army Corps of Engineers, Waterways Experiment Station, Vicksburg, Miss., 1955.

27. "Design of Pavement with High Quality Structural Layers," G. M. Hammitt, *Research in Airport Pavements,* Special Report 175, Transportation Research Board, Washington, 1978.

28. *Economic Analysis of Airport Pavement Rehabilitation Alternatives,* Rep. FAA-RD-81-78, Federal Aviation Administration, Washington, 1981.

29. "Effect of Dynamic Loads on Airport Pavements," R. H. Ledbetter, *Research in Airport Pavements,* Special Report 175, Transportation Research Board, Washington, 1978.

30. *ELSYM5—Computer Program for Determining Stresses and Deformation in a Five Layer Elastic System,* G. Ahlborn, University of California, Berkeley, 1972.

31. "Equivalent Passages of Aircraft with Respect to Fatigue Distress of Flexible Airfield Pavements," J. A. Deacon, *Proceedings,* Association of Asphalt Paving Technologists, 1971.

32. *Field Survey and Analysis of Aircraft Distribution on Airport Pavements,* Rep. FAA-RD-74-36, Federal Aviation Administration, Washington, 1975.

33. *Flexible Airfield Pavements,* Tech. Manual TM 5-825-2, U.S. Army Corps of Engineers, A.G. Publication Center, St. Louis, Mo., 1978.

34. *Full-Depth Asphalt Pavements for General Aviation,* Information Series, no. IS-154, The Asphalt Institute, Lexington Ky., 1973.

35. *Geomechanics Computing Programme,* N.1, Computer Programmes for Circle and Strip Loads on Layered Anisotropic Media, W. J. Harrison, C. M. Gerrard, and L. J. Wardel, Division of Applied Geomechanics, SCIRO, Australia, 1972.

36. *Guidelines and Procedures for Maintenance of Airport Pavements,* Advisory Circular AC 150/5380-6, Federal Aviation Administration, Washington, 1982.

37. *Hot Mix Asphalt Paving Handbook,* Advisory Circular AC 150/5370-14, Federal Aviation Administration, Washington, 1991.

38. "Influence Charts for Rigid Pavements," G. Pickett and G. K. Ray, *Transactions,* vol. 116, pp. 49–73, American Society of Civil Engineers, New York, 1951.

39. "Layered Systems under Normal Surface Loads," M. G. Peutz, H. P. M. van Kempen, and A. Jones, *Highway Research Record,* no. 228, Highway Research Board, Washington, 1968.

40. *Measurement, Construction, and Maintenance of Skid Resistant Airport Pavement Surfaces,* Advisory Circular AC 150/5320-12B, Federal Aviation Administration, Washington, 1991.

41. *Methodology for Determining, Isolating and Correcting Runway Roughness,* Rep. FAA-RD-75-110-II, Federal Aviation Administration, Washington, 1977.
42. *Micro-PAVER, Concept and Development Airport Pavement Management System,* Rep. DOT/FAA/PM-87-8, Federal Aviation Administration, Washington, 1987.
43. *Micro-PAVER, Pavement Management System,* Advisory Circular AC 150/5000-6, Federal Aviation Administration, Washington, 1987.
44. "Modern Pavement Evaluation Techniques," B. A. Vallerga and R. G. Lee, *Airport Challenges of the Future,* American Society of Civil Engineers, New York, 1973.
45. *Multiple-Wheel Heavy Gear Load Pavements Tests,* vol. 1: *Basic Report,* R. G. Ahlvin, Tech. Rep. AFWL-TR-70-113, AFWL, Kirtland Air Force Base, N. Mex., 1971.
46. "New Formulas for Stresses in Concrete Pavements of Airfields," H. M. Westergaard, *Transactions,* vol. 113, American Society of Civil Engineers, New York, 1948.
47. "Numerical Computation of Stresses and Strains in a Multiple-Layer Asphalt Pavement System," H. Warren and W. L. Dieckmann, internal unpublished report, Chevron Research Corporation, 1963.
48. *Offpeak Construction of Airport Pavements Using Hot-Mix Asphalt,* Advisory Circular AC 150/5370- 13, Federal Aviation Administration, Washington, 1990.
49. *Pavement Computer Aided Structural Engineering (PCASE),* Pavement design and analysis computer programs, U.S. Army Corps of Engineers, St. Louis, Mo., 1992.
50. *Pavement Design for Frost Conditions,* Tech. Manual TM 5-818-2, U.S. Army Corps of Engineers, A.G. Publication Center, St. Louis, Mo., 1965.
51. "Pavement Evaluation and Design for Jumbo Jets," B. A. Vallerga and B. F. McCullough, *Transportation Engineering Journal,* vol. 95, no. TE4, American Society of Civil Engineers, New York, 1969.
52. *Pavement Management System,* Advisory Circular AC 150/5380-7, Federal Aviation Administration, Washington, 1988.
53. *Pavement Response to Aircraft Dynamic Loads,* Rep. FAA-RD-74-39-II, Federal Aviation Administration, Washington, 1974.
54. "Performance of Concrete Pavements Subjected to Wide-Body Jet Aircraft Loading," R. L. Hutchinson, *Proceedings,* Annual Meeting on Roadways and Airport Pavements, Paper SP 51-8, American Concrete Institute, Detroit, Mich., 1975.
55. *Principles of Pavement Design,* E. J. Yoder and M. W. Witczak, 2d ed., Wiley, New York, 1975.
56. *Reinforced Concrete Pavements,* R. D. Bradbury, Wire Reinforcement Institute, Washington, 1938.
57. *Relative Pavement Bearing Strength Requirements of Aircraft,* R. C. O'Massey, Tech. Paper, no. 780568, Douglas Aircraft Company, McDonnell-Douglas Corporation, Long Beach, Calif., 1978.
58. *Review of Soil Classification Systems Applicable to Airport Pavement Design,* Rep. FAA-RD-73-169, Federal Aviation Administration, Washington, May 1974.
59. *Rigid Pavements for Airfields,* Tech. Manual TM 5-825-3, U.S. Army Corps of Engineers, A.G. Publication Center, St. Louis, Mo., 1988.
60. *Standards for Specifying Construction of Airports,* Advisory Circular AC 150/5370-10A with Changes 1 through 4, Federal Aviation Administration, Washington, 1992.
61. *Standardized Method of Reporting Airport Pavement Strength—PCN,* Advisory Circular AC 150/5335-5, Federal Aviation Administration, Washington, 1983.
62. *Structural Design of Asphalt Pavements for Heavy Aircraft,* J. M. Edwards and C. P. Valkering, Shell International Petroleum Company Limited, London, 1970.
63. *Structural Design of Pavements for Light Aircraft,* FAA-RD-76-179, Federal Aviation Administration, Washington, 1976.
64. *The ACN/PCN System for Reporting Airfield Bearing Strength,* R. C. O'Massey, Douglas Aircraft Company, McDonnell-Douglas Corporation, Long Beach, Calif., 1979.
65. "The Application of Elastic Theory to Flexible Pavements," A. C. Whiffin and N. W. Lister, *Proceedings,* 1st International Conference on the Structural Design of Asphalt Pavements, University of Michigan, Ann Arbor, 1963.

66. *Thickness Design—Asphalt Pavements for Air Carrier Airports,* Manual Series, no. MS-11, The Asphalt Institute, Lexington, Ky., 1987.
67. *Thickness Design for Concrete Pavements,* Concrete Information Series, Portland Cement Association, Skokie, Ill., 1966.
68. *Use of Nondestructive Testing Devices in the Evaluation of Airport Pavements,* Advisory Circular AC 150/5370-11 with Change 1, Federal Aviation Administration, Washington, 1980.

13

Airport Lighting, Marking, and Signing

This chapter is devoted principally to outlining the visual requirements of the pilot and the various lighting, marking, and sign systems in use or proposed to meet these requirements. No attempt has been made to describe in detail the hardware or its installation.

The emphasis in this chapter is on the following facilities:

1. Approach lighting
2. Runway threshold lighting
3. Runway edge lighting
4. Runway centerline and touchdown zone lights
5. Runway approach slope indicators
6. Taxiway edge and centerline lighting
7. Runway and taxiway marking systems
8. Taxiway guidance sign systems

These facilities provide the following functions:

1. Ground to air visual information required during landing
2. The visual requirements for takeoff and landing
3. The visual guidance for taxiing

Requirements for Visual Aids

Since the earliest days of flying, pilots have used ground references for navigation when approaching an airport, just as officers on ships at sea have used landmarks on shore when approaching a harbor.

Pilots need visual aids in good weather as well as bad and during the day as well as at night.

In the daytime there is adequate light from the sun and sky, so artificial lighting is not usually required. But it is necessary to have adequate contrast in the field of view and to have a suitable pattern of brightness so that the important features of the airport can be identified and oriented with respect to the position of the aircraft in space. These requirements are met almost automatically during the day when the weather is clear. The runway for conventional aircraft always appears as a long, narrow strip with straight sides and is free of obstacles. It can therefore be easily identified from a distance or by flying over the field. Therefore, the perspective view of the runway and other identifying reference landmarks are used by pilots as visual aids for orientation when they are approaching the airport to land. Experience has demonstrated that the horizon, runway edges, runway threshold, and centerline of the runway are the most important elements for pilots to see. These features are shown schematically in Figs. 13-1 and 13-2. To enhance the visual information during the day, the runway is painted with a standard marking pattern, as is discussed later. The key elements in this pattern are the threshold, the centerline, the edges, and multiple parallel lines to increase the perspective and to define the plane of the surface.

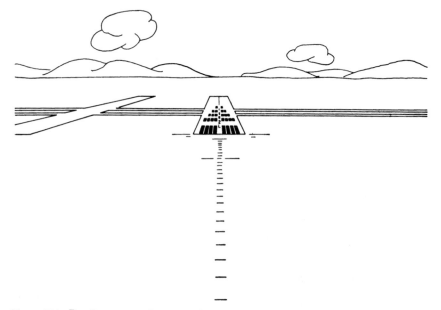

Figure 13-1 Daytime approach as seen from a height of 200 ft, 3500 ft from the runway threshold.

Figure 13-2 Daytime approach as seen from a height of 80 ft, 800 ft from the runway threshold.

During the day when visibility is poor and at night, the visual information is reduced by a significant amount compared to the clear-weather daytime scene. It is therefore essential to provide visual aids which will be as meaningful to pilots as possible.

The aircraft landing operation

An aircraft approaching a runway in a landing operation can be visualized as a sequence of operations involving a transient body suspended in a three-dimensional grid that is approaching a fixed two-dimensional grid. While in the air, the aircraft can be considered as a point mass in a three-dimensional orthogonal coordinate system, in which it has translation along three coordinate directions and rotation about three axes. If the three coordinate axes are aligned horizontal, vertical, and parallel to the end of the runway, the directions of motion can be described as lateral, vertical, and forward. The rotations are normally called *pitch, yaw,* and *roll* for the horizontal, vertical, and parallel axes, respectively. During a landing operation, pilots must control and coordinate all six degrees of freedom of the aircraft so as to bring it into coincidence with the desired approach or reference path to the touchdown point on the runway. To do this, pilots need translation information regarding the aircraft's alignment, height, and distance; rotation information regarding pitch, yaw, and roll; and information concerning the rate of descent and the rate of closure

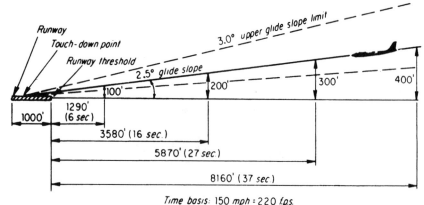

Figure 13-3 Glide slope, height, distance, and time relationships.

with the desired path. The glide path, height, time, and distance relationships during a typical landing are shown in Fig. 13-3.

Alignment guidance

Pilots must know where their aircraft are with respect to lateral displacement from the centerline of the runway. Most runways are 75 to 200 ft wide and 3000 to 12,000 ft long. Thus any runway is a long, narrow ribbon when it is first seen from several thousand feet away. The predominant alignment guidance comes from longitudinal lines that constitute the centerline and edges of the runway. All techniques, such as painting, lighting, or surface treatment, that develop contrast and emphasize these linear elements are helpful in providing alignment information.

Height information

Estimation of the height above ground from visual cues is one of the most difficult judgments for pilots. It is simply not possible to provide good height information from an approach lighting system. Consequently the best source of height information is the instrumentation in the aircraft. However, use of these instruments requires the availability of an instrument landing system on the ground. Many airports have no instrument landing system at all, and at others only some runways have this equipment. Consequently two types of ground-based visual aids defining the desired glide path have been developed. These are known as the *visual-approach slope indicator* (VASI) and the *precision-approach path indicator* (PAPI). However, these visual aids can be used only when the visibility is reasonably good.

How much the pilot sees on the ground

Several parameters influence how much a pilot can see on the ground. One is the *cockpit cutoff angle*. This is the angle between the longitudinal axis of the fuselage and an inclined plane below which the view of the pilot is blocked by some part of the aircraft, indicated by α in Fig. 13-4. Normally the larger the angle α, the more the pilot can see of the ground. Also important is the *pitch angle* ß of the fuselage axis during the approach to the runway. Few aircraft approach a runway with the fuselage angle horizontal; they are pitched either up or down. The larger the angle ß (in a pitch-up attitude), the larger the angle α must be to have adequate over-the-nose vision. Approach speed has a profound influence on ß. As an example, for some aircraft ß can be decreased by about 1° with each 5-kn increase in speed above the reference approach speed.

In Fig. 13-4, VR is the visual range or the maximum distance a pilot can see at some height above the runway h. The horizontal segment of the ground that a pilot can see is H. According to Fig. 13-4,

$$H = VR \cos \theta - h \cot (\alpha - \text{ß}) \tag{13-1}$$

and

$$\sin \alpha = \frac{h}{VR} \tag{13-2}$$

Note from Eq. (13-1) that for a fixed value of VR the ground segment H increases as the height h of the eyes of a pilot above the ground decreases. Typical values of α range from 11° to 16°, and typical values of ß are ± 0.5°.

It has been found through experience that 3 s is approximately the minimum reaction time for a pilot to cause the aircraft to react after sighting a visual aid [28]. If a minimum of 3 s is necessary for perception, pilot action, aircraft response, and checking the response, and if

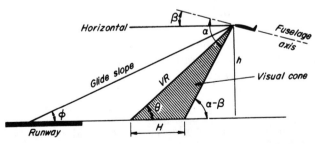

Figure 13-4 Visual parameters: ϕ = glide slope angle; α = cockpit cutoff angle; ß = pitch angle; VR = visual range; H = horizontal segment of visual range; h = height of glide slope above the runway; and θ = angle formed by VR with the horizontal.

the approach speed of the aircraft is 150 mi/h (220 ft/s), then the minimum horizontal segment on the ground should not be less than 660 ft. Using Eq. (13-1) with the glide slope angle ϕ of 2.5° and a value of $\alpha - \beta$ of 12° results in H being about 200 ft when h is 200 ft. However, when h is 100 ft, H is 687 ft. Consequently, a pilot cannot derive visual guidance from approach lights until the aircraft reaches a height of 100 ft above the runway.

Approach Lighting

Because of the difference in the slant visual range discussed in the preceding section, the approach lights need to meet different photometric requirements from the threshold and runway lights. In general, much higher intensity is required in the *approach lighting system* (ALS), especially in the outermost lighting units. Also, special identification information, such as high-brightness flashing lights, may be warranted when visibility is extremely poor.

Studies of visibility in fog [3] have shown that for a visual range of 2000 to 2500 ft it would be desirable to have as much as 200,000 candelas (cd) available in the outermost approach lights where the slant range is relatively long. Under these same conditions, the optimum intensity of the approach lights near the threshold should be on the order of 100 to 500 cd. A transition in the intensity of the light directed toward the pilot is highly desirable, to provide the best visibility at the greatest possible range and to avoid glare and the loss of contrast sensitivity and visual acuity at short range.

Sequenced flashing high-intensity lights are available for airport use and are installed as supplements to the standard ALS at those airports where very low visibilities occur frequently. These lights operate from the stored energy in a capacitor which is discharged through the lamp in approximately 5 ms and may develop as much as 30 million cd of light. They are mounted in the same pedestals as the light bars. The lights are sequence-fired, beginning with the unit farthest from the runway. The complete cycle is repeated every 2 s. This results in a brilliant ball of light continuously moving toward the runway. Since the very bright light can interfere with the eye adaptation of the pilot, condenser discharge lamps are usually omitted in the 1000 ft of the ALS nearest the runway.

System configurations

The configurations which have been adopted are the Calvert system [3], shown in Fig. 13-5, which is widely used in Europe and other parts of the world; the ICAO category II and category III system, shown in Fig. 13-6; and several system configurations which have been adopted

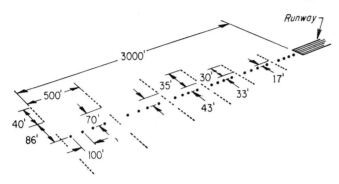

Figure 13-5 Calvert system of approach lights.

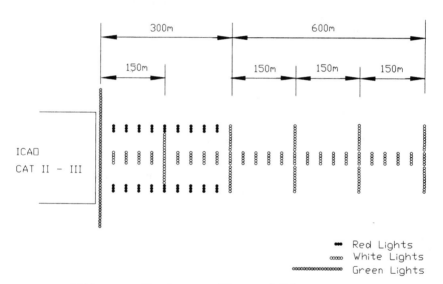

Figure 13-6 ICAO category II and category III approach light systems.

by the FAA in the United States, shown in Fig. 13-7. The FAA publishes criteria for the establishment of the ALSs [13] and other navigation facilities at airports [6]. Approach lights are normally mounted on frangible pedestals of varying height to improve the perspective of the pilot in approaching a runway. Details concerning the requirements in the United States are contained in the references [4].

The Calvert system is 3000 ft in length. In this system, developed by E. S. Calvert in Great Britain, there are a line of single-bulb lights spaced on 100-ft centers along the extended runway centerline and six transverse crossbars of lights of variable length spaced on 500-ft centers. The length of the transverse rows diminishes as the aircraft gets nearer to the runway threshold. Roll guidance in the Calvert system is provided principally by the transverse rows of lights. The details of the layout of the Calvert system are shown in Fig. 13-5.

For operations in very poor visibility, category II or category III, there is a common internationally accepted ICAO system which applies only to the inner 300 m of the system closest to the runway threshold. The remaining 600 m is unaffected, and consequently it can be the Calvert system, the U.S. system, or another system. The category II and category III system adopted by the ICAO, shown in Fig. 13-6, consists of two lines of red bars on each side of the runway centerline and a single line of white bars on the runway centerline, both at 30-m intervals and both extending out 300 m from the runway threshold. In addition, there are two longer bars of white light at a distance of 150 m and 300 m from the runway threshold as well as a long threshold bar of green light at the runway threshold. The ICAO also recommends that the longer bars of white light be placed 450, 500, and 750 m from the runway threshold if the runway centerline lights extend out that distance, as shown in Fig. 13-6.

The systems currently approved by the FAA for installation in the United States consist of a high-intensity ALS with sequenced flashing lights (ALSF-2), which is required for category II and category III precision-instrument approaches, a high-intensity ALS with sequenced flashing lights (ALSF-1), and three medium-intensity ALSs (MALSR, MALS, MALSF).

In each of these systems, there is a long transverse crossbar located 1000 ft from the runway threshold to indicate the distance from the runway threshold. In these systems roll guidance is provided by crossbars of white light 14 ft in length, placed at either 100- or 200-ft centers on the extended runway centerline. The 14-ft crossbars consist of closely spaced five-bulb white lights to give the effect of a continuous bar of light.

The high-intensity ALS is 2400 ft long (some are 3000 ft long) with various patterns of light located symmetrically about the extended runway centerline and a series of sequenced high-intensity flashing lights located every 100 ft on the extended runway centerline for the outermost 1400 ft. In the high-intensity ALSs, the 14-ft crossbars of five-bulb white light are placed at 100-ft intervals, and in the medium-intensity ALSs these crossbars of white light are placed at 200-ft intervals—both for a distance of 2400 ft from the runway threshold on the extended runway centerline. The high-intensity ALSs have a long crossbar of green light at the edge of the runway threshold. In the ALSF-2 system, shown in Fig. 13-7a, two additional crossbars consisting of three-bulb white light are placed symmetrically about the runway centerline at a distance of 500 ft from the runway threshold, and two additional three-bulb red light crossbars are placed symmetrically about the extended runway centerline at 100-ft intervals

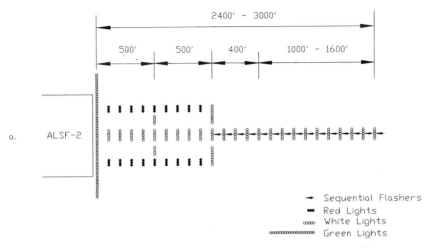

Figure 13-7 FAA approach light systems: (*a*) High-intensity approach light system ALSF-2.

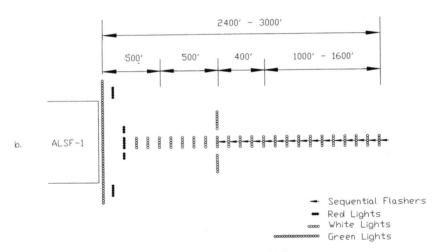

Figure 13-7 (*b*) High-intensity approach light system ALSF-1.

for the inner 1000 ft, to delineate the edges of the runway surface. In the ALSF-1 system, shown in Fig. 13-7*b,* two additional crossbars consisting of five-bulb red light are placed symmetrically about the runway centerline at a distance of 100 ft from the runway threshold, to delineate the edge of the runway, and two additional three-bulb red light crossbars are placed symmetrically about the extended runway centerline at 200 ft from the runway threshold.

The MALSR system, shown in Fig. 13-7*c,* is a 2400-ft medium-intensity ALS with *runway alignment indicator lights* (RAILs). The inner 1000 ft of the MALSR is the medium-intensity ALS portion of

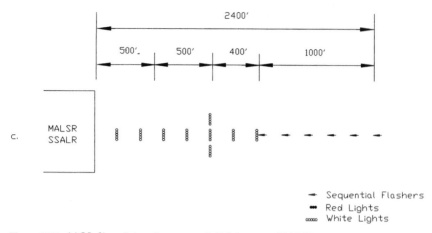

Figure 13-7 (c) Medium-intensity approach light system MALSR.

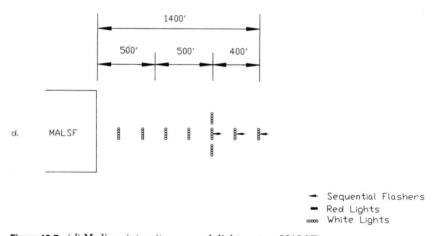

Figure 13-7 (d) Medium-intensity approach light system MALSF.

the system, and the outer 1400 ft is the RAIL portion of the system. The system has sequential flashing lights for the outer 1000 ft. It is recommended for category I precision-instrument approaches. The MALSF system, shown in Fig. 13-7d, is a 1400-ft medium-intensity ALS with runway alignment indicators lights. The system has sequential flashing lights for the outer 400 ft. The *simplified short approach lighting system* (SSALR) has the same configuration as the MALSR system.

At smaller airports where precision-instrument approaches are not required, a medium-intensity approach lighting system with sequential flashers (MALSF) or with sequenced flashers (MALS) is adequate. The system is only 1400 ft long compared to 2400 ft for a precision-instrument approach system. It is therefore much more economical, an important factor at small airports. The MALSF, similar to the MALSR shown in Fig. 13-7d, is a short-approach medium-intensity ALS, but the sequenced flashers replace the runway alignment indicator lights, and these are provided only in the outermost 400 ft of the 1400-ft system, to improve pilot recognition of the runway approach in areas where there are distracting lights in the vicinity of the airport. The MALS system does not have the runway alignment indicator lights or the sequential flashers.

At international airports in the United States, the 2400-ft ALSs are often extended to 3000 ft to conform to international specifications.

Visual-Approach Slope Aids

A larger percentage of landing accidents have occurred in good-visibility weather and have been attributed to poor ground reference data which caused difficulty in judging height.

Visual-approach slope indicator

As an aid in defining the desired glide path in relatively good weather conditions, an optical reference device located on the ground adjacent to the sides of a runway has been developed [11]. This device, the *visual-approach slope indicator* (VASI), has been adopted in the United States and was developed in Great Britain. The FAA publishes establishment criteria for VASI systems at airports [11].

There are a number of different VASI configurations depending on the desired visual range, the type of aircraft, and whether large widebodied aircraft will be using the runway, as shown in Table 13-1. Each group of lights transverse to the direction of the runway is referred to as a *bar* irrespective of the fact that it may be on one side or on both sides of the runway. A bar is made up of one, two, or three light units, referred to as *boxes*. Thus the VASI-12 system, shown in Table 13-1, is a two-bar system consisting of twelve boxes. The bar nearest to the runway threshold is referred to as the *downwind bar*, and the bar farthest from the runway threshold is referred to as the *upwind bar*. If pilots are on the proper glide path, the downwind bar appears white and the upwind bar appears red; if pilots are too low, both bars appear red; and if pilots are too high, both bars appear white.

To accommodate large wide-bodied aircraft where the height of the eye of the pilot is much greater than in smaller jets, a third upwind bar is added, as represented by the VASI-6 and VASI-16 systems in Table 13-1. For wide-bodied aircraft, the middle bar becomes the downwind bar, and the third bar is the upwind bar. In other words, pilots of large wide-bodied aircraft ignore the bar closest to the run-

TABLE 13-1   Configuration of Visual-Approach Slope Indicators

Type	Schematic	Day VFR range, nmi	Comments
VASI-16		5	All aircraft including long large-body turbojet; used at major airports requiring maximum boldness of signal
VASI-12		5	All aircraft except long large-body turbojet; used at major airports requiring maximum boldness of signal
VASI-6		4	For runways with B-747 or C5A operations; two visual glide paths to serve all aircraft types
VASI-4		4	Normally installed on primary runways
VASI-2		3	Normally installed on secondary and utility runways
SAVASI		11/2	Option for VASI-2; for VFR basic utility runways only

way threshold and use the other two bars for visual reference. The location of the lights for VASI-6 systems is shown in Fig. 13-8.

The more common systems in use in the United States are the VASI-2, VASI-4, VASI-12, and VASI-16. VASI systems are particularly useful on runways that do not have an instrument landing system or for aircraft not equipped to use an instrument landing system.

Precision-approach path indicator

The FAA presently allows the use of another type of visual-approach indicator called the *precision-approach path indicator* (PAPI) [20]. This system gives more precise indications to the pilot of the approach path of the aircraft and utilizes only one bar as opposed to the minimum of two bars required by the VASI system. A schematic diagram of the PAPI system is shown in Fig. 13-9. The system consists of a unit with four lights on either side of the approach runway. By utilizing the color scheme indicated on Fig. 13-9, the pilot is able to ascertain five approach angles relative to the proper glide slope, compared to three with the VASI system. One problem with the VASI system has been the lack of an immediate transition from one color

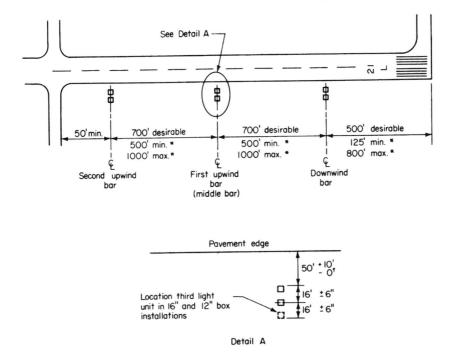

Figure 13-8 Location of visual-approach slope indicators: Upwind and downwind light units are located the same distance from the runway edge; center optical aperture of all light units is within 1 ft of runway centerline grade. Longitudinal tolerances to be used only to avoid taxiways, runways, etc., or to achieve a desirable height due to terrain. Lateral tolerances to be used to avoid ditches, catch basins, utility holes, etc. (*Federal Aviation Administration*).

indication to another, resulting in shades of colors. The PAPI system solves this problem by providing an instant transition from one color indication to another as a reaction to the descent path of the aircraft. An advantage of the system is that it is a one-bar system as opposed to the two-bar VASI system. This results in greater operating and maintenance cost economies and eliminates the need for the pilot to look at two bars to obtain glide slope indications.

Threshold Lighting

During the final approach for landing, pilots must make a decision to complete the landing or "execute a missed approach." The identification of the threshold is a major factor in pilot decisions to land or not to land. For this reason, the region near the threshold is given special lighting consideration. The threshold is identified at large airports by a complete line of green lights extending across the entire width of the runway, as shown earlier in Fig. 13-7, and at small airports by four green lights on each side of the threshold. When the lights are

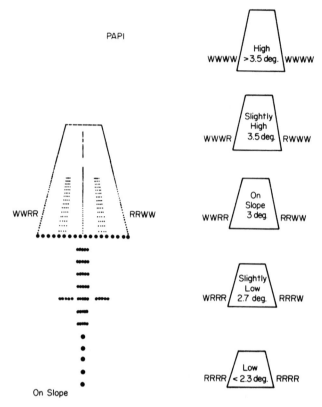

Figure 13-9 Precision-instrument approach path indicator (*Federal Aviation Administration*).

extended across the entire width of the runway, they must be semi-flush types. The lights on either side of the runway threshold may be elevated. Threshold lights in the direction of landing are green, but in the opposite direction these lights are red, to indicate the end of the runway.

Runway Lighting

After crossing the threshold, pilots must complete a touchdown and rollout on the runway. The runway visual aids for this phase of landing should be designed to give pilots information on alignment, lateral displacement, roll, and distance. The lights should be arranged to form a visual pattern that pilots can easily interpret.

At first, night landings were made by floodlighting the general area. Various types of lighting devices were used, including automobile headlights, arc lights, and search lights. Boundary lights were added to outline the field and to mark hazards such as ditches and fences. Gradually, preferred landing directions were developed, and special lights were used to indicate these directions. Floodlighting was then restricted to the preferred landing directions, and runway edge lights were added along the landing strips. As experience grew, the runway edge lights were adopted as visual aids on a runway. This was followed by the use of runway centerline and touchdown zone lights for operations in very poor visibility.

Runway edge lights

Recommended standards for the design and installation of runway edge lighting systems are published by the FAA [21] and are contained in ICAO Annex 14 [1]. These light fixtures are usually elevated units. Semiflush lights, however, are permitted. Each unit has a specially designed lens which projects two main light beams down the runway. Elevated runway lights are mounted on frangible fittings and project no more than 30 in above the surface on which they are installed. They are located along the edge of the runway not more than 10 ft from the edge of the full-strength pavement surface. The longitudinal spacing is not more than 200 ft. Runway edge lights are white, except that in the last 2000 ft of an instrument runway in the direction of aircraft operations, these lights are yellow to indicate a caution zone. A typical layout of high-intensity runway lights is shown in Fig. 13-10. If the runway threshold is displaced but the area that is displaced is usable for takeoffs and taxiing, then the runway edge lights in the displaced area in the direction of aircraft operations are red, as shown in Fig. 13-11.

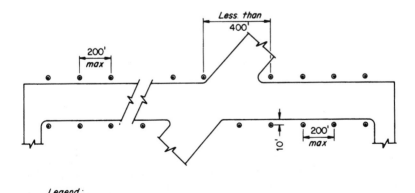

Legend:

◎ --- 360° white, except for the last 2,000' of the instrument runway

Figure 13-10 High-intensity runway edge lights (*Federal Aviation Administration*).

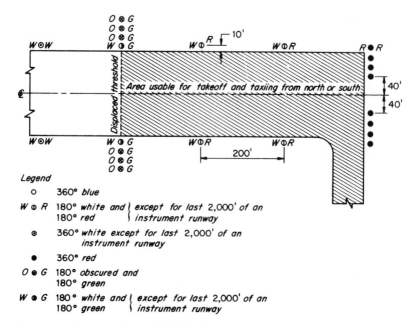

Legend

o   360° blue

W ⊕ R   180° white and ⎰ except for last 2,000' of an
         180° red   ⎱ instrument runway

⊛   360° white except for last 2,000' of an
     instrument runway

●   360° red

O ● G   180° obscured and
         180° green

W ● G   180° white and ⎰ except for last 2,000' of an
         180° green   ⎱ instrument runway

Figure 13-11 Runway edge and threshold lights for a displaced threshold (*Federal Aviation Administration*).

Runway centerline and touchdown zone lights

As an aircraft proceeds over the approach lights, pilots are looking at relatively bright light sources on the extended runway centerline. Over the runway threshold, pilots continue to look along the centerline, but the principal source of guidance—namely, the runway edge lights—has moved far to each side in their peripheral vision. The result is that the central area appears excessively black, and pilots are flying virtually blind, except for the peripheral reference information. Attempts to eliminate this "black hole" by increasing the intensity of runway edge lights have proved ineffective. To reduce the black-hole effect and provide adequate guidance during very poor visibility conditions, runway centerline and touchdown zone lights have been installed in the pavement. These lights are usually installed only at those airports which are equipped for instrument operations. These lights are required for category II and category III runways and for category I runways used for landing operations below a 2400-ft *runway visual range* (RVR). Runway centerline lights are required on runways used for takeoff operations below 1600-ft RVR. Although not required, runway centerline lights are recommended for category I runways greater than 170 ft wide or when used by aircraft with approach speeds over 140 kn. The FAA publishes recommended standards for the design of these lighting systems [14].

The touchdown zone lights are white, consist of a three-bulb bar on either side of the runway centerline, and extend 3000 ft from the runway threshold or one-half the runway length if the runway is less than 6000 ft long. They are spaced at intervals of 100 ft, with the first light bar 100 ft from the runway threshold, and are located 36 ft on either side of the runway centerline, as shown in Fig. 13-12. The centerline lights are spaced at intervals of 50 ft. They are normally offset a maximum of 2 ft from the centerline, to avoid the centerline paint line and to prevent the nose gear of the aircraft from riding over the light fixtures. These lights are also white, except for the last 3000 ft of runway in the direction of aircraft operations, where they are color-coded. For the last 1000 ft, centerline lights are red, and for the next 2000 ft they are alternated red and white, as shown in Fig. 13-13.

When there are displaced thresholds, the centerline lights extend into the displaced threshold area. If the displaced area is not used for takeoff operations, or if the displaced area is used for takeoff operations and is less than 700 ft long, the centerline lights are blanked out in the direction of landing. For displaced thresholds more than 700 ft long or for displaced areas used for takeoffs, the centerline lights in the displaced area must be capable of being shut off during landing operations.

Runway end identifier lights

Runway end identifier lights (REILs) are installed at airports where there are no approach lights to provide pilots with positive visual identification of the approach end of the runway. Criteria for the establishment of these facilities are published by the FAA [10]. The system consists of a pair of synchronized white flashing lights located on each side of the runway threshold and is intended for use when there is adequate visibility.

Taxiway Lighting

Either after a landing or on the way to take off, pilots must maneuver the aircraft on the ground on a system of taxiways to and from the terminal and hangar areas. At a large airport, the taxiway system can be very complex; consequently, adequate lighting aids should be provided for taxiing at night and during the day when visibility is very poor.

The following overall guidelines should be followed in determining the lighting, marking, and signing visual aid requirements for taxiways:

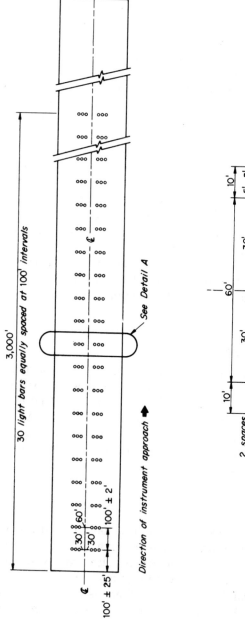

Figure 13-12 Runway touchdown zone lights (*Federal Aviation Administration*).

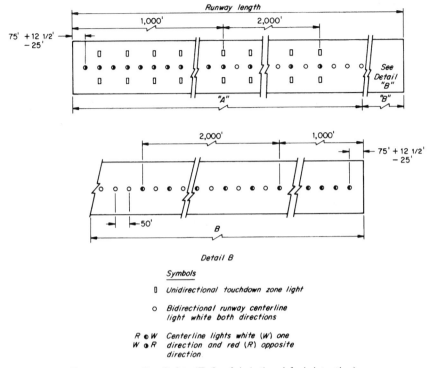

Figure 13-13 Runway centerline lights (*Federal Aviation Administration*).

1. To avoid confusion with runways, taxiways must be clearly identified.

2. Runway exits need to be readily identified. This is particularly true for high-speed runway exits so that pilots can locate these exits 1200 to 1500 ft before the turnoff point.

3. Adequate visual guidance along the taxiway must be provided.

4. Specific taxiways must be readily identified.

5. The intersections between taxiways, the intersections between runways and taxiways, and runway-taxiway crossings must be clearly marked.

6. The complete taxiway route from the runway to the apron and from the apron to the runway should be easily identified.

Two types of lights are used for the designation of taxiways. One type delineates the edges of taxiways [21], and the other type delineates the centerline of the taxiway [27]. Taxiway edge lights are blue whereas taxiway centerline lights are green.

Taxiway edge lights

Elevated bidirectional lights are usually located at intervals of not more than 200 ft on either side of the taxiway. The exact spacing is influenced by the physical layout of the taxiways. Closer spacing is required on curves. Light fixtures are located not more than 10 ft from the edge of full-strength pavement surfaces. The lights cannot extend more than 30 in above the pavement surface. The spacing of lights along a curve is shown in Fig. 13-14. Entrance points to run-

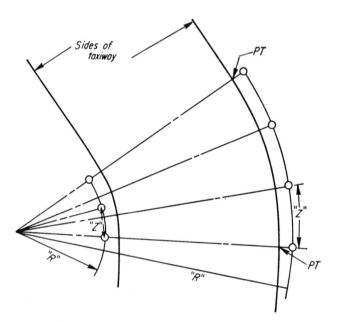

Radius "R" of curve in feet	Dimension "Z" in feet	Radius "R" of curve in feet	Dimension "Z" in feet
15	20	300	80
25	27	400	95
50	35	500	110
75	40	600	130
100	50	700	145
150	55	800	165
200	60	900	185
250	70	1000	200 *max*

Notes:

1. For radii not listed, "Z" spacing shall be determined by linear interpolation.

2. "Z" is the chord length

Figure 13-14 Typical taxiway lighting layout on straight and curved sections (*Federal Aviation Administration*).

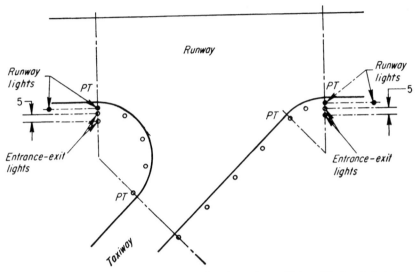

Figure 13-15 Location of entrance and exit taxiway lights (*Federal Aviation Administration*).

ways and exit points from them are lighted as shown in Fig. 13-15. With the advent of large wide-bodied jet aircraft, the thrust of the outboard engines tends to impinge on the taxiway lights, in some cases causing damage to the fixtures. Pilots have indicated that taxiway edge lights do not provide sufficient guidance, especially with respect to lateral displacement.

Taxiway centerline lighting

Research and experience have demonstrated that guidance from centerline lights is superior to that from edge lights, particularly in low-visibility conditions. The spacing of the lights on curves and tangents is given in Table 13-2 [27].

For normal exits, the centerline lights are terminated at the edge of the runway. At taxiway intersections the lights continue across the intersection. For long-radius high-speed exit taxiways, the taxiway lights extend onto the runway from a point 200 ft back from the point of curvature PC of the taxiway to the point of tangency (PT) of the central curve of the taxiway (see Chap. 9). Within these limits, the spacing of lights is 50 ft. These lights are offset 2 ft from the runway centerline lights and are gradually brought into alignment with the centerline of the taxiway.

Where the taxiways intersect with runways and aircraft are required to hold short of the runway, several yellow lights spaced at 5-ft intervals are placed transversely across the taxiway.

TABLE 13-2   Taxiway Lighting Spacing

	Maximum longitudinal spacing, ft	
	1000-ft RVR* and above	Below 1000-ft RVR
Curves		
Centerline radius of curve, ft		
125–399	25	12.5
400–1199	50	25
1200 to straight	100	50
Very long-radius exits	50	50
Tangents		
Taxiway length, ft		
Up to 350	25	
351–450	50	
451–500	75	
500 or greater	100	

*RVR = runway visual range.

SOURCE: Federal Aviation Administration.

Runway and Taxiway Marking

To aid pilots in guiding the aircraft on the runways and taxiways, pavements are marked with lines and numbers. These markings are of benefit primarily during the day and at dusk; at night, lights are used to guide pilots in landing and maneuvering at the airport. White is used for all markings on runways, and yellow is used on taxiways and aprons. The FAA has developed a comprehensive plan for marking runways and taxiways [17]. Similarly, the ICAO recommendations for marking are contained in Annex 14 [2]. The FAA and ICAO requirements are quite similar.

Runways

The FAA has grouped runways for marking purposes into three classes: (1) visual runways; (2) non-precision-instrument runways; and (3) precision-instrument runways. The visual runway is a runway with no straight-in instrument approach procedure and is intended solely for the operation of aircraft using visual-approach procedures. The non-precision-instrument runway has an existing precision-instrument approach procedure utilizing air navigation facilities with only horizontal guidance (typically a VOR station) for which a straight-in non-precision-instrument approach procedure has been approved. A precision-instrument runway has an existing precision-instrument approach procedure utilizing a precision-instrument landing system (ILS or MLS).

The end of each runway is marked with a number which indicates the magnetic azimuth (clockwise from magnetic north) of the runway in the direction of operations. The marking is given to the nearest 10° with the last digit omitted. Thus a runway in the direction of an azimuth of 163° would be marked as runway 16, and this runway would be in the approximate direction of south-south-east. Therefore, the east end of an east-west runway would be marked 27 (for 270° azimuth), and the west end of an east-west runway would be marked 9 (for a 90° azimuth). If there were two parallel runways in the east-west direction, e.g., these runways would be given the designation 9L-27R and 9R-27L, to indicate the direction of each runway and their position (L for left and R for right) relative to each other in the direction of aircraft operations. If a third parallel runway existed in this situation, it would be given the designation 9C-27C, to indicate its direction and position relative (C for center) to the other runways in the direction of aircraft operations. When there are more than three parallel runways, e.g., four parallel runways, one pair is marked with the magnetic azimuth to the nearest 10° while the other pair is marked with the magnetic azimuth to the next nearest 10°. Therefore, if there were four parallel runways in the east-west direction, one pair would be designated as 9L-27R and 9R-27L, and the other pair could be designated as either 10L-28R and 10R-28L or 8L-26R and 8R-26L.

Figures 13-16 and 13-17 have been reproduced from the FAA publication [17] to show the essential markings on runways. The amount of marking required on a runway depends on whether the runway is used for VFR or IFR conditions. The required markings for a visual runway are the runway designation, centerline, threshold, fixed distance markings (on runways 4000 ft or longer used by turbine-powered aircraft), and holding positions for taxiway-runway intersections. The required markings for a non-precision-instrument runway are the runway designation, centerline, threshold, fixed distance markings (on runways 4000 ft or longer used by turbine-powered aircraft), and holding positions for runway-taxiway intersections and instrument landing system (ILS) critical areas. The required markings for a precision-instrument runway are the runway designation, centerline, threshold, fixed distance markings, touchdown zone, side stripes, and holding positions for runway-taxiway intersections and ILS critical areas.

Runway designation markings consist of numbers and letters indicating the direction of the runway, as discussed earlier. These markings are white, are 60 ft tall, and, depending upon the number or letter used, vary from 5 ft wide for the numeral 1 to 23 ft wide for the numeral 7. When more than one number or letter is needed to desig-

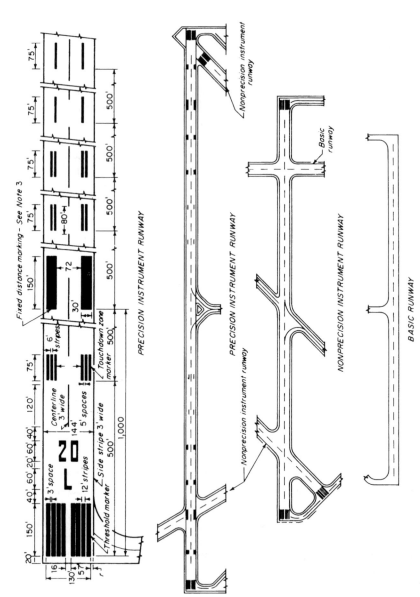

Figure 13-16 Typical runway markings (*Federal Aviation Administration*).

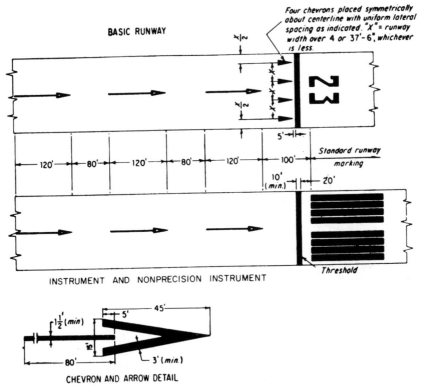

Figure 13-17 Displaced threshold marking (*Federal Aviation Administration*).

nate the runway, the spacing between the designators is normally 15 ft. The sizes of the runway designator markings are proportionally reduced only when necessary due to space limitations on narrow runways, and these designation markings should be no closer than 2 ft from the edge of the runway or the runway edge stripes.

Runway threshold markings are white; consist of two groups of four longitudinal stripes 150 ft long, 12 ft wide, spaced 3 ft apart; and are located symmetrically about the runway centerline. The center spacing between each group of four stripes is 16 ft. For runways less than 150 ft wide, the length of the stripes remains the same but the width and spacing of the stripes are proportionally changed. For runways wider than 150 ft, the width and spacing of the stripes may be proportionally increased, or additional stripes may be added.

Runway centerline markings are white, are located on the centerline of the runway, and consist of a line of uniformly spaced stripes and gaps. The stripes are 120 ft long, and the gaps are 80 ft long. Adjustments to the lengths of stripes and gaps, where necessary to accommodate runway length, are made near the runway midpoint. The minimum width of stripes is 12 in for visual runways, 18 in for non-precision-instrument runways, and 36 in for precision-instrument runways.

Runway touchdown zone markings are white and consist of groups of one, two, and three rectangular bars symmetrically arranged in pairs about the runway centerline. These markings begin 500 ft from the runway threshold. The bars are 75 ft long, 6 ft wide, with 5-ft spaces between them, and are longitudinally spaced at distances of 500 ft along the runway. The inner stripes are placed 36 ft on either side of the runway centerline. For runways less than 150 ft wide, the width and spacing of the stripes may be proportionally reduced. On shorter runways, those pairs of markings which would extend to within 900 ft of the runway midpoint are eliminated.

The fixed distance markings (sometimes called the *aiming point*) are placed on all runways 4000 ft or longer which are used by turbine-powered aircraft. The marking consists of two bold stripes 60 ft long and 30 ft wide, spaced 72 ft apart symmetrically about the runway centerline and beginning 1000 ft from the threshold, as shown in Fig. 13-16.

Runway side stripes, consisting of continuous white lines along each side of the runway, provide contrast with the surrounding terrain or delineate the edges of the full-strength pavement. The maximum distance between the outer edges of these markings is 200 ft, and these markings must be a minimum of 3 ft wide for precision-instrument runways and at least as wide as the centerline stripes on other runways.

At some airports it is desirable or necessary to "displace" the runway threshold on a permanent basis. A *displaced threshold* is one which has been moved a certain distance from the end of the runway. Most often this is necessary to clear obstructions in the flight path on landing. The displacement reduces the length of runway available for landings, but takeoffs can use the entire length of runway. The FAA requires that displaced thresholds be marked as shown in Fig. 13-17.

To prevent erosion of the soil, many airports provide a paved *blast pad* 150 to 200 ft long adjacent to the runway end. Similarly, some airport runways have a *stopway,* which is designed only to support aircraft during rare aborted takeoffs and is not designed as a full-strength pavement. Since these paved areas are not designed to support aircraft and yet may have the appearance of being so designed, markings are needed to show this. The markings for blast pads and stopways are shown in Fig. 13-18. Likewise the area adjacent to the edge of the runway may have a paved shoulder that is not capable of supporting aircraft. These areas are marked with a 3-ft-wide stripe, as shown in Fig. 13-19. Yellow is used for these types of markings.

Figure 13-18 Runway blast pad or stopway marking (*Federal Aviation Administration*).

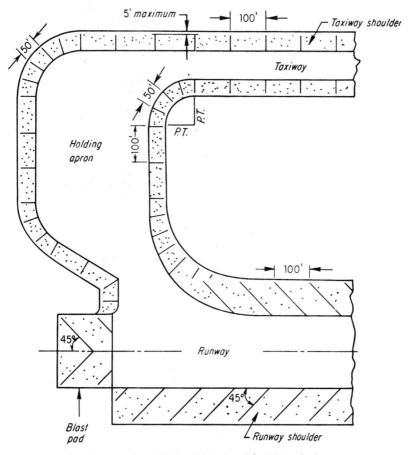

Figure 13-19 Shoulder marking (*Federal Aviation Administration*).

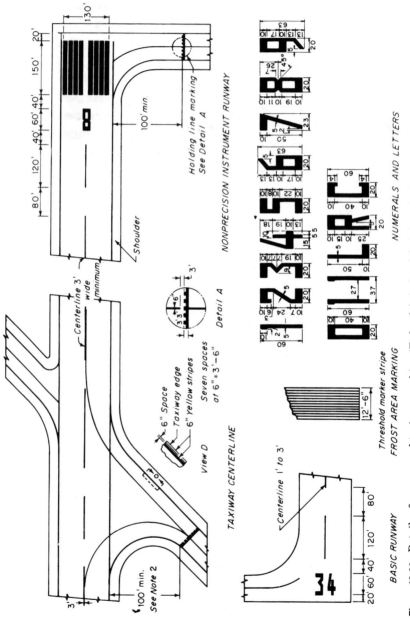

Figure 13-20 Details of runway and taxiway marking (*Federal Aviation Administration*).

Taxiways

Taxiways are marked as shown in Fig. 13-20. These are essential features of the taxiway marking system recommended by the FAA [25]:

1. The centerline of the taxiway is marked with a single, continuous 6-in yellow line.

2. On taxiway curves, the taxiway centerline marking continues from the straight portion of the taxiway at a constant distance from the outside edge of the curve.

3. At taxiway intersections which are designed for aircraft to travel straight through the intersection, the centerline markings continue straight through the intersection.

4. For taxiway intersections where there is an operational need to hold aircraft, a dashed yellow holding line is placed perpendicular to and across the centerline of both taxiways.

5. At the intersection of a taxiway with a runway end, the centerline stripe of the taxiway terminates at the edge of the runway.

6. At the intersection between a taxiway and a runway where the taxiway serves as an exit from the runway, the taxiway marking is usually extended onto the runway into the vicinity of the runway centerline marking. The taxiway centerline marking is extended parallel to the runway centerline marking a distance of 200 ft beyond the point of tangency (PT). The taxiway curve radius should be large enough to provide clearance to the taxiway edge and the runway edge of at least one-half the width of the taxiway.

7. For a taxiway crossing a runway, the taxiway centerline marking may continue across the runway, but it must be interrupted for the runway markings.

8. When the edge of the full-strength pavement of the taxiway is not readily apparent, or when a taxiway must be outlined as it is established on a large paved area such as an apron, the edge of the taxiway is marked with two continuous 6-in-wide yellow stripes 6 in apart.

9. When a taxiway intersects a runway or a taxiway enters an ILS critical area, a holding line is placed across the taxiway, as shown in Fig. 13-20. The holding line for a taxiway intersecting a runway consists of two solid lines of yellow stripes and two broken lines of yellow stripes placed perpendicular to the centerline of the taxiway and across the width of the taxiway. The solid lines

TABLE 13-3   Location Distances for Holding-Position Markings

Aircraft approach category	Airplane design group	Perpendicular distance from runway centerline to intersecting runway-taxiway centerline, ft	
		Visual non-precision-instrument	Precision-instrument
A, B	I, II Small airplanes only	125	175
	I, II, III	200*	250*
	IV	250	250†
C, D	I, II, III, IV	250‡	250†,‡
	V	250‡	280‡

*For airplane design group III only, increase by 1 ft for each 100 ft above 5100 ft above mean sea level.

†For airplane design group IV only, increase by 1 ft for each 100 ft above mean sea level.

‡For aircraft approach category D only, increase by 1 ft for each 100 ft above mean sea level.

SOURCE: Federal Aviation Administration [25].

are always placed on the side on which the aircraft is to hold. The holding line for an ILS critical area consists of two solid lines placed perpendicular to the taxiway centerline and across the width of the taxiway joined with three sets of two solid lines symmetric about and parallel to the taxiway centerline. These holding lines are located the minimum distance from the centerline of the runway as indicated in Table 13-3.

10. When the taxiway shoulder has the same appearance as the load-bearing pavement, it is marked as shown in Fig. 13-19.

Closed runways and taxiways

When runways or taxiways are permanently or temporarily closed to aircraft, yellow crosses are placed on these trafficways. For permanently closed runways, the threshold, runway designation, and touchdown markings are obliterated, and crosses are placed at each end and at 1000-ft intervals. For temporarily closed runways, the runway markings are not obliterated, and the crosses are usually a temporary type and are placed only at the runway ends. For permanently closed taxiways, a cross is placed on the closed taxiway at each entrance to the taxiway. For temporarily closed taxiways, barricades with orange and white markings are normally erected at the entrances.

Surface guidance and control

To assist air traffic control and pilots in maintaining the safe movement of aircraft on the ground in poor-visibility conditions and dense

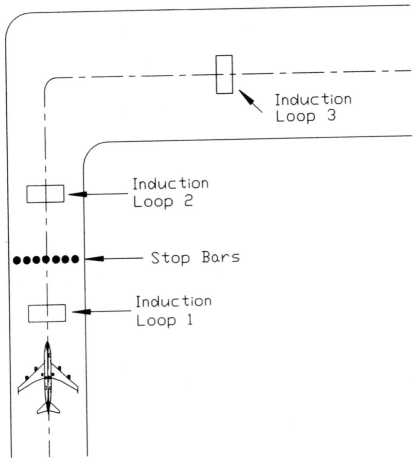

Figure 13-21 Typical induction loop stop bar system (*International Civil Aviation Organization*).

traffic situations, a system of surface movement control devices is being implemented at many airports. The most common system is a system of stop bars placed across taxiways to supplement taxiway holding position and intersection markings [2]. This type of system is very helpful in preventing runway incursions.

The stop bars are controlled in the air traffic control tower. An example of a stop bar system is shown in Fig. 13-21. The stop bars are placed across the taxiway at the point where aircraft are to stop. The stop bars typically consist of semiflush lights showing the color red in the intended direction of travel, and they are spaced at 10-ft intervals across the taxiway. In some cases, the operation of the stop bars is controlled by induction loop sensors embedded in the pavement. These sensors consist of three induction loops: a loop on the taxiway to detect the presence of an aircraft approaching the stop bar, a loop beyond the stop bar to indicate that an aircraft has passed the stop bar, and a loop on the runway pavement beyond the threshold to indicate the presence of an aircraft. The taxiway centerline lights remain lighted only up to the first loop. When clearance is given for the aircraft to proceed beyond the stop bar, the stop bar lights are switched off by the air traffic controller. This also illuminates the taxiway centerline lights beyond the stop bar. Once the aircraft passes over the second loop, the stop bar is automatically activated again. When the aircraft passes the third loop, the portion of the taxiway centerline lights between the stop bar and the runway is automatically turned off.

Runway Distance-Remaining Signs

Recently, runway distance-remaining signs, sometimes called "distance-to-go" signs, have been installed at many airports. These signs are placed on the side of a runway and provide the pilot with information on how much runway is left during takeoff or landing operations. These signs are placed at 1000-ft intervals along the runway in a descending sequential order. Normally, these signs consist of white numerals on a black background.

The FAA recommends that the signs be configured in one of three ways [25]. The preferred method of configuration, and the most economical, is to place double-faced signs on only one side of the runway. In this configuration the signs should be placed on the left side of the most frequently used direction of the runway. The signs may be placed on the right side of the runway when necessary due to required runway-taxiway separations or due to conflicts between intersecting runways or taxiways. The second method is to provide a set of single-faced signs on either side of the runway to indicate the distance remaining when the runway is used in both directions. The advantage of this configuration is that the distance remaining is more accurately reflected when the runway length is not an even multiple of 1000 ft. The third method uses double-faced signs on both sides of the runway. The advantage of this method is that the runway dis-

TABLE 13-4 Sign Sizes and Location Distance for Distance-Remaining Signs

FAA sign size	Maximum sign height, in			Distance from pavement edge, ft
	Legend	Face	Installed	
4	40	48	60	50–75
5	25	30	42	20–35

SOURCE: Federal Aviation Administration [25].

tance is displayed on both sides of the runway in each direction, which is helpful when a sign on one side needs to be omitted because of a clearance conflict. In the preferred configuration and the second configuration, when the runway distance is not an even multiple of 1000 ft, one-half of the excess distance is added to the distance on each sign on each runway end. For example, if the runway length available is 8250 ft, the last sign is located 1000 ft plus 125 ft from the end of the runway. A tolerance of $\pm$ 50 ft is allowed for the placement of runway distance-remaining signs. These signs should be illuminated anytime the runway edge lights are illuminated. The recommended sizes and placement of these signs are given in Table 13-4.

Taxiway Guidance Sign System

The primary purpose of a taxiway guidance sign system is to aid pilots in taxiing aircraft at an airport. At controlled airports, the signs supplement the instructions of the air traffic controllers and aid the pilot in complying with those instructions. The sign system also substantially aids the air traffic controller by simplifying instructions for taxiing clearances and for the routing and holding of aircraft. At locations not served by air traffic control towers, or for aircraft without radio, the sign system provides guidance to the pilot to major destination areas in the airport.

The taxiway guidance sign system consists of four basic types of signs: *mandatory instruction* signs, which indicate that aircraft should not proceed beyond a point without positive clearance; *location* signs, which indicate the location of an aircraft on the taxiway or runway system and the boundaries of critical airfield surfaces; *direction* signs, which identify the paths available to aircraft at intersections; and *destination* signs, which indicate the direction to a particular destination.

The efficient and safe movement of aircraft on the surface of an airport requires that a well-designed, properly thought-out, and standardized taxiway guidance sign system be provided at the airport. The system must provide the pilot with the capacity to readily determine the designation of any taxiway on which the aircraft is located, readily identify routings to a desired destination on the airport property, indicate mandatory aircraft holding positions, and identify the boundaries for aircraft approach areas, ILS critical areas, runway safety areas, and obstacle-free zones. It is virtually impossible, except for holding-position signs, to completely specify the locations and types of signs required on a taxiway system at a particular airport owing to the wide variation in the types of functional layouts for airports. The ICAO publishes recommendations concerning surface movement guidance and control systems [2, 15]. The FAA recommends that the following guidelines be followed in designing a taxiway guidance sign system [25]:

1. A holding-position sign and taxiway location sign should be installed at the holding position on any taxiway that provides access to a runway.

2. A holding-position sign should be installed on any taxiway at the boundary of the ILS critical area or the runway approach area when it is necessary to protect the navigational signal, airspace, or safety area for a runway. This sign should be placed at the entrance to and the exit from such areas.

3. A holding-position sign should be installed on any runway where that runway intersects another runway.

4. A sign array consisting of taxiway direction signs should be installed prior to each intersection between taxiways if an aircraft would normally be expected to turn at or hold short of the intersection. The direction sign in the array should include a sign panel, consisting of a taxiway designation and an arrow, for each taxiway that an aircraft would be expected to turn onto or hold short of. A taxiway location sign should be included as part of the sign array unless it is determined to be unnecessary. If an aircraft normally would not be expected to turn at or hold short of the intersection, the sign array is not needed unless the absence of guidance would cause confusion.

5. A runway exit sign identifying the exit taxiway should be installed along each runway for each normally used runway exit.

6. Destination signs may be substituted for direction signs at the intersection between taxiways or for runway exit signs at uncontrolled airports.

7. Standard highway stop signs should be installed on ground vehicle roadways at the intersection of each roadway with a runway or taxiway. For roadway intersections with taxiways, a standard highway yield sign may be used instead of the stop sign.

8. Additional signs should be installed on the airfield where necessary to eliminate confusion or to provide confirmation of location.

Developing taxiway designations

The first step in designing a proper taxiway guidance sign system is the development of a simple and logical method for designating taxiways. The FAA recommends that the following guidelines be followed [25]:

1. Keep it simple and logical.

2. Letters of the alphabet should only be used for designating taxiways. Optimally, the designation of taxiways should start at one end

of the airport and continue to the opposite end of the airport in a sequential manner. When there are more taxiways than letters of the alphabet, double-letter designations may be used. Numbers alone should never be used for taxiway designations since they can be confused with runway designations. The letters I and O should not be used since they could be confused with a runway number designation. The letter X should not be used for a taxiway designation since it might be confused with a designation for a closed taxiway or runway.

3. For a major taxiway with numerous stub exits, the short segments of the taxiway may be designated by the taxiway letter and an exit number, where the number is sequential along the taxiway route.

4. All separate and distinct taxiway segments should be designated. No separate or distinct taxiway segment should have the same designation as another taxiway.

5. Taxiway designations should not be changed when there is no significant change in the direction or continuity of the taxiway. If a change is found necessary, the change should be made only at a taxiway intersection and should be appropriately signed.

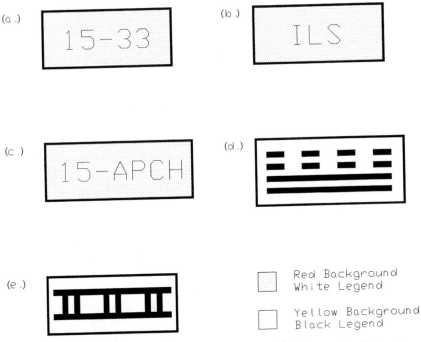

Figure 13-22 Mandatory instruction signs and boundary area signs: (*a*) Runway safety area holding-position sign; (*b*) ILS holding-position sign; (*c*) runway approach area holding-position sign; (*d*) runway safety area, runway obstacle-free zone, and runway approach area boundary location sign; (*e*) ILS critical area boundary location sign.

6. Taxiways should not be designated by direction of travel or location. Designations such as *inner, outer,* and *parallel* should be avoided.

Types of taxiway signs

Mandatory instruction signs. Mandatory instruction signs denote an entrance to a runway, critical area, or prohibited area. They are used for holding-position signs for runway-taxiway and runway-runway intersections (Fig. 13-22a), ILS critical areas (Fig. 13-22b), and runway approach areas (Fig. 13-22c) and to designate areas where entry by aircraft is prohibited. These signs have white inscriptions on a red background and are installed on the left side of the runway or taxiway. In some cases, runway-taxiway intersections require a sign on both sides of the taxiway. This includes situations on taxiways which are at least 150 ft wide, where the painted holding line extends across an adjacent holding bay, where the painted holding line markings do not extend straight across the taxiway, and where the painted holding line markings are located a short distance from an intersection with another taxiway. Generally arrows are not permitted on mandatory instruction signs unless arrows are necessary at the taxiway-runway-runway intersections to indicate directions to these runways. For runway designation use, these signs normally contain both designations of the runway; the designation on the left is for the runway to the left, and that on the right is for the runway to the right.

At controlled airports, aircraft and ground vehicles are required to hold at these points unless cleared by air traffic control. At uncontrolled airports, these signs are intended to be informational, and progression beyond these signs is permitted only after appropriate precautions have been taken.

Location signs. Location signs are used to identify the runway or taxiway on which an aircraft is located (Figs. 13-23a and 13-23b). These signs consist of a yellow inscription and border on a black background. Location signs are also used to identify the boundary of the runway safety area or obstacle-free zones (Fig. 13-22d), or the ILS critical area (Fig. 13-22e) for a pilot exiting a runway. In the latter cases, the signs consist of a black inscription and border on a yellow background, and the inscription on the sign is the same as relevant holding line marking.

Direction signs. Direction signs are used to indicate the directions of other taxiways leading out of an intersection. These signs are used as taxiway direction signs (Fig. 13-23c) and runway exit signs. The signs have black inscriptions and borders on a yellow background and

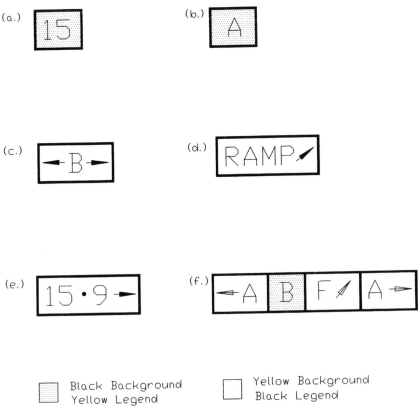

(a.)

(b.)

(c.)

(d.)

(e.)

(f.)

Black Background
Yellow Legend

Yellow Background
Black Legend

Figure 13-23 Runway and taxiway location signs, destination signs, and taxiway sign array: (*a*) Runway location sign; (*b*) taxiway location sign; (*c*) taxiway direction sign; (*d*) inbound destination sign; (*e*) outbound destination sign; (*f*) taxiway sign array consisting of three taxiway direction signs and one taxiway location sign.

always contain arrows. The arrows are oriented in the approximate direction of the turn required. These signs should not be collocated with holding-position signs and should not be located between the holding line and the runway. Signs used to indicate the direction of taxiways on the opposite side of the runway should be located on the opposite side of the runway. Runway exit signs should be located prior to the exit on the side of the runway on which the aircraft is expected to exit. If the taxiway crosses the runway and the aircraft could be expected to exit on either side, then a runway exit sign should be installed on both sides of the runway.

Destination signs. Destination signs have black inscriptions and black borders on a yellow background and always contain arrows. These signs indicate the general direction to a remote location at the

TABLE 13-5   Typical Legends of Taxiway Destination Signs

General parking, servicing, and loading areas	RAMP or RMP
Areas specifically set aside for aircraft parking	PARK
Areas where aircraft are fueled or serviced	FUEL
Gate position at which aircraft are loaded or unloaded	GATE
Areas set aside for itinerant aircraft	VSTR
Areas set aside for military aircraft	MIL
Areas set aside for freight or cargo handling	CRGO
Areas set aside for handling international flights	INTL
Hangar or hangar area	HGR
ILS critical area	ILS

airport, such as an inbound destination (Fig. 13-23*d*), and they are generally not required where taxiway direction signs are used. Outbound destination signs are used to identify directions to the takeoff runways. These routes normally begin at the entrance to a taxiway from the apron area. More than one runway number may be used, separated by a dot, if the route is common to more than one runway (Fig. 13-23*e*). Inbound destination signs are often used to indicate the general direction to major airport facilities such as passenger terminal aprons, cargo areas, military aprons, or general aviation facilities. These signs should consist of a minimum of three letters to avoid confusion with taxiway guidance signs. The typical legends found on taxiway destination signs are given in Table 13-5.

Information signs. Other types of signs may be needed on the airfield which are not part of the taxiway guidance systems described above. These signs are called *information signs,* and they might be used, e.g., to indicate a noise abatement procedure to a pilot ready to take off on a specific runway. These signs should have black inscriptions on a yellow background, and they are not required to be lighted.

Taxiway ending sign. The sign system does not provide a sign to indicate that a taxiway does not continue beyond an intersection. A frangible, retroreflective sign should be installed on the far side of the intersection if normal visual cues such as marking and lighting are inadequate.

Signing conventions. The FAA recommends the following signing conventions [25]:

1. Signs should be placed on the left side of the taxiway as viewed by the pilot of an approaching aircraft. If signs are installed on both sides of the taxiway at the same location, the sign faces should be identical. Signs may be placed on the right side of the taxiway when necessary to meet clearance requirements or where it is

impractical to install them on the left side because of terrain features or conflicts with other objects.

2. Some signs may be installed on the back of other signs, even though this may result in the sign being on the right side of the taxiway. Signs which may be installed in this manner include the following:

   a. Runway safety area, obstacle-free zone area, and runway approach area boundary signs may be installed on the back of taxiway-runway intersection and runway approach area holding-position signs.

   b. ILS critical-area boundary signs may be installed on the back of ILS critical-area holding-position signs.

   c. Taxiway location signs, when installed on the far side of the intersection, may be installed on the back of direction signs.

   d. Taxiway location signs may be installed on the back of holding-position signs.

   e. Destination signs may be installed on the back of direction signs on the far side of intersections when the destination referred to is straight ahead.

3. Taxiway location signs installed in conjunction with holding-position signs for taxiway-runway intersections should always be installed outboard of the holding-position sign.

4. Location signs are normally included as part of a direction sign array located prior to the taxiway intersection. Except for the intersection of two taxiways, the location sign is placed in the array so that the designations for all turns to the left would be located to the left side of the sign and designations for all turns to the right or straight ahead to the right of the location sign (Fig. 13-23f).

5. All direction signs have arrows. Arrows on signs should be oriented toward the approximate direction of the turn. Each designation appearing in the array of direction signs should have only one arrow, except when the taxiway intersection comprises only two taxiways, the direction sign for the taxiway may have two arrows.

6. Destination signs should be located in advance of intersections and should not be collocated with other signs. These may also be

TABLE 13-6   Location Distances and Sizes for Taxiway Guidance Signs

FAA sign size, ft	Sign height, in			Perpendicular distance from taxiway runway edge to sign, ft
	Legend	Face	Installed	
1	12	18	30	10–20
2	15	24	36	20–35
3	18	30	42	36–60

SOURCE: Federal Aviation Administration [25].

installed on the far side of the intersection when the taxiway does not continue and direction signs are provided prior to the intersection.

7. Information signs should not be collocated with mandatory, location, direction, or destination signs.

8. Each designation and its associated arrow included in the array of direction signs or destination signs should be delineated from the other designations in the array by a black vertical border.

Sign size and location

Taxiway guidance signs are available in three heights, as indicated in Table 13-6. The choice of a particular size sign involves several factors including effectiveness, aircraft clearance, jet blast, and snow removal operations. Normally, the larger the sign and the closer it is located to the runway or taxiway edge, the more effective it is. However, aircraft clearance requirements and jet blast effects necessitate smaller signs when they are located near the pavement edges, whereas effectiveness necessitates larger signs when they are located at farther distances. The effects of snow removal operations on the signs should be considered in the choice of sign size and location. The sign must provide 12 in of clearance between the top of the sign and any part of the most critical aircraft using or expected to use the airport when the wheels of the aircraft are at the defined pavement edge. The distances shown in Table 13-3 should be used in determining runway holding positions. All signs in an array should be the same size and at the same height above the ground. In determining sign locations with respect to intersecting runways, the clearance requirements to other moving aircraft, as given in Table 13-7, should be used. For signs installed at holding positions, the signs should be in line with the holding line markings within a tolerance of 10 ft.

TABLE 13-7   Distances for Taxiway Intersection Markings
from Centerline of Crossing Taxiway

Airplane design group	Distance, ft
I	44.5
II	65.5
III	93.0
IV	129.5
V	160.0
VI	193.0

SOURCE: Federal Aviation Administration [25].

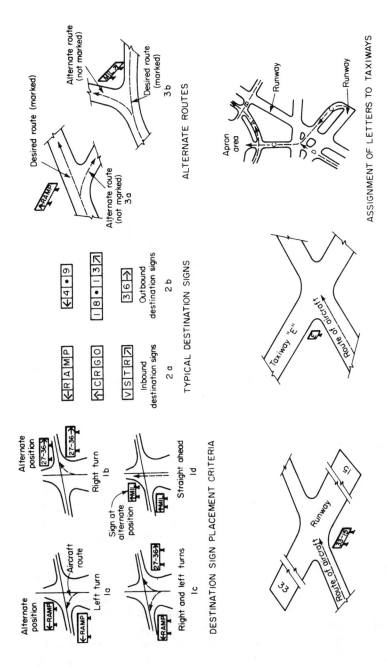

ALTERNATE ROUTES

ASSIGNMENT OF LETTERS TO TAXIWAYS

TYPICAL DESTINATION SIGNS

TYPICAL TAXIWAY INTERSECTION SIGN

DESTINATION SIGN PLACEMENT CRITERIA

TYPICAL RUNWAY INTERSECTION SIGN

Figure 13-24 Taxiway sign locations.

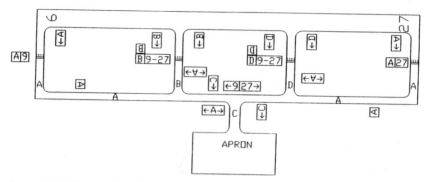

Figure 13-25 Illustration of sign requirements and placement at an airport (*Federal Aviation Administration*).

Where there is no operational need for taxiway holding line markings, the signs may be installed in the area from the taxiway point of tangency to the location where the holding line markings would be installed [25]. Typical locations for taxiway guidance signs are shown in Fig. 13-24. An illustration of the required signs and their placement for a basic airport layout is given in Fig. 13-25 [25].

Sign operation

Holding-position signs for runways, ILS critical areas, approach areas, and their associated taxiway location signs should be illuminated when the associated runway lights are illuminated. Other taxiway signs should be illuminated when the associated taxiway lights are illuminated.

Retroreflective Marking of Runway and Taxiway Centerlines

Installation of retroreflective markings is not mandatory. However, it is quite economical, especially at airports where lights cannot be justified because of the volume or nature of air traffic [12]. The marking is very similar to that used successfully on highways for many years.

Related Visual Aids

Beacons are lighted to mark an airport. They are designed to produce a narrow horizontal and vertical beam of high-intensity light which is rotated about a vertical axis so as to produce approximately 12 flashes per minute for civil airports and 18 flashes per minute for military airports [28]. The flashes with a clearly visible duration of at least

0.15 s are arranged in a white-green sequence for land airports and a white-yellow sequence for landing areas on water. Military airports use a double white flash followed by a longer green or yellow flash to differentiate them from civil airfields. The beacons are mounted on top of the control tower or similar high structure in the immediate vicinity of the airport.

Obstructions are identified by fixed, flashing, or rotating red lights or beacons. All structures that constitute a hazard to aircraft in flight or during landing or takeoff are marked by obstruction lights having a horizontally uniform intensity duration and a vertical distribution design to give maximum range at the lower angles (1.5° to 8°) from which a colliding approach would most likely come. The criteria for determining which structures need to be lighted are published by the FAA [18, 19].

Visibility

Throughout this chapter and elsewhere in the text, reference has been made to poor-visibility conditions; categories I, II, and III; and runway visual range (RVR). However, these terms were never defined. Various definitions of visibility are used in aviation, and they are all briefly described here.

Meteorological visibility and ceiling

Meteorological visibility is defined differently depending on whether it is day or night. At night it is the distance from which a human can see a 25-cd light. During the day it is the distance at which a black circular target subtended by a visual angle of 1° can be seen. Meteorological visibility is also usually associated with the height of the underside of a dense cloud above the airport surface. This height is referred to as the *ceiling*. Thus the term 300-½ means a 300-ft ceiling and ½-mi visibility.

Runway visual range

While various definitions of visibility have been used by meteorologists, visibility on a runway is being reported more and more in terms of the RVR [22]. The RVR is the distance over which the high-intensity runway edge lights can be seen by pilots. The RVR is determined by transmissometer measurements at the side of the runway near the threshold, and the calibration is correlated with the intensity setting of the edge lights and the time of day or night.

The transmissometer required to measure RVR consists of a light source with a narrow-beam projector and a receiver having a narrow acceptance angle. The distance between the units is usually 500 ft. The baseline between the two units should be parallel with the runway, and the light source and the receiver should be about 15 ft above the ground.

For category II and category III operations, two transmissometer measurements are made, one near the threshold and the other about midway down the length of the runway. The latter location provides useful information for aircraft taking off as well as for landing aircraft during landing roll. The limitation of the transmissometer is that it measures visibility on the ground, whereas pilots approaching a runway like to know the visibility on the approach path.

Visibility categories

For landing operations primarily, visibilities on the runway have been classified into three categories, namely, I, II, and III. Category III is further subdivided into a, b, and c. The three groups are defined in terms of the RVR and decision height. The *decision height* is the lowest height above the runway where pilots make the decision to continue the landing or to abort. It is based on the ability of pilots to obtain guidance from visual cues on the ground rather than from instruments in the cockpit. If pilots are unable to see a sufficient number of visual cues at the decision height, the landing must abort.

The visual categories are specified in Table 13-8. For category IIIa operations, an automatic landing capability, i.e., the aircraft touching down on the runway, is required. For category IIIb operations, an automatic rollout capability in addition to an automatic landing capability is required. For Category IIIc operations, the taxiing portion of the landing must also be automatic.

TABLE 13-8   RVR Minima and Decision Heights

ILS category	Lowest minima, ft	
	RVR	Decision height
Non-precision	2400	250
Precision, category I	1800	200
Precision, category II	1200	100
Precision, category IIIa	700	0
Precision, category IIIb	150	0
Precision, category IIIc	0	0

SOURCE: Federal Aviation Administration.

References

1. *Aerodromes, Annex 14 to the Convention on International Civil Aviation,* vol. 1: *Aerodrome Design and Operations,* 1st ed., International Civil Aviation Organization, Montreal, Canada, July 1990.
2. *Aerodrome Design Manual,* pt. 4: *Visual Aids,* 2d ed., International Civil Aviation Organization, Montreal, Canada, 1983.
3. "Airport Approach, Runway and Taxiway Lighting Systems," E. C. Walter, *Journal of the Air Transport Division,* vol. 84, no. AT1, American Society of Civil Engineers, New York, June 1958.
4. *Airport Design,* Advisory Circular AC 150/5300-13, Federal Aviation Administration, Washington, 1989.
5. *Airport Miscellaneous Lighting Visual Aids,* Advisory Circular AC 150/5340-21, Federal Aviation Administration, Washington, 1971.
6. *Airway Planning Standard Number One—Terminal Air Navigation Facilities and Air Traffic Control Services,* Order 7031.2B, Federal Aviation Administration, Washington, 1976.
7. "Aviation Ground Lighting for All-Weather Operation," M. Latin, *Airport Forum,* vol. 7, no. 1, February 1977.
8. *Comparison Between ICAO Annex 14 Standards and Recommended Practices and FAA Advisory Circulars,* Doc. D6-58344, Boeing Commercial Airplane Company, Seattle, 1979.
9. *Economy Approach Lighting Aids,* Advisory Circular AC 150/5340-14B with Changes 1 and 2, Federal Aviation Administration, Washington, 1970.
10. *Establishment Criteria for Runway End Identification Lights (REIL),* Rep. FAA-ASP-79-4, Federal Aviation Administration, Washington, 1979.
11. *Establishment Criteria for Visual Approach Slope Indicator (VASI),* Rep. FAA-ASP-76-2, Federal Aviation Administration, Washington, 1977.
12. *FAA Specification L-853, Runway and Taxiway Retroreflective Markers,* Advisory Circular AC 150/5345-39B, Federal Aviation Administration, Washington, 1980.
13. *Installation Criteria for the Approach Lighting System Improvement Program (ALSIP),* Rep. FAA-ASP-78-5, Federal Aviation Administration, Washington, 1978.
14. *Installation Details for Runway Centerline and Touchdown Zone Lighting Systems,* Advisory Circular AC 150/5340-4C with Changes 1 and 2, Federal Aviation Administration, Washington, 1978.
15. *Manual of Surface Movement Guidance and Control Systems,* 1st ed., Doc. 9476, International Civil Aviation Organization, Montreal, Canada, 1986.
16. *Marking and Lighting of Unpaved Runways,* V. F. Dosch, NAFEC Tech. Letter, Rep. NA-78-34-LR, Federal Aviation Administration, Technical Center, Atlantic City, N.J., 1978.
17. *Marking of Paved Areas on Airports,* Advisory Circular AC 150/5340-1F, Federal Aviation Administration, Washington, 1987.
18. *Obstruction Marking and Lighting,* Advisory Circular AC 70/7460-1H, Federal Aviation Administration, Washington, 1991.
19. *Proposed Construction or Alteration of Objects That May Affect Navigable Airspace,* Advisory Circular AC 70/7460-2I, Federal Aviation Administration, Washington, 1988.
20. *Precision Approach Path Indicator (PAPI) Systems,* Advisory Circular AC 150/5345-28D, Federal Aviation Administration, Washington, 1985.
21. *Runway and Taxiway Edge Lighting System, Advisory Circular AC 150/5340-24, Federal Aviation Administration, Washington,* 1975.
22. Runway Visual Range (RVR), Advisory Circular AC 97-1A, Federal Aviation Administration, Washington, 1977.
23. *Segmented Circle Airport Marker System,* Advisory Circular AC 150/5340-5B Federal Aviation Administration, Washington, 1984.
24. *Specifications for Taxiway and Runway Signs,* Advisory Circular AC 150/5345-44E, Federal Aviation Administration, Washington, 1991.
25. *Standards for Airport Sign Systems,* Advisory Circular AC 150/5340-18C, Federal Aviation Administration, Washington, 1991.

26. *Taxiway Centerline Lighting System,* Advisory Circular AC 150/5340-19, Federal Aviation Administration, Washington, 1968.
27. "The Theory of Visual Judgements in Motion and Its Application to the Design of Landing Aids for Aircraft," E. S. Calvert, *Transactions of the Illuminating Engineering Society,* vol. 22, no. 10, London, 1957.
28. "Working Papers," A. E. Jenks, *International Air Transport Association,* Special Meeting on Visual Aids to Flare and Landing, Amsterdam, Netherlands, Nov. 14–22, 1955.

14

Airport Drainage

An adequate drainage system for the removal of surface and subsurface water is vital for the safety of aircraft and for the longevity of the pavements. Improper drainage results in the formation of puddles on the pavement surface, which can be hazardous to aircraft taking off and landing. Poor drainage can also result in the early deterioration of pavements. Flat longitudinal and transverse grades and wide pavement surfaces often pose difficulties in making provision for adequate drainage at airports.

The material in this chapter is principally concerned with estimating the amounts of surface and subsurface runoff and not with the hydraulics of pipes or details of installation. These latter items are adequately covered in texts on hydraulics and literature provided by pipe manufacturers.

The FAA and the Corps of Engineers have developed most of the information on airport drainage in the United States. We have drawn freely from the publications of these two agencies.

Purpose of Drainage

The functions of an airport drainage system are as follows:

1. Interception and diversion of surface and groundwater flow originating from lands adjacent to the airport

2. Removal of surface runoff from the airport

3. Removal of subsurface flow from the airport

In very few cases will the natural drainage on a site be sufficient by itself to satisfy these functions; consequently artificial drainage must be installed.

Design Storm for Surface Runoff

The selection of the severity of the storm which the drainage system should accommodate involves economic consideration. An extremely severe storm occurring very infrequently would undoubtedly cause some damage if the system were designed for a storm of lesser severity. However, if serious interruptions in traffic are not anticipated, a system designed for the larger storm may not be economically justified. Taking these factors into account, the FAA recommends that for civil airports the drainage system be designed for a storm whose probability of occurrence is once in 5 years [1]. The design should, however, be checked with a storm of lesser frequency (10 to 15 years) to ascertain if serious damage or interruption of traffic would result from such a storm. Drainage for military airfields is based on a 2-year storm frequency [7]

Ordinarily no ponding is permitted on paved surfaces, but in the intervening areas ponding is permitted, provided it will not result in undesirable saturation of the subgrades underneath the pavements.

Determining the Intensity-Duration Pattern
for the Design Storm

The determination of the amount of rainfall which can be expected at the site of the airport is the first step in the design of a drainage system. Rainfall intensity is expressed in inches per hour for various durations of a particular storm. The expected frequency of occurrence is also an important factor to consider. The severity of storms is related to their frequencies; a storm which is expected to occur once in 100 years will be more severe than one having a frequency of occurrence of once in 5 years.

David L. Yarnell of the U.S. Department of Agriculture conducted extensive investigations concerning rainfall intensities, durations, and frequencies throughout the United States [15]. West of the 105th meridian, where the Yarnell information is not as complete, the National Weather Service has compiled rainfall data which appear in Refs. 12 to 14.

Yarnell developed rainfall intensities for 5-, 10-, 15-, 30-, 60-, and 120-min durations for a storm which can be expected to occur once in 5 years and the intensities for a 1-h duration for storms whose expected frequencies of occurrence are once in 2, 5, 10, 25, 50, and 100 years. The intensities for a duration of 1 h for frequencies of 2, 5, 10, 25, 50, and 100 years are shown in Fig. 14-1.

The 1-h intensity does not by itself portray the intensity-duration pattern of a storm. The Corps of Engineers made extensive studies of rainfall patterns in the United States and found that irrespective of

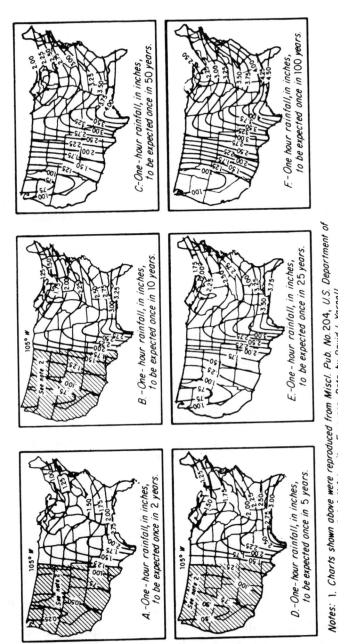

Notes: 1. Charts shown above were reproduced from Miscl. Pub. No. 204, U.S. Department of Agriculture, Rainfall Intensity-Frequency Data by David L.Yarnell.

2. For the Western part of the U.S. West of the 105th Meridian, see One Hour Rainfall Data in Weather Bureau Technical Paper No. 24.

Figure 14-1 One-hour rainfall intensities for the United States (Corps of Engineers).

frequency, the intensity-duration patterns of storms were largely governed by their 1-h intensities. That is, two storms of different frequency of occurrence whose 1-h intensities are equal will have similar intensity-duration patterns. This is shown in Fig. 14-2. For example, if the 1-h intensities of storms whose frequencies were 5, 10, or 15 years were all exactly 2.0 in/h, the intensity-duration patterns would be expected to follow the pattern indicated by the curve labeled 2.0.

If the drainage system is to be designed for a storm whose expected frequency of occurrence is once in 5 years and if detailed data concerning the intensity-duration pattern are nonexistent, the pattern can be approximated from Fig. 14-2, provided the 1-h intensity is known. It goes without saying that if sufficient rainfall data are available at an airport site, the intensity-duration frequency data should be developed from this information rather than from other sources. Rarely, however, does a site have such complete rainfall information.

Determining the Amount of Runoff by the FAA Procedure

The FAA analysis of airport surface drainage revolves about the solution of the *rational method* expression

$$Q = CIA \qquad (14\text{-}1)$$

where Q = runoff from given drainage basin, ft³/s
  C = ratio of runoff to rainfall
  I = rainfall intensity for time of concentration of runoff, in/h
  A = drainage area, acres

Examples and charts illustrating the FAA procedure for design have been taken largely from the FAA [1].

Time of concentration

The *time of concentration* is defined as the time taken by water to reach the drain inlet from the most remote point in the tributary area. The *most remote point* refers to the point from which the time of flow is the greatest. The time of concentration is usually divided into two components: inlet time and time of flow. The *inlet time* is the time required for water to flow overland from the most remote point in the drainage area to the inlet. The *time of flow* is the time taken by the water to flow from the drain inlet through the pipes to the point in the system under consideration. Sometimes the inlet time will be the time of concentration; at other times the time of concentration will be the sum of the inlet time and time of flow.

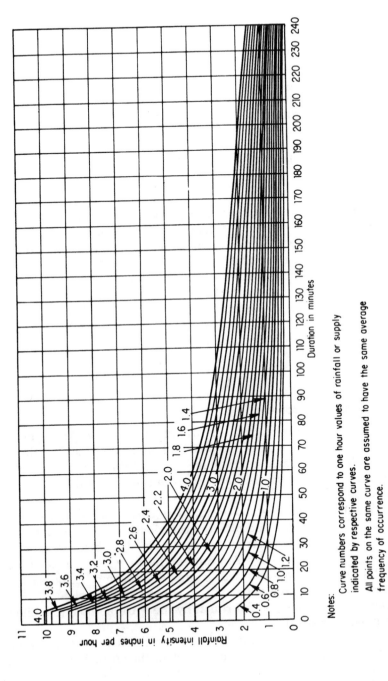

Notes:

Curve numbers correspond to one hour values of rainfall or supply indicated by respective curves.

All points on the same curve are assumed to have the same average frequency of occurrence.

Figure 14-2 Rainfall intensity-duration curves (*Corps of Engineers*).

The time of flow can be computed by the use of well-established hydraulic formulas. The inlet time is obtained largely empirically from the relationship

$$D = kT^2 \qquad (14\text{-}2)$$

Where D = distance, ft
T = time, min
k = dimensional empirical factor which is dependent on slope, roughness of terrain, extent of vegetative cover, and distance to drain inlet

Inlet times can be estimated from Fig. 14-3.

Coefficient of runoff

Application of the rational method requires the exercise of considerable judgment on the part of the engineer. The runoff rate is variable from storm to storm and varies even during a single period of precipitation. The coefficient of runoff depends on antecedent storm conditions, slope and type of surface, and extent of the drainage area. The range of values suggested by the FAA is indicated in Table 14-1.

For drainage basins consisting of several types of surfaces with different infiltration characteristics, the weighted runoff coefficient should be computed in accordance with

$$C = \frac{A_1 C_1 + A_2 C_2 + A_3 C_3}{A_1 + A_2 + A_3} \qquad (14\text{-}3)$$

Typical example—No ponding

In order that the worst conditions attendant upon the design storm may be used in the design of the pipe system, a separate duration of storm is selected for each subdrainage area tributary to a drain inlet. The duration of the storm is made equal to the sum of the inlet time and time of flow.

Each reach of pipe must be designed to carry the discharge from the inlet at its upstream end plus the contribution from all preceding inlets. For economy of construction, the grade of each reach is determined largely by topography. A minimum mean velocity on the order of 2.5 ft/s should be maintained to provide scouring action so that reduction of the pipe area due to silting will not be a problem.

To clarify the computation of runoff by the FAA method, the following example is presented.

The intensity-duration rainfall pattern for a 5-year-frequency storm at the site of the proposed airport is shown in Fig. 14-4. The layout of the drains on a portion of the airport is shown in Fig. 14-5. Design data for establishing inlet times and coefficients of runoff for

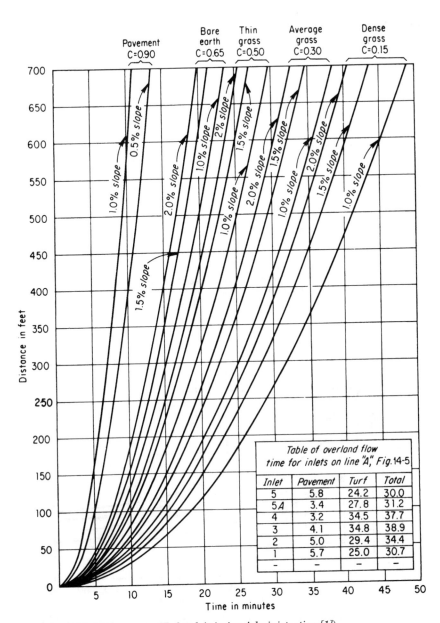

Figure 14-3   Inlet time curves (*Federal Aviation Administration [1]*).

drainage areas tributary to drain line *A,* shown in Fig. 14-5, are tabulated in Table 14-2. It is assumed that the coefficients of runoff for pavement and for turf are 0.90 and 0.30, respectively.

From these data, inlet times have been computed by the use of Fig. 14-3 on the basis that the slope of the pavement is 1 percent and the

TABLE 14-1   Coefficients of Runoff C

Type of surface	Factor C
For all watertight roof surfaces	0.75–0.95
For asphalt runway pavements	0.80–0.95
For concrete runway pavements	0.70–0.90
For gravel or macadam pavements	0.35–0.70
For impervious soils (heavy)*	0.40–0.65
For impervious soils with turf*	0.30–0.55
For slightly pervious soils*	0.15–0.40
For slightly pervious soils with turf*	0.10–0.30
For moderately pervious soils*	0.05–0.20
For moderately pervious soils with turf*	0.00–0.10

*For slopes from 1 to 2 percent.

SOURCE: Federal Aviation Administration [1].

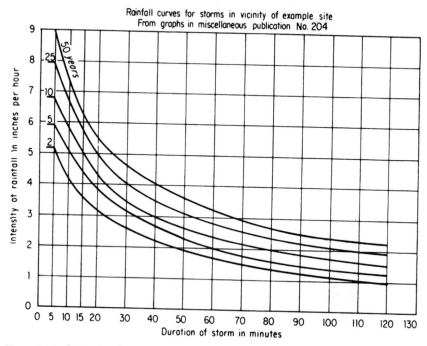

Figure 14-4 Intensity-duration rainfall pattern for design storm (*Federal Aviation Administration [1]*).

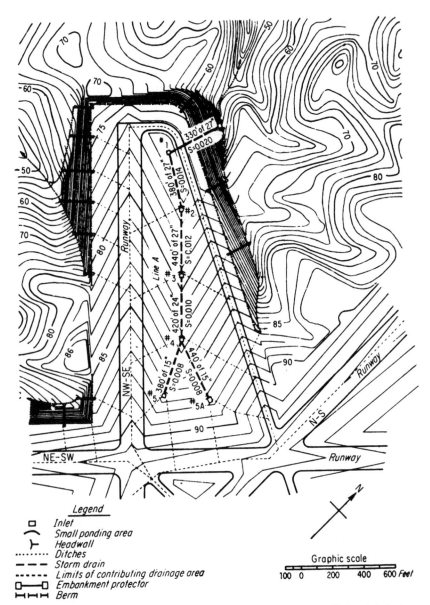

Figure 14-5 Portion of airport showing drainage design details (*Federal Aviation Administration [1]*).

TABLE 14-2   Design Data for Line A in Fig. 14-5

	Tributary area to inlets, acres				Distance remote point to inlet, ft				
	Pave-ment	Turf	Both	Subtotal	Pave-ment	Turf	Total	Line segment	Length, ft
Inlets									
5	1.27	1.05	2.32	2.32	200	340	540	5-4	380
5A	1.02	1.86	2.88	2.88	70	450	520	5A-4	440
4	1.40	7.35	8.75	13.95	60	690	750	4-3	420
3	0.78	6.46	7.24	21.19	100	700	800	3-2	440
2	0.83	4.56	5.39	26.58	150	500	650	2-1	380
1	1.14	3.70	4.84	31.42	190	360	550	1-outlet	330
Outlet	—	—	—	—	—	—	330		
Total	6.44	24.98	31.42						

Weighted Average for C

To inlet 5:

$$\frac{1.27}{2.32}(0.90) = 0.49$$

$$\frac{1.05}{2.32}(0.30) = \underline{0.14}$$

$$C = 0.63$$

To inlet 5A:

$$\frac{1.02}{2.88}(0.90) = 0.32$$

$$\frac{1.86}{2.88}(0.30) = \underline{0.19}$$

$$C = 0.51$$

To inlet 4:

$$\frac{1.40}{8.75}(0.90) = 0.14$$

$$\frac{7.35}{8.75}(0.30) = \underline{0.25}$$

$$C = 0.39$$

To inlet 3:

$$\frac{0.78}{7.24}(0.90) = 0.10$$

$$\frac{6.46}{7.24}(0.30) = \underline{0.27}$$

$$C = 0.37$$

To inlet 2:

$$\frac{0.83}{5.39}(0.90) = 0.14$$

$$\frac{4.56}{5.39}(0.90) = \underline{0.25}$$

$$C = 0.39$$

To inlet 1:

$$\frac{1.14}{4.84}(0.90) = 0.21$$

$$\frac{3.70}{4.84}(0.30) = \underline{0.23}$$

$$C = 0.44$$

SOURCE: Federal Aviation Administration [1].

slope of the turfed area is 1.5 percent. The inlet times for this specific problem are shown in Fig. 14-3. The computations for runoff, assuming no ponding, are shown in Table 14-3.

Typical example—Ponding

In the design of an airfield drainage system, ponding may be used to effect a reduction in the cost of installation. Ponding is simply a means of providing temporary storage of runoff prior to its entry into the underground system. For purposes of design computation, the ponded volume may be assumed to be an inverted pyramid or a truncated pyramid, the height of which is the depth of water above the inlet at any stage. The area of the base of the pyramid is taken as the surface area of the pond. If ponding were permitted, the layout of

TABLE 14-3 Drainage System Design Data

Inlet	Line segment	Length of segment, ft	Inlet time, min	Flow time, min	Time of concentration, min	Runoff coefficient C	Rainfall intensity I	Tributary area A, acres	Remarks
5A	5A-4	440	31.2	1.8	31.2	0.51	3.10	2.88	n = 0.015
5	5-4	380	30.0	1.6	30.0	0.63	3.15	2.32	
4	4-3	420	37.7	1.1	37.7	0.39	2.80	8.75	See accumulated runoff computed below.
3	3-2	440	38.9	1.0	38.9	0.37	2.70	7.24	Accumulated runoff adjustment negligible.
2	2-1	380	34.4	0.8	39.9	0.39	2.65	5.39	
1	1-outlet	330	30.7	0.5	40.7	0.44	2.60	4.84	
Outlet									

Calculation of Example

Maximum flow from inlets 5 and 5A will reach inlet 4 in 31.6 and 33.0 min, respectively. All inlet 4 subarea will be contributing to the system only after 37.7 min. Flow from inlets 5 and 5A must be adjusted for 37.7-min time of concentration. Adjusted time of concentration for inlets 5 and 5A (that is, the inlet time for end-of-line structures) equals the time of concentration to inlet 4 less the flow time through the respective pipe segments. For inlet 5, adjusted time of concentration = 37.7 − 1.6 = 36.1 min. For inlet 5A, adjusted time of concentration = 37.7 − 1.8 = 35.9 min. By using these adjusted times of concentration, an intensity of rainfall of 2.85 in/h is obtained (slight time difference cannot be read from curves).

Applying these data in the formula $Q = CIA$:

Adjusted flow from inlet 5 = $0.63 \times 2.85 \times 2.32$ = 4.16

Adjusted flow from inlet 5A = $0.51 \times 2.85 \times 2.88$ = 4.18

Flow into inlet 4 from inlets 5 and 5A in 37.7 min = 8.34

Flow from inlet 4 subarea = 9.56

Accumulated flow entering inlet 4 in 37.7 min = 17.90 ft³/s

SOURCE: Federal Aviation Administration [1].

TABLE 14-3 (Continued)

	Runoff Q ft³/s	Accumulated runoff, ft³/s	Velocity of drain, ft/s	Size of pipe, in	Slope of pipe, ft/ft	Capacity of pipe, ft³/s	Invert elevation	Remarks
Inlet								
5A	4.55	4.55	4.0	15	0.008	5.0	81.52	$n = 0.015$
5	4.60	4.60	4.0	15	0.008	5.0	81.04	
4	9.56	17.90	6.2	24	0.010	20.0	78.00	See accumulated runoff computed below.
3	7.23	25.13	7.4	27	0.012	30.0	73.80	Accumulated runoff adjustment negligible.
2	5.57	30.70	8.0	27	0.014	33.0	68.52	
1	5.54	36.24	9.5	27	0.020	37.5	63.20	
Outlet							56.60	

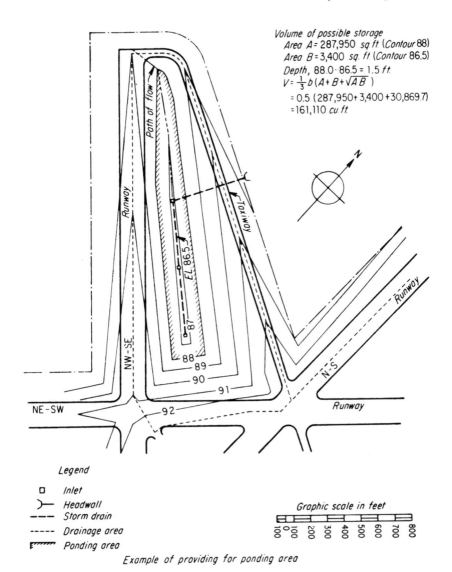

Volume of possible storage
Area A = 287,950 $sq.ft.$ (Contour 88)
Area B = 3,400 $sq.ft.$ (Contour 86.5)
Depth, 88.0 - 86.5 = 1.5 $ft.$
$V = \frac{1}{3}b(A+B+\sqrt{AB}\ )$
= 0.5 (287,950 + 3,400 + 30,869.7)
= 161,110 $cu.ft.$

Legend

□   Inlet
⟩—   Headwall
---   Storm drain
-----   Drainage area
▨▨▨   Ponding area

Graphic scale in feet

100 0 100 200 300 400 500 600 700 800

Example of providing for ponding area

Figure 14-6   Layout of drainage for ponding (*Federal Aviation Administration [1]*).

the drainage system might be as shown in Fig. 14-6. The most remote point to one of the inlets is 950 ft, comprising 100 ft of pavement and 850 ft of turf. The time of concentration is estimated at 4 + 54 = 58 min. The complete drainage area is 31.42 acres, of which 6.44 acres is paved. Assuming that the coefficients of runoff for pavement and turf are 0.90 and 0.30, respectively, the combined C is 0.423. From Fig. 14-4 the rainfall intensities for durations of 5, 10, 15, 20, 30, 60, 90,

TABLE 14-4   Volume of Runoff—Ponding

Time, min	Intensity* I	$Q = CIA$, ft³/s	Volume $V = CIAt$, ft³
5	5.80	77.1	23,100
10	4.96	65.9	39,600
15	4.33	57.5	51,800
20	3.95	52.5	63,000
30	3.18	42.3	76,100
60	2.00	26.6	95,700
90	1.62	21.5	116,300
120	1.26	16.7	120,600
180	0.87	11.6	125,000

*Hourly intensities from Fig. 14-4.

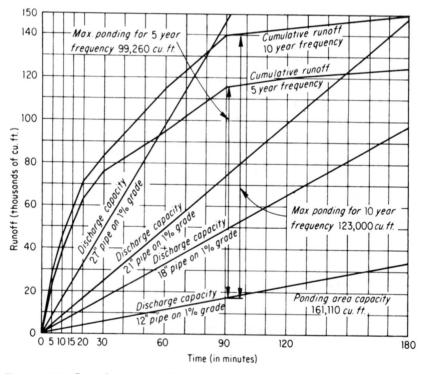

Figure 14-7 Cumulative runoff for ponding in Fig. 14-6 (*Federal Aviation Administration [1]*).

120, and 180 min are obtained, and the volumes of runoff are computed as shown in Table 14-4.

To visualize the effects of ponding, a comparison is made of the discharge capacity of tentative drainage pipes and the cumulative runoff for the design storm frequency. This comparison is best made as a plot of runoff on the ordinate axis and time on the abscissa. An example of such a plot is shown in Fig. 14-7. The discharge capacity for each of four selected pipe sizes is shown as a straight line. These discharge curves were computed for an assigned slope and roughness coefficient n for each pipe by use of the Manning formula. For this example the pipes were assumed to be concrete with a roughness coefficient n of 0.015 and laid on a 1.0 percent slope. The discharge in cubic feet per second multiplied by 3600 s is the discharge capacity ordinate in cubic feet at the 60-min abscissa in Fig. 14-7. Each discharge capacity curve must pass through the origin of coordinates, and one point as determined above will define the straight-line relationship.

The significance of the cumulative runoff and discharge capacity curves as plotted in Fig. 14-7 is that the difference in ordinates (cumulative runoff minus discharge capacity) represents the amount of ponding at any instant after the beginning of the storm. The maximum amount of ponding is determined by scaling the largest difference between the cumulative runoff curve and the discharge capacity curve.

It is considered essential that all ponding area edges be kept at least 75 ft from the edges of pavements. In this example, this would mean that the pond should not reach a level above elevation 88.0. The storage capacity below this elevation is 161,000 ft³. If a 12-in-diameter pipe were used, the maximum ponding would amount to 99,260 ft³, considerably less than the available 161,100 ft³. For practical consideration a pipe of lesser diameter is not recommended.

Although not shown in this text, computations were also made for a 10-year-frequency storm. With a 12-in-diameter pipe such a storm would develop a pond of 123,000 ft³, still less than the available capacity of 161,000 ft³.

Determining the Amount of Runoff by the Corps of Engineers Procedure

For determining runoff, the Corps of Engineers uses a relationship for overland flow developed by R. E. Horton [16]. This relationship, as modified by the Corps of Engineers, is as follows:

$$q = (\sigma \tan h^2)\left[0.922t\left(\frac{\sigma}{nL}\right)^{1/2} S^{1/4}\right] \tag{14-4}$$

where q = rate of overland flow at lower end of elemental strip of turfed, bare, or paved surface, in/h of ft³/s per acre of drainage area

Q = total discharge from a drainage area, ft³/s; Q equals product of q and drainage area in acres

S = slope of surface or hydraulic gradient, absolute, i.e., 1 percent = 0.01

t = time or duration, min; time from beginning of supply (storm); total time $t = t_c + t_d$

t_c = duration of supply which produces maximum rate of outflow from a drainage area but not in a pipe

t_d = time water flows in pipe

σ = rate of supply, or rainfall in excess of rate of infiltration, in/h

L = effective length of overland or channel flow, ft

n = retardance coefficient

The term t_c is nothing more than the time of concentration for the drainage area under consideration. The term L, the effective length, represents the length of overland sheet flow from the most remote point in the drainage area to the drain inlet, measured in a direction parallel to the maximum slope, before the runoff has reached a defined channel or ponding basin, plus the length of flow in a channel if one is present. If ponding is permitted, L is measured from the most remote point in the drainage area to the mean edge of the pond.

The term n is referred to as the *retardance coefficient.* Typical coefficients are given in Table 14-5.

When a drainage area is composed of two or three types of surfaces, an average retardance coefficient must be computed. For example, if a drainage area consists of 4 acres of average grass cover and 2 acres of pavement, the average retardance coefficient is equal to

$$\frac{4(0.40) + 2(0.02)}{6} = 0.27$$

Infiltration rate

Use of the Horton formula requires an estimate of the amount of rainfall which is absorbed in the ground and which therefore does not

TABLE 14-5   Retardance Coefficients

Surface	Value of n
Smooth pavements	0.02
Bare packed soil free of stone	0.10
Sparse grass cover, or moderately rough bare surface	0.30
Average grass cover	0.40
Dense grass cover	0.80

SOURCE: Corps of Engineers [7].

appear as runoff. This is referred to as *infiltration* and is expressed as a rate in inches per hour. Thus, the intensity of rainfall (in inches per hour) less the infiltration rate is equal to the rate of runoff or the rate of supply σ in the formula for runoff.

The infiltration rate is dependent largely on the structure of the soil cover, moisture content, and temperature of the air. The infiltration rate is not constant throughout the duration of the storm, but is assumed so in the computations. It is felt that such an assumption is reasonable, especially when the soil is near saturation.

The infiltration rate for paved surfaces is usually assumed to be zero. Infiltration rates for other types of surfaces and soil cover must be estimated from experience. A value of 0.5 in/h has been suggested for turfed areas. Thus, if the rainfall intensity on a turfed area were 2.0 in/h, the rate of supply σ would be 1.5 in/h.

Standard supply curves

By use of Eq. (14-4) maximum rates of runoff q for rates of supply σ of 0.8, 1.0, 1.6, and 1.8 in/h are shown in Figs. 14-8 and 14-9. Maximum rates of runoff are also shown for rates of supply of 0.4, 0.6, 1.2, 1.4, 2.0, 2.2, 2.4, 2.6, 2.8, 3.0, 3.2, and 3.4 in/h [7].

Maximum rates of runoff for the curve labeled *supply curve no. 1.0* (Fig. 14-8) were obtained in the following manner. From Fig. 14-2 the intensities of runoff for various durations corresponding to the curve labeled 1.0 are obtained. These intensities are entered as σ in Eq. (14-4), and L is varied to produce the family of curves shown in Fig. 14-8. The curve labeled σ is supply curve no. 1.0, obtained from Fig. 14-2. The dotted line labeled t_c represents the maximum rate of runoff q which would occur from an elemental area with various effective lengths L. For example, the maximum rate of runoff from an area whose effective length L is 60 ft is 2.0 ft³/s. Multiplying this rate by the drainage area yields the maximum total discharge Q.

Figures 14-8 and 14-9 were prepared for $n = 0.40$ and $S = 1$ percent. If these charts are to be used for other cases, the actual effective L for the area under study must be converted in terms of L for $n = 0.40$ and $S = 1$. A conversion chart is shown in Fig. 14-10. For example, if the actual $n = 0.30$ and $S = 2$ percent and the effective length L is 400 ft, then the equivalent effective L for $n = 0.40$ and $S = 1$ percent is 140 ft.

Typical example—No ponding

In the Corps of Engineers procedure, a reach of drain pipe is always designed for a storm whose duration is equal to the time of concentration for the drainage area above the pipe. The time of concentration

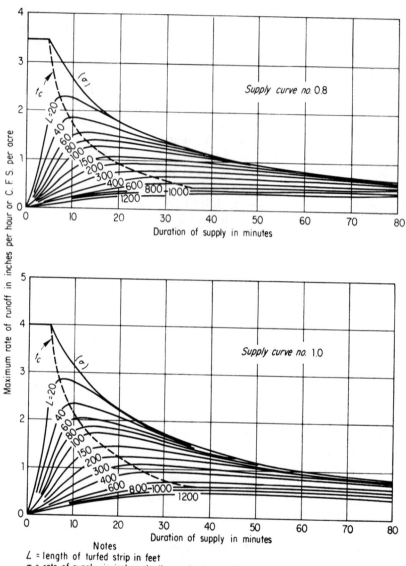

Notes

L = length of turfed strip in feet
σ = rate of supply, in inches depth per hour.
t_c = critical time of runoff in minutes, assuming
surface storage as negligible.

Figure 14-8 Standard supply curves, 0.8 and 1.0 in/h (*Corps of Engineers*).

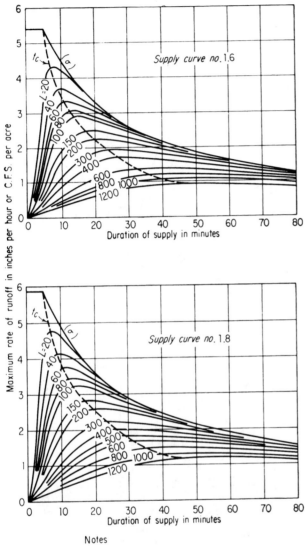

Notes

L = length of turfed strip in feet.

σ = rate of supply, in inches depth per hour.

t_c = critical time of runoff in minutes, assuming surface storage as negligible.

Figure 14-9 Standard supply curves, 1.6 and 1.8 in/h (*Corps of Engineers*).

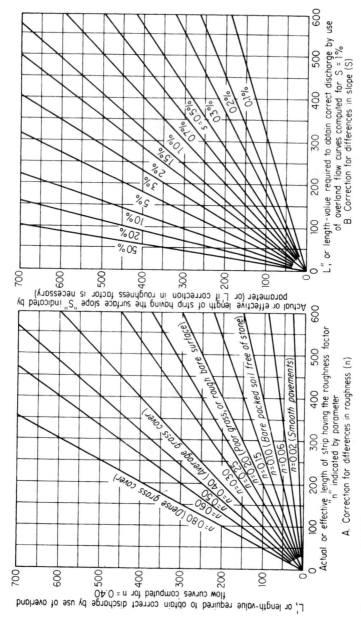

Figure 14-10 Modification in L required to compensate for difference in n and S (*Corps of Engineers*).

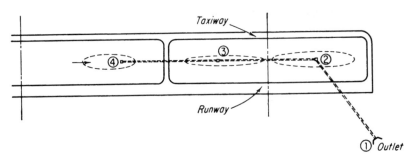

Figure 14-11 Portion of airport showing drainage layout (*Corps of Engineers*).

corresponds to the time necessary to produce maximum flow into a particular inlet (which is the same as the time necessary for water to reach an inlet from the most remote point in the area) plus the flow time in the pipe.

To clarify the computation of runoff by the Corps of Engineers procedure, the following example is presented.

Consider the drainage areas shown in Fig. 14-11. The 1-h intensity of the design storm is assumed to be 2.0 in/h. The infiltration rate for the turfed areas is assumed to be 0.5 in/h. The retardance coefficient for the pavement is $n = 0.02$, and for the turfed area $n = 0.40$. The drainage areas, retardance coefficients (referred to as *roughness factors*), and actual effective lengths L are shown in Table 14-6. Values of L and S were obtained from a grading plan of the area. The equivalent L's are obtained from Fig. 14-10. Column 14, labeled *adopted for selecting diagrams,* designates the nearest whole number which can be identified on the supply curves (Figs. 14-8 and 14-9). The standard supply curve to be used for the example is obtained by weighting the supply curves for the paved and turfed areas. For example, for inlet 4, the paved area is 5.97 acres and the supply curve is 2.0 in/h; the turfed area is 26.81 acres and the supply curve is 1.5 in/h. The weighted supply curve is equal to

$$\frac{5.97(2) + 26.81(1.5)}{5.97 + 26.81} = 1.6$$

In columns 20 and 21, the critical inlet time t_c (the time that will produce the maximum discharge) and the corresponding rates of runoff are listed. These values are obtained from Fig. 14-9. In columns 23 and 24, additional rates of runoff for arbitrarily selected times are listed. This is done to facilitate computation for various times of concentration for the several points along a drainage system.

The next step is to compute the volumes of runoff into inlets 4, 3, and 2. The computations are shown in Table 14-7. Obviously the dura-

TABLE 14-6   Airfield Drainage—Drain Inlet Capacities

	Supply curve nos.	
For paved areas	2.0	Drainage section assuming
For bare areas		
For turfed areas	1.5	

	Drainage area (DA), acres				Permissible ponding						Length L, ft		
Inlet no.	Paved, n = 0.02	Unpaved Bare	Unpaved Turf n = 0.40	Total	Depth at inlet, ft	Pond area, 1000 ft²	Volume 1000 ft³	Volume, ft³/acre DA	Average roughness factor n	Average slope S, %	Actual or effective length, ft	Equivalent L for n = 0.40 and S = 1%	L adopted for selecting diagrams
1	2	3	4	5	6	7	8	9	10	11	12	13	14
4	5.97		26.81	32.78					0.33	2.0	575	330	330
3	5.69		25.54	31.23					0.33	2.8	575	280	280
2	5.69		25.54	31.23					0.33	2.8	575	280	280

SOURCE: Corps of Engineers [7].

TABLE 14-7   Airfield Drainage—Size and Profile of Underground Storm Drains

	Supply curve nos.	
For paved areas	2.0	Drainage section
For bare areas		
For turfed areas	1.5	

Point of design					Critical runoff time to produce maximum flow in underground drain					Rate of	
	Distance, ft					Drain time, min					
Inlet or junction	From main outlet	From preceding inlet	Critical inlet	t_c, min	Assumed velocity in pipe, ft/s	From preceding inlet	Accumulation total	Approximate t_c' (col. 5 + col. 8)	Adopted t_c', min	4	3
1	2	3	4	5	6	7	8	9	10	11	12
4	4805	—	4	24	3.0			24	25	59.0	
3	3155	1650	4	24	3.0	9.2	9.2	33	30	59.0	62.5
2	1505	1650	4	24	3.0	9.2	18.4	42	40	55.8	56.2

SOURCE: Corps of Engineers [7].

no ponding of runoff

Standard supply curve no. × DA					Drain inlet capacity			Critical contribution to system		
Paved areas	Unpaved areas		Total	Weighted supply curve (col. 18 + col. 5)	t_c, min	q_d ft³/s/ acre	Q_d, ft.³/s (col. 21 × col. 5)	t_c', min	q_d, ft³/s/ acre	Q_d, ft³/s (col. 24 × col. 5)
	Bare	Turf								
15	16	17	18	19	20	21	22	23	24	25
11.94		40.22	52.16	1.6	24	1.8	59.0	30	1.8	59.0
								40	1.7	55.8
11.38		38.31	49.69	1.6	23	2.0	62.5	25	2.0	62.5
								30	2.0	62.5
								40	1.8	56.2
11.38		38.31	49.69	1.6	23	2.0	62.5	25	2.0	62.5
								30	2.0	62.5
								40	1.8	56.2

assuming no ponding of runoff

inflow into underground drains, ft³/s, corresponding to adopted value of t_c' (col 10)

Inlet																	Total
2																	
13	14	15	16	17	18	19	20	21	22	23	24	25	26	27	28	29	30
																	59.0
																	121.5
56.2																	168.2

TABLE 14-8   Airfield Drainage—Drain Inlet Capacities Required to Limit Ponding to Permissible Volumes

Supply curve nos.													
For paved areas		20									Drainage		
For bare areas													
For turfed areas		1.5											

	Drainage area (DA), acres				Permissible ponding						Length L, ft		
		Unpaved											
Inlet	Paved, $n = $ 0.02	Bare	Turf $n = $ 0.40	Total	Depth at inlet, ft	Pond area, 1000 ft²	Volume 1000 ft³	Volume, ft³/acre DA	Average roughness factor n	Average slope S, %	Actual or effective length, ft	Equivalent L for $n = 0.40$ and $S = 1\%$	L adopted for selecting diagrams
1	2	3	4	5	6	7	8	9	10	11	12	13	14
4	5.97		26.81	32.78	3.0	138	206	6,292	0.33	2.0	525	300	300
3	5.69		25.54	31.23	1.73	145	125	4,016	0.33	2.8	340	200	200
2	5.69		25.54	31.23	2.73	270	368	11,800	0.33	2.8	340	200	200

*Not required when appreciable ponding is permissible.

SOURCE: Corps of Engineers [7].

tion of a storm necessary to provide the maximum rate of runoff into inlet 4 is equal to 24 min. The pipe from inlet 4 to inlet 3 is designed for a storm of this duration. At inlet 3 the time of concentration is 24 min plus the flow time in the pipe from inlet 4 to inlet 3 (9.2 min). The pipe from inlet 3 to inlet 2 would be designed for a storm of 33.2-min duration. Enter Fig. 14-9 (supply curves 1.6) with 33 min as the abscissa, and read the rates of runoff for effective lengths L of 280 ft (inlet 3) and 330 ft (inlet 4). Multiply these rates by their respective drainage areas. According to the computations at inlet 3, the area directly tributary to it contributes 62.5 ft³/s, and the area tributary to inlet 4 contributes 59.0 ft³/s. Thus the pipe from inlet 3 to inlet 2 should be designed for a capacity of 59.0 + 62.5 = 121.5 ft³/s. The same process would be repeated for the design of the pipe from inlet 2 to the outlet.

It should be emphasized that the duration of the storm for the analysis of a particular point along the drainage system always corresponds to the time of concentration above this point. Had the inlet time t_c for the area directly tributary to inlet 3 been larger than the sum of the inlet time for the area tributary to inlet 4 plus the flow time to inlet 3, the former would have established the duration of the storm for the design of the pipe from inlet 3 to inlet 2.

Typical example—Ponding

If ponding is permissible, the first step is to establish the limits of the ponding area. From a grading and drainage plan, the volumes in the section east side of airfield

Standard supply curve no. × DA				Weighted supply curve (col. 18 ÷ col. 5)	Drain inlet capacity			Critical contribution to system		
Paved areas	Unpaved areas		Total		t_c, min	q_d ft³/s/ acre	Q_d, ft.³/s (col. 21 × col. 5)	t_c', min	q_d, ft³/s/ acre	Q_d, ft³/s (col. 24 × col. 5)
	Bare	Turf								
15	16	17	18	19	20	21	22	23	24	25
11.94		40.21	51.15	1.6	*	0.52	17.05			
11.38		38.31	49.69	1.6	*	0.52	16.24			
11.38		38.31	49.69	1.6	*	0.52	16.24			

various ponds can be computed. These volumes are then expressed in terms of cubic feet per acre of drainage area, as shown in column 9 of Table 14-8. The actual and equivalent L values are determined in the same manner as for the case of no ponding, with one exception. The actual L is measured to the mean edge of the pond rather than to the drain inlet. The actual and equivalent effective lengths are listed in columns 12 and 13.

The Corps of Engineers has developed charts which yield drain inlet capacities to prevent ponds from exceeding certain specified volumes. Typical charts are shown in Fig. 14-12 and 14-13. The volumes are computed for various supply curves (Fig. 14-2), assuming the slope of the basins forming the drainage areas is 1 percent. The supply curves represent the intensity-duration pattern for storms whose 1-h intensities correspond to the supply curve numbers. The volumes of runoff for a specific supply curve are computed in a manner similar to the procedure used by the FAA. The cumulative volumes of runoff are compared with the various capacities of drain inlets to arrive at the volumes of storage shown in Figs. 14-12 and 14-13. Since the volumes of runoff depend on L and S, charts must be prepared for a wide range of L values. Figures 14-12 and 14-13 show drain inlet capacities for L equal to 100, 200, 300, and 400 ft. Additional charts have been prepared for $L = 0, 40, 600, 800, 1000,$ and 1200 ft [7].

The physical significance of the charts may be described by reference to the following example. Suppose that L for a large drainage area is 100 ft and that the runoff pattern corresponds to supply curve 2. Assume that the maximum permissible ponding is 300 ft³/acre of drainage area. From Fig. 14-12 a pipe which has a capacity of 1.0 ft³/s

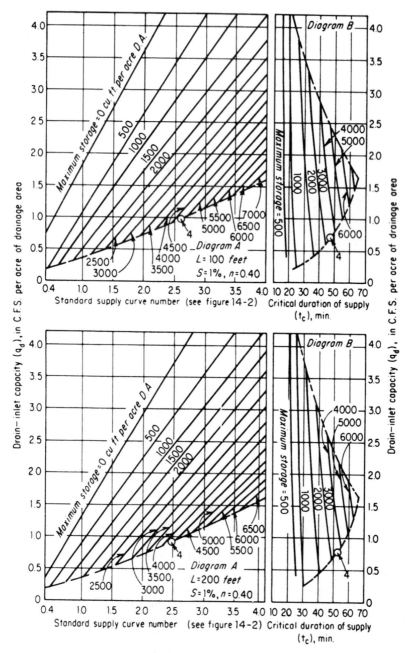

Figure 14-12   Drain inlet capacity versus maximum surface storage. C.F.S. = cubic feet per second (*Corps of Engineers*).

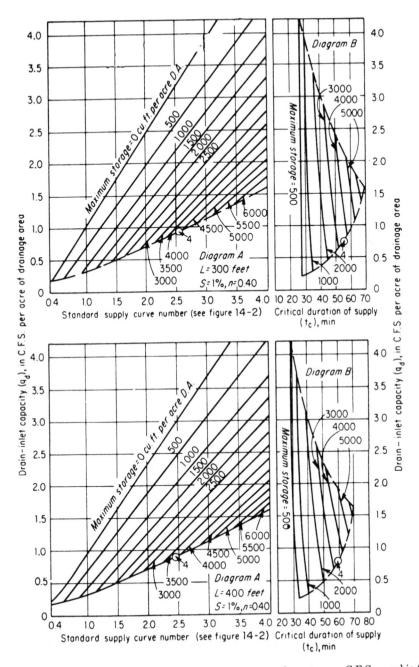

Figure 14-13 Drain inlet capacity versus maximum surface storage. C.F.S. = cubic feet per second (*Corps of Engineers*).

per acre of drainage area would be adequate to prevent the pond from exceeding a volume of 3000 ft^3 during any part of the storm. The dashed lines labeled 4 are equal to rates of supply corresponding to a duration of 4 h. Although smaller drain inlets are possible, it is felt that the sizes corresponding to a duration of 4 h are about the minimum from a practical standpoint.

The required drain inlet capacities for the drainage layout in Fig. 14-11 were obtained from Figs. 14-12 and 14-13 and are tabulated in Table 14-8. Note that the time of concentration is not a factor in these computations.

Layout of Surface Drainage

A finished grade contour map of the runways, taxiways, and aprons is extremely helpful for the layout of a storm drain system. Several trial drainage layouts may be necessary before the most economical system can be selected. The grades of the storm drain should be such as to maintain a minimum mean velocity on the order of 2.5 ft/s to provide sufficient scouring action to avoid silting. To maintain an adequate cross section for flow at all times, the diameter of the storm drain should not be less than 12 in.

Water from a drainage area is collected into the storm drain by means of inlets. The inlet structure consists of a concrete box, the top of which is covered with a grate made of cast iron, cast steel, or reinforced concrete. The grates must support aircraft wheel loads and should therefore be designed for contact pressures for the aircraft which will be served by the airport.

On long tangents, drain inlets are usually placed at intervals varying from 200 to 400 ft. The location of the inlets depends on the configuration of the airport and on the grading plan. Normally, if there is a taxiway parallel to the runway, the inlets are placed in a valley between runway and taxiways, as indicated in Fig. 14-11. If there is no parallel taxiway, the drains are placed near the edge of the runway pavement or at the toe of the slope of the graded area.*

On aprons, inlets are usually placed in the pavement proper. This is the only way a large apron area can be drained. All grates should be securely fastened to the frames so that they will not be jarred loose with the passage of traffic (see Fig. 14-14).

Adequate depths of cover should be provided over the pipes so that the pipes can support traffic. The recommended minimum depths of cover are shown in Table 14-9.

---

*The FAA recommends that the inlets not be closer than 75 ft to the edge of the pavement.

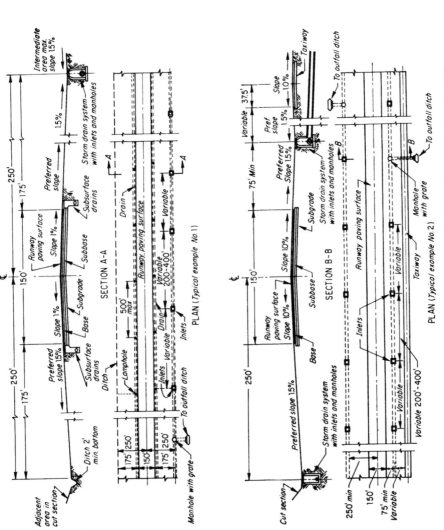

Figure 14-14 Recommended pavement drainage sections (*Federal Aviation Administration [1]*).

TABLE 14-9   Recommended Minimum Depth of Cover for Pipe, ft

Kind of pipe	Flexible Pavement															
	Pipe cover															
	Nominal diameter of pipe, in								Nominal diameter of pipe, in							
	12	18	24	30	36	42	48	60	12	18	24	30	36	42	48	60
	Wheel load 15,000 lb								Wheel load 30,000 lb							
Clay sewer pipe	2.5	3.0	3.0	3.5	3.5				3.0	3.5	4.0	4.5	4.5			
Clay culvert pipe	1.5	1.5	1.5	2.0	2.0				2.5	3.0	3.0	3.0	3.0			
Concrete sewer pipe	2.5	3.0	3.0						3.0	3.5	4.0					
Concrete sewer pipe (extra-strength)	1.5	1.5	2.0						2.5	3.0	3.0					
Reinforced-concrete culvert pipe																
Class I								3.0								4.5
Class II	2.5	2.5	2.5	2.5	2.5	2.5	2.5	2.5	3.0	3.0	3.0	3.0	3.0	3.5	3.5	3.5
Class III	2.0	2.0	2.0	2.0	2.0	2.0	2.0	2.0	2.5	2.5	2.5	2.5	2.5	3.0	3.0	
Class IV	1.5	1.5	1.5	1.5	1.5	1.5	1.5	1.5	2.0	2.0	2.0	2.0	2.0	2.0	2.0	2.0
Class V	1.0	1.0	1.0	1.0	1.0	1.0	1.0	1.0	1.5	1.5	1.5	1.5	1.5	1.5	1.5	1.5
Corrugated metal pipe, gauge no.																
16	1.0	1.5	1.5						1.5	2.0	2.0					
14	1.0	1.0	1.0	1.5	2.0				1.0	1.0	1.5	2.0	2.5			
12	1.0	1.0	1.0	1.0	1.5	1.5	2.0		1.0	1.0	1.5	1.5	2.0	2.5	2.5	
10			1.0	1.0	1.0	1.0	1.5	1.5			1.0	1.5	1.5	2.0	2.0	2.5
8					1.0	1.0	1.0	1.0					1.0	1.5	1.5	2.0
	Wheel load 45,000 lb								Wheel load 60,000 lb							
Clay culvert pipe	3.0	3.5	3.5	4.0	4.0				3.5	4.0	4.5	4.5	5.0			
Concrete sewer pipe (extra-strength)	3.0	3.5	3.5						3.5	4.0	4.5					
Reinforced-concrete culvert pipe																
Class I																
Class II	3.0	3.5	3.5	4.0	4.0	4.5	4.5		3.5	4.0	4.5					
Class III	2.5	3.0	3.0	3.5	3.5	3.5	4.0	4.0	3.0	3.5	4.0	4.0	4.5			
Class IV	2.0	2.5	2.5	2.5	2.5	2.5	3.0	3.0	2.5	3.0	3.0	3.0	3.5	3.5	3.5	4.0
Class V	1.5	1.5	2.0	2.0	2.0	2.0	2.5	2.5	2.0	2.0	2.5	2.5	2.5	3.0	3.0	3.0
Corrugated metal pipe, gauge no.																
16	2.0	2.5	3.0						2.5	3.0	3.5					
14	1.5	2.0	2.5	3.0	3.0				2.0	2.5	3.0	3.5	4.0			
12	1.5	2.0	2.0	2.5	2.5	3.0	3.5		1.5	2.0	2.5	3.0	3.5	4.0	4.0	
10			1.5	2.0	2.0	2.5	3.0	3.5			2.0	2.5	3.0	3.5	3.5	4.0
8					1.5	2.0	2.5	3.0					2.5	3.0	3.0	3.5

As a guide for the design of storm drains, the coefficient of roughness n for various types of pipes and open channels is listed in Table 14-10.

Subsurface Drainage

The functions of subsurface drainage are to (1) remove water from a base course, (2) remove water from the subgrade beneath a pavement,

	Flexible Pavement															
	Pipe cover															
Kind of pipe	Nominal diameter of pipe, in								Nominal diameter of pipe, in							
	12	18	24	30	36	42	48	60	12	18	24	30	36	42	48	60
	Wheel load 75,000 lb								Wheel load 100,000 lb							
Reinforced-concrete culvert pipe																
Class I																
Class II																
Class III	3.0	4.0	4.5	4.5	5.0				3.5	4.5	5.0	5.0	5.5			
Class IV	3.0	3.0	3.5	3.5	4.0	4.0	4.5	5.0	3.5	3.5	4.0	4.0	4.5	5.0	5.0	5.5
Class V	2.5	2.5	3.0	3.0	3.0	3.5	4.0	4.0			3.5	3.5	3.5	4.0	4.5	4.5
Corrugated metal pipe, gauge no.																
16	3.0	3.5	4.0						3.5	4.0	4.5					
14	2.5	3.0	3.5	4.0	4.5				3.0	3.5	4.0	4.5	5.0			
12	2.5	3.0	3.0	3.5	4.0	4.5	5.0		3.0	3.0	3.5	4.0	4.5	5.0	5.5	
10			2.5	3.0	3.5	4.0	4.5	5.0			3.0	3.5	4.0	4.5	5.0	5.5
8					3.0	3.5	4.0	4.5					3.5	4.0	4.5	5.0

Cover depths measured from top of flexible pavement or unsurfaced areas to top of pipe. Cover for pipe in areas not used by aircraft shall be in accordance with cover requirements for 15,000-lb wheel loads.

Rigid Pavement
Pipe placed under rigid pavements shall have a minimum cover, measured from the bottom of the slab, of 1.0 ft.

Note: The recommended minimum depth of cover for pipe does not provide protection against freezing conditions in seasonal freezing areas.

SOURCE: Federal Aviation Administration [1].

and (3) intercept, collect, and remove water flowing from springs or pervious strata.

Base drainage is normally required (1) where frost action occurs in the subgrade beneath a pavement, (2) where the groundwater is expected to rise to the level of the base course, and (3) where the pavement is subject to frequent inundation and the subgrade is highly impervious.

Subgrade drainage is desirable at locations where the water may rise beneath the pavement to less than 1 ft below the base course.

Intercepting drainage is highly desirable where it is known that subsurface waters from adjacent areas are seeping toward the airport pavements.

Methods for draining subsurface water

Base courses are usually drained by installing subsurface drains adjacent to and parallel to the edges of the pavement. The pervious material in the trench should extend to the bottom of the base course, as shown in Fig. 14-15. The center of the drainpipe should be placed a minimum of 1 ft below the bottom of the base course.

Subgrades are drained by pipes installed along the edges of pavement and in some instances, where the groundwater is extremely high, underneath the pavements. The center of the subsurface drain should be placed no less than 1 ft below the level of the groundwater. When subgrade drains are installed along the edges of the pavement, they may also serve for draining the base course.

Intercepting drainage can be accomplished by means of open ditches well beyond the pavement areas. If this is not practical, then subdrains can be used.

Types of pipe

The following types of pipe have been used for subdrainage:

1. Perforated metal, concrete, or vitrified clay pipe. The joints are sealed. The perforations normally extend over about one-third of the circumference of the pipe. The perforated area is usually placed adjacent to the soil.

2. Bell-and-spigot pipes are laid with the joints open. Vitrified clay, cast iron, and plain concrete are used in the manufacture of bell-and-spigot pipes.

3. Porous concrete pipe collects water by seepage through the concrete wall of the pipe. This type of pipe is laid with the joints sealed.

4. Skip pipe manufactured of both vitrified clay and cast iron is a special type of bell-and-spigot pipe with slots at the bells.

5. Farm tile is made of clay or concrete with the ends separated slightly to permit the entrance of water. This type of pipe is rarely used on airport projects.

Pipe sizes and slopes

Experience has shown that a 6-in-diameter drain is adequate, unless extreme groundwater conditions are encountered. If desired, the flow may be estimated by means of the available theories for soil drainage [6]. These theories require knowledge of the effective porosity and coefficient of permeability of the soil which is being drained, as well as the head on the pipe and the distance which the water must flow to reach the drain. Rarely is theory relied on to compute pipe sizes.

The recommended minimum slope for subdrains is 0.15 ft in 100 ft. A minimum thickness of 6 in of filter material should surround the drain. The gradation of the filter material is discussed in succeeding paragraphs.

TABLE 14-10 Coefficients of Roughness *n*

	n
Pipe	
Clay and concrete	
Good alignment, smooth joints, smooth transitions	0.013
Less favorable flow conditions	0.015
Corrugated metal	
100% of periphery smoothly lined	0.013
Paved invert, 50% of periphery paved	0.018
Paved invert, 25% of periphery paved	0.021
Unpaved, bituminous-coated or noncoated	0.024
Open channels	
Paved	0.015–0.020
Unpaved	
Bare earth, shallow flow	0.020–0.025
Bare earth, depth of flow over 1 ft	0.015–0.020
Turf, shallow flow	0.06–0.08
Turf, depth of flow over 1 ft	0.04–0.06

SOURCE: Federal Aviation Administration [1].

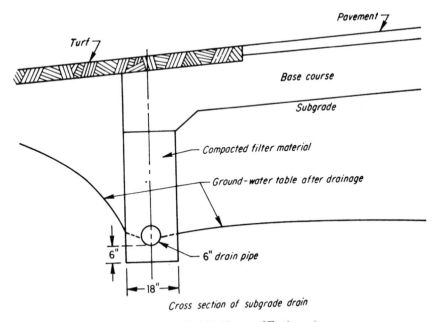

Cross section of subgrade drain

Figure 14-15 Subgrade subdrainage details (*Corps of Engineers*).

Utility holes and risers

For cleaning and inspection, utility holes and risers are often installed along the drains. The Corps of Engineers recommends that utility holes be placed at intervals of not more than 1000 ft, with one riser approximately midway between the holes [6]. The function of the riser is to be able to insert a hose for flushing the system. The function of a utility hole is to permit inspection of the pipes.

Gradation of filter material

The term *filter material* applies to the granular material which is used as backfill in the trenches where subdrains are placed. To permit free water to reach the drain, the filter material must be many times more pervious than the protected soil. Yet if the filter is too pervious, the particles of soil to be drained will move into the filter material and clog it.

On the basis of some general studies conducted by K. Terzaghi, the Corps of Engineers has developed an empirical design for filter material which has been substantiated by tests [9]. The criteria for selecting the gradation of the filter material are as follows:

1. To prevent clogging of a perforated pipe with filter material, the following requirement must be satisfied:

$$\frac{85\% \text{ size of filter material*}}{\text{Diameter of perforation}} > 1$$

2. To prevent the movement of particles from the protected soil into the filter material, the following conditions must be satisfied:

$$\frac{15\% \text{ size of filter material}}{85\% \text{ size of protected soil}} \leqq 5$$

and

$$\frac{50\% \text{ size of filter material}}{50\% \text{ size of protected soil}} \leqq 25$$

3. To permit free water to reach the pipe, the following condition must be fulfilled:

---

*This means that 85 percent (by weight) is finer than the specified size.

$$\frac{15\% \text{ size of filter material}}{15\% \text{ size of protected soil}} \geq 5$$

A typical example of design is shown in Fig. 14-16. Concrete sand has proved to be a satisfactory filter material for the majority of fine soils which are drainable. A single gradation of filter material is preferred for simplicity of construction.

Filter materials tend to segregate as they are placed in trenches. To minimize this tendency, the material should not have a coefficient of uniformity greater than 20. For the same reason, filter materials should not be skip-graded. Filter materials should always be placed in a moist state. The presence of moisture tends to reduce segregation.

Drainability of soils

Certain types of soils, such as gravelly sands, sand, and sandy loams, are usually self-draining and require very little, if any, subsurface drainage. Subsurface drainage can be effective for draining clay loams, sandy clay loams, and certain silty loams. The amount of sand in these soils largely determines how drainable they are. For soils containing a high percentage of silt and clay, subsurface drainage becomes very problematic.

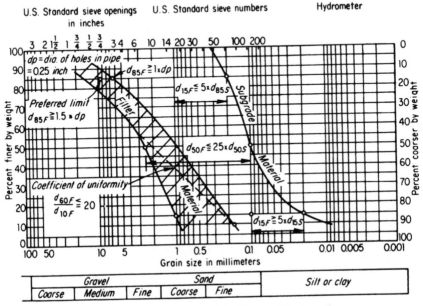

Figure 14-16 Design example for filter materials (*Corps of Engineers*).

References

1. *Airport Drainage,* Advisory Circular AC 150/5320-5B, Federal Aviation Administration, Washington, 1970.
2. *Conduits, Culverts, and Pipes,* Engineering Manual EM 1110-2-2902, Department of the Army, Washington, 1969.
3. "Design of Drainage Facilities for Military Airfields," G. A. Hathaway, *Transactions,* American Society of Civil Engineers, New York, 1949.
4. *Drainage and Erosion Control—Drainage for Areas Other than Airfields,* Tech. Manual TM 5-820-4, Department of the Army, Washington, 1965.
5. *Drainage and Erosion Control—Structures for Airfields and Heliports,* Tech. Manual TM 5-820-3, Department of the Army, Washington, 1965.
6. *Drainage and Erosion Control—Subsurface Drainage Facilities for Airfield Pavements,* Tech. Manual TM 5-820-2, Department of the Army, and Air Force Manual AFM 88-5, Department of the Air Force, Washington, 1979.
7. *Drainage and Erosion Control—Surface Drainage Facilities for Airfields and Heliports,* Tech. Manual TM 5-820-1, Department of the Army, and Air Force Manual AFM 88-5, Washington, 1987.
8. *Drainage of Asphalt Pavement Structures,* Manual Series MS-15, The Asphalt Institute, College Park, Md., 1984.
9. *Filter Experiments and Design Criteria,* Tech. Memo 3-360, U.S. Army Corps of Engineers, Waterways Experiment Station, Vicksburg, Miss., 1953.
10. *On-Site Stormwater Management: Applications for Landscape and Engineering,* B. Ferguson and T. H. Debo, 2d ed., Van Nostrand and Reinhold, New York, 1990.
11. "Pavement Subsurface Drainage Systems," H. H. Ridgeway, *Synthesis of Highway Practice,* No. 96, Transportation Research Board, Washington, 1982.
12. *Precipitation Frequency Atlas of the Western United States,* J. F. Miller, National Weather Service, Washington, 1973.
13. *Rainfall Frequency Atlas of the United States,* Technical Paper 40, U.S. Weather Service, Washington, 1961.
14. *Rainfall Intensities for Local Drainage Design in the United States,* Tech. Paper 24, pts. 1 and 2, U.S. Weather Service, Washington, 1953, 1954.
15. *Rainfall-Intensity Frequency Data,* D. L. Yarnell, Miscellaneous Publication 204, Department of Agriculture, Washington, 1935.
16. "The Interpretation and Application of Runoff Plat Experiments with Reference to Soil Erosion Problems," R. E. Horton, *Proceedings,* vol. 3, Soil Science Society of America, Madison, Wis., 1938.
17. *Urban Hydrology for Small Watersheds,* 2d ed., U.S. Soil Conservation Service, Washington, 1986.

15

Environmental and Economic Assessment

The current concern for an assessment and understanding of the environmental, ecological, and sociological consequences of development actions has resulted in the emergence of a holistic approach to planning. This approach views all actions as being undertaken in a single system and examines the consequences of these actions in terms of the entire system. Traditionally, proposals for transportation facilities have been evaluated in terms of sound engineering and technological principles, economic criteria, and benefits to the users and community. Policy decisions today, however, are being made with a more complete awareness of the impacts of these decisions on both users and nonusers from economic, social, environmental, and ecological viewpoints.

Airports must be planned in a manner which ensures their compatibility with the environs in which they exist. Many serious compatibility problems which presently exist in the vicinity of airports represent a serious confrontation between two important characteristics of urban economics: the need for airports to meet transportation needs and the continuing demand for community expansion. Airport planning must be done within the context of a comprehensive regional plan. The location, size, and configuration of an airport must be coordinated with the existing and planned patterns of development in a community, by considering the effect of airport operations on people, ecological systems, water resources, air quality, and the other areas of community concern [9].

This chapter presents an overview of the factors which must be considered to assess and evaluate the impact of airport development decisions in the context of a systems approach to planning.

Policy Considerations

In the United States, the overall basis for policies related to the consideration of the environmental, ecological, and social impacts of airport development is rooted in the National Environmental Policy Act of 1969 (Public Law 91-190). The policy of the Department of Transportation (DOT) [43] is

> To integrate national environmental objectives into the missions and programs of the Department and to:
>
> 1. Avoid or minimize adverse environmental effects wherever possible;
> 2. Restore or enhance environmental quality to the fullest extent practicable;
> 3. Preserve the natural beauty of the countryside and public park and recreational lands, wildlife and waterfowl refuges, and historic sites;
> 4. Preserve, restore, and improve wetlands;
> 5. Improve the urban physical, social and economic environment;
> 6. Increase access to opportunities for disadvantaged persons; and
> 7. Utilize a systematic, interdisciplinary approach in planning and decision making which may have an impact on the environment.

To implement this policy, the FAA has established an environmental assessment and consultation process which provides the relevant officials, policymakers, and the public with an understanding of the potential environmental consequences of proposed actions and ensures that the decision-making process includes environmental assessments as well as economic, technological, and other factors relevant to the decision. The FAA requires that environmental impact statements and negative declarations serve to document and record compliance with this policy and reflect a thorough study of all relevant environmental factors using a systematic, comprehensive, and interdisciplinary approach. The National Environmental Policy Act (NEPA) requires (Public Law 91-190)

> All agencies of the Federal government to include in every recommendation or report on proposals for legislation and other major Federal actions affecting the quality of the human environment, a detailed statement on:
>
> 1. The environmental impact of the proposed action;
> 2. Any adverse environmental effects which cannot be avoided should the proposal be implemented;
> 3. Alternatives to the proposed action;
> 4. The relationship between local short-term uses of man's environment and the maintenance and enhancement of long-term productivity; and
> 5. Any irreversible and irretrievable commitments of resources which would be involved in the proposed action should it be implemented.

Complementing this overall policy statement, the FAA also established an *aviation noise abatement policy* (Public Law 96-193 and Public Law 101-508) to significantly reduce the adverse impacts of aviation noise on existing land uses and to achieve a substantial degree of noise compatibility between airports and their environs. The FAA has endorsed coordinated actions between aircraft operators and owners, the FAA, airport owners and sponsors, and the community. The FAA proposed several actions to achieve airport noise control and land-use compatibility, including source noise reductions through aircraft retrofit and replacement, modifications of landing and takeoff procedures, and compatibility plans which have the objective of containing severe noise impacts within airport controlled areas.

Noise is the most apparent impact of the airport upon the community, but consideration is required for all those social, economic, environmental, and ecological factors which are influenced by airport activity. These factors may be grouped into four categories: pollution factors, social factors, ecological factors, and engineering and economic factors [22]. The pollution factors include air and water quality, noise, and construction impacts. The social factors include land development, displacement and relocation of businesses and residences, parks and recreational areas, historic places and archaeological resources which may be impacted, areas which are unique because of natural or scenic beauty, and the consistency of the proposed development with local planning. The ecological factors include the impact on wildlife and waterfowl, flora and fauna, endangered species, and wetlands or coastal zones. The engineering and economic factors include a consideration of flood hazards, costs of construction and operation, benefits of implementation, and energy and natural resource use.

The FAA has identified the requirements for environmental assessment (EA) reports, environmental impact statements (EISs), and findings of no significant impact (FONSI) for various types of projects, and the FAA has also categorically excluded certain types of projects from the requirements of a formal environmental assessment [2, 43]. Table 15-1 lists a breakdown of the type of environmental study required for some common airport planning actions.

The general format for an environmental study consists of a statement of need for the proposal, an inventory of problems and issues, an identification of constraints and opportunities, an identification of the improvement components including physical and nonphysical entities, measures to increase benefits and reduce harm, a discussion of the alternatives and their impacts, and the manner and degree of community and public agency involvement in the process [2, 36, 43].

TABLE 15-1   Environmental Study Requirements of Airport Development Project*

Typical actions normally requiring environmental assessment:
Airport location
New runway
Major runway extension
Runway strengthening to permit use by noisier aircraft
Major expansion of terminal or parking facilities
Establishment or relocation of instrument landing system
Land acquisition
    Required for facility modifications
    Relocation of business or residences
    Affecting historical, recreational, or archaeological resources
    Affecting wetlands, coastal zones, or floodplains
    Affecting endangered or threatened species
Typical actions normally requiring environmental impact statement:
Adoption of new national airport system plan if criteria are substantially different
    from former plan
First-time airport location or airport layout planned
New runways capable of air carrier traffic in metropolitan areas

*See Federal Aviation Administration [43].

SOURCE: Federal Aviation Administration [2].

Pollution Factors

Air quality

Many of the larger, more densely populated urban areas are facing serious difficulties associated with the emission of dangerous gaseous and particulate matter into the atmosphere from industrial processes, combustion, and transportation. Air pollution affects the public welfare including the personal comfort and health of people, and it causes damage to soil, water, vegetation, wildlife, and animals, deterioration of property and erosion of property values, and a reduction in visibility resulting in losses of aesthetic appeal and increased hazards in transportation. *Air pollution* is the introduction of foreign substances or compounds into the air or the alteration of the concentrations of naturally occurring elements. Hub airports with a considerable volume of commercial jet aircraft traffic may contribute substantially to this problem.

Air quality is defined by the concentration level of six pollutants for which standards have been adopted, namely, carbon monoxide, hydrocarbons, nitrogen oxides, sulfur dioxide, suspended particulates, and photochemical oxidants. The standards are specified in the Clean Air Act and consist of two categories: primary standards related to health and secondary standards related to welfare.

The amount of a particular pollutant produced by an aircraft is a function of the type of engines and mode of operation of the aircraft

[17]. An analysis must include a consideration of aircraft idling at the gate and runway threshold, engine power run-ups, taxiing, takeoff, climb-out, approach, and landing. The dispersion of the pollutants is studied through the use of either emission models or diffusion models. The emission model assumes a uniform dispersion of the pollutants within the atmosphere of concern, whereas the diffusion model uses emissions or emission rates together with physical and meteorological conditions to determine concentrations of pollutants. A study of the air quality impacts for an airport project requires a determination of ambient air quality; local meteorological conditions; the mix, number, and paths of aircraft using the airport; and the emission rate of the aircraft in different operating modes. It also requires knowledge of the operating characteristics and volume of ground transportation modes providing access to and services at the airport, as well as the point sources of pollution occasioned by normal operation of an airport. A flowchart of the interaction of those factors normally considered in an air quality study at an airport is shown in Fig. 15-1. The results of an air quality study are typically displayed on maps which show the before and after concentration of pollutants in the area of the airport, together with charts indicating the level of compliance with air quality standards.

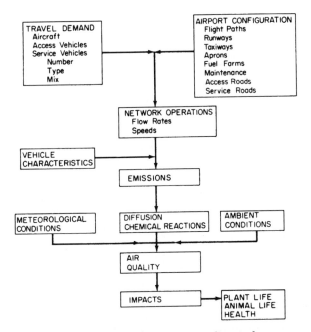

Figure 15-1 Flowchart illustrating air quality study process for airports.

Water quality

Water is one of the most valuable resources on earth. Not only is it essential for the maintenance of life itself, but also it is used by people in nearly all daily activities. As the population has grown, so has the demand for water, and today that need is so great that in many areas of the world the demand has outpaced the supply. The construction and operation of airport facilities can contribute to the degradation of water quality and reduction of the quantity of groundwaters or surface waters. Water quality can be affected by the addition of soluble or insoluble organic or inorganic materials into rivers, streams, and aquifers, resulting in a water source which is inadequate to support aquatic life and other uses such as fishing, swimming, and water supply needs. Changes in the cover, composition, and topography of the ground in the vicinity of airport sites can cause changes in the amount, peaking, routing, and filtration of runoff and the recharge area of aquifers. Construction-related activities may cause the introduction of materials and wastes into streams and water sources, increases in the volumes of sanitary wastes and water supply demand, and increases in storm water management systems.

A water quality study for an airport facility should address both the direct and indirect effects of the project on water quality [8, 21]. The direct effects include soil erosion, the amount and composition of runoff from the facility, infiltration, spills, turbidity, and the quantities of water supply and sewage disposal needs. Indirect effects include the accelerated weathering of exposed geologic and construction materials, disruption of nutrient cycles for the support of life, and the extraction of construction materials which may alter natural filtering, the degree of imperviousness of soils, and water storage capacity. Typically, a water quality study will identify the source and receptors of pollutants as well as the amount of degradation that the introduction of pollutants will cause. The study will also address the impact on the quantity of water sources through a determination of flow rates, flow and recharge areas, permeability, infiltration, and flow interruptions. Construction measures utilized to minimize degradation of water quality and supply include the construction of check dams, sediment traps, berms, dikes, channels, and slope drains; sodding and seeding; brush barriers; and paving. Wastewater management plans are usually prepared integrally with a review of this impact area [8, 21, 30].

Aircraft and airport noise

The effects of aircraft noise on communities surrounding airports present a serious problem to aviation. Since commercial jet transport operations began in 1958, the public reaction to aircraft noise has been vigorous. Because of these reactions much has been learned about the generation and propagation of noise and about human reactions to noise. On the basis of this knowledge, procedures have been developed which permit the planner to estimate the magnitude and extent of noise from airport operations and to predict community response. Several of these procedures are outlined here. The impact of aircraft noise on a community is dependent upon several factors, including the magnitude of the sound, duration of the sound, flight paths used during takeoff and landing, number and types of operations, operating procedures, aircraft mix, runway system utilization, time of day and season, and meteorological conditions. The response of communities to exposure to aircraft noise is a function of the land and building use, type of building construction, distance from the airport, ambient noise level, and community attitudes [9, 45].

Quantifying aircraft noise. Like most other environmental issues, aircraft noise has many dimensions. Most of these dimensions relate to the reaction of people to aircraft noise. These reactions relate to the sound level, the varying sensitivity of the human ear to different frequencies or pitches of sound, the frequency of aircraft noise intrusions, the time of day of these intrusions, and the number of intrusions over a period such as a day. Given this range of dimensions, it is not surprising that several metrics of aircraft noise have been developed over the years. While some metrics were built into the first sound meters over half a century ago, most have been developed since the introduction of the first transport-category turbojet aircraft in the late 1950s.

Many metrics have been developed over the years to describe aircraft noise. Some of the more common ones are presented in the following paragraphs. The goal of these metrics is to quantify aircraft noise in a manner which relates the physical aspects of sound to human assessments of loudness and noisiness. These metrics are the basis of most noise analyses conducted at airports throughout the United States and elsewhere. In addition, there are other similar noise metrics with specialized purposes, and these are also discussed: *effective perceived noise level* (EPNL), *composite noise rating* (CNR), and *noise exposure forecast* (NEF).

Sound pressure and sound pressure level. All sounds come from a sound source such as a musical instrument, a voice speaking, or an airplane passing overhead. Sound energy radiated by such sources is transmitted through the air in sound waves, which are tiny pressure fluctuations just above and below atmospheric pressure. These pressure fluctuations, called *sound pressures,* impinge on the ear, creating audible sound. Sound pressures are quantified by the root mean square (rms) value, i.e., the square root of the average squared pressure fluctuation over some brief period (about 1 s for aircraft noise purposes)

$$p_{\mathrm{rms}} = \left(\frac{1}{T} \int_{t=0}^{T} p(t)^2 \, dt \right)^{1/2} \tag{15-1}$$

where p_{rms} = rms sound pressure
$p(t)$ = deviation from atmospheric pressure at time t
T = averaging time, 1 s for airport noise purposes

The human auditory system is sensitive to a very wide range of rms sound pressures. The loudest sounds people can hear without pain have about 1 million times the rms sound pressure of the faintest sounds people can hear. Equally remarkable is the way the auditory system perceives *changes* in loudness. To a first approximation, equal *percentage* changes in rms sound pressure are perceived as equal changes in loudness. Hence, at high rms sound pressures, larger absolute changes in rms sound pressure are required to make a noticeable difference in loudness than at low rms sound pressures. The smallest difference in rms sound pressure that the human auditory system can detect is about 10 percent.

For these reasons a logarithmic scale, or *decibel scale,* is well suited for quantifying sound in a manner which relates to human perception. In its logarithmic form, rms sound pressure is called the rms *sound pressure level* (SPL). The sound pressure level is the logarithm of the ratio of two squared pressures, with the numerator containing the pressure of the sound source of interest and the denominator containing a reference pressure, as shown in Eq. (15-2). The units of SPL level are decibels (dB).

$$L_p = 10 \log \frac{p_{\mathrm{rms}}^2}{p_0^2} \tag{15-2}$$

where L_p = rms sound pressure level
p_{rms} = rms sound pressure
p_0 = reference pressure of 20×10^{-6} N/m² or 2.90×10^{-9} lb/in²
$\log$ = logarithm to base 10

The value of p_0 has been chosen to approximate the lowest rms sound pressure that a healthy young adult can hear. Substituting this barely audible rms sound pressure for p_{rms} in Eq. (15-2) produces a sound pressure level of 0 dB. In contrast, an rms sound pressure 1 million times greater produces a sound pressure level of 120 dB. Most sounds in our day-to-day environment have sound pressure levels on the order of 30 to 100 dB. Two useful rules of thumb for comparing SPLs are that (1) on average, people perceive a 6- to 10-dB increase in sound pressure level as a doubling of subjective loudness and (2) changes of less than 2 or 3 dB are not readily detectable outside a laboratory environment.

The A-weighted sound level. Another important attribute of sound is its frequency, or *pitch*. For a pure tone this is the number of times per second that the sound pressure oscillates about atmospheric pressure. The units of frequency are hertz (Hz), but they may also be referred to as cycles per second in references predating the adoption of hertz as an international standard. Virtually all sounds contain energy across a broad range of frequencies. Even the single note of a musical instrument contains a fundamental frequency plus a number of overtones.

The normal frequency range of hearing for a young adult extends from a low of 16 Hz to a high of about 16,000 Hz. However, the human auditory system is not equally sensitive across this entire range. Frequencies in the range of 2000 to 4000 Hz sound louder than lower or higher frequencies heard at the same rms sound pressure level. Thus, it is possible for two different sounds with the same sound pressure level to sound different in loudness.

For this reason the A-*weighted sound level* (A level) was developed. Incorporated in almost every commercially available sound-level meter, a standardized A-weighting filter adds gain or attenuation to different frequencies in a manner approximating the sensitivity of the human ear. The frequency response of the filter has a ± 3-dB effect in the midfrequency range between 500 and 10,000 Hz and increasing attenuation outside this range. Although the A-weighting filter is only an approximation of a complex physiological process, one sound judged louder than another will generally have a higher A-weighted sound level. Similarly, two sounds judged equally loud will generally have nearly the same A-weighted sound levels. A range of commonly encountered A-weighted sound levels is shown in Fig. 15-2.

For environmental assessment purposes, the A-weighted sound level represents a significant improvement over the overall (unweighted) sound pressure level. Unweighted sound pressure levels are rarely, if ever, used in environmental analyses. All federal agencies dealing with community noise, including transportation, have adopted the A-weighted sound level as the basic unit for analysis of environmental impact.

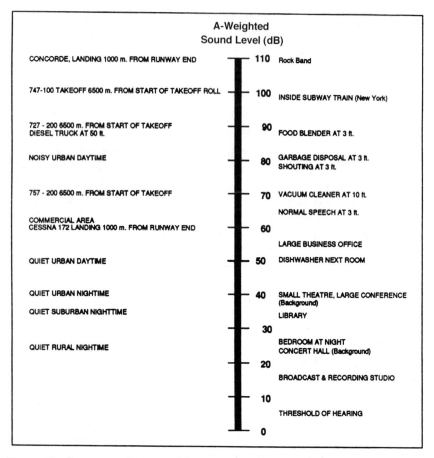

Figure 15-2 Common environmental A-weighted sound levels in decibels (*Harris Miller Miller Hanson*).

The A-weighted sound levels are measured in decibels, as are unweighted sound pressure levels and several other metrics discussed in this chapter. Thus, the measurement units themselves do not identify the quantity being reported. To avoid ambiguity, the quantity—in this case the A-weighted sound level—should always be reported along with the units. An example is an A-weighted sound level of 85 dB. [Although they do not meet current acoustical terminology standards, A-weighted sound levels may be reported in the literature as dBA, dB(A), or simply A-weighted.]

Maximum A-weighted sound level. In addition to sound *level*, another important dimension to environmental sound is its variation over time. A distant highway with relatively steady traffic, e.g., produces a fairly continuous background sound level with moment-to-moment variations of only a few dBs. In contrast, an aircraft pass-by produces a distinct, transient noise event. During an aircraft pass-by, the sound level emerges out of the fluctuating background environment, continues to increase until the aircraft passes the observer, and then decreases to blend in with the background noise as the aircraft recedes into the distance. Figure 15-3 illustrates this phenomenon.

For reporting as well as comparison purposes, it is desirable to use a single number for describing the sound level of such a noise event. A convenient metric is the maximum A-weighted sound level. This value is easy to measure, requiring an observer to simply note the maximum reading on a sound-level meter. It is also easy to describe since most people can relate to the loudest part of a noise event. In Fig. 15-3, the maximum A-weighted sound level is 85 dBA.

Sound exposure level. While being a very useful metric of aircraft noise events, the maximum sound level does not address the time element, or *duration,* of the event. During the late 1960s and early 1970s, several psychoacoustic listening studies were conducted to investigate how people assessed the relative noisiness of noise events with differing durations. All other things being equal, it was found that increased duration resulted in greater perceived noisiness. On average, the studies determined that people were willing to trade a

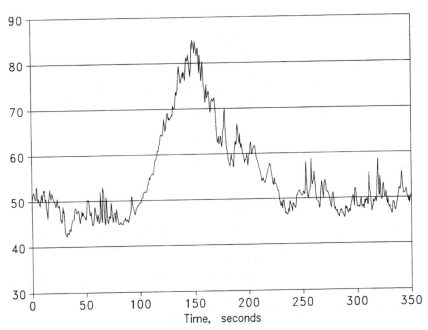

Figure 15-3 Typical A-weighted sound-level time history of an aircraft pass-by (*Harris Miller Miller Hanson*).

doubling of duration for a 3-dB reduction in maximum A-level sound. This finding supported a simple model for subjective noisiness: Noise events with equal time-integrated A-weighted sound energy are rated as equally noisy.

Developed to address this finding, the *sound exposure level* (SEL) is the total A-weighted sound energy contained in the noise event. Its description as a continuous integral is shown in Eq. (15-3). The units are decibels. Theoretically, this integral could approach infinity as T becomes large. If integration is over the top 20 dB, the computation will be only 0.1 dB less than the theoretical maximum.

$$L_{AE} = 10 \log \left(\frac{1}{T_0} \int_{t=0}^{T} 10^{L_{At}/10} \, dt \right) \tag{15-3}$$

where L_{AE} = sound exposure level
T_0 = 1 s to maintain a dimensionless argument for logarithm
L_{AT} = continuous A-weighted sound-level function describing the noise event time history. Limits of t from 0 to T are sufficient to encompass the top 10 to 20 dB of noise event.

For measurement purposes, the continuous integral presents two difficulties: (1) A continuous, mathematical function for the A-weighted sound-level time history is never known, and (2) the time limits of integration are nebulous since there is no precisely defined beginning or end to an aircraft noise event which slowly emerges from, and then blends back into, a time-varying background. These difficulties are circumvented by using discrete samples of A-weighted sound level, the summation approximation of Eq. (15-4), and empirically derived guidelines for the limits of i from 1 to N.

$$L_{AE} = 10 \log \left(\frac{1}{T_0} \sum_{i=1}^{N} 10^{L_{A,i}/10} \, \Delta t \right) \tag{15-4}$$

where $L_{A,i}$ is the instantaneous ith A-weighted sound level measured every 0.5 s and Δt is 0.5 s. The limits of i from 1 to N are sufficient to perform the summation over at least the top 10 dB of the noise event.

An accepted sampling interval Δt is 0.5 s. If the summation starts, i equal to 1, with the first sample to come within 20 dB of the maximum value and continues until the last sample, i equal to N, is within 20 dB of the maximum, then approximation will be about 0.1 dB lower than the theoretical value. If the summation is performed over only the top 10 dB of the time history, then the discrepancy will be less than 1 dB. This discrete summation approximation to the integral is used in all sound-level meters and monitoring devices. The use of Eq. (15-4) in computing the SEL is illustrated next.

Example Problem 15-1   The following sample of A-weighted sound levels was measured at 0.5-s intervals during an aircraft flyover: 64.5, 66.7, 67.1, 69.2, 71.3, 73.2, 74.1, 75.6, 77.8, 79.1, 78.6, 77.2, 75.7, 74.5, 72.6, 71.1, 69.7, 68.6, 68.0, and 66.4 dB.

To determine the SEL of this noise event, we must substitute into Eq. (15-4), obtaining

$$L_{AE} = 10 \log \left[\frac{1}{1} \, (10^{64.5/10} + 10^{66.7/10} + \cdots + 10^{66.4/10})(0.5) \right]$$

$$= 10 \log 253{,}734{,}091 = 84.0 \text{ dB}$$

Observe that even though none of the individual sound events had an A-weighted sound level in excess of 79.1 dB, the effect of the duration of the noise event leads to a numerically higher value for the SEL.

Because of the sound-level durations involved with typical aircraft pass-bys, the SEL will always be numerically larger than the maximum A-weighted sound level of the event. For most aircraft overflights, the difference is on the order of 7 to 12 dB. Factors affecting this difference are the aircraft speed (the greater the speed, the smaller the difference) and the distance to the aircraft at its closest point of approach to the observer (the greater the distance, the greater the difference).

Equivalent steady sound level.   The preceding discussion focused on measures of sound associated with individual events. It is frequently necessary, however, to quantify sound levels over longer periods such as an hour, several hours, or even a day. Such needs arise when one is tracking diurnal patterns or describing cumulative exposure over intermediate exposure periods (such as during school or office hours) or over a 24-h day. In contrast to the energy *summation* metrics used for individual noise events, energy *average* metrics are used for longer periods. One such metric is the *equivalent steady sound level* (QL). (Although it does not meet current acoustical terminology standards, the equivalent steady sound level may be reported in the literature as the *energy average sound level* L_{eq} or LEQ.) The units are decibels. Mathematically, the equivalent steady sound level is the sound pressure level shown in Eq. (15-2) calculated by using a long-term rms sound pressure [T in Eq. (15-1) equals the period of interest]. As a practical matter, the equivalent steady sound level is almost always calculated from a time series of A-weighted sound levels acquired with sound-level meter readings taken at 0.5-s intervals or less. Equation (15-5) shows the manner in which the equivalent steady sound level is computed from a discrete time series of data.

$$L_{eq} = 10 \log \left(\frac{1}{T} \sum_{i=1}^{N} 10^{L_{A,i}/10} \Delta t \right) \tag{15-5}$$

where L_{eq} = equivalent steady sound level
  $L_{A,i}$ = instantaneous ith A-weighted sound level measured every 0.5 s
  T = time period of interest (for example, 1 h)
  Δt = 0.5 s or less typically
  N = $T/\Delta t$, with T and Δt in same units

An equivalent and computationally more efficient manner of expressing Eq. (15-5) is

$$L_{eq} = 10 \log \left(\frac{1}{N} \sum_{i=1}^{N} 10^{L_{A,i}/10} \right) \tag{15-6}$$

where $L_{A,i}$ is the instantaneous ith A-weighted sound level measured every 0.5 s and N is total number of sound-level samples.

The computational process described in Eqs. (15-5) and (15-6) does not make any distinction between sources; i.e., it accumulates sound levels produced by both aircraft and nonaircraft sources. When the equivalent sound level is computed in this manner, it is called the *total* equivalent sound level. It is often useful, however, to know only the aircraft component. The aircraft component can be calculated from the SELs of individual events from

$$L_{eq} = 10 \log \left(\frac{1}{T} \sum_{j=1}^{M} 10^{L_{AE,j}/10} \right) \tag{15-7}$$

where $L_{AE,j}$ = sound exposure level produced by jth aircraft pass-by during time period
  T = time period of interest (say, 1 h), s
  M = number of aircraft noise events during period T

Functionally, this equation accumulates all the aircraft sound energy from multiple events, then spreads it out uniformly over the time period by dividing by the length of the period (not just the length of time that aircraft were present).

The computation of the hourly average sound level is illustrated in this example problem.

Example Problem 15-2 The following SELs for four aircraft flyovers were measured in a 1-h period: 84.0, 89.1, 90.2, and 86.6 dBA.

To compute the hourly average sound level, we must substitute into Eq.

(15-7), obtaining

$$L_{eq} = 10 \log \left[\frac{1}{3600} \left(10^{84.0/10} + 10^{89.1/10} + 10^{90.2/10} + 10^{86.6/10}\right) \right]$$

$$= 10 \log 713{,}339 \approx 58.5 \text{ dBA}$$

Observe that even though the SEL of each aircraft flyover was greater than 58.5 dB, the averaging process reduces the hourly average SEL.

Experience has shown that the concept of an *average* sound level is often misinterpreted by the affected public as an underreporting or understatement of their noise environment. Their concern is that the metric does not report the *total* noise energy over the period. As can be seen in Eqs. (15-5), (15-6), and (15-7), this metric, as well as other average metrics, does indeed include all the noise energy. Each and every noise event, no matter how high or low the sound level, increases the value of the metric. Viewed another way, the average value is the total noise energy adjusted by the constant $10 \log T$.

The local community component of the equivalent steady sound level is also a frequently reported statistic which serves as a basis of comparison for the aircraft component. It may be estimated by using a variant of Eq. (15-6) which accumulates sound levels only during subintervals of the total period when no aircraft are present. It is an *estimate* because there is no way of knowing the community sound-level contribution during periods when aircraft are present. Equation (15-8) shows the basic summation process. Each summation in the equation represents a nonaircraft subinterval.

$$L_{eq} = 10 \log \left[\frac{1}{N} \left(\sum_{i=1}^{N_1} 10^{L_{A,i}/10} + \sum_{i=1}^{N_2} 10^{L_{A,i}/10} + \cdots + \sum_{i=1}^{N_n} 10^{L_{A,i}/10} \right) \right] \quad (15\text{-}8)$$

where
$L_{A,i}$ = instantaneous ith A-weighted sound level measured every 0.5 s

N_1, N_2, N_n = number of sound-level samples in each subinterval containing no aircraft noise

N = total number of samples, which is equal to $N_1 + N_2 + \cdots + N_n$

This metric is referred to as the *hourly average sound level* when 1 h of averaging time is used. (Although it does not meet current acoustical terminology standards, the hourly average sound level may be reported in the literature as the *hourly noise level, HNL, hourly* L_{eq}, or 1-h L_{eq}.) In airport applications, hourly average sound levels may be used for plotting and visualizing diurnal trends. And 8- and 24-h periods are referred to as 8- and 24-h average sound levels. The symbol L_{eq} is generic, referring to any arbitrary period. To avoid ambiguity, the subscript is replaced by the appropriate time frame. Thus, the symbols $L_{1\text{ h}}$, $L_{8\text{ h}}$, and $L_{24\text{ h}}$ are used for 1-, 8-, and 24-h periods, respectively.

Day-night average sound level. As the name implies, the day-night average sound level DNL is a metric used to describe sound exposure over a 24-h period. The units are decibels. Computationally it is identical to the 24-h average sound level with one important difference. The DNL incorporates a time-of-day weighting which adds 10 dB to sound levels occurring between 10 p.m. and 7 a.m. Although periodically the magnitude of the weighting becomes a topic of discussion within the scientific community, the intent is to account for a presumed increase in human sensitivity to noise during nighttime hours. While the formal definition is a continuous integral, Eq. (15-9) shows the formula for computing the total (aircraft plus community sources) DNL from discrete samples of the A-weighted sound level.

$$L_{dn} = 10 \log \left(\frac{1}{86,400} \sum_{i=1}^{N} 10^{(L_{A,i} + W_i)/10} \, \Delta t \right) \qquad (15\text{-}9)$$

where L_{dn} = day-night average sound level for 1 day
           $L_{A,i}$ = instantaneous ith A-weighted sound level, measured every 0.5 s
    86,400 = number of seconds in 1 day
           W_i = time-of-day weighting for ith A-weighted sound level (0 dB if occurred between 7 a.m. and 10 p.m.; 10 dB if between 10 p.m. and 7 a.m.)
           Δt = 0.5 s or less typically; units must be seconds
           N = 86,400/Δt

The aircraft component of DNL may be computed from SELs of individual events by using Eq. (15-10).

$$L_{dn} = 10 \log \left(\frac{1}{86,400} \sum_{j=1}^{M} 10^{(L_{AE,j} + W_j)/10} \right) \qquad (15\text{-}10)$$

where $L_{AE,j}$ = sound exposure level produced by jth aircraft pass-by during day
          W_j = time-of-day weighting for the jth aircraft pass-by (0 dBA if between 7 a.m. and 10 p.m.; 10 dBA if between 10 p.m. and 7 a.m.)
         M = number of aircraft noise events during 24-h period

The application of this equation to determine the DNL of several aircraft flyovers at various times during the day is illustrated next.

Example Problem 15-3 The following SELs of five aircraft flyovers were measured over the course of a 24-h period: 81.2 dBA at 6:03 a.m., 95.1 dBA at 10:32 a.m., 79.2 dBA at 2:15 p.m., 88.8 dBA at 7:33 p.m., and 71.2 dBA at 10:05 p.m.

To compute the DNL, we must substitute into Eq. (15-10). The sound exposure levels at 6:03 a.m. and 10:05 p.m. must be increased by the time-of-day weighting of 10 dBA since these flyovers occurred between 10:00 p.m. and 7:00 a.m.

$$L_{dn} = 10 \log \left[\frac{1}{86,400} (10^{91.2/10} + 10^{95.1/10} + \cdots + 10^{81.2/10}) \right]$$

$$= 10 \log 63,978.85 = 48.1 \text{ dBA}$$

To find the aircraft which has the greatest and least contribution to the day-night average sound level, the value of the quantity $10^{(L_{AE,j} + W_j)/10}$ must be evaluated for each aircraft. Clearly, by adding the time-of-day weighting, we see that the aircraft flyover at 10:32 a.m. is the greatest contributor and that the aircraft flyover at 2:15 p.m. is the least contributor to the day-night average sound exposure level.

A useful rule of thumb for estimating the contribution of DNL to a single daytime (7 a.m. to 10 p.m.) noise event may be obtained by simplifying Eq. (15-10) for the condition where M is equal to 1. The approximation shown in Eq. (15-11) is accurate to within 0.5 dB.

$$L_{dn} \approx L_{AE} - 50 \qquad (15\text{-}11)$$

where L_{AE} is the sound exposure level of a single aircraft pass-by. The use of Eqs. (15-10) and (15-11) to compute the DNL of a single daytime noise event is illustrated now.

Example Problem 15-4 Let us determine the DNL produced by a single daytime noise event with an SEL of 105 dBA.

Using Eq. (15-10), we have

$$L_{dn} = 10 \log \left[\frac{1}{86,400} (10^{105.0/10}) \right]$$

$$= 10 \log 366,004 = 55.6 \text{ dBA}$$

Using Eq. (15-11), we have

$$L_{dn} \approx 105 - 50 \approx 55 \ dBA$$

If this noise event were added to the noise events in Example Problem 15-3, we would find that the DNL was increased to 56.3 dBA, or there would be an increase of 0.7 dB.

Environmental reporting criteria often require annual average values of DNL. Both airport and atmospheric factors contribute to day-to-day variability in the DNL observed at a particular location near the airport. In cases where average values must be computed from measurements, the averaging must be done on a sound-energy basis.

Equation (15-12) shows the formula for computing the annual average value $L_{dn,\text{ann}}$.

$$L_{dn,\text{ann}} = 10 \log \left(\frac{1}{365} \sum_{i=1}^{365} 10^{L_{dn,\,i}/10} \right) \tag{15-12}$$

where $L_{dn,i}$ is the DNL for the ith day of the year. This equation assumes that 365 individual DNL values are to be used in the averaging process. For conditions where the number of days differs from 365 (leap years, missing data, etc.), the available number of data points should be used in the summation, and the number 365 should be replaced by the actual number of data points used.

Representative values of DNL range from a low of 40 to 45 dB in extremely quiet, isolated locations to highs of 80 or 85 dB immediately adjacent to a busy truck route or just off the end of a runway at an active military air base. The Environmental Protection Agency (EPA) identified this measure as the most appropriate means of evaluating community (including aircraft) noise in 1974 [28]. Most other public agencies dealing with noise exposure, including the FAA, the Department of Defense, and the Department of Housing and Urban Development (HUD), also have adopted DNL in their guidelines and regulations.

Time-above threshold level. The preceding metrics quantify noise exposure in terms of sound level or sound energy. An alternate descriptor uses duration, or time, as the basic metric. The metric is *time above* (TA), defined as the length of time that the A-weighted sound level exceeds a specified *threshold* level over a given period. Typically TA is reported as the number of minutes per day that the A-weighted sound level exceeds values of 55, 65, 75, 85, 95, and 105 dB. The historical appeal of TA has generally been its simplicity; that is, TAs are arithmetically additive. For single noise events, TAs can be arithmetically added to compute hourly TAs, and hourly TAs can be arithmetically added to find the 24-h values. Proponents of TA argue that the arithmetic addition process allows easy-to-understand assessments of major contributors to 24-h totals. TA may be required for some environmental analyses. However, at present there are no accepted criteria or land-use compatibility guidelines using TA.

Other single-event sound-level metrics. The *perceived noise level* (PNL) and *effective perceived noise level* (EPNL) are quantities similar to the A level and sound exposure level, respectively. The PNL and EPNL were developed specifically to correlate with subjective response to aircraft sound. The perceived noise level, in units of PNdB, is a quantity which varies from moment to moment, just as the A-weighted sound level does. As a general rule, the PNL is approximately 13 dB greater than the A-weighted sound level.

The effective perceived noise level, in units of EPNdB, is a single-event metric which sums the PNL in a manner similar to the way SEL sums the A level. However, EPNL also incorporates a tone correction (TC) adjustment to account for the increased subjective noisiness of sounds containing discrete frequency tones (like those produced by turbofan engine compressor blades). The formula for computing EPNL is [39]

$$L_{\text{EPN}} = 10 \log \left(\frac{1}{T_0} \sum_{i=1}^{N} 10^{(L_{\text{PN},i} + TC_i)/10} \Delta t \right) \qquad (15\text{-}13)$$

where L_{EPN} = effective perceived noise level
  $L_{\text{PN},i}$ = instantaneous ith perceived noise level, measured every 0.5 s
  TC_i = instantaneous ith tone correction
  Δt = 0.5 s
  T_0 = 10 s. Limits of i from 1 to N are sufficient to sum over top 10 dB of noise event

Both the PNL and the tone correction are computed from sound pressure levels measured in individual one-third-octave bands from 50 to 10,000 Hz. The magnitude of the tone correction ranges from 0 to 6 dB depending upon the frequency where the tone occurs and upon the sound pressure level of the tone relative to the broadband noise in the same frequency range. As a general rule, the EPNL is about 3 dB greater than SEL, but it can be more if very noticeable pure tones are present or less at very large distances.

Because of the complexity involved in measurement—sophisticated frequency analyses and nonlinear amplitude adjustments are required—PNL and EPNL are not used in the United States for routine environmental analyses. Their current use is limited to aircraft airworthiness certification under FAR part 36 [39].

Other 24-h sound-level metrics. The *community noise equivalent level* (CNEL) adopted in California airport noise standards was actually a forerunner of DNL. The computation procedure is virtually identical to that for DNL. Equations (15-9) and (15-10) can be used to compute CNEL, with the only difference being the use of three weighting periods instead of two. For 7 a.m. to 7 p.m. the weighting is 0 dB, for 7 p.m. to 10 p.m. the weighting is 4.77 dB (the actual weighting is a factor of 3 in sound energy and 10 log 3, or 4.77 dB), and for 10 p.m. to 7 a.m. the weighting is 10 dB. The only difference between the two metrics is the approximately 5-dB weighting during the three evening hours from 7 p.m. to 10 p.m. Numerically, CNEL is always greater than DNL. but from a practical standpoint this difference is rarely more than 1 dB.

Before the adoption of DNL, two other descriptors of daily noise exposure were used to quantify noise impacts around airports. Neither is still in active use in the United States today. The *composite noise rating* (CNR) was one of the first 24-h metrics to embody individual aircraft sound levels, their frequency of occurrence, and their time of day in a single-number rating. Predating the development and use of personal computers by more than two decades, CNR calculations were performed by using a handbook procedure published in 1963 under a joint effort by the U.S. Air Force and the FAA. The CNR used the maximum PNL as the single-event sound-level descriptor.

The *noise exposure forecast* (NEF) was developed shortly thereafter in 1967, and it quickly replaced CNR. The NEF uses EPNL as the single-event sound-level descriptor and a sound-energy summation process similar to DNL. Equation (15-14) shows the formula for computing NEF. Because of the computational complexities involved in their underlying single-event metrics, both CNR and NEF fell into disuse with the adoption of DNL.

$$\text{NEF} = 10 \log \left(\sum_{j=1}^{M} 10^{(L_{\text{EPN},j} + W_j)/10} \right) - 88 \qquad (15\text{-}14)$$

where $L_{\text{EPN},j}$ = EPNL produced by jth aircraft pass-by during day
$\quad\quad W_j$ = time-of-day weighting for jth aircraft pass-by (0 dB if between 7 a.m. and 10 p.m.; 12 dB if between 10 p.m. and 7 a.m.)
$\quad\quad 88$ = adjustment factor designed to shift metric to lower numeric range not occupied by any other then-current 24-h metric

Because of differences in frequency weightings, differences in accounting for the duration of individual events, and differences in the evening and nighttime weightings, there is no exact functional relationship among these three metrics. Within ± 3 dB, however, the relationship shown in Eq. (15-15) has been found to be valid. Thus, for DNL and CNEL values of 65, an NEF value of 30, and a CNR value of 100, all indicate approximately the same degree of noise exposure, within ± 3 dB.

$$L_{dn} \approx \text{NEF} + 35 \approx \text{CNR} - 35 \qquad (15\text{-}15)$$

Aircraft noise effects and land-use compatibility.   The effects of noise on people can be classified into one of two categories, namely, behavioral effects and health or physiological effects. Behavioral effects are associated with activity interference. These effects include annoyance and the interference with communication, mental activity, rest, and sleep. Health effects are those that produce hearing loss or nonauditory effects such as cardiovascular disease and hypertension.

Various federal agencies have developed guidelines for assessing the compatibility of noise with land uses, including the EPA, HUD, and the FAA. All the guidelines are based on the day-night average sound level (DNL) and were designed to protect public health and welfare, but also take into account the feasibility of controlling noise [28, 44].

Speech interference. One of the primary effects of aircraft noise is the tendency to drown out or *mask* speech, making it difficult or impossible to carry on a normal conversation without interruption. The sound level of speech decreases as the distance between talker and listener increases. As the level of speech decreases in the presence of background noise, it becomes harder and harder to hear. Figure 15-4 presents typical distances between talker and listener for satisfactory

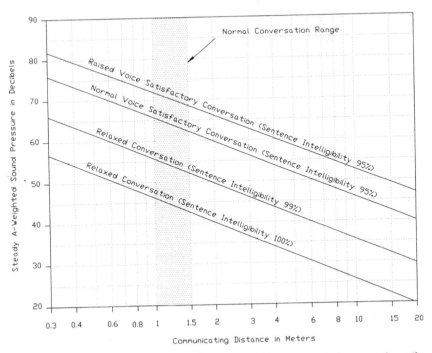

Figure 15-4 Maximum distances outdoors over which conversation is satisfactorily intelligible outdoors in a steady noise environment (*adapted from Environmental Protection Agency [28]*).

outdoor conversations in the presence of different steady A-weighted background sound levels for three degrees of vocal effort, namely, raised, normal, and relaxed. As the background level increases, the individuals either must talk louder or must get closer together to continue the conversation.

As indicated in the figure, *satisfactory conversation* does not always require hearing every word—95 percent intelligibility is acceptable for many conversations. This is because a few unheard words can be inferred when they occur in a familiar context. However, for relaxed conversation, people have higher expectations of hearing speech and require complete, 100 percent intelligibility. Any combination of talker-listener distances and background noise that falls below the bottom line in Fig. 15-4, thus ensuring 100 percent intelligibility, represents an ideal environment for outdoor speech communication and is considered necessary for acceptable indoor conversation as well.

One implication of the relationships in Fig. 15-4 is that for typical communication distances of 3 or 4 ft (1 to 1.5 m), acceptable outdoor conversations where 95 percent intelligibility is acceptable can be carried on in a normal voice as long as the background A-weighted sound level is less than about 65 dB. In other situations, where greater speech intelligibility is required, background levels must be lower. For example, indoors, where 100 percent intelligibility is desired, the background A-weighted sound level must be less than about 45 dB. If the noise exceeds either of these levels, as might occur when an aircraft passes overhead, intelligibility is lost unless vocal effort is increased or communication distance decreased.

A second implication of these relationships is that an acceptable A-weighted background level of 60 to 65 dB outdoors does not guarantee an acceptable background level indoors. This is because most housing construction typically provides about 15 dB of sound attenuation from outside to inside the building when windows are open. Thus, only if the outdoor A-weighted sound level is 60 dB or less will there be a reasonable chance of the resulting indoor sound level affording acceptable conversation inside.

Sleep interference. The disruptive effects of noise on sleep can be a concern to communities exposed to aircraft overflights during nighttime hours. Over the past two decades, many investigations of noise-induced sleep disruption have been conducted worldwide. The functional relationship most often evaluated has been the probability of a sleep disruption created by a single noise intrusion of a given sound level. Four major research review studies [10, 32, 37, 46] all support the same general finding that increased sound exposure level (SEL) results in higher probabilities of sleep disruption. Currently, however, there are no guidelines or acceptability criteria for assessing the cumulative impact of many aircraft noise intrusions of various SELs over the course of a nighttime sleeping period.

The aforementioned reviews provide some insight as to why such guidelines and criteria have not been forthcoming. Data acquisition methodologies, the treatment of mitigating variables, and the choice of both dose and disruption metrics are far from standardized, even for studies limited to short-term, transient noise events similar to those produced by aircraft overflights.

Despite all this variability, first-order dose-response curves have been developed which attempt to relate the probability of an arousal, either a sleep stage change or an actual awakening, to the SEL of a single noise event [11, 26]. At present, caution should be exercised in drawing inferences from such curves. Among other things, the preponderance of underlying data generally represents laboratory listening conditions. Thus the curves could be expected to better predict reaction to new and unfamiliar sounds rather than to older, more familiar sounds. When compared with in-laboratory studies, the limited data available from in-home investigations using familiar sound sources suggest that arousal probabilities may be on the order of only one-eighth those observed from unfamiliar sources. In addition to these adaptation issues, uncertainty still remains on important questions such as the cumulative effects of multiple noise intrusions, the effects of noise on falling asleep as opposed to awakening, and the extent to which sleep deprivation represents a quantifiable physiological problem.

Community annoyance. Social survey data have long made it clear that individual reactions to noise vary widely for a given 24-h average sound level. As a group, however, the aggregate response of people to factors such as speech and sleep interference and desire for an acceptable environment is predictable and relates well to measures of cumulative noise exposure such as DNL. Figure 15-5 shows the most widely recognized relationship between day-night average sound level and the percentage of people highly annoyed, regardless of the noise source. Based on data from 18 surveys conducted worldwide of the attitudes of people toward noise, the curve indicates that the relationship between group reaction and 24-h exposure is quantifiable. The curve shows that at DNLs as low as 55 dB, approximately 5 percent of the people will still be highly annoyed with their noise environment. The percentage increases more rapidly as the DNL increases above 65 dB [49].

Separate work by the EPA suggests that overall community reaction to a noise environment is also dependent on the level of the intruding noise compared to the level of the existing noise. Research was conducted to determine the relationship between intruding noise

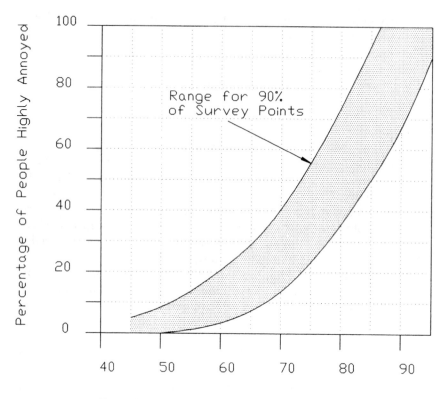

Day-Night Average Sound Level (decibels)

Figure 15-5 Percentage of people highly annoyed as a function of day-night average sound level (*adapted from Schultz, Journal of the Acoustical Society of America [49]*).

level and community reaction for 55 cases of community noise intrusion where reactions were known [16]. The data were normalized to the same set of conditions so that the cases were comparable. In particular, the conditions were adjusted to an existing noise environment, without the intruding noise, of about 60 dB. The data show that sporadic complaints occurred when the intruding equivalent noise level was between about 59 and 65 dB, and widespread complaints occurred when the intruding noise fell between about 63 and 75 dB.

The implication of this research is that complaints may begin to occur when the aircraft DNL is approximately equal to the background DNL, and that widespread complaints start to occur when the aircraft DNL exceeds the background DNL by 3 to 5 dB. Such a conclusion provides some assistance in anticipating what community reaction could be to a change in the noise level of an intruding source, such as an airport. If the change is likely to result in an increase above existing noise levels of 3 to 5 dB, some community reaction may be expected [16].

Noise-induced hearing loss. Hearing loss is measured as *threshold shift*. The term *threshold* refers to the quietest sound a person can hear. When a threshold shift occurs, the sound must be louder before it can be heard. For hundreds of years it has been known that excessive exposure to loud noises can lead to noise-induced temporary threshold shifts, which in time can result in permanent hearing impairment. With a threshold shift of 25 dB a person could correctly understand only about 90 percent of the sentences spoken in a conversational level at a 3-ft (1-m) distance in a quiet room [42].

Research over the last 40 years on industrial and military populations provides an understanding of the development of noise-induced hearing loss and its relationship to noise level, spectral content, and length of exposure. Detailed international criteria have been developed that identify maximum noise exposures which do not produce noise-induced hearing loss in any segment of the population exposed [1]. The Occupational Safety and Health Administration (OSHA) regulation [41] cites a maximum permissible A-weighted sound exposure of 90 dB for 8 h.

It is extremely unlikely that aircraft noise around airports could ever produce hearing loss. For example, it would take continuous exposure to more than 1000 overflights per day with an SEL of 100 dB each to produce a time-weighted average sound level of 85 dB. If this occurred 5 days per week for 40 years, and if people were exposed to this outdoors without any attenuation from buildings, then the resultant noise exposure would start to produce a *noise-induced permanent threshold shift* (NIPTS) of less than 10 dB in the most sensitive 10 percent of the population.

Nonauditory health effects. Concern is often raised that noise has adverse effects on human health other than hearing. In spite of considerable worldwide research, however, there is no unambiguous scientific evidence to relate quantitatively any noise environment with the origin of or contribution to any clinical nonauditory disease. Even the most recent research, conducted at levels above the limits for conservation of hearing, failed to give consistent results. Most authoritative reviews agree with the World Health Organization's environmental health criteria document on noise [33], which states that "research on this subject has not yielded any positive evidence, so far, that disease is caused or aggravated by noise exposure, insufficient to cause hearing impairment." For practical noise control considerations, the present status of our knowledge means that by using criteria that prevent noise-induced hearing loss, minimize speech and sleep disruption, and minimize community reactions and annoyance, any effects on health will also be prevented. In general, these guidelines should not be regarded as identifying levels of exposure that are desirable, but rather as a balancing of what is desirable with what is feasible.

Noise and land-use compatibility guidelines. Based on the relationships between noise and the collective response of people to their environment, the DNL has become accepted as a standard for evaluating community noise exposure and as a decision-making aid regarding the compatibility of alternative land uses.

In their application to airport noise in particular, DNL projections have two principal functions:

1. To provide a means for comparing existing noise conditions with those that might result from the implementation of noise abatement procedures or from forecast changes in airport activity

2. To provide a quantitative basis for identifying and judging potential noise impacts

Both these functions require the application of objective criteria. Government agencies dealing with environmental noise have devoted significant attention to this issue and have developed noise and land-use compatibility guidelines to help federal, state, and local officials with the noise evaluation process.

In FAR part 150 [6], which defines procedures for developing airport noise compatibility programs, the FAA has established DNL as the official cumulative noise exposure metric for use in airport noise analyses, and it has developed guidelines for noise and land-use compatibility evaluation. Table 15-2 presents these guidelines.

The guidelines represent a compilation of extensive scientific research into noise-related activity interference and attitudinal response. However, reviewers of DNL contours should recognize the highly subjective nature of response to noise and the special circumstances that can either increase or decrease the tolerance of an individual. For example, a high nonaircraft background or ambient noise level, such as from ground traffic, can reduce the significance of aircraft noise. Alternatively, residents of areas with unusually low background noise levels may find relatively low levels of aircraft noise very annoying. Response may also be affected by expectation and experience. People often get used to a level of noise exposure that guidelines suggest may be unacceptable, and similarly, changes in exposure may generate a response that is far greater than the guidelines might suggest.

Finally, the cumulative nature of DNL means that the same level of

TABLE 15-2 FAA Noise and Land-Use Compatibility Guidelines

	Yearly day-night average sound level DNL, dB					
	Below 65	65–70	70–75	75–80	80–85	Over 85
Residential use						
Residential other than mobile homes and transient lodgings	Y*	N	N	N	N	N
Mobile home park	Y	N	N	N	N	N
Transient lodgings	Y	N	N	N	N	N
Public use						
Schools	Y	N	N	N	N	N
Hospitals and nursing homes	Y	25	30	N	N	N
Churches, auditoriums, and concert halls	Y	25	30	N	N	N
Government services	Y	Y	25	30	N	N
Transportation	Y	Y	Y	Y	Y	Y
Parking	Y	Y	Y	Y	Y	N
Commercial use						
Offices, business and professional	Y	Y	25	30	N	N
Wholesale and retail—building materials, hardware, and farm equipment	Y	Y	Y	Y	Y	N
Retail trade—general	Y	Y	25	30	N	N
Utilities	Y	Y	Y	Y	Y	N
Communication	Y	Y	25	30	N	N
Manufacturing and production						
Manufacturing, general	Y	Y	Y	Y	Y	N
Photographic and optical	Y	Y	25	30	N	N
Agriculture (except livestock) and forestry	Y	Y	Y	Y	Y	Y
Livestock farming and breeding	Y	Y	Y	N	N	N
Mining and fishing, resource production and extraction	Y	Y	Y	Y	Y	Y
Recreational						
Outdoor sports arenas and spectator sports	Y	Y	Y	N	N	N
Outdoor music shells, amphitheaters	Y	N	N	N	N	N
Nature exhibits and zoos	Y	Y	N	N	N	N
Amusements, parks, resorts, and camps	Y	Y	Y	Y	Y	Y
Golf courses, riding stables, and water recreation	Y	Y	25	30	N	N

*Note:

Y (yes) Land use and related structures are compatible without restrictions.

N (no) Land use and related structures are not compatible and should be prohibited.

25, 30, or 35 Land use and related structures generally compatible; measures to achieve outdoor-to-indoor noise level reduction of 25, 30, or 35 dB must be incorporated into design and construction of structure.

There are special provisions pertaining to many of the compatibility designations that are not included here; refer to FAR part 150 [6] for details.

SOURCE: Federal Aviation Administration [36].

noise exposure can be achieved in an essentially infinite number of ways. For example, a large increase in relatively quiet flights can counterbalance a smaller reduction in relatively noisy operations, with no net change in DNL. The increased frequency of operations can annoy residents, despite the apparent unchanged status quo of the noise. With these words of caution in mind, the guidelines of the FAA for compatible land use can be combined with DNL contours indicating points of equal exposure to identify the potential types and locations of land uses and the degree of their incompatibility. Note that, by these guidelines, all land uses are considered compatible with aircraft day-night average sound levels below 65 dB. This does not mean that people will not complain or otherwise be disturbed by aircraft noise at lower levels, as has been shown earlier; nor does it preclude individual communities or other jurisdictions from adopting lower standards to meet local needs.

Determining the extent of the problem

The extent of a potential or ongoing airport noise problem is generally quantified in one of two ways:

1. Prediction using computer-based simulation models

2. Measurement through portable or permanent monitoring systems

The simulation models produce maps depicting contours of equal sound level such as DNL. Measurements are used to provide or confirm input to the simulation models as well as to confirm model predictions at specific ground locations.

The integrated noise model and NOISEMAP. Two computer-based simulation models are currently used in the United States. Both produce maps showing contours of equal day-night average sound level. Developed by the FAA, the *integrated noise model* (INM) is most often used for civil airports [29]. NOISEMAP, developed by the U.S. Air Force, is generally used for military airbases but is also used for civil and joint-use airports. The FAA has approved both models for use in airport noise studies. Both models run on personal computers, but due to the highly computation-intensive nature of the programs, state-of-the-art processors are recommended. The two models require the same basic input parameters, but their formats differ.

Use of either model requires inputs in two principal categories: aircraft noise and performance data and aircraft operational data. The major difference between the two categories of input is that the first is generally not airport-dependent while the second is airport-specific and must be individually developed for each airport.

Aircraft noise and performance data. The INM uses a standard, internal noise and performance database containing a large number of aircraft types. The model uses the noise data to determine the SEL of specific aircraft types as a function of thrust and distance from the observer. The performance data used by the model define the length of the takeoff roll, climb rate, speed, and thrust management for both departures and arrivals.

Aircraft operational data. The INM also requires operational input data specific to the airport under study. These data are often difficult to obtain because they are not routinely collected by either the airport or the FAA. To address this problem, airports are beginning to develop specific data collection procedures for this specific purpose. Operational inputs describe activity at the airport by using average values during the period of interest, and they include the following:

1. Physical description of the airport runways, including any displaced takeoff or landing thresholds

2. Runway utilization percentages

3. Number of aircraft operations by aircraft type for all noise-significant aircraft types in the fleet mix

4. Day-night split of operations by aircraft type

5. Flight corridor descriptions

6. Flight corridor utilization percentages

Output of noise models. Both the INM and NOISEMAP produce output in two forms, namely, contour maps of equal day-night average sound level and detailed tabular analyses for user-specified ground locations. Figure 15-6 shows an example of a DNL contour map. A typical map shows contours from 60 to 80 dB at 5-dB intervals. For presentation purposes, these contours are superimposed graphically on a good-quality base map.

In addition to DNL contours, SEL contours can be helpful in addressing issues of sleep and speech interference and for analyzing the effects of noise abatement procedures, such as proposed noise abatement flight tracks. Graphical comparisons of SEL contours of

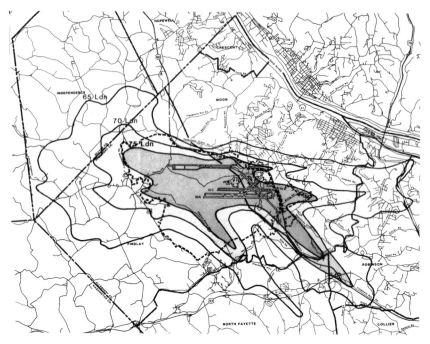

Figure 15-6   Noise exposure map for Greater Pittsburgh International Airport (*Aviation Planning Associates, Inc. [7]*).

various aircraft types can also provide powerful images for comparing noise emissions of differing aircraft types.

Tabular listings for user-specified ground locations show not only the predicted DNL but also the SEL and DNL contribution of individual aircraft by runway and flight corridor. This information is invaluable to understanding the major contributors to the total DNL. It can also be used to compare the model predictions with data from noise-monitoring locations. Such comparisons often provide the basis for fine-tuning model inputs as well as promoting public confidence in the computer model and the contours it produces.

Aircraft noise and operations monitoring.   Many civil aviation airports in the United States have installed aircraft noise monitoring systems to assist in managing airport-community relations. The first systems, installed 20 or more years ago, performed strictly sound-level monitoring. Current technology systems have evolved into complete noise monitoring systems capable of providing information on both aircraft sound levels and aircraft operations. The primary uses of airport noise and operations monitoring systems are to help establish and monitor compliance with noise abatement procedures, verify trends in overall fleet noise, and provide input and validation data for computer-based airport noise simulation models.

When people complain about aircraft noise, the complaint is often followed by a reference to some operational characteristic of the aircraft which differed from their expectations. For example, "they're not supposed to fly directly over my house," "that aircraft flew too low," or "they never used to use that runway so often." While admittedly anecdotal, such informative complaints can be extremely helpful in pinpointing the operational source of the complaint and in starting a process of noise mitigation and community education.

The operations side of the monitoring system provides airport managers with the tools to verify the underlying cause of complaints, determine the extent of identified problems, and provide an objective basis for seeking solutions. The primary source of information for modern systems is data routinely collected by the FAA with their automated radar terminal system (ARTS). The ARTS retains information sufficient to reconstruct the three-dimensional flight trajectory, with the aircraft type, airline, flight number, and type of operation (departure or arrival) for every commercial aircraft movement. Modern, computer-based operations monitoring systems access these data, provide extensive on-line data storage capacity, and embody sophisticated database management systems for retrieving, sorting, and reporting the enormous volumes of data they acquire.

Useful, long-term summary statistics from operations monitoring systems include the percentage of runway utilization, with breakdowns by departures and arrivals and by aircraft type, and overall traffic counts, with breakdowns by aircraft type and by time of day. Detailed presentations of actual aircraft flight tracks, such as those shown in Fig. 15-7, are extremely helpful for examining noise abatement alternatives. The data from operations monitoring systems are also one of the few objective data sources for preparing accurate and defensible airport noise contours.

The sound-level monitoring side of the system consists of a number of remote microphones located in the community surrounding the airport and a central processing site usually located at airport administrative offices. Microphones are located on top of 7-m-high poles, and the microphone signal is processed in real time *at the pole* to compute and store most all sound-level metrics of interest. Data are then transmitted digitally from the microphone site to the central station via a modem and voice-grade telephone lines.

Finding solutions

Table 15-3 presents a matrix of aircraft-related noise problems and

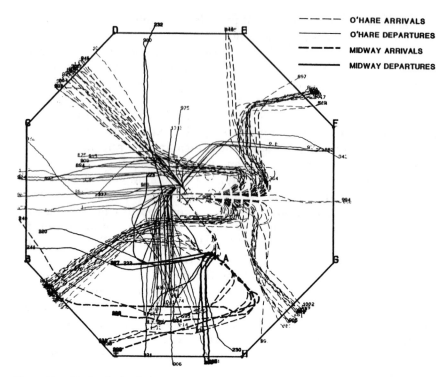

Figure 15-7 Radar-derived air carrier departure flight tracks in Chicago area terminal airspace (*Landrum and Brown Aviation Consultants [14]*).

potential solutions [36]. In general, the solutions seek to increase distance between the aircraft and noise-sensitive elements of the community, reduce noise levels at the source, or reduce the numbers of noise events in noise-sensitive areas. Some specific solutions require FAA expertise and approval; hence the involvement of the agency should be sought at the earliest possible opportunity. Details of some solutions are discussed in the following paragraphs.

Noise barriers.   Noise barriers offer opportunities for controlling ground-based noise sources such as takeoff and landing roll, taxiway and apron movements, aircraft power-backs, *auxiliary power units* (APUs), and maintenance engine runs. To be effective, the barriers must break the line of sight between the noise source and the receiver. Hence, they provide no benefit once the aircraft is airborne and is visible above the barrier. Maximum effectiveness is achieved when a barrier is close to either the source or the receiver, rather than halfway between them.

TABLE 15-3   **Matrix of Noise Control Actions**

If you have this problem with noise from: / Consider these actions		Taxiing	Departure	Approach	Landing roll	Training flights	Maintenance	Ground equipment
Airport plan	Changes in runway location, length, or strength	●	●	●	●	●		
	Displaced thresholds			●		●		
	High-speed exit taxiways	●		●				
	Relocated terminals	●					●	●
	Isolating maintenance run-ups or use of test stand noise suppressors and barriers	●					●	●
Airport and airspace use	Preferential or rotational runway use*	●	●	●	●	●		
	Preferential flight track use or modification to approach and departure procedures.*		●	●		●		
	Restrictions on ground movement of aircraft*	●						
	Restrictions on engine run-ups or use of ground equipment						●	●
	Limitations on number or types of operations on types of aircraft	●	●	●	●	●	●	●
	Use restrictions Rescheduling Move flights to another airport	●	●	●	●	●	●	●
Aircraft operation	Raise glide slope angle or intercept*			●		●		
	Power and flap management*		●	●		●		
	Limited use of reverse thrust*				●			
Land use	Land or easement acquisition	●	●	●	●	●	●	●
	Joint development of airport property	●	●	●	●	●	●	●
	Compatible-use zoning	●	●	●	●	●	●	●
	Building code provisions and sound insulation of buildings	●	●	●	●	●	●	●
	Real property noise notices		●	●	●	●	●	●
	Purchase assurance		●	●	●	●	●	●

TABLE 15-3   Matrix of Noise Control Actions (Continued)

If you have this problem with noise from: / Consider these actions	Taxiing	Departure	Approach	Landing roll	Training flights	Maintenance	Ground equipment
Noise program management — Noise-related landing fees	●	●	●	●	●		
Noise monitoring			●	●		●	●
Establish citizen complaint mechanism							
Establish community participation program	●	●	●	●	●	●	●

*These are examples of restrictions that involve the FAA's responsibility for safe implementation. They should not be set in place unilaterally by the airport operator.

SOURCE: Federal Aviation Administration [36].

Typical barriers are walls, earth berms, or wall-berm combinations. Long buildings, such as the terminal itself, also make effective barriers. Blocking the line of sight to APUs and low engine aircraft such as the Boeing 737 usually requires barriers of only modest height, assuming flat terrain. Blocking the line of sight to high tail-mounted engines, such as those on the DC-10 or L-1011, presents a greater challenge. Barriers just blocking the line of sight generally provide about 5 dB of noise reduction. Higher barriers provide more.

For maintenance runups, a barrier is often in the form of a pen or series of walls. A pen surrounds the aircraft as closely as possible but allows entry through the front. It also contains a blast shield to prevent engine exhaust damage to the barrier. Complete enclosures, often referred to as *hush houses,* feature doors, roofs, and exhaust-silencing treatment. They are used where large amounts of noise reduction are required.

Along the runway sideline, especially in the vicinity of the start of the takeoff roll, barriers are most effectively placed near the residences they are meant to protect. Obstruction height clearance requirements usually preclude placing barriers close enough to the runway to be effective in these locations.

Barrier performance can be degraded by temperature inversions and winds with a component blowing in the direction of source to receiver. This is especially true if the barrier cannot be located as close as desired to the source or receiver. Under these atmospheric conditions, refracted sound travels a higher curved path from the source to receiver, and sound attenuation is reduced or eliminated under extreme conditions such as in high wind.

Sound insulation. Sound insulation of structures, such as residences and schools, attempts to improve the environment indoors through treatment of the structure itself. FAA funding criteria for sound insulation projects seek a 5-dB transmission loss improvement and a day-night average sound level (DNL) goal of 45 dB indoors. Windows are usually the weak link in the sound attenuation properties of structures. With windows open, the noise reduction properties of other parts of the structure are largely irrelevant, and a noise reduction up to 15 dB is all that can be expected. With windows closed, noise reduction is greater, but the additional reduction is dependent on the extent of

1. Any remaining air gaps such as around windows and doors and through attic and basement vents
2. The thickness and number of panes of glazing
3. The weight of exterior doors
4. The weight of roofing and walls

Cost-effective sound insulation programs can achieve 25 to 35 dB of noise reduction through attention to air gaps (caulking around door and window frames, insulation of walls and attics, sound-absorbing material around attic vents and soffits), window treatment (replacement of jalousie or poorly fitting windows and use of double-strength or double-pane glass in the form of special acoustical windows or storm windows), and doors (replacement of hollow core with solid-core units). To be effective during the summer months, central air conditioning must also be part of a basic noise insulation package so proper ventilation can be achieved with the windows closed.

Enhancing roof and wall weight can provide additional benefit once the aforementioned items are no longer the weak link. However, the cost of ensuring that other elements are not the weak link, such as installing triple instead of double glazing and sophisticated air duct treatment, added to the cost of the structural enhancements themselves generally increases the cost significantly.

Preferential runway system. The preferential runway concept is based on optimizing runway utilization under wind, weather, demand, and airport layout constraints to minimize population impacts by taking advantage of uneven population distribution around the airport. Preference is given, weather permitting, to those runways for which arrivals or departures affect the fewest people. Considerable effort can be devoted to determining which runway flight track combinations create the least noise impact and to developing with the FAA a workable plan that can be implemented. Monitoring the effectiveness of any preferential runway-use program is important; it can also entail significant effort to develop and implement.

Noise abatement departure procedures and flight tracks. The FAA has developed a recommended noise abatement takeoff procedure involving power settings and profile characteristics for turbojet-powered aircraft with maximum certificated gross takeoff weights in excess of 75,000 lb [34]. Most domestic airlines have incorporated this procedure or an equivalent in their flight manuals. The National Business Aircraft Association (NBAA) has also developed and recommended noise abatement procedures for turbojet business aircraft. The objectives of the NBAA program are to ensure that jet aircraft noise abatement procedures are safe, standardized, and uncomplicated while at the same time being effective at reducing noise levels in the community. Noise abatement departure procedures can also include use of specific headings and turns to avoid populated areas. The INM may be used to assess the effectiveness of such procedures.

Noise abatement flight paths can offer significant opportunities for noise abatement where distribution of incompatible land uses is uneven. Typically, noise abatement flight paths are designed to avoid the noise-sensitive areas and to route air traffic over less sensitive areas. Implementation of these kinds of flight paths will also require extensive interaction with the FAA. Again, the INM may be useful in assessing the noise impacts of various flight tracks.

Airport use restrictions. Noise-based airport use restrictions address noise control through reductions in the average noisiness of the aircraft that use the airport. Use restrictions have come under court challenge, especially by the FAA, as unduly restrictive of interstate commerce. In general, the courts have found restrictions of an airport to be legal if they are

1. Reasonable in the circumstances of the particular airport
2. Carefully tailored to the local needs and to community expectations
3. Based upon data which support the need and rationale for the restriction
4. Not unduly restrictive of interstate commerce

Several types of restrictions can be considered. Curfews or other nighttime-use restrictions are designed to reduce or eliminate noisy operations during late-night hours when people may be particularly sensitive to noise. Such restrictions can have large DNL benefits relative to the number of aircraft operations affected because of the 10-dB penalty added to noise between 10:00 p.m. and 7:00 a.m. when the DNL is computed. Aircraft operators may react by canceling operations by restricted aircraft types, switching to quieter aircraft types, or rescheduling.

Full curfews—eliminating all nighttime flights—have been found to be *overbroad* and to impose *undue burden* on interstate commerce, and they are often viewed as *arbitrary and capricious*. The overbroad issue has to do with the fact that a full curfew may deny access to the airport by users who, in fact, could operate quietly at night without significant disruption to sleep. A full curfew might have interstate commerce implications because of nighttime activity to and from out-of-state destinations. The arbitrary and capricious test has to do with whether a use restriction can be justified in terms of its noise benefits. Perhaps the most important point to be made is that a detailed quantitative noise analysis should be developed to provide justification for any noise-based use restriction.

Use restrictions can also be based on FAA noise certification categories. These categories are identified in FAR part 36 and discussed later. These restrictions limit the use of the airport based on the noise certification stage of the aircraft. For example, an airport may adopt a restriction that limits its use to stage 3 aircraft at night.

As part of the certification process, specific noise levels are measured for each aircraft [38]. Use restrictions can be based on these specific levels. For example, an airport could prohibit nighttime departures by aircraft with certified noise levels exceeding 108 EPNdB.

Use restrictions do not have to be based on certified noise levels. Certified noise levels may not be available for some older transport aircraft or for many general aviation aircraft. Certified noise levels may also be deemed unrepresentative of the sound levels produced under actual local operating conditions. In such cases, it may be preferable to set limits based on other published data [25] or on the noise levels measured at the airport itself.

Noise-based landing fees provide an economic incentive to discourage the operation of noisier aircraft, especially during noise-sensitive times of the day. Noise-based landing fees are proportional in some way to the noise produced by the aircraft. For example, an operator may be charged more for a takeoff by a stage 2 aircraft than for the same operation by a stage 3 aircraft. Alternatively, the fee may be higher for a night departure than for a day departure by the same aircraft. To be effective, however, the fee structure must be set high enough to affect airport user decision making.

Noise regulations

Federal aviation noise regulations are set forth in a number of forms. The highest forms of regulations are those set forth in various parts of Title 14 of the *Code of Federal Regulations* (14 CFR). This section of the federal code is called the *Federal Aviation Regulations* (FAR). The FAA also publishes orders and advisory circulars. Orders are procedures to which FAA staff must adhere in performing their responsibilities. To the extent that the FAA approves actions by others in the aviation industry (airports, airlines, etc.), the orders apply to them as well. Advisory circulars are printed documents which provide useful guidance and information often related to the FARs or to FAA orders.

FAR part 36.   FAR part 36 sets noise standards that aircraft must meet to obtain type and airworthiness certificates for operation in the United States [39]. They were first promulgated in 1969 for application to civil subsonic turbojets and large (over 12,500 lb) propeller-driven aircraft; the government subsequently amended the regulation to address civil supersonic aircraft, small (not over 12,500 lb) propeller aircraft, and rotary-wing aircraft such as helicopters. FAR part 36 also prescribes the procedures for aircraft manufacturers and others to use in measuring aircraft noise for certification purposes. The FAA publishes companion advisory circulars which present measurement results [25, 31, 38].

In 1977, the certification limits were made more stringent, leading to the classification of aircraft into three groups known as stages. *Stage 1* aircraft were flying before the regulation was initially adopted and were never required to meet level limits when they were first issued. *Stage 2* aircraft met the original (1969) noise emission limits but not the revised (1977) limits. *Stage 3* aircraft are those newest, quietest types that must meet the revised limits.

The regulation requires that aircraft meet gross-weight-based noise limits at three locations. Figure 15-8 shows the required measurement locations for turbojet and large propeller aircraft. These are under the takeoff path 6500 m from brake release, under the approach path 2000 m from runway threshold, and along the flight track sideline 450 m from the runway centerline (650 m for older turbojet aircraft). The sideline measurement is at the point of maximum sideline noise. In practice, this is normally to the side of takeoff, not landing.

Figures 15-9 through 15-11 show the original stage 2 noise limits and the lower stage 3 limits for these three locations. Shown with the

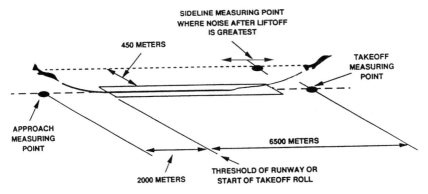

Figure 15-8 FAR part 36 noise measurement locations (*Federal Aviation Administration [39]*).

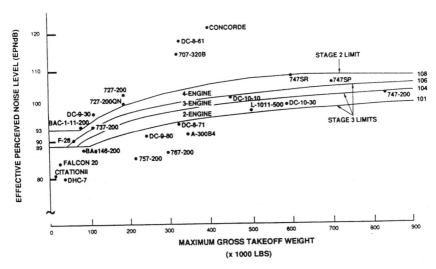

Figure 15-9 FAR part 36 certification levels for takeoff noise (*Federal Aviation Administration [38]*).

limits are several examples of the actual certificated levels for a variety of different aircraft. Note that some of the quieter types include the McDonnell-Douglas DC-8-70 series, DC-9-80s (also known as the MD-80), and the Boeing 757-200 and 767-200. FAR part 36 certification noise levels are published and regularly updated in advisory circulars [25, 31, 38]. A further discussion of these noise limits is contained in Chap. 3.

FAR part 91. FAR part 91 limits civil aircraft operations in the United States based on FAR part 36 certification status. The noise

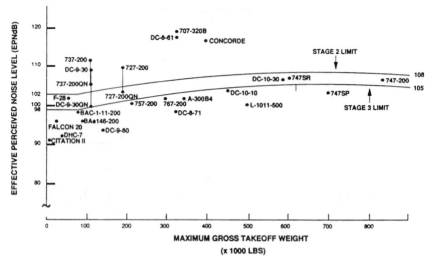

Figure 15-10   FAR part 36 certification levels for approach noise (*Federal Aviation Administration [38]*).

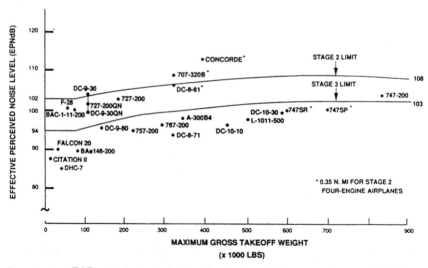

Figure 15-11   FAR part 36 certification levels for sideline noise (*Federal Aviation Administration [38]*).

elements of FAR part 91 were first adopted in 1977. This regulation prohibits operation of civil subsonic turbojet aircraft with maximum weights over 75,000 lb, unless they were certificated under FAR part 36 stage 2 or stage 3 limits. FAR part 91 has led to the elimination and ongoing prohibition of all stage 1 operations in the United States in civil subsonic turbojets over 75,000 lb but does not affect aircraft of

75,000 lb or less which are primarily corporate aircraft.

In 1990, the federal government enacted the Airport Noise and Capacity Act of 1990 (Public Law 101-508). This act called for the FAA to develop a national aviation noise policy and regulations to implement the policy. This was accomplished, in part, by amending FAR part 91 to require the phased elimination of stage 2 operations in civil subsonic turbojets over 75,000 lb by the end of 1999 with limited waivers through 2003. This will leave only stage 3 aircraft operating in the air carrier fleet in the near future.

The interim compliance schedule of FAR part 91 requires that aircraft operators remove 25 percent of their stage 2 airplanes by the end of 1994, 50 percent by 1996, and 75 percent by 1998; or, alternatively, phase-in stage 3 airplanes to achieve a fleet mix with 55 percent stage 3 by 1994, 65 percent by 1996, and 75 percent by 1998. Aircraft operators must submit annual reports showing their progress toward meeting the terms of the phase-in phase-out program and plan for future conversion. The balance of the FAA national noise policy required by the Airport Noise and Capacity Act is embodied in FAR part 161, which is discussed later.

FAR part 150. The Aviation Safety and Noise Abatement Act of 1979 (Public Law 96-193) required the FAA to establish regulations that set forth national standards for identifying airport noise and land-use incompatibilities and to develop programs to eliminate them. The FAA promulgated these regulations as FAR part 150 [6].

FAR part 150 prescribes specific standards and systems for

1. Measuring noise

2. Estimating cumulative noise exposure by using computer models

3. Describing noise exposure including instantaneous noise levels, single-event levels, and cumulative exposure

4. Coordinating noise compatibility program development with local land-use planning officials and other interested parties

5. Documenting the analytical process and developing the compatibility program

6. Submitting documentation to the FAA

7. FAA and public review processes

8. FAA approval or disapproval of the submission

A full FAR part 150 submission consists of two basic elements, namely, a *noise exposure map* (NEM) and its associated documentation, and a *noise compatibility program* (NCP). It is possible, however, to submit only the NEM. In addition to these elements, a critical ingredient to a successful FAR part 150 program is a thorough and effective public involvement program.

Noise exposure map. The NEM describes the airport layout and operation, aircraft-related noise exposure, land uses in the airport environs, and the resulting noise-related land-use compatibility situation. It addresses the year of submission and the fifth year from submission. It includes graphical depiction of existing and future noise exposure resulting from aircraft operations and of land uses in the airport environs. Documentation must accompany the NEM that describes the data collection and analysis undertaken in its development. The basic output of the map development is identification of existing and potential future noise and land-use incompatibilities. FAR part 150 includes a table presenting noise and land-use compatibility guidelines, shown earlier in Table 15-2.

Noise compatibility program. Following development of an NEM, which essentially defines the extent of noise and land-use incompatibility, the airport proprietor may elect to develop an NCP. In so doing, the airport proprietor must consider all potential compatibility measures, including the airport layout, operational and use alternatives, and land-use alternatives. FAR part 161, discussed below, further regulates the evaluation and adoption of airport use restrictions. The ultimately developed program is essentially a list of the actions that the airport proprietor proposes to take to minimize existing and future noise and land-use incompatibilities. The NCP documentation must recount the development of the program, including a description of all measures considered, the reasons that individual measures were accepted or rejected, how measures will be implemented and funded, and predicted effectiveness of individual measures and the overall program.

Following FAA acceptance of the NEM and approval of the NCP submissions, the airport operator may apply for FAA funding of program implementation. Official FAA approval of the NCP does not eliminate requirements for formal environmental assessment of any proposed actions pursuant to requirements of the National Environmental Policy Act (NEPA) [2, 43]. However, acceptance of the submission is a prerequisite to application for funding of implementation actions.

Public involvement. At every stage of the FAR part 150 planning process opportunities exist to apprise airport neighbors, user groups, and local officials of project alternatives and to solicit their comments, criticism, and support. A key objective is to utilize the broadest possible definition of the *airport public,* which includes more than just the residents of areas around the airport. A balanced discussion of issues must include representatives of the aviation industry in local, state, regional, and national agencies with jurisdiction; the business community; and individual airport users. Successful public information programs include the following elements:

1. Regular advisory committee meetings
2. Technical and other subcommittee meetings as required
3. Informational newsletters tailored for public distribution
4. Informal workshops open to the general public
5. A final public hearing
6. Briefings to local public officials

FAR part 161. The second major element of the national noise policy enacted through the Airport Noise and Capacity Act of 1990 is FAR part 161. It establishes requirements that an airport operator must meet prior to promulgating any airport noise or access restriction on the use of stage 2 or stage 3 aircraft [40].

Under FAR part 161, an airport proprietor may impose restrictions on stage 2 aircraft operations as long as two conditions are met: First, the proprietor must prepare an analysis of the anticipated costs and benefits of the proposed restriction. Second, the proprietor must provide proper notice of the restriction to the public and to affected parties. The cost-benefit analysis required by FAR part 161 must include

1. An analysis of the anticipated or actual costs and benefits of the proposed restriction
2. A description of alternative restrictions which were considered
3. A description of the alternative measures considered that do not involve airport restrictions
4. A comparison of the costs and benefits of the alternative measures considered

FAR part 161 imposes substantial impediments to local restrictions on stage 3 aircraft. No local stage 3 restriction may become effective unless it has been submitted to and approved by the FAA. The process for FAA review and approval has three principal elements [5]:

1. The collection and analysis of data to justify the restriction and to explain its environmental and economic impact
2. The notification of the public and allowance of adequate time for comment on the proposed restriction
3. The submission of the restriction for FAA review and approval

An airport cannot implement a stage 3 aircraft restriction unless the airport complies with each of the above elements.

FAR part 161 stipulates that both types of restrictions may be developed through the FAR part 150 process. The benefit of using FAR part 150 as a mechanism for developing a rule restricting airport access is the availability of federal funding. Since such federal funding may be used for the preparation of a FAR part 150 NCP, this funding can be used for the preparation of any noise or access restriction included in the FAR part 150 study. The major disadvantage of submitting stage 2 restrictions to the FAA as part of the FAR part 150 submission is that a formal submission will invoke the approval process, which is otherwise not necessary under FAR part 161. Stage 3 restrictions, on the other hand, require FAA review whether or not they are developed during a FAR part 150 process.

Construction impacts

The construction of facilities at airports can have temporary and long-term impacts on the community and travelers. Those factors of primary concern during the construction process include soil erosion, water and air quality, noise due to construction equipment and methods, source and quantity of construction materials, disruption and relocation of businesses and residences, continued operation of existing facilities both on and off the airport during the construction process, and interference with other construction projects [23].

A review of the environmentally sensitive areas and facilities should be undertaken to identify those which will be subject to impact and the likely duration and consequences of these activities. The location and quantity of cut and fill must be identified, and methods to minimize the effect of construction activities on soil erosion during construction should be identified. Procedures for the handling of construction materials and wastes to minimize the introduction of particulate matter into the air and water resources are required. Those land uses which will be adversely affected by noise from construction activities should be identified, and the optimal routing of construction vehicles and timing of activities should be chosen to minimize damage. Obvious positive effects of a major construction activity are the increases in employment and payroll for personnel associated with the project and the purchase of materials and supplies from local firms which support the local economy. However, certain businesses and residences in the vicinity of the project may be subject to disruption due to the possible rerouting and congestion of vehicular traffic and restrictions on access to land uses.

The study of the direct and indirect socioeconomic impacts on the community should include an identification of the location, timing, and amount of impact throughout the entire construction period.

Social Factors

Land development

The development of an airport can cause a change in the pattern of land-use activity both in the vicinity of the airport and in the geographic region. These land development activities may cause changes in the level of economic activity and population growth and demography. The existence of an airport generally induces land development in the vicinity of the airport which may be measured in terms of the density of industrial, commercial, retail, agricultural, and residential use. It usually impacts the nature, magnitude, and operating patterns of other transportation modes providing access to the airport and those land uses associated with its development. The presence of the airport may induce industrial or commercial activities to move into or expand within the region due to increased accessibility to markets and materials.

An airport will affect land development as a function of its direct economic impact on the region. The following land development activities should be studied [2, 43]:

1. Plant relocation from outside the region which will require construction activities

2. Increases in the production or sales of existing business enterprises requiring new or expanded capital facilities

3. Increased expenditures in tourist and recreational facilities resulting in requirements for new or expanded facilities and increases in retail sales

4. Expansion of agricultural markets resulting in increased productivity and resource utilization

5. Increased demand for specialized facilities for such activities as business or convention centers

6. Expansion of commercial and financial markets resulting in a demand for additional facilities

The study of the land development impacts requires an analysis of those factors which influence industrial, commercial, and residential location decisions, including accessibility to raw materials, labor, and markets; the costs of production and transportation; and those quality-of-life and community factors associated with such decisions. The analysis is usually conducted through an examination of historical trends for the area and similar locations and through the use of surveys and economic models to predict land development. The study should identify the nature and extent of existing zoning ordinances in the vicinity of the airport and should recommend those changes necessary to accommodate the likely development in a compatible manner. The study should also identify those requirements for regional policy decisions needed to stimulate overall land development in accordance with the stated objectives of the communities affected.

Displacement and relocation

The construction or expansion of an airport often creates a need for additional land; the relocation of residences, businesses, and community facilities; a disruption in business activity and community character and cohesiveness; the impairment of community service functions; and an increased demand for public services. The assessment is directed toward determining the type, extent, characteristics, and effects of displacement and relocation and mitigation measures to minimize adverse consequences.

The boundaries of the areas affected are determined from area maps. The community structure is usually defined in terms of population demography, growth, and density; housing and business characteristics such as the type, distribution, condition, value, occupancy, and vacancy levels; open land; recreational resources; and community services. The availability of relocation resources for those land uses required for airport and ancillary activities is identified. The changes in demand for public services are quantified. Comparisons of the relative impact of project alternatives usually attempt to address the following items [43]:

1. The nature, location, and extent of the displacement of homes, businesses, and community facilities

2. The creation of physical barriers or divisions

3. The impairment of mobility, accessibility, community services, and community facilities

4. The disruption of homes, businesses, and community facilities during construction

5. The nature, availability, adequacy, and compatibility of relocation resources

6. The nature, location, and extent of land-use changes

7. The aesthetic appeal of the design for the facility and surrounding environment

Parks, recreational areas, historical places, archaeological resources, and natural and scenic beauty

Particular attention is required to determine the impact upon parks, recreational areas, open spaces, cultural and historic places, archaeological resources, and natural and scenic beauty. The analyst must identify the type of facilities which will be impacted, their size and utilization, and those measures which can be implemented to preserve the nature, character, compatibility, and accessibility of the facilities. It is particularly important to document the relative impact of project alternatives on these types of facilities and to avoid the acquisition of such lands for the implementation of a project alternative.

The assessment can be performed on the basis of a participatory evaluation with those groups that possess expertise or interest in this impact area. Suggested generalized evaluation criteria might include [2, 43]:

1. The existence, nature, and extent of any physical alteration to the facilities

2. The degree of conformity of the planned facilities with the existing environment

3. The disruption of access

4. The disruption of the ambient environment

5. The compatibility of access-induced development with the facilities

Consistency with local planning

The planning and design of airports can have significant effects on the economy, land use, infrastructure, and nature of community development. The planning effort must be carried on in an environment which is compatible and coordinated with other local planning efforts and guidelines. Care must be taken from the inception of planning to assess the impact of project alternatives and operations on the goals and objectives of communities and to identify those facets of the project which may present conflicts between existing plans and community goals. Modifications to airport project proposals which will be in conformity with local policies and plans must be explicitly examined. Policies required to preserve the overall development objectives of the community should be identified, and mechanisms to implement such policies should be proposed.

The assessment of impacts in this area requires an identification of and coordination with those federal, state, and local agencies which have a concern or jurisdiction in matters related to airport and community development actions; the delineation of clear statements of goals and objectives; the integration of airport plans with local comprehensive land-use, economic, and transportation development plans; and the establishment of a continuing dialogue on issues related to these plans. The presentation of the results of planning efforts and development recommendations in public forums, with mechanisms for citizen participation in the planning and review processes, together with timely and well-documented responses to community concerns is essential.

Ecological Factors

Wildlife, waterfowl, flora, fauna, and endangered species

Consideration of the impact of airport development on changes in the natural state of land and waterways is essential to protect ecosystems. Living and nonliving elements, plants, and animals all interact on land and in water to produce highly interdependent aquatic and terrestrial ecosystems. The relationship between species and the ecosystem is essential to maintain the life support system for wildlife, waterfowl, flora, fauna, and endangered species [20, 51, 52]. Of particular importance are vegetation, plant, and animal life. The principal impacts which could occur are the loss of or injury to the organisms or the loss or degradation of the ecosystem.

The use of land for airport development creates disturbances and disruptions to flora and fauna. The specific project elements often include the clearing and grubbing of land areas, changes in the composition and nature of the topography, and interferences with watershed patterns. Thus airports can destroy the natural habitat and feeding grounds of wildlife and eliminate or reduce flora essential to the maintenance of the ecological balance in the area. Particular hazards may be presented to birds and aircraft due to striking birds, and care must be exercised in choosing airport sites to avoid natural migration routes and land which attracts birds. The protection of endangered species in the United States is legislated through the Endangered Species Act of 1973 (Public Law 93-205). Lists of endangered species are published [20, 51, 52]. Refer also to state and location regulations in this regard.

The assessment techniques used include identification of important aquatic and terrestrial organisms present in the area and determination of the life support systems required for different species. An analysis is performed to determine the impacts on vegetation requirements, food chains, and habitats of these species, as well as their tolerance to air and water pollution. Care must be taken in the case of aquatic species to examine the effects of soil erosion, flooding, and sedimentation on streambeds where food chains, spawning grounds, and habitats exist.

Wetlands and coastal zones

Improperly planned or operated drainage systems at airports can contaminate streams, lakes, and waterways. The normal operation of an airport results in potential contamination through aircraft and ground vehicle washing, servicing, and fueling; airport and aircraft maintenance; and terminal services. In the construction phase of a project, there is a high potential for contamination through clearing, grubbing, pest control, and changes in topography. Changes in the natural drainage patterns of the area are very common due to the nature of airport development projects. Preservation of recharge areas and stream flows, the elimination of flooding and sedimentation problems, and the preservation of the quality and routing of water resources are all vital to the maintenance of water quality and protection of ecosystems.

Engineering and Economic Factors

Flood hazards

The flood hazard potential of any development is a necessary consideration since alterations in the topography, cover, and soil characteristics on the property are inevitable. The storage capacity of local rivers, streams, canals, and groundwater areas can be exceeded due to changes in the magnitude and paths of runoff from storms and high rainfall or thawing events. The potential for flooding is analyzed by evaluating the characteristics of the ground surface, soil materials, topography, and floodplains; the historical frequency and intensity of storms; and storm water drainage and retention facilities. The methods for conducting such analyses are discussed in Chap. 14.

If it is found that the project design increases the potential for on-or off-site flooding, then those areas subject to these effects are identified and the mechanisms required to alleviate the hazards incorporated into the project design. The construction of new or increased-capacity

TABLE 15-4   Unit-Cost Indices for Space and Special Equipment at Airport
Terminal Buildings

Location	Area category	Area type	Cost unit	Benchmark index per unit		
				Shell	Tenant	Total cost
Terminal	Type A: Passenger-handling facilities	1. Lobbies	ft²	1.00	NA*	
		2. Waiting rooms				
		3. Circulation				
		4. Rest rooms				
		5. Counter areas			2.00	3.00
		6. Baggage claim facilities, including claim device			0.50	1.50
		7. Service and storage areas			NA	
Terminal	Type B: Airline/tenant operations space, partly finished	1. Customer service offices	ft²	0.65	0.40	1.05
		2. Agent supervisor offices, checkout, and agent lounge			0.50	1.15
		3. Toilets			1.55	2.20
		4. VIP/PR rooms			1.10	1.75
		5. Lost and found			0.72	1.37
Terminal or connector	Type C: Airline operations space, lower level unfinished	1. Offices	ft²	0.60	0.55	1.15
		2. Tire shop (including equipment)			0.62	1.22
		3. Storerooms			0.30	0.90
		4. Ready and lunch rooms			0.43	1.03
		5. Lockers			0.20	0.80
		6. Toilets			1.80	2.40
		7. Planning center and load planning			0.85	1.45
Connector	Type D: Passenger-handling	1. Corridors	ft²	0.80	NA	
		2. Rest rooms				
		3. Service and storage area				
		4. Boarding areas			0.26	1.06

*NA = not applicable.

These cost data have been compiled for new terminal and concourse construction, and they may not be applicable for remodeling or small additions to existing facilities. They do not include installed equipment, such as loading bridges, claim devices, ramp utilities, and so on.

SOURCE: Federal Aviation Administration [50].

storm sewers and impounding areas, channels, and dikes is most commonly indicated. Changes in the elevation of facilities and the slope and cover of the ground surface at the site can also be of considerable benefit in reducing flood hazards.

Costs of construction and operation

All engineering planning studies consider the capital, maintenance, and operating costs of all feasible alternatives as an integral part of the planning process. For airport projects, these costs include land acquisition; purchases of land leases and covenants to protect aircraft operations and environmental quality; facility construction; operating, maintenance, and administrative costs. Typically, construction costs are derived from quantity takeoffs of materials which are related to locally appropriate cost indices for the various construction items [13, 18, 19]. The overall cost factor is computed by a concept of benchmark cost indices for various components of the airport [50]. A tabulation of these indices is given in Table 15-4. The capital costs usually include materials, supplies, labor, and engineering.

The construction costs for various items are normally assessed at different times, and therefore, for comparative and evaluative purposes, these are brought back to some common point in time so that value may be properly attributed to the construction needs. The operating, maintenance, and administrative costs are usually annualized. These costs are normally estimated through comparisons with similar installations, historical trends, and economic influences. The costs of capital and the required coverage, as discussed in Chap. 2, are also included to arrive at the total program cost.

An overview of the categories and items typically found in airport projects for which capital, operating, maintenance, and administrative cost estimates are required includes the following:

1. Airfield facilities
   a. Runways, taxiways, and aprons
   b. Fueling and fixed power systems; crash, fire, and rescue units
   c. Air traffic control facilities, lighting and navigational aids
2. Terminal building facilities
   a. Terminal buildings and connectors
   b. General aviation servicing buildings and hangar areas
   c. Boarding devices, mechanical and electrical systems
   d. Communications and security systems
   e. Air cargo buildings
   f. Maintenance and administrative buildings
   g. Furnishings
3. Access facilities
   a. Roadways, drives, and curb frontage
   b. Rental-car, limousine, and transit areas
   c. Parking lots and garages
   d. Graphics, signage, and lighting
4. Infrastructure facilities
   a. Landscaping and drainage
   b. Utilities including water supply, sewage disposal, power supply systems
   c. Land acquisition

For the purposes of evaluation, these costs are usually related to passenger and aircraft traffic characteristics such as the cost per enplaned passenger or cost per air carrier operation. The costs are also allocated among the various users and nonusers of the project, normally tenants, concessionaires, airlines, general aviation, cargo businesses, and federal, state, and local government agencies as appropriate.

Economic benefits and fiscal requirements

The evaluation of the economic and financial feasibility of the project requires an identification of both the level and the allocation of benefits and costs of the project, as well as a revenue analysis performed for the various cost centers. Both direct and indirect benefits can accrue to the users of the airport and to the community in which the airport is located. Generally, user benefits include reductions in delay, fuel consumption, time, and other operating and maintenance costs. These can usually be derived relative to dollar value. Nonuser and community benefits take the form of increased economic activity, rises in employment, and purchases of goods and services. These can also be evaluated through classical economic techniques [2, 8, 21].

Although it may be possible to justify expenditures from an economic standpoint, it may not be possible to generate or capture the value of these benefits from a revenue viewpoint. A revenue analysis seeks to identify the revenue requirements by cost center and the level of revenue required to cover project costs. Normally, the costs are allocated to facility users; and rents, rate and charges, and concession agreements are negotiated on the principle that all users should pay their fair and proportionate share of the costs of providing, maintaining, operating, and administering the facilities they use.

Various indices are used to determine the reasonableness of revenue requirements including the percentage of revenue generated which is paid for terminal rents or landing fees and the revenue required per enplaned passenger. A comparison of the bonding capacity of the government units concerned with the project is vital to determine the influence of the project on other public revenue requirements.

Energy and natural resources

The use of new technology in power generation systems at airports, the efficient layout of apron areas and taxiing routes, the improvements in the capacity of runway systems, the installation of navigational aids, and the effective uses of construction materials can substantially reduce the costs and resource use of energy. The impact of airport design elements on energy consumption should be examined in detail. Typically, comparisons between existing and planned systems, and between alternative systems for proposed facility modifications,

Vis-à-vis fuel consumption of aircraft and terminal systems may yield essential information concerning their feasibility and merit.

Summary

Although the incentive for the study of environmental, sociological, and ecological factors in the evaluation of engineering projects was initially provided through national, state, and local legislation, the state of the art has evolved to the point that a better and more complete understanding of the short- and long-term implications of these projects is leading to more efficient engineering designs. True, the costs of planning have increased because of the need to study a great number of criteria in the evaluation of planning proposals. But the potential for the overall reduction in the real *costs* of these proposals on the long-term requirements of society through the use of comprehensive planning approaches has also been increased.

References

1. *Acoustic Determination of Occupational Noise Exposure and Estimation of Noise-Induced Hearing Impairment,* Publication 1999, International Organization for Standardization, Geneva, Switzerland, 1990.
2. *Airport Environmental Handbook,* Order 5050.4A, Federal Aviation Administration, Washington, 1985.
3. *Airport Landscaping for Noise Control Purposes,* Advisory Circular AC 150/5320-14, Federal Aviation Administration, Washington, 1978.
4. *Airport Master Plans,* Advisory Circular AC 150/5070-6A, Federal Aviation Administration, Washington, 1985.
5. *Airport Noise: A Guide to the FAA Regulations under the Airport Noise and Capacity Act,* Cutler and Stanfield and Harris Miller Miller and Hanson, Inc., Lexington, Mass., January 1992.
6. *Airport Noise Compatibility Planning,* pt. 150, Federal Aviation Regulations, Federal Aviation Administration, Washington, 1991.
7. *Airport Noise Compatibility Study, Greater Pittsburgh International Airport,* Aviation Planning Associates, Cincinnati, Ohio, 1986.
8. *Airport Planning and Environmental Assessment,* Notebook Series, four volumes, DOT P 5600.5, Department of Transportation, Washington, 1978.
9. *Airport Planning Manual,* pt. 2: *Land Use and Environmental Control,* 2d ed., Doc. 9184-AN/902, International Civil Aviation Organization, Montreal, Canada, 1985.
10. *Analysis of the Predictability of Noise-Induced Sleep Disturbance,* K. S. Pearsons, D. S. Barber, and B. G. Tabachnick, Rep. HSD-TR-89-029, U.S. Air Force, Washington, 1989.
11. "Applied Acoustical Report: Criteria for Assessment of Noise Impacts on People," L. S. Finegold, C. S. Harris, and H. E. Von Gierke, submitted to *Journal of Acoustical Society of America,* New York, 1992.
12. *Aviation Noise Effects,* S. J. Newman, Rep. FAA-EM-85-2, Federal Aviation Administration, Washington, 1985.
13. *Building Construction Cost Data,* R. S. Means Company, Inc., Duxbury, Mass. (annual).
14. *Chicago Delay Task Force Technical Report,* vol. 1: *Chicago Airport/Airspace Operating Environment,* Landrum and Brown Aviation Consultants, Chicago, April 1991.
15. *Citizen Participation in Airport Planning,* Advisory Circular AC 150/5050-4,

Federal Aviation Administration, Washington, 1975.
16. *Community Noise,* Rep. DOT-NTID300.3, Wyle Laboratories, Office of Noise Abatement and Control, Environmental Protection Agency, Washington, 1971.
17. *Compilation of Air Pollution Emission Factors,* Rep. AP-42, Environmental Protection Agency, Washington, periodically revised.
18. *Dodge Guide for Estimating Public Works Construction Costs,* McGraw-Hill Information Systems Company, New York (annual).
19. *Dodge Manual for Building Construction Pricing and Scheduling,* McGraw-Hill Information Systems Company, New York (annual).
20. *Endangered and Threatened Wildlife and Plants,* Fish and Wildlife Service, Department of the Interior, Washington (updated periodically).
21. *Environmental Assessment Notebook Series,* Rep. DOT P5600.4, seven volumes, Department of Transportation, Washington, 1975.
22. "Environmental Considerations in Airport Planning," C. V. Robart, Course notes for airport planning and design, short course, University of California, University Extension, Berkeley, June 1977.
23. *Environmental Impact Assessment Report for the Expansion of the Existing Terminal Complex at the Fort Lauderdale–Hollywood International Airport,* Aviation Division, Broward County Department of Transportation, Fort Lauderdale, Fla., 1980.
24. *Environmental Protection, Annex 16 to the Convention on International Civil Aviation,* vol. 1: *Aircraft Noise,* 2d ed., International Civil Aviation Organization, Montreal, Canada, 1988.
25. *Estimated Airplane Noise Levels in A-Weighted Decibels,* Advisory Circular AC 36-3F, Federal Aviation Administration, Washington, 1990.
26. *Federal Agency Review of Selected Noise Analysis Issues,* Federal Interagency Committee on Noise (FICON), Washington, 1992.
27. *General Operating and Flight Rules,* pt. 91, Federal Aviation Regulations, Federal Aviation Administration, Washington, 1992.
28. *Information on Levels of Environmental Noise Requisite to Protect Public Health and Welfare with an Adequate Margin of Safety,* Environmental Protection Agency, Arlington, Va., 1974.
29. *INM Integrated Noise Model Version 3, User's Guide,* Rep. FAA-EE-81-17, Office of Environment and Energy, Federal Aviation Administration, Washington, 1982.
30. *Management of Airport Industrial Waste,* Advisory Circular AC 150/5320-15, Federal Aviation Administration, Washington, 1991.
31. *Measured or Estimated (Uncertified) Airplane Noise Levels,* Advisory Circular AC 36-2C, Federal Aviation Administration, Washington, 1986.
32. *Measures of Noise Level: Their Relative Accuracy in Predicting Objective and Subjective Responses to Noise during Sleep,* J. S. Lucas, Rep. No. EPA-600/1-77-010, Environmental Protection Agency, Washington, 1977.
33. *Noise,* Environmental Health Series, no. 12, World Health Organization, Geneva, Switzerland, 1980.
34. *Noise Abatement Departure Profiles,* Advisory Circular AC 91-53, Federal Aviation Administration, Washington, 1978.
35. *Noise Certification Handbook,* Advisory Circular AC 36-4B, Federal Aviation Administration, Washington, 1988.
36. *Noise Control and Compatibility Planning for Airports,* Advisory Circular AC 150/5020-1, Federal Aviation Administration, Washington, 1983.
37. "Noise-Induced Sleep Disturbances and Their Effect on Health," B. Griefahn and A. Muzet, *Journal of Sound and Vibration,* vol. 59, no. 1, 1978.
38. *Noise Levels for Certified and Foreign Aircraft,* Advisory Circular AC 36-1F, Federal Aviation Administration, Washington, 1992.
39. *Noise Standards: Aircraft Type and Airworthiness Certification,* pt. 36, Federal Aviation Regulations, Including Changes 1 to 21, Federal Aviation Administration, Washington, 1991.
40. *Notice and Approval of Airport Noise and Access Restrictions,* pt. 161, Federal Aviation Regulations, Federal Aviation Administration, Washington, 1991.
41. "Occupational Noise Exposure; Hearing Conservation Amendment," *Federal*

Register 48(46), Occupational Safety and Health Administration, Washington, 1983.

42. *Physiological, Psychological and Social Effects of Noise,* K. D. Kryter, Reference Publication 1115, National Aeronautics and Space Administration, Washington, 1984.

43. *Policies and Procedures for Considering Environmental Impacts,* Order 1050.1D, Federal Aviation Administration, Washington, 1986.

44. *Public Health and Welfare Criteria for Noise,* Environmental Protection Agency, Arlington, Va., 1973.

45. *Recommended Method for Computing Noise Contours around Airports,* Circular-205 AN1/25, International Civil Aviation Organization, Montreal, Canada, 1988.

46. "Research on Noise-Disturbed Sleep since 1973," B. Griefahn, *Proceedings,* The Third International Congress on Noise as a Public Health Problem, ASHA Rep. 10, Freiburg, West Germany, 1980.

47. *Special Report: Summary of Significant Provisions of the Final FAA Noise Rule,"* Airports Association Council International-North America, Washington, 1991.

48. *Study of Soundproofing Public Buildings near Airports,* Rep. DOT-FAA- AEQ-77-9, Wyle Laboratories, Federal Aviation Administration, Washington, 1977.

49. "Synthesis of Social Surveys on Noise Annoyance," T. J. Schultz, *Journal of the Acoustical Society of America,* vol. 64, no. 2, 1978.

50. *The Apron and Terminal Building Planning Report,* Rep. FAA-RD-75-191, Federal Aviation Administration, Washington, July 1975.

51. *Threatened Wildlife of the United States,* Fish and Wildlife Service, Department of the Interior, Washington (updated periodically).

52. *United States List of Endangered Fauna,* Fish and Wildlife Service, Department of the Interior, Washington (updated periodically).

Metric Conversion of U.S. Customary System Units

English	Metric
1 ft	0.3048 m
1 in	0.0254 m = 2.54 cm
1 lb	0.4536 kg
1 ton (2000 lb)	907.2 kg
1 nmi (nautical)	1852 m = 1.852 km
1 mi (statute)	1609 m = 1.609 km
degrees (°) Fahrenheit	degrees (°) Celsius = (°F − 32)$\frac{5}{9}$
1 gal (U.S. liquid)	0.003785 m^3
1 acre	4046.8 m^2
1 mi^2 (statute)	2,589,988 m^2
1 ft^2	0.0929 m^2
1 ft/s	0.3048 m/s

Problems for Solution

1. The characteristics of a new aircraft are given in Table B-1. In domestic route service, the applicable regulations require only a 0.75-h fuel reserve for this aircraft. The aircraft has an average route speed of 500 mi/h and an average fuel burn of 13 lb/mi.
   a. Plot payload versus range for this aircraft.
   b. Determine the takeoff weight of this aircraft for a 1200-mi trip at maximum payload.
2. An aircraft has the performance characteristics given in Table B-2 in normal route operation. In normal use this aircraft will consume 11,000 lb of fuel in 1 h of flight. The normal cruising speed is 550 mi/h. Generally, alternative destinations are located within 100 mi of planned destinations, and regulations require

TABLE B-1 Preliminary Aircraft Characteristics, lb

Maximum ramp weight	141,000
Maximum structural takeoff weight	140,000
Maximum structural landing weight	128,000
Zero-fuel weight	112,000
Operating empty weight	73,274
Fuel capacity	39,162
Maximum structural payload	38,726
Maximum passenger capacity	27,000
Maximum cargo hold capacity	12,500

TABLE B-2   Aircraft Performance Characteristics, lb

Maximum ramp weight	191,000
Maximum structural takeoff weight	190,000
Maximum structural landing weight	160,000
Zero-fuel weight	140,000
Operating empty weight	100,000
Fuel capacity	57,000
Maximum structural payload	40,000
Maximum passenger capacity	39,000
Maximum cargo hold capacity	15,000

sufficient fuel to fly to the planned destination, fly to an alternate destination, and fly for 1.25 h at the alternate destination.

 a. Find the maximum passenger payload in passengers and lb which may be carried on this aircraft if 15,000 lb of cargo is also carried.

 b. If a planned route distance is 2000 mi, determine the number of lb of fuel that must be boarded at the gate for this trip.

 c. If the aircraft carries 160 passengers and no cargo, except for passenger baggage, on a planned 2000-mi trip and if an incident occurs 30 min into the flight, requiring a landing, determine the amount of additional time that the aircraft must remain aloft prior to being able to land safely without exceeding its maximum allowable landing weight.

 d. Plot payload versus route range for this aircraft.

3. A turbine-powered aircraft has these runway performance characteristics: For a normal takeoff, the liftoff distance is 6500 ft, and the aircraft reaches a height of 35 ft above the end of the runway at a distance of 7000 ft from the runway threshold. For an engine-failure takeoff, the liftoff distance is 7200 ft, and the aircraft reaches a height of 35 ft above the end of the runway at a distance of 7900 ft from the runway threshold. For an aborted takeoff, the accelerate-stop distance is 9700 ft. For landing, the stop distance is 4600 ft. Determine the minimum FAR runway length requirements for this aircraft, indicating the length of the full-strength pavement, the length of the stopway, and the length of the clearway if the runway is to be used by this aircraft in both directions.

4. An aircraft has these runway performance characteristics: For a normal takeoff, the liftoff distance is 8000 ft, and the aircraft reaches a height of 35 ft above the end of the runway at a distance of 8700 ft from the runway threshold. For an engine-failure takeoff, the liftoff distance is 10,000 ft, and the aircraft reaches a height of 35 ft above the end of the runway at a distance of 11,000 ft from the runway threshold. For an aborted takeoff, the acceler-

ate-stop distance is 10,500 ft. For landing, the stop distance is 5000 ft. Determine the minimum FAR runway requirements for this aircraft if the runway is to be used in both directions.

5. An airline is considering the use of a particular aircraft on a non-stop route of 1230 mi between two airports. The nearest alternate airport at either destination is 260 mi away. The time fuel reserves are included in the item listed as typical operating empty weight plus time fuel reserves in Table B-3. Table B-4 indicates the takeoff requirements for this aircraft. The airport under study has a runway of 8000 ft available with a maximum difference in centerline elevation of 15 ft. The airport elevation is 500 ft above mean sea level (AMSL), and the mean daily highest temperature in the hottest month is 90°F.

   a. Find the maximum allowable takeoff weight of this aircraft at the study airport.

TABLE B-3   Landing Runway Requirements

	Maximum Allowable Landing Weight, 1000 lb						
	Airport elevation, ft						
Temp., °F	0	1000	2000	3000	4000	5000	6000
80	363.5	363.5	363.5	363.5	363.5	363.5	363.5
85	363.5	363.5	363.5	363.5	363.5	363.5	359.8
90	363.5	363.5	363.5	363.5	363.5	363.5	350.6
95	363.5	363.5	363.5	363.5	363.5	354.6	341.7
100	363.5	363.5	363.5	363.5	358.4	345.5	332.8

	Landing Runway Length, 1000 ft						
	Airport elevation, ft						
Weight, 1000 lb	0	1000	2000	3000	4000	5000	6000
280	5.44	5.57	5.71	5.85	6.00	6.15	6.31
290	5.59	5.73	5.87	6.02	6.17	6.33	6.49
300	5.75	5.89	6.04	6.19	6.35	6.51	6.68
310	5.90	6.05	6.20	6.36	6.52	6.69	6.86
320	6.06	6.21	6.37	6.53	6.70	6.87	7.05
330	6.21	6.37	6.54	6.70	6.87	7.05	7.23
340	6.37	6.53	6.69	6.86	7.03	7.21	7.40
350	6.51	6.66	6.83	7.00	7.19	7.37	7.57
360	6.64	6.81	6.98	7.16	7.35	7.54	7.75
370	6.77	6.96	7.14	7.33	7.52	7.71	7.92

Maximum structural takeoff weight = 408,900 lb
Maximum structural landing weight = 363,500 lb
Typical operating weight plus time fuel reserves = 263,300 lb
Average fuel consumption = 30 lb/mi
Typical maximum passenger load = 54,000 lb
Maximum structural payload = 99,660 lb
Fuel capacity = 21,762 gal

TABLE B-4   Takeoff Runway Requirements

Temp., °F	Airport elevation, ft						
	0	1000	2000	3000	4000	5000	6000
	Maximum Allowable Takeoff Weight, 1000 lb						
80	408.9	400.6	392.9	382.8	369.1	356.2	343.9
85	408.9	400.6	386.0	372.3	359.4	347.1	335.4
90	403.6	388.8	374.9	362.0	349.7	338.1	326.9
95	391.1	377.3	364.2	352.0	340.3	329.2	318.4
100	379.6	366.4	354.1	342.4	331.2	320.4	309.9
	Reference Factor						
80	46.0	48.2	50.5	53.4	57.0	61.0	65.4
85	46.3	48.5	51.8	55.2	59.0	63.1	67.8
90	47.2	50.3	53.6	57.2	61.1	65.5	70.4
95	48.8	52.0	55.4	59.2	63.3	68.0	73.2
100	50.5	53.8	57.4	61.4	65.8	70.7	76.3

Takeoff Runway Length, 1000 ft

Weight, 1000 lb	Reference factor				
	40	50	60	70	80
320	4.23	5.28	6.39	7.50	8.56
330	4.44	5.58	6.78	7.98	9.09
340	4.67	5.89	7.18	8.46	9.63
350	4.90	6.21	7.60	8.95	10.16
360	5.13	6.55	8.03	9.45	10.68
370	5.38	6.89	8.46	9.94	
380	5.63	7.25	8.91	10.44	
390	5.89	7.63	9.36	10.92	
400	6.17	8.01	9.82		
410	6.45	8.40	10.29		

b. If the aircraft takes off from the other airport with a takeoff weight of 338,000 lb with the maximum allowable payload and minimum fuel, find the maximum allowable payload at the origin airport as well as the landing weight and the landing runway requirements at the destination airport.

6. An airport has an 8000-ft runway available. The maximum difference in centerline elevation is 10 ft. The airport elevation is 3500 ft AMSL, and the mean daily high temperature in the hottest month is 87.5°F. An airline wishes to use a particular aircraft from this airport. The planned route distance is 700 mi, and the nearest alternate destination is 200 mi from the planned destination. The time fuel reserves are included in the item listed as typical operating empty weight plus time fuel reserves in Table B-5. The takeoff characteristics of this aircraft are given in Table B-6.

TABLE B-5   Landing Runway Requirements

Maximum Allowable Landing Weight, 1000 lb

Temp., °F	Airport elevation, ft						
	0	1000	2000	3000	4000	5000	6000
70	103.0	101.5	97.9	94.3	90.0	87.6	84.4
75	103.0	101.5	97.9	94.3	90.9	87.6	84.4
80	103.0	101.5	97.9	94.3	90.9	87.6	84.4
85	103.0	101.1	97.5	94.0	90.6	87.3	84.1
90	102.8	99.1	95.5	92.0	88.6	85.4	88.2
95	100.5	97.0	93.5	90.0	86.7	83.5	80.4
100	98.2	94.7	91.3	88.0	84.8	81.6	78.5

Landing Runway Length, 1000 ft

Weight, 1000 lb	Airport elevation, ft						
	0	1000	2000	3000	4000	5000	6000
70	3.92	4.02	4.11	4.20	4.30	4.40	4.50
75	4.15	4.24	4.34	4.44	4.54	4.64	4.75
80	4.37	4.46	4.56	4.67	4.78	4.89	5.00
85	4.58	4.69	4.79	4.90	5.02	5.14	5.26
90	4.80	4.91	5.02	5.13	5.26	5.38	5.51
95	5.01	5.13	5.25	5.37	5.50	5.63	5.77
100	5.23	5.35	5.47	5.60	5.74	5.88	6.02
105	5.44	5.57	5.70	5.84	5.98	6.12	6.27

Maximum structural takeoff weight = 114,900 lb
Maximum structural landing weight = 103,000 lb
Typical operating weight plus time reserves = 66,620 lb
Average fuel consumption = 16 lb/mi
Typical maximum passenger load = 26,000 lb
Maximum structural payload = 28,740 lb
Fuel capacity = 7500 gal

    a. Determine if there are any takeoff or landing weight restrictions on this aircraft based upon available runway length at this airport for such a planned flight.

    b. Determine the required runway length if any takeoff restrictions based upon runway length are to be eliminated.

    c. Find the maximum allowable takeoff weight from this airport and from the other airport so that the maximum allowable landing weight is not exceeded at this airport under these conditions.

   7. A new runway is being planned for Denver International Airport. The runway is to be used by an aircraft with the characteristics given in Tables B-3 and B-4. The longest nonstop trip planned from this airport with this aircraft is to be Denver-Pittsburgh, a route distance of 1300 mi, with the listed alternate being

TABLE B-6   Takeoff Runway Requirements

Temp., °F	Airport elevation, ft						
	0	1000	2000	3000	4000	5000	6000
Maximum Allowable Takeoff Weight, 1000 lb							
70	114.9	110.4	105.8	101.5	97.4	93.4	89.6
75	114.9	110.4	105.8	101.5	97.4	93.4	89.6
80	114.9	110.4	105.8	101.5	97.4	93.4	89.6
85	114.9	109.8	105.3	101.0	96.9	93.0	89.2
90	111.8	107.1	102.7	98.5	94.5	90.7	87.0
95	108.9	104.4	100.1	96.0	92.1	88.4	84.9
100	105.9	101.6	97.5	93.6	89.8	86.2	82.8
Reference Factor							
70	54.1	58.1	62.6	67.5	73.0	79.1	85.8
75	54.5	58.6	63.1	68.1	73.7	79.9	86.7
80	54.9	59.1	63.7	68.8	74.4	80.6	87.5
85	55.7	59.9	64.6	69.8	75.5	81.9	88.9
90	57.6	62.1	67.0	72.4	78.4	85.0	92.3
95	59.6	64.1	69.2	74.9	81.2	88.2	95.7
100	61.7	66.3	71.6	77.6	84.1	91.4	99.2

Takeoff Runway Length, 1000 ft

Weight, 1000 lb	Reference factor					
	50	60	70	80	90	100
75	4.00	4.04	4.62	5.18	5.76	6.34
80	4.00	4.53	5.21	5.88	6.56	7.25
85	4.26	5.09	5.90	6.69	7.48	8.30
90	4.75	5.73	6.67	7.59	8.53	9.49
95	5.31	6.45	7.54	8.61	9.69	10.80
100	5.95	7.26	8.51	9.74	10.98	12.25
105	6.66	8.16	9.59	10.98	12.39	13.83
110	7.46	9.16	10.77	12.34	13.92	15.55
115	8.35	10.26	12.07	13.83	15.58	

Philadelphia, a route distance of 1575 mi. The mean maximum daily temperature at Denver is 95°F, and the airport elevation is 5330 ft AMSL. The maximum difference in centerline elevation is 20 ft. The maximum structural takeoff weight of the aircraft is 408,900 lb.

a. Determine the required takeoff and landing runway length for this aircraft at Denver.

b. Determine the actual takeoff and landing weight of this aircraft at Pittsburgh for the trip to and from Denver.

8. The total air carrier passenger traffic between two airports from 1970 through 1990 is given in Table B-7. A new carrier began ser-

TABLE B-7   Total Annual Air Carrier Passenger Traffic
between Airports, 1970–1990

Year	Passengers, all airlines
1970	198,128
1975	259,317
1980	295,780
1985	340,717
1990	360,670

TABLE B-8   New Airline's Market Share, 1989–1991

Year	Quarter	Percentage of total passengers
1989	First	0.40
	Second	1.11
	Third	2.80
	Fourth	5.70
1990	First	10.10
	Second	12.90
	Third	15.20
	Fourth	17.30
1991	First	19.00
	Second	19.40
	Third	19.90
	Fourth	20.20

vice in the first quarter of 1989 between the airports and has been steadily increasing its share of the market, as shown in Table B-8.

a. Determine the total market share which the new airline may hope to capture, assuming all services remain the same.

b. Determine the total number of passengers that the new airline may expect to serve in this market in the year 2000.

c. If an average load factor of 0.48 is required to cover costs and there are a total of 12 round-trips per day by the airlines with aircraft having an average passenger capacity of 125, determine whether the route is profitable for the airlines collectively in the fourth quarter of 1991 and for the new airline if it offers 3 round-trips per day.

9. At present, there is air service from Epsilon City to the cities of Alpha, Beta, Gamma, and Delta. The current passenger traffic between Epsilon City and the other cities, together with total disposable income and travel time, is given in Table B-9.

a. Plot a curve relating enplaned passengers per unit of disposable income versus travel time for travel from Epsilon City to each of the above cities.

TABLE B-9   Intercity Travel and Socioeconomic Data (1992) for Airline Service
from Epsilon City

Destination city	Disposable income, $ millions	Average travel time, h	Enplaned passengers
Alpha	$1000	4.0	60,000
Beta	1000	2.0	120,000
Gamma	875	1.0	300,000
Delta	1000	0.5	720,000

TABLE B-10   General Aviation Aircraft Registration and Income
Data

Year	Registered aircraft Country	State	Per capita income for the state, $
1970	140,000	2,900	5,750
1975	152,000	3,300	7,700
1980	168,000	3,650	9,400
1985	181,000	3,870	10,500
1990	195,000	4,195	11,600

    b. Using the above curve, determine the expected current market for passengers on a new route between Epsilon City and Sigma City if it is known that the total disposable income of Sigma City is $1.3 billion and the route will have a travel time of 1.5 h.

10. General aviation aircraft registration and per capita income data for 1970 through 1990 are contained in Table B-10.

    a. Develop a top-down forecast of the number of aircraft registrations in the state for the years 1995 and 2000.

    b. Develop an econometric model relating aircraft registration in the state to per capita income in the state. Give the equation relating the variables as well as the standard error of the estimate of the equation.

    c. Develop a bottom-up forecast of the number of aircraft registrations in the state in 1995 if the per capita income is expected to be $12,300. State the most probable range in the forecast.

    d. Compare the top-down and bottom-up forecasts of the number of aircraft registrations in the state for the year 1995.

11. The historical pattern of airline passenger growth at Beta City Airport and the simultaneous pattern of population and economic growth are given in Table B-11.

    a. Develop a top-down forecast for the 1997 enplaned passengers at Beta City Airport.

    b. Develop an econometric model of enplaned passenger growth at Beta City Airport. Also prepare a bottom-up forecast of the enplaned passengers at Beta City Airport in 1997 if it is

TABLE B-11   Socioeconomic and Passenger Data for Beta City

Year	Enplaned passengers, millions Beta	United States	Population of Beta, millions	Per capita disposable income of Beta, $
1976	0.50	94	0.125	8,500
1979	0.80	135	0.160	9,500
1982	1.00	165	0.200	10,500
1985	1.20	185	0.222	11,000
1988	1.30	200	0.235	12,000
1991	1.45	220	0.255	13,500

TABLE B-12   Wind Data for Daytime Visual Meteorological Conditions

Direction	0–3	4–12	13–15	16–18	19–24	25–31	32+	Total
N	0.4	3.6	1.0	0.5	0.2	0.0	0.0	5.7
NNE	0.3	2.5	0.6	0.2	0.1	0.0	0.0	3.7
NE	0.4	3.3	0.6	0.2	0.1	0.0	0.0	4.6
ENE	0.3	2.7	0.5	0.2	0.1	0.0	0.0	3.8
E	0.5	2.7	0.3	0.1	0.0	0.0	0.0	3.6
ESE	0.3	1.6	0.2	0.1	0.0	0.0	0.0	2.2
SE	0.4	2.3	0.3	0.1	0.0	0.0	0.0	3.1
SSE	0.5	4.9	0.6	0.1	0.0	0.0	0.0	6.1
S	0.7	11.8	1.9	0.7	0.2	0.0	0.0	15.3
SSW	0.4	6.0	1.7	1.1	0.6	0.1	0.1	10.0
SW	0.4	3.1	0.5	0.3	0.2	0.0	0.0	4.5
WSW	0.3	2.2	0.4	0.2	0.1	0.0	0.0	3.2
W	0.3	2.6	0.7	0.4	0.3	0.1	0.0	4.4
WNW	0.2	3.0	1.4	1.2	1.0	0.4	0.0	7.2
NW	0.2	4.7	2.5	2.0	1.3	0.3	0.0	11.0
NNW	0.2	5.2	2.0	1.4	0.7	0.1	0.0	9.6
Calm	2.0							
Total	7.8	62.2	15.2	8.8	4.9	1.0	0.1	100.0

Wind speed, mi/h

expected that the per capita disposable income will be $15,000 and the population will be 310,000 in Beta City in 1997.

12. Wind data for daylight hours for visual meteorological conditions for an airport are given in Table B-12.

   a. Plot the wind rose for daylight hours in visual meteorological conditions.

   b. Determine the orientation of the main runway at this airport based upon daylight VMC if the crosswind component to the runway cannot exceed 13 mi/h and tailwinds must be less than 4 mi/h.

   c. Determine the percentage of time that this runway may be utilized under VMC.

d. Determine the orientation of a crosswind runway at this airport to maximize wind coverage under VMC.

13. At present, a 2000-ft paved runway exists at an airport. The approach threshold to runway 9 is located at the physical end of the runway. A highway is located 250 ft from the physical end of the runway along the extended centerline of the runway. The elevation of the end of the runway is 120 ft AMSL. The elevation of the centerline of the highway is 131 ft AMSL. Regulations specify that the approach surface must have a 17-ft vertical clearance above the centerline of the highway. The runway is designed to have an approach surface slope of 20:1. Determine whether the threshold is properly located or must be displaced. If it must be displaced, find the amount of displacement required and the effective length of runway available from the threshold to the end of runway 9.

14. A sketch of the obstruction surfaces defined in FAR part 77 is shown in Fig. B-1. The obstruction surfaces are to be specified for approaches to a 6000-ft runway designed for use by transport aircraft for non-precision-instrument operations with visibility minima greater than 0.75 mi. The elevation of the runway is 350.0 ft AMSL. Determine the coordinates of points *A, B, C, D, E, F,* and *G,* shown in Fig. B-1.

15. A sketch of the obstruction surfaces defined in FAR part 77 is shown in Fig. B-1. The obstruction surfaces are to be specified for approaches to a 10,000-ft runway designed for use by transport aircraft for precision-instrument operations. The elevation of the runway is 642.5 ft AMSL.

a. Determine the coordinates of points *A, B, C, D, E, F,* and *G,* shown in Fig. B-1.

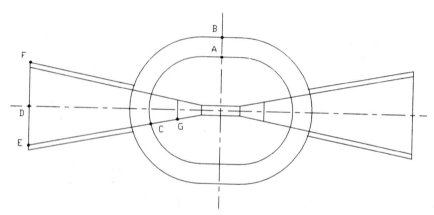

Figure B-1 Schematic diagram of FAR part 77 surfaces.

b. Determine whether an object which is 280 ft high located at a point 12,000 ft from the midpoint of the runway along the extended runway centerline and offset a distance of 5000 ft from the extended runway centerline penetrates the FAR part 77 obstruction surfaces.

c. Determine the maximum elevation of a tower located at a point 2000 ft from the midpoint of the runway measured along the runway centerline and offset a distance of 8000 ft from the runway centerline, if it is not to penetrate the FAR part 77 obstruction surfaces.

16. The pattern of hourly aircraft demand on the average day of the peak month at an airport is given in Table B-13. The fleet mix and operating cost per hour in the peak hour consist of 5 percent type A aircraft at $180, 25 percent type B at $300, 60 percent type C at $1400, and 10 percent type D at $3000. The existing runway system provides for an hourly capacity of 52 operations. The runway system is known to behave as a single-server queueing system with a Poisson arrival distribution and a uniform service distribution with a variance of zero.

a. Find the average daily delay in minutes to aircraft at this airport in the peak month.

b. Find the average daily cost of delay at this airport in the peak month.

c. Find the capacity required at this airport if the peak-hour delay on the average day in the peak month is to be limited to 8 min per operation.

17. An airport has the scheduled aircraft demand shown in Table B-14. A parallel-runway system is being operated at the airport.

a. Assume that the above runway system can be modeled by using a Poisson arrival and general service distribution with a constant service rate. One runway services only arrivals at a rate of 30 aircraft per hour, and the other services only departures at a rate of 30 aircraft per hour.

(1) Find the average daily delay in minutes and the average daily queue length for arriving aircraft.

TABLE B-13   Hourly Aircraft Demand

Hour	Demand	Hour	Demand	Hour	Demand	Hour	Demand
00:00	12	06:00	16	12:00	58	18:00	67
01:00	9	07:00	43	13:00	55	19:00	51
02:00	7	08:00	54	14:00	69	20:00	43
03:00	4	09:00	36	15:00	73	21:00	38
04:00	4	10:00	39	16:00	75	22:00	27
05:00	4	11:00	62	17:00	71	23:00	19

TABLE B-14   Average Hourly Runway Arrival Demand

Hour	Arrivals	Departures	Hour	Arrivals	Departures
7:00	5	13	15:00	17	18
8:00	7	12	16:00	30	25
9:00	10	13	17:00	35	30
10:00	12	22	18:00	21	23
11:00	17	21	19:00	18	14
12:00	18	21	20:00	20	15
13:00	15	19	21:00	16	7
14:00	12	18	22:00	15	6

    (2) Find the average daily delay in minutes and the average daily queue length for departing aircraft.

    (3) Estimate the largest average hourly delay in minutes and the largest hourly queue length for arriving aircraft.

    (4) What arrival runway service rate is required if arrivals are to be delayed an average of 4 min daily?

   b. Assume that the above runway system can be modeled by using a Poisson arrival and exponential service distribution. One runway services only arrivals at a rate of 30 aircraft per hour, and the other services only departures at a rate of 30 aircraft per hour.

    (1) Find the average daily delay in minutes and the average daily queue length for arriving aircraft.

    (2) Find the average daily delay in minutes and the average daily queue length for departing aircraft.

    (3) Estimate the largest average hourly delay in minutes and the largest hourly queue length for arriving aircraft.

    (4) What arrival runway service rate is required if arrivals are to be delayed an average of 4 min daily?

   c. Assume that the total aircraft demand can be modeled by using a deterministic model and that the runways jointly service arrivals and departures at a rate of 40 aircraft per hour.

    (1) Find the time period when aircraft are delayed.

    (2) Find the largest number of aircraft delayed at any one time during the day and the delay to these aircraft.

    (3) Find the average daily delay to all aircraft using the airport on this day.

18. The pattern of typical daily aircraft demand for a runway system at an airport is given in Table B-15. The runway system can service 40 operations per hour. The fleet mix and operating cost per hour in the peak hour consist of 10 percent type A aircraft at $140, 30 percent type B at $480, 50 percent type C at $2000, and 10 percent type D at $4500. Determine the period of time when

TABLE B-15   Runway System Aircraft Demand

Hour	Operations	Hour	Operations
00:00	5	12:00	25
01:00	5	13:00	25
02:00	5	14:00	20
03:00	5	15:00	35
04:00	5	16:00	55
05:00	5	17:00	60
06:00	10	18:00	50
07:00	20	19:00	35
08:00	30	20:00	20
09:00	25	21:00	15
10:00	30	22:00	10
11:00	25	23:00	5

TABLE B-16   Hourly Runway System Demand in Aircraft

Hour	Aircraft	Hour	Aircraft	Hour	Aircraft
07:00	29	12:00	91	17:00	56
08:00	29	13:00	100	18:00	56
09:00	35	14:00	47	19:00	79
10:00	53	15:00	53	20:00	65
11:00	62	16:00	50	21:00	56

aircraft delays exist, the total daily delay to aircraft, and the total daily cost of delay, using a deterministic model.

19. The hourly demand for use of the runway system at an airport on a typical day is given in Table B-16. It may be assumed that the arrival distribution is Poisson and that the service distribution is exponential.

   a. Plot the average daily delay as a function of the runway capacity.

   b. Determine the required capacity of the runway system if the average daily delay is not to exceed 4 min.

20. An airport has a daily demand of 2050 aircraft. In the busiest part of the day, the average hourly demand is 128 operations. The average hourly capacity available on the runway system is 130 aircraft. Assume that the runway system may be modeled by a single-server model with a Poisson arrival distribution and an exponential service distribution.

   a. Find the average hourly delay during the day.

   b. Find the average hourly delay during the busiest period of the day.

   c. Find the required runway system capacity to limit the average hourly delay during the busiest period of the day to 15 min.

d. Find the average demand rate required during the busiest period of the day to limit the average hourly delay during the busiest period of the day to 15 min with an average hourly runway system capacity of 130 operations.

21. A runway is to service arrivals and departures. The common approach path is 6 mi long for all aircraft. During a particular interval of time, the runway is serving only two types of aircraft: a Spartan-100 (S-100) which travels the common approach path in 210 s and has an arrival runway occupancy time of 30 s, and a Spartan-300 (S-300) which travels the common approach path in 180 s and has an arrival runway occupancy time of 40 s. The air traffic separation rules in effect require a minimum of 120 s between consecutive departures, a minimum 3-mi separation between consecutive arrivals, and a departure can be released if the incoming arrival is at least 2 mi from the threshold. During the period to be analyzed, an ordered arrival queue of S-300, S-100, and S-300 aircraft and an ordered departure queue of S-100, S-100, and S-300 aircraft are to be serviced by the runway.

a. Using the space-time diagram, analyze the capacity of this runway.

    (1) Draw the space-time diagram to service these aircraft, assuming that the first arrival is at the entry gate at time zero.

    (2) Find the required time to service the above arrival and departure queue.

    (3) Estimate the hourly capacity of the runway to service a repeating pattern of aircraft as described.

    (4) Estimate the hourly capacity of the runway to service a repeating pattern of aircraft as described if the departure-departure spacing is reduced to 90 s.

b. Analyze the capacity of this runway, assuming operations occur in an error-free environment.

    (1) Find the capacity of the runway to service only arrivals.

    (2) Find the capacity of the runway to service only departures.

    (3) Find the probability of releasing a departure after each arrival and the capacity of the runway to service mixed operations in the case where arrivals are given priority over departures.

    (4) Find the required interarrival time if one departure is to be released after each arrival and the resulting capacity of the runway to service mixed operations.

    (5) Find the required interarrival time if two departures are to be released after each arrival and the resulting capacity of the runway to service mixed operations.

c. Solve part *b* of this problem, assuming operations occur in a position error context if the standard deviation of the position error is 20 s and the minimum-separation rules can be violated 10 percent of the time.

22. A runway is to service arrivals and departures. The common approach path is 6 mi long for all aircraft. During a particular interval the runway is serving only two types of aircraft: a T-10, which travels the common approach path in 180 s and has an arrival runway occupancy time of 30 s, and a T-30, which travels the common approach path in 150 s and has an arrival runway occupancy time of 40 s. The air traffic separation rules in effect require a minimum of 90 s between consecutive departures, a minimum 3-mi separation between consecutive arrivals, except when the leading arrival is a T-30 and the trailing is a T-10, for which a 4-mi minimum separation is required. A departure can be released if the incoming arrival is at least 2 mi from the threshold. During the period to be analyzed, an ordered arrival queue of T-30, T-10, T-30, and T-30 aircraft and an ordered departure queue of T-10, T-30, and T-30 aircraft are to be serviced by the runway.

a. Using the space-time diagram, analyze the capacity of this runway.

(1) Draw the space-time diagram to service these aircraft, assuming that the first arrival is at the entry gate at time zero.

(2) Find the required time to service the above arrival and departure queue.

(3) Estimate the hourly capacity of the runway to service a repeating pattern of aircraft as described.

b. Analyze the capacity of this runway, assuming operations occur in an error-free environment.

(1) Find the capacity of the runway to service only arrivals.

(2) Find the capacity of the runway to service only departures.

(3) Find the probability of releasing a departure after each arrival and the capacity of the runway to service mixed operations in the case where arrivals are given priority over departures.

(4) Find the required interarrival time if one departure is to be released after each arrival and the resulting capacity of the runway to service mixed operations.

(5) Find the required interarrival time if two departures are to be released after each arrival and the resulting capacity of the runway to service mixed operations.

c. Solve part *b* of this problem, assuming operations occur in a position error context if the standard deviation of the position error is 30 s and the minimum-separation rules may be violated 5 percent of the time.

23. A single runway at an airport is capable of servicing mixed operations. Aircraft operate in an error-free environment. The aircraft consist of 70 percent type A aircraft with an approach speed of 120 mi/h and an arrival runway occupancy time of 50 s, and 30 percent type B aircraft with an approach speed of 90 mi/h and an arrival runway occupancy time of 40 s. The common approach path for arrivals is 5 mi long. The air traffic separation rules in effect require a minimum of 3.5 mi between consecutive arrivals. A minimum time is required between consecutive departures of 90 s when the trailing departure is a type A aircraft, 120 s when the leading departure is a type A aircraft and the trailing departure is a type B aircraft, and 60 s when both the leading and trailing departures are type B aircraft. A departure can be released if the incoming arrival is at least 2 mi from the threshold.

   a. Find the capacity of the runway to service only arrivals.

   b. Find the capacity of the runway to service only departures.

   c. Find the capacity of the runway to service mixed operations while maintaining minimum interarrival separations.

   d. Find the capacity of the runway system to service mixed operations if a departure is to be inserted between a pair of arrivals at least 50 percent of the time.

   e. Find the capacity of the runway to service mixed operations if two departures are always to be inserted between a pair of arrivals.

24. A single runway at an airport is used to service only arrival aircraft. The aircraft fleet using the runway comprises two types. The type A aircraft has an approach speed of 120 mi/h and consists of 30 percent of the fleet. The type B aircraft has an approach speed of 90 mi/h and consists of 70 percent of the fleet. The common approach path to the runway system is 5 mi long. The standard deviation of the position error of aircraft is 40 s. Minimum-separation rules may be violated 10 percent of the time. The minimum separation distance between arrivals is 3 mi when the trailing arrival is a type A aircraft or when the leading and trailing arrivals are both type B aircraft and 4 mi when the leading arrival is type A and the trailing arrival is type B.

   a. Find the hourly capacity of the runway to service only arrivals.

   b. Find the actual distance maintained between aircraft when the leading aircraft is type A and the trailing aircraft is type A.

   c. If two parallel runways existed at the airport, what would be the arrival-only hourly capacity of this runway system if only type A aircraft use one runway and only type B aircraft use the other runway?

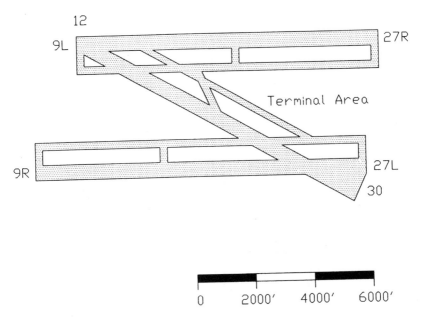

Figure B-2 Runway layout plan.

TABLE B-17   Historical Runway-Use Configurations at Airport

Use configuration		Hourly capacity		Use, %	
		VFR	IFR	VFR	IFR
A	A 27R D 27R	55	53	9.0	4.0
B	A&D 27R A 27L	111	105	31.5	6.0
C	A&D 27R A 30	77	59	22.5	0.0
D	A 30 D 27L			27.0	0.0

25. The runway system at an airport is shown in Fig. B-2. The historical runway-use configurations at this airport are given in Table B-17. The aircraft fleet mix at this airport consists of 3 percent type A, 7 percent type B, 75 percent type C, and 15 percent type D aircraft in all weather conditions. During the peak hour, the arrivals and departures are about equal. Fig. B-3 is included for assistance in the solution of this problem.

   a. Compute the VFR hourly capacity of runway-use configuration D.

   b. Compute the weighted hourly capacity of the runway system at this airport.

26. The runway system at any airport is given in Fig. B-4. The historical pattern of runway use is given in Table B-18. The aircraft

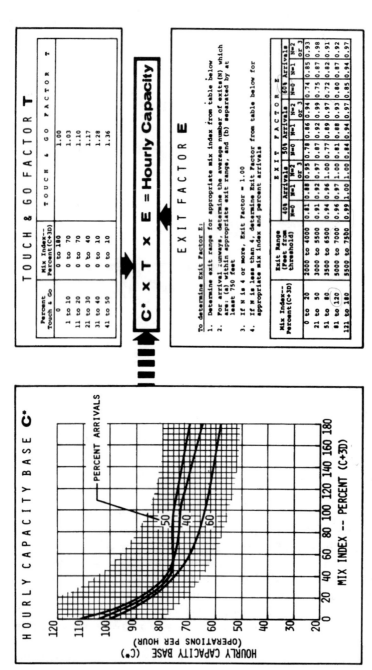

Figure B-3 Hourly capacity of intersecting runway-use configuration with arrivals on one runway and departures on the other runway under visual flight rules.

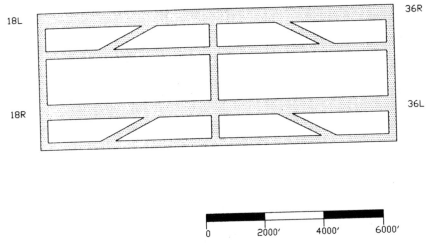

Figure B-4   Runway layout plan.

TABLE B-18   Runway-Use Patterns at Airport

Runway-use configuration	Arrival	Departure	Use, %	
			VFR	IFR
A	36L	36L	48.0	19.0
B	36L	36R	22.0	0.0
C	18R	18L	11.0	0.0

mix using the airport consists of 5 percent type A, 15 percent type B, 60 percent type C, and 20 percent type D aircraft. During the peak hour, the percentage of arrivals is 60 percent. The level of annual operations is 200,000. Figures B-5 through B-7 are abstracted from the FAA capacity handbook for use in the solution of this problem.

a. Find the hourly capacity of runway-use configuration A in both VFR and IFR.

b. Find the hourly capacity of runway-use configurations B and C in VFR.

c. Find the weighted hourly capacity of the airport.

d. Estimate the level of annual delay at the airport in aircraft-minutes.

27. The existing runway system at an airport consists of two closely spaced parallel runways whose centerlines are separated by 1000 ft. The future annual demand on the airport is expected to be 232,000 operations. The aircraft use and hourly operating cost data for aircraft at the airport consist of 10 percent type A at $500, 50 percent type B at $1500, 30 percent type C at $3000, and

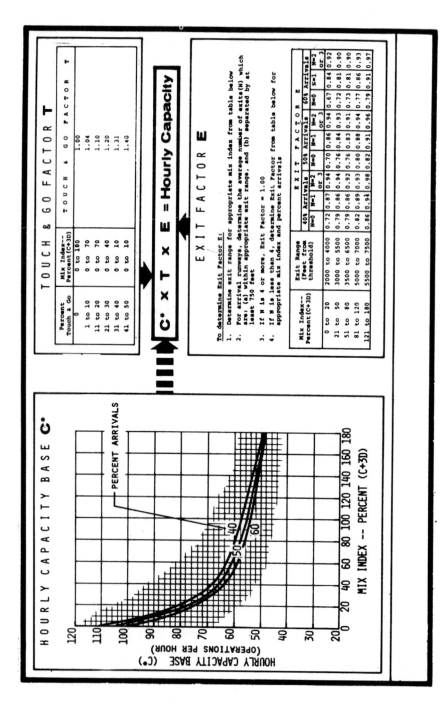

Figure B-5 Hourly capacity of single-runway-use configuration with both arrivals and departures under visual flight rules.

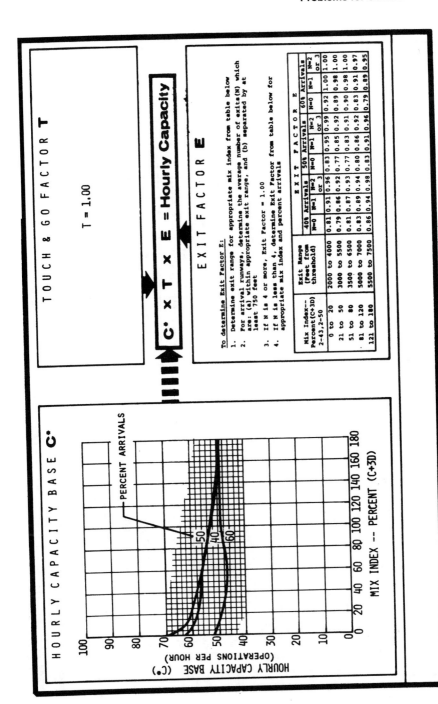

Figure B-6 Hourly capacity of single-runway-use configuration with both arrivals and departures under instrument flight rules.

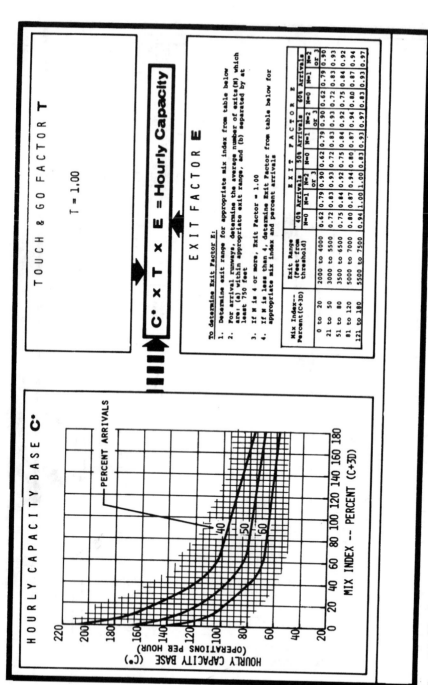

Figure B-7 Hourly capacity of parallel-runway-use configuration with arrivals on one runway and departures on the other runway under visual flight rules.

10 percent type D at $5000. A third runway is to be added to the airport to minimize aircraft operating costs.

a. Determine if the third runway should be another parallel runway with a centerline spaced 3000 ft from the centerline of an existing parallel runway, an intersecting runway, or an open-V runway.

b. Determine the annual costs of delay for the choice made in part *a.*

28. A rough sizing of the airport site is to be made for an airport with parallel runways to be operated in an IFR environment with arrivals on one runway and departures on the other runway. The thresholds of the runways are not staggered. The airside system is to consist of parallel taxiways serving each runway, the terminal complex on one side of the parallels and the property boundary on the other side. The design aircraft is to be a Boeing 767-200. The terminal is to have parallel concourses perpendicular to the runway centerlines, with nose-in aircraft parking. The top of the concourse will be 28 ft high at the end. The runways are 10,000 ft long. A schematic layout of the above facility is shown in Fig. B-8.

a. Determine the separation requirements between runway centerlines, between the runway centerline and a parallel taxiway centerline, between parallel taxiway centerlines, between the runway and the terminal building line, between the runway and the property line, and between concourse faces. Solve this, using the Boeing 767-200 as the design aircraft.

b. Determine the area of the apron surface, in acres, between the concourses if six aircraft are parked on each terminal face.

c. Determine the property area, in acres, between the property line and the terminal building line from the ends of the runways.

d. Determine the distance between the runway centerline and the aircraft holding line for takeoffs.

29. The layout of the runway, taxiway, taxilane, and apron area at an airport with a gate arrivals concept is shown in Fig. B-9. The gate design aircraft are the DC-10-10 for gate A, Boeing 747-200 for gate B, Boeing 767-200 for gate C, Boeing 727-200 for gate D, and Boeing 737-200 for gate E. The minimum separation between runway and taxiway centerlines is 400 ft. Runways are 150 ft wide, and taxiways are 75 ft wide. The apron area is the area required for parking aircraft including the required separations between wing tips, between aircraft and fixed obstructions, and between aircraft and movable obstructions. Geometric design criteria require that the most demanding aircraft in an aircraft design group be used to determine separation criteria.

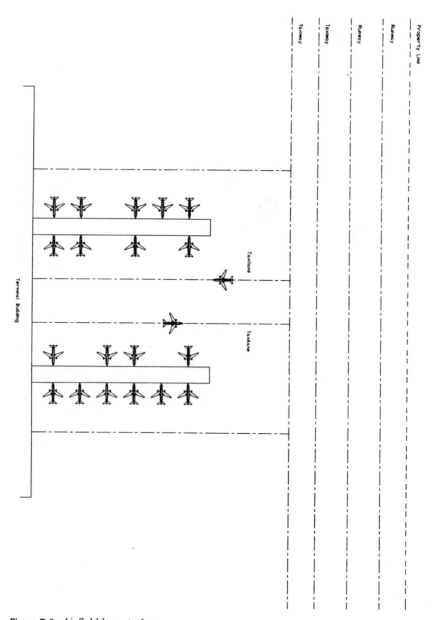

Figure B-8 Airfield layout plan.

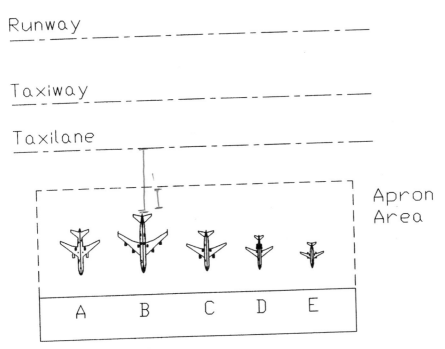

Runway

Taxiway

Taxilane

Apron
Area

A   B   C   D   E

Figure B-9   Airfield layout plan.

a. Determine the minimum distance between the runway center-line and the face of the terminal building if a single taxiway, a single taxilane, and the apron parking area are included.

b. Determine the length and width of the apron area.

30. A terminal system is to be placed between two parallel 10,000-ft runways which are to accommodate simultaneous arrivals in IFR conditions. Each of the runways is to be serviced by a dual parallel-taxiway system. The design aircraft is the Boeing 747-400. Assume that the wing tips of the Boeing 747-400 are located 60 percent of the distance along the centerline of the fuselage. An apron area with two circular satellite buildings, each accommodating eight gates, is to be constructed for the design aircraft. The apron area is to be serviced by dual parallel taxilanes between the satellites.

a. Lay out the apron area, and locate the apron area relative to the satellite buildings.

b. Compare the relative sizes of the apron area for the above satellite configuration and a single rectangular pier with eight aircraft parked on either side of the pier.

31. An analysis of the total aircraft demand at an airport in the current year indicates that the expected traffic on the peak-month

average day is 1200 operations. The peak hour at present sees an average of 50 arrivals and 65 departures. In the design year, air carrier traffic is expected to be 73 percent of the total airport traffic. Peak-hour passenger load factors are expected to be 0.90, and the average aircraft seating capacity in the peak hour is expected to be 190 passengers. Through passengers are expected to be 5 percent of inbound passengers, and connecting passengers are expected to be 25 percent of the deplaning passengers. The present level of annual aircraft demand is 300,000 operations, and this is expected to grow to 350,000 annual operations in the design year. The percentage of connecting passengers passing through security is expected to be 40 percent.

a. Estimate the total peak-hour operations in the design year.
b. Estimate the total peak-hour enplaning, deplaning, connecting, originating, and terminating passengers in the design year.
c. Estimate the number of peak-hour passengers for the design of ticketing facilities, security, boarding devices, baggage claim, and the deplaning curb front in the design year.

32. An airport is expected to have 200,000 annual aircraft operations in the design year. The airport services 80 percent general aviation and 20 percent air carrier traffic. Air carrier aircraft are expected to have an average capacity of 86 passengers and are the only aircraft which use the terminal facilities at the airport. During the peak hour, it is expected that the inbound and outbound load factors will be about 0.45. Deplaning passengers are expected to be 65 percent of the inbound passengers; connections, 10 percent of the deplaning passengers. It is also expected that in the peak hour, inbound air carrier operations will be about 60 percent of total air carrier operations. Ground transportation vehicles are expected to have average occupancies of 2.4 passengers per vehicle.

a. Estimate the total number of air carrier operations on the peak-month average day during the peak hour in the design year.
b. Estimate the total number of originating, terminating, enplaning, deplaning, and connecting passengers during the peak hour on the peak-month average day in the design year in the terminal building.
c. Estimate the total number of vehicles on the ground access system in the vicinity of the terminal building during the peak hour on the peak-month average day in the design year.
d. Estimate the number of passengers who use the baggage claim area in the terminal building during the peak hour on the peak-month average day in the design year if 80 percent of the terminating passengers are expected to have baggage to claim.

33. The peak-hour originating passenger demand at the ticket counters at an airport is expected to be 300 passengers per hour. A ticket counter typically services a passenger in 2.5 min. The ticket counter can be modeled as a single-server system in which the arrival process is Poisson and the service process is constant.
   a. Determine the number of ticket counters which must be provided if the demand splits up evenly among each ticket counter and if the average passenger delay at the ticket counters cannot exceed 10 min.
   b. Determine the number of ticket counters which must be provided if the total demand is queued in one line, if the first passenger in line goes to the first available processor, and if the average passenger delay at the ticket counters cannot exceed 10 min.

34. The arrival distribution of passengers at the initial originating passenger-processing positions during a 40-min design period is given in Table B-19. These positions consist of ticketing, baggage check-in, and gate information positions. Performance and use data indicating the type of passenger processing required at these positions and the average service time are given in Table B-20. Five passenger-processing positions exist, and each is capable of providing all the above services. Assume that the arrival distribution is Poisson and that the service rate is constant.
   a. Find the average passenger service time at the initial passenger-processing position.
   b. Find the average delay to passengers during the design period.
   c. Find the average queue length during the design period.

TABLE B-19   Passenger Arrival Distribution for Design Period

Period	Passengers arriving in period	Period	Passengers arriving in period
9:00–9:05	4	9:20–9:25	28
9:05–9:10	8	9:25–9:30	19
9:10–9:15	14	9:30–9:35	8
9:15–9:20	26	9:35–9:40	2

TABLE B-20   Facility Passenger Servicing Data

Type of service required	Passengers requiring service, %	Average service time, s
Ticket only	20	80
Baggage only	15	60
Ticket and baggage	45	150
Information only	20	30

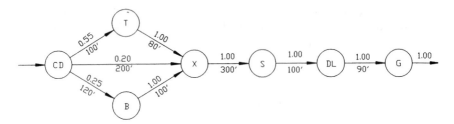

Figure B-10 Link-and-node diagram for enplaning system.

TABLE B-21   Characteristics of Passenger-Processing System

Component	Number of servers	Service rate per person, s
Curb-front doors (CD)	3	5
Ticket counters (T)	6	120
Baggage check-in (B)	2	50
Security (X)	2	15
Seat selection (S)	4	30
Boarding devices (G)	3	6

35. An originating passenger-processing network is to be analyzed during the peak hour. The peak-hour demand is 300 passengers. The passengers proceed from the curb front, to the ticketing and baggage check-in area, to security, to seat selection, to the departure lounge, and then to the boarding device. The distribution of passengers and the walking distance in ft between these processors are shown in Fig. B-10. There is an average of 0.25 visitor per passenger, and 40 percent of visitors proceed beyond security. Average walking speeds are 2 ft/s. The characteristics of the passenger-processing components are given in Table B-21. The system exhibits Poisson arrival and constant service distributions, and the passenger demand is split equally between each processor of a particular type. Delays at the departure lounge are not considered in this analysis. Assume that the period during which demand exceeds capacity is limited to 10 min.
   a. Draw a network which shows the number of persons per hour who proceed between processing components.
   b. Determine the average passenger-processing time during the peak hour.
   c. Determine the number of ticketing processing positions and the number of baggage check-in positions required if the average line length at each is not to exceed 5 passengers in the peak hour.

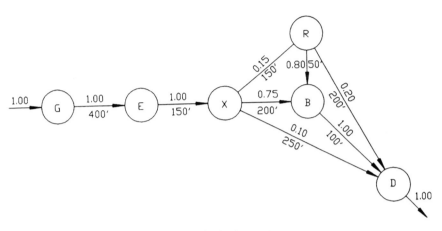

Figure B-11 Link-and-node diagram for deplaning system.

36. The terminating passenger-processing network at an airport is to be analyzed during the peak hour. Passengers proceed from the boarding device, to an escalator, to the baggage claim area, to the central car area, and to the curb front. The distribution of passengers and the walking distances between processors are shown in Fig. B-11. It takes 10 min from the arrival of the aircraft until the first piece of baggage is placed on the baggage claim device, and the baggage is placed on the device at a rate of 10 bags per minute. Each passenger has an average of 1.7 pieces of checked baggage. There are 200 terminating passengers in the peak hour. There is an average of 0.3 visitor per passenger, and 50 percent of the visitors wait at the gate for the passengers. Passenger walking speeds are 3 ft/s. The characteristics of the passenger-processing network are given in Table B-22. The system exhibits Poisson arrival and exponential service distributions, and the passenger demand is split evenly between each processor of a particular type. The demand on any processor cannot exceed its capacity for more than 15 min.

TABLE B-22   Characteristics of Passenger-Processing System

Component	Number of servers	Service, persons/min
Boarding device (G)	4	10.0
Escalator (E)	2	15.0
Security exit (X)	2	12.0
Baggage claim devices (B)	2	
Rental-car checkout (R)	6	0.5
Curb-front doors (D)	4	6.0

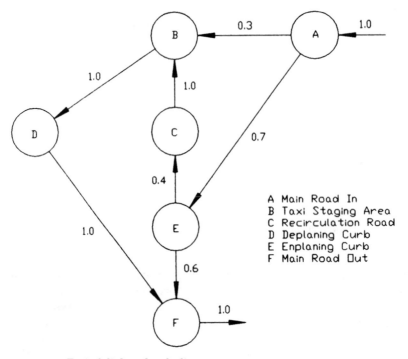

Figure B-12 Taxicab link-and-node diagram.

    a. Draw a network which shows the number of persons per hour who proceed between processing components.

    b. Compute the average passenger delay at the baggage claim devices.

    c. Determine the average passenger-processing time during the peak hour.

37. The flow of taxicabs on the airport roadway system is shown on the link-and-node diagram in Fig. B-12. It is expected that 236 taxicabs will enter the airport on the main road into the airport during the peak hour on the design day. Taxicabs are considered to be 20 ft long and to have dwell times at the enplaning curb of 85 s and at the deplaning curb of 140 s.

    a. Write the transition matrix for taxicab flow on airport property.

    b. Using the transition matrix, determine the total number of taxicabs which will use the taxicab staging area, the enplaning and the deplaning curbs, and the recirculation roadway during the peak hour. Verify the results, using the network flow iteration approach.

    c. Determine the lengths of deplaning and enplaning curb front required for taxicabs in the peak hour, if the maximum ratio of demand to capacity is 0.65.

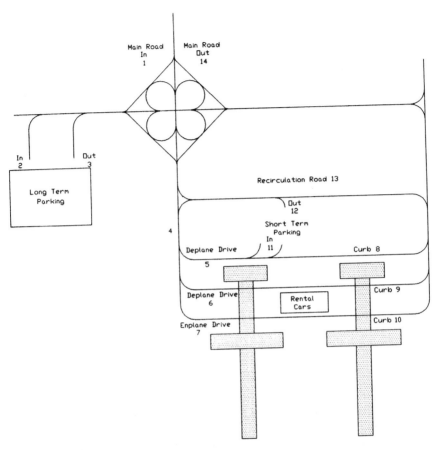

Main Road
In
1

Main Road
Out
14

In
2

Out
3

Long Term
Parking

Recirculation Road 13

Out
12

Short Term
Parking
In
11

Deplane Drive

4

Curb 8

5

Deplane Drive
6

Rental
Cars

Curb 9

Enplane Drive
7

Curb 10

Figure B-13 Ground access system layout.

38. The plan of the terminal-area ground access system at an airport is shown in Fig. B-13. The traffic flows and traffic splits for private automobiles at this airport are shown in Fig. B-14. The total number of private automobiles entering the airport on Main Road during the peak hour in 1990 is estimated to be 1000.

   a. Determine the total peak-hour flow of the private automobiles on the Terminal Road In, Enplane Drive, Recirculation Road, and Main Road leaving the airport in 1990. Show the results of using the transition matrix approach.

   b. Determine the accumulation of vehicles during the peak hour in 1990 in both the long-term and short-term parking facilities.

   c. If the dwell time of private automobiles at the enplaning curb averages 1.8 min and the length of a private automobile slot is 25 ft, find the length of enplaning curb front required for private automobiles in 1990. Assume that 70 percent of the inside

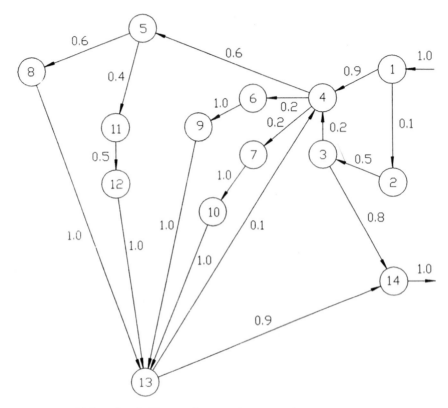

Figure B-14 Link-and-node diagram for ground access system.

lane and 30 percent of the outside lane are available for enplaning curb-front vehicles, and the level-of-service criterion is a demand/capacity ratio of 0.65.

39. The August 1992 schedule of an airline at any airport is given in Table B-23.

 a. Draw a ramp chart and determine the number of gates required at this airport if all aircraft may use all gates. The criteria of the airline for ramp scheduling stipulate that the aircraft may be scheduled into a given gate only if there is a 15-min time gap between aircraft departures and aircraft arrivals being assigned to that gate.

 b. Compute the average gate occupancy time and the gate utilization factor, considering only aircraft which both arrive and depart during the period from 06:00 to 24:00.

 c. Determine the peak hour for the gate occupancy and the number of gates occupied by aircraft during the peak hour with the gate system used by the airline at this airport.

 d. The airline expects to add 10 more flights to this airport by

TABLE B-23   Airline Flight Schedule, August 1992

Flight	Time Arrival	Departure	Aircraft	Flight	Time Arrival	Departure	Aircraft
DL 109	09:00	09:30	L-1011	DL 711	17:45	18:15	767
DL 120	08:15	08:45	L-1011	DL 719	15:30	16:15	767
DL 134	16:45	17:30	L-1011	DL 721/784	23:30	07:45	767
DL 136/143	17:00	09:00	L-1011	DL 735	19:15	20:00	767
DL 147	09:00	09:45	L-1011	DL 739	08:45	09:30	767
DL 148	13:00	13:30	L-1011	DL 744	10:30	11:00	767
DL 154	08:15	08:45	L-1011	DL 746/732	21:30	07:00	767
				DL 752	08:15	08:45	767
DL 201	13:30	15:45	72S	DL 787	22:15	22:45	767
DL 239	09:00	09:45	72S				
DL 259	18:00	18:30	72S	DL 931	17:00	17:45	D8S
DL 260	19:30	20:15	72S	DL 964/926	19:15	07:15	D8S
DL 262	19:15	20:00	72S	DL 968/904	23:30	08:45	D8S
DL 278	06:15	07:00	72S				
DL 281	22:15	22:45	72S	DL 1163	09:00	10:00	73S
DL 283	22:15	23:00	72S	DL 1173	14:00	15:45	73S
DL 286	08:15	08:45	72S	DL 1184	06:15	07:00	73S
DL 339	13:00	13:45	72S				
DL 354	15:15	15:45	72S	DL 1520	08:15	09:30	D9S
DL 355/559	22:15	07:30	72S	DL 1533/1672	14:00	16:00	D9S
DL 359	22:15	22:45	72S	DL 1565/1580	06:45	09:30	D9S
DL 362	13:00	13:30	72S	DL 1566	15:15	15:45	D9S
DL 404	15:15	16:00	72S	DL 1567	10:30	11:30	D9S
DL 418	19:30	20:00	72S	DL 1577	08:00	08:45	D9S
DL 454	20:15	20:45	72S	DL 1614	19:30	20:30	D9S
DL 456	22:15	22:45	72S	DL 1626	19:30	20:15	D9S
DL 464	19:15	20:00	72S	DL 1629	20:30	22:00	D9S
DL 482	11:15	11:45	72S	DL 1631	13:00	13:45	D9S
DL 485	22:00	22:45	72S	DL 1637	13:15	13:45	D9S
DL 505	09:00	09:30	72S	DL 1644/1543	23:15	07:00	D9S
DL 537	13:00	13:30	72S	DL 1654	15:15	16:00	D9S
DL 546	10:30	11:00	72S	DL 1662	20:30	21:30	D9S
DL 586	18:15	18:45	72S	DL 1677	21:30	22:00	D9S
DL 604	19:00	06:45	72S	DL 1689	18:00	18:45	D9S
DL 611	09:00	09:30	72S	DL 1697/1523	22:00	08:15	D9S
DL 648	19:30	20:00	72S				

August 1996. Using the random numbers given in Table B-24, simulate the arrival time and gate occupancy time of these 10 flights, and schedule them into the ramp chart above. Simplify the simulation by using flight arrivals in 1-h and gate occupancy times in 15-min increments. From the simulation, estimate flight arrival times and gate occupancy times to the nearest 15 min. From the simulation results, determine whether any additional gates are required for the August 1996 schedule. How many total gates are required for the August 1996 schedule at this airport?

TABLE B-24   Random Numbers for Flight Arrival and Gate
Occupancy Times

| Flight no. | Random numbers | |
	Arrival time	Gate occupancy time
1	2451	7309
2	3050	8030
3	3119	0738
4	5152	8142
5	0583	6420
6	7889	9816
7	4971	4127
8	3638	0632
9	9845	6379
10	9016	6555

TABLE B-25   March 1992 Airline Schedule at Airport for Gate Assignments in
Main Terminal Building

| Flight | Time | | | Flight | Time | | |
	Arrival	Departure	Aircraft		Arrival	Departure	Aircraft
AL 715	11:15	12:00	D9S	NW 234	13:45	15:00	72S
AL 256	12:45	13:30	72S	NW 298/205	19:15	09:45	72S
AL 390	16:15	17:00	73S				
AL 353/310	20:45	08:00	73S	PA 221	11:30	12:45	72S
				PA 922	12:30	16:30	D10
EA 218	07:00	08:00	72S	PA 266	19:00	19:30	72S
EA 559	09:45	10:15	AB3				
EA 831	09:45	10:30	D95	RC 844	12:00	12:30	D95
EA 888	10:45	11:30	D95	RC 880	16:30	17:15	D95
EA 541	11:00	12:00	AB3				
EA 804	12:00	13:00	D95	TW 296	10:15	11:30	72S
EA 588	12:00	14:15	AB3	TW 204	13:15	14:30	72S
EA 293	13:15	13:45	72S	TW 610	15:00	16:00	L-10
EA 133	14:45	15:45	727				
EA 214	15:30	16:30	72S	UA 300	11:30	14:00	73S
EA 258	16:30	17:30	72S	UA 288	18:15	19:00	72S
EA 239	16:45	18:00	72S	UA 241/270	21:30	09:30	72S
EA 871	18:30	19:30	D95				
EA 699	18:45	19:30	L-10				
EA 822/814	20:30	08:30	D95				
EA 890	21:00	21:45	D95				
EA 400/430	20:30	09:30	757				
EA 751/740	23:45	06:45	D95				

40. The March 1992 airline schedule for those airlines which are operating at an airport is given in Table B-25. Under the gate-use strategy gates are shared by Allegheny (AL) and Northwest (NW) airlines, and by Republic (RC), Trans World (TW) and United (UA) airlines. Gates are used exclusively by Eastern (EA) and Pan Am (PA) airlines. Flight numbers for wide-bodied aircraft range from 500 to 699, and from 900 to 999.

   a. Construct the March 1992 ramp chart for the various gate uses. A minimum gate free time of 15 min is required between flights at any gate, to allow for flight schedule irregularities.

   b. Determine the number of gates required for the March 1992 schedule based upon the current gate-use strategy.

   c. Determine the average gate occupancy time by wide-bodied aircraft, narrow-bodied aircraft, and all aircraft at this station.

   d. Determine the gate utilization factor for each gate.

41. The following sample of A-weighted sound levels was measured at 0.5-s intervals during an aircraft flyover: 61.3, 64.2, 66.5, 68.7, 69.1, 73.2, 77.3, 78.2, 78.9, 77.6, 76.8, 74.1, 72.6, 70.2, 68.7, 65.5, 64.6, 63.1, 62.7, 61.6, 61.3, and 61.0 dB. Determine the sound exposure level (SEL) of this noise event.

42. The following SELs for six aircraft flyovers were measured in a 1-h period: 87.0, 92.1, 90.7, 92.5, 83.1, and 89.6 dB. Compute the hourly average sound level.

43. The following sound exposure levels (SEL) of 10 aircraft flyovers were measured over the course of a 24-h period: 81.2 dB at 6:30 a.m., 85.1 dB at 9:00 a.m., 89.2 dB at 10:15 a.m., 78.8 dB at 10:30 a.m., 75.2 dB at 12:30 p.m., 87.2 dB at 2:30 p.m., 88.1 dB at 3:30 p.m., 91.2 dB at 6:15 p.m., 76.8 dB at 10:30 p.m., and 77.2 dB at 11:00 p.m. Compute the day-night average SEL.

44. An aircraft produces an SEL of 95 dBA.

   a. Determine the day-night average SEL produced by a single daytime noise event with this aircraft.

   b. Determine the day-night average SEL produced by a single nighttime noise event with this aircraft.

45. The runway system at an airport consists of three parallel runways. The centerlines of the first and second runways are separated by 2500 ft, and the centerlines of the second and third runways are separated by 4300 ft. A fourth parallel runway is being considered which would have a centerline spacing from the third runway of 2500 ft. The traffic mix and the aircraft operating cost per h consist of 3.8 percent type A aircraft at $96, 6.1 percent type B aircraft at $412, 58.6 percent type C aircraft at $1400, and 31.5 percent type D aircraft at $2900. The projected annual demand is 600,000 aircraft operations. Find the range in annual benefits due to delay reduction if the fourth runway is constructed.

46. An airline is considering expansion of its terminal facilities to effect time savings for its passengers. The total capital construction costs of the project are estimated to be $10 million. Construction funds are required at the beginning of the year in which construction is started, and construction will take 1 year. Planners estimate that 1.5 million passengers a year will use the facility and that these passengers will experience an average time savings of 20 min each. Passengers value time at the rate of $6 per hour. No increase in maintenance or operating costs is expected. The discount rate is 8 percent.

    a. Plot net present value versus time for 1983 through 1993.

    b. Determine the earliest year in which the project is economically justified, if ever.

47. An airport presently consists of a single runway, and at some future time it may become necessary to construct a parallel runway capable of handling simultaneous approaches. The traffic mix and the aircraft operating cost per hour consist of 10 percent type A aircraft at $120, 20 percent type B aircraft at $500, 50 percent type C aircraft at $1850, and 20 percent type D aircraft at $2800. The annual aircraft demand is expected to rise from 50,000 in 1992, to 61,000 in 1995, to 97,000 in 2000, to 146,000 in 2005, and to 213,000 in 2010. Construction of the new runway and associated taxiway will cost $28 million, and the annual maintenance cost is expected to be $120,000. The discount rate is 8 percent.

    a. Plot net present value versus time from now to the year 2010.

    b. Assuming the runway has no salvage value, analyze the net present value of the costs and benefits of the new runway, and determine the earliest time the runway should be constructed, if ever.

48. An airport improvement project is expected to cost $100 million. The project cost is to be recovered through landing fees charged to aircraft. The aircraft demand is expected to rise from 300,000 to 400,000 annual operations over a 20-year capital recovery period. The fleet mix is expected to remain constant and consist of 30 percent MD-87, 40 percent Boeing 737-300, 10 percent Boeing 757-200, 10 percent Boeing 767-200, and 10 percent DC-10-30. A 1.25 coverage factor is required, and the bonds will be retired at an annual interest rate of 6 percent. Earnings on the capital recovery fund are expected to be 6% annually. Determine the average landing fee which must be charged to each aircraft to recover the capital costs of the project.

49. An airport is considering the imposition of a passenger facility charge on enplaned passengers to raise capital for airport improvements. The number of annual enplaned passengers is

expected to increase uniformly from 300,000 to 400,000 over a 10-year period for which the passenger facility charge will be imposed. The annual interest rate is 8 percent. Determine the amount of capital which might be financed at present for a $1, $2, and $3 passenger facility charge.

50. The construction of a new runway at an airport will cost $30 million, to be paid with capital development bonds over a 20-year period. The runway is expected to decrease the average aircraft delay from 9 to 5 min. The average annual demand is expected to be 200,000 operations, and the average operating cost of an aircraft is $1500 per hour. The discount rate is 10 percent. Compare the annual benefits to the annual costs of this runway construction project.

Index

AAAE (*see* American Association of Airport Executives)
AAS (*see* Advanced automation system)
Above mean sea level (AMSL), 145, 149
Accelerate-stop distance (DAS), 113
Access interface system, terminal, 431–432, 449
ACF (*see* Area control facility)
ACI (*see* Airports Council International)
ACN (*see* Aircraft classification number)
ADAP (*see* Airport Development Aid Program)
ADF (*see* Arrival delay factor)
ADI (*see* Arrival delay index)
ADO (*see* Airports District Office)
Advanced air traffic management system study, 175
Advanced automation system (AAS), 176
AERA (*see* Automated en route air traffic control system)
Aerospace Industries Association of America (AIA), 41
AFSS (*see* Automated flight service station)
Agencies and organizations, 37–42
  (*See also* specific agency names)
AIA (*see* Aerospace Industries Association of America)
AIP (*see* Airport Improvement Program)
Air cargo:
  domestic, 7
  worldwide, 14
Air carriers:
  commuter and regional, 7–9
  general aviation and, 16

Air carriers (*Cont.*):
  passenger traffic, 4, 14
  U.S. air cargo traffic, 7
  U.S. passenger traffic, 3–6
Air commerce, worldwide, 13–16
Air Commerce Act of 1926, 21–22, 144
Aircraft:
  angled nose-in parking, 510
  angled nose-out parking, 510
  characteristics related to airport design, 77–141
  datum length, 411, 413
  dimensional terms, 83
  fueling, 512–514
  general aviation, 80–81
  grounding facilities, 515
  nose-in parking, 509–510
  parallel parking, 510
  parking, 507–510
  short-haul passenger, 80–81
  STOL (*see* Short takeoff and landing)
  transport, 78–79, 82, 84–110
Aircraft classification number (ACN), 627–629
Aircraft Owners and Pilots Association (AOPA), 41
Airfield:
  components, 301
  geometric design, 363–429
  separation requirements related to runways, 383
Air Line Pilots Association (ALPA), 41
Airline:
  activities in terminal area, 464
  deregulation, 9–13, 33–35
Airline Deregulation Act of 1978, 33–35

ABOUT THE AUTHORS

ROBERT HORONJEFF (deceased) was an internationally known consultant on airport design and professor of transportion engineering at the University of California, Berkeley.

FRANCIS X. McKELVEY is professor of civil engineering at Michigan State University. He has also served as a consultant on transportation and airport planning to federal, state, and local agencies, as well as to private firms in the United States and abroad. He played a major role in the planning and design of the Fort Lauderdale-Hollywood International Airport.

$$Tij \quad \begin{bmatrix} 106 & IRA \\ A_0 & 90 \end{bmatrix} \begin{bmatrix} 0.25 & 0.25 \\ 0.25 & 0.25 \end{bmatrix}$$

0.5T   ?.